T0274183

Analytical Mechanics

Analytical mechanics is the foundation of many areas of theoretical physics, including quantum theory and statistical mechanics, and has wide-ranging applications in engineering and celestial mechanics. This introduction to the basic principles and methods of analytical mechanics covers Lagrangian and Hamiltonian dynamics, rigid bodies, small oscillations, canonical transformations and Hamilton-Jacobi theory. This fully up-to-date textbook includes detailed mathematical appendices and addresses a number of advanced topics, some of them of a geometric or topological character. These include Bertrand's theorem, proof that action is least, spontaneous symmetry breakdown, constrained Hamiltonian systems, non-integrability criteria, KAM theory, classical field theory, Lyapunov functions, geometric phases and Poisson manifolds. Providing worked examples, end-of-chapter problems and discussion of ongoing research in the field, it is suitable for advanced undergraduate students and graduate students studying analytical mechanics.

Nivaldo A. Lemos is Associate Professor of Physics at Federal Fluminense University, Brazil. He was previously a visiting scholar at the Massachusetts Institute of Technology. His main research areas are quantum cosmology, quantum field theory and the teaching of classical mechanics.

Analytical Mechanics

Analytical Mechanics

NIVALDO A. LEMOS

Fluminense Federal University

CAMBRIDGE
UNIVERSITY PRESS

University Printing House, Cambridge CB2 8BS, United Kingdom

One Liberty Plaza, 20th Floor, New York, NY 10006, USA

477 Williamstown Road, Port Melbourne, VIC 3207, Australia

314–321, 3rd Floor, Plot 3, Splendor Forum, Jasola District Centre, New Delhi – 110025, India

79 Anson Road, #06–04/06, Singapore 079906

Cambridge University Press is part of the University of Cambridge.

It furthers the University's mission by disseminating knowledge in the pursuit of education, learning, and research at the highest international levels of excellence.

www.cambridge.org
Information on this title: www.cambridge.org/9781108416580
DOI: 10.1017/9781108241489

© 2004, Editora Livraria da Física, São Paulo

English translation © Cambridge University Press 2018

Revised and enlarged translation of Mecânica Analítica by Nivaldo A. Lemos originally published in Portuguese by Editora Livraria da Física, São Paulo 2004

English edition first published 2018

Printed in the United Kingdom by TJ International Ltd. Padstow Cornwall

A catalogue record for this publication is available from the British Library.

Library of Congress Cataloging-in-Publication Data
Names: Lemos, Nivaldo A., 1952– author.
Title: Analytical mechanics / Nivaldo A. Lemos, Fluminense Federal University.
Other titles: Mecânica analítica. English
Description: Cambridge, United Kingdom ; New York, NY : Cambridge University Press, 2018. | Includes bibliographical references and index.
Identifiers: LCCN 2018011108| ISBN 9781108416580 (hb) | ISBN 1108416586 (hb)
Subjects: LCSH: Mechanics, Analytic.
Classification: LCC QA805 .L4313 2018 | DDC 531.01/515–dc23
LC record available at https://lccn.loc.gov/2018011108

ISBN 978-1-108-41658-0 Hardback

To my wife
Marcia
and my daughters
Cintia, Luiza and Beatriz

Contents

Preface   *page* xi

1 Lagrangian Dynamics   1
  1.1   Principles of Newtonian Mechanics   1
  1.2   Constraints   7
  1.3   Virtual Displacements and d'Alembert's Principle   11
  1.4   Generalised Coordinates and Lagrange's Equations   17
  1.5   Applications of Lagrange's Equations   24
  1.6   Generalised Potentials and Dissipation Function   28
  1.7   Central Forces and Bertrand's Theorem   31
  Problems   37

2 Hamilton's Variational Principle   42
  2.1   Rudiments of the Calculus of Variations   42
  2.2   Variational Notation   49
  2.3   Hamilton's Principle and Lagrange's Equations   50
  2.4   Hamilton's Principle in the Non-Holonomic Case   56
  2.5   Symmetry Properties and Conservation Laws   65
  2.6   Conservation of Energy   71
  2.7   Noether's Theorem   73
  Problems   78

3 Kinematics of Rotational Motion   86
  3.1   Orthogonal Transformations   86
  3.2   Possible Displacements of a Rigid Body   92
  3.3   Euler Angles   95
  3.4   Infinitesimal Rotations and Angular Velocity   96
  3.5   Rotation Group and Infinitesimal Generators   102
  3.6   Dynamics in Non-Inertial Reference Frames   103
  Problems   109

4 Dynamics of Rigid Bodies   112
  4.1   Angular Momentum and Inertia Tensor   112
  4.2   Mathematical Interlude: Tensors and Dyadics   114
  4.3   Moments and Products of Inertia   119

4.4	Kinetic Energy and Parallel Axis Theorem	120
4.5	Diagonalisation of the Inertia Tensor	122
4.6	Symmetries and Principal Axes of Inertia	126
4.7	Rolling Coin	129
4.8	Euler's Equations and Free Rotation	131
4.9	Symmetric Top with One Point Fixed	139
	Problems	145

5   Small Oscillations   149
5.1	One-Dimensional Case	149
5.2	Anomalous Case: Quartic Oscillator	154
5.3	Stationary Motion and Small Oscillations	161
5.4	Small Oscillations: General Case	163
5.5	Normal Modes of Vibration	165
5.6	Normal Coordinates	170
5.7	Mathematical Supplement	176
	Problems	178

6   Relativistic Mechanics   183
6.1	Lorentz Transformations	183
6.2	Light Cone and Causality	188
6.3	Vectors and Tensors	191
6.4	Tensor Fields	194
6.5	Physical Laws in Covariant Form	196
6.6	Relativistic Dynamics	199
6.7	Relativistic Collisions	204
6.8	Relativistic Dynamics in Lagrangian Form	207
6.9	Action at a Distance in Special Relativity	209
	Problems	211

7   Hamiltonian Dynamics   215
7.1	Hamilton's Canonical Equations	215
7.2	Symmetries and Conservation Laws	220
7.3	The Virial Theorem	221
7.4	Relativistic Hamiltonian Formulation	224
7.5	Hamilton's Equations in Variational Form	226
7.6	Time as a Canonical Variable	228
7.7	The Principle of Maupertuis	234
	Problems	237

8   Canonical Transformations   242
8.1	Canonical Transformations and Generating Functions	242
8.2	Canonicity and Lagrange Brackets	248

8.3	Symplectic Notation	250
8.4	Poisson Brackets	254
8.5	Infinitesimal Canonical Transformations	258
8.6	Angular Momentum Poisson Brackets	262
8.7	Lie Series and Finite Canonical Transformations	264
8.8	Theorems of Liouville and Poincaré	268
8.9	Constrained Hamiltonian Systems	272
	Problems	281

9 The Hamilton-Jacobi Theory   288

9.1	The Hamilton-Jacobi Equation	288
9.2	One-Dimensional Examples	291
9.3	Separation of Variables	294
9.4	Incompleteness of the Theory: Point Charge in Dipole Field	299
9.5	Action as a Function of the Coordinates	303
9.6	Action-Angle Variables	306
9.7	Integrable Systems: The Liouville-Arnold Theorem	312
9.8	Non-integrability Criteria	316
9.9	Integrability, Chaos, Determinism and Ergodicity	318
9.10	Prelude to the KAM Theorem	320
9.11	Action Variables in the Kepler Problem	323
9.12	Adiabatic Invariants	325
9.13	Hamilton-Jacobi Theory and Quantum Mechanics	329
	Problems	332

10 Hamiltonian Perturbation Theory   338

10.1	Statement of the Problem	338
10.2	Generating Function Method	341
10.3	One Degree of Freedom	342
10.4	Several Degrees of Freedom	345
10.5	The Kolmogorov-Arnold-Moser Theory	348
10.6	Stability: Eternal or Long Term	353
10.7	Lie Series Method	354
	Problems	358

11 Classical Field Theory   360

11.1	Lagrangian Field Theory	360
11.2	Relativistic Field Theories	365
11.3	Functional Derivatives	367
11.4	Hamiltonian Field Theory	371
11.5	Symmetries of the Action and Noether's Theorem	374
11.6	Solitary Waves and Solitons	378
11.7	Constrained Fields	381
	Problems	385

Appendix A   Indicial Notation   392
Appendix B   Frobenius Integrability Condition   398
Appendix C   Homogeneous Functions and Euler's Theorem   406
Appendix D   Vector Spaces and Linear Operators   408
Appendix E   Stability of Dynamical Systems   421
Appendix F   Exact Differentials   426
Appendix G   Geometric Phases   428
Appendix H   Poisson Manifolds   433
Appendix I   Decay Rate of Fourier Coefficients   440
References   442
Index   452

Preface

It is no exaggeration to say that analytical mechanics is the foundation of theoretical physics. The imposing edifice of quantum theory was erected on the basis of analytical mechanics, particularly in the form created by Hamilton. Statistical mechanics and the quantum field theory of elementary particles are strongly marked by structural elements extracted from classical mechanics. Moreover, the modern development of the theories of chaos and general dynamical systems has given rise to a renaissance of classical mechanics that began in the middle of the twentieth century and is ongoing today. Thus, the study of virtually any branch of contemporary physics requires a solid background in analytical mechanics, which remains important in itself on account of its applications in engineering and celestial mechanics.

For more than two decades, intermittently, the author has taught analytical mechanics to undergraduate physics students at Universidade Federal Fluminense in Niterói, Brazil. The present book, fruit of the author's teaching experience and ceaseless fascination with the subject, aims at advanced undergraduate or graduate students who have previously taken a course in intermediate mechanics at the level of Marion & Thornton (1995) or Taylor (2005). As to mathematical prerequisites, the usual courses on calculus of one and several variables, ordinary differential equations and linear algebra are sufficient.

Analytical mechanics is much more than a mere reformulation of Newtonian mechanics, and its methods have not been primarily conceived to help solve the equations of motion of specific mechanical systems. To make this clear, symmetry properties, invariance and general structural features of mechanics are highlighted in the course of the presentation.

As remarked by Vladimir I. Arnold, numerous modern mathematical theories, with which some of the greatest names in the history of mathematics are associated, owe their existence to problems in mechanics. Accordingly, without neglecting the physical aspects of the discipline but recognising its eminently mathematical character, it is trusted that the level of mathematical precision maintained throughout the exposition will be found satisfactory. The main compromise in this aspect is the use of infinitesimal quantities, which seems advisable when first exposing students to the notions of virtual displacement and continuous families of transformations. Auxiliary mathematical results are kept out of the main text except when they naturally fit the development of the formalism. Theorems of more frequent and general use are stated and some proved in the appendices. With regard to the theorems stated without proof, reference is made to mathematical texts in which proofs can be found. The progressive replacement of heuristic arguments with rigorous mathematical justifications is in line with the trend of contemporary theoretical physics, whose mathematical language is becoming increasingly sophisticated.

The mathematical apparatus of analytical mechanics is very rich and grants students the first contact with concepts and tools that find extensive application in the most varied fields of physics. Notions such as linear operator, eigenvalue, eigenvector, group, Lie algebra, differential form and exterior derivative arise naturally in a classical context, where circumstances are more easily visualisable and intuition is almost always a safe guide. The high degree of generality of the formalism is conducive to enhancing the student's abstraction ability, so necessary for the proper understanding of contemporary physical theories.

The presentation mainly follows the traditional approach, which still seems the most suitable for a first course in analytical mechanics, and for which the author is heavily indebted to acclaimed classics such as Goldstein (1980) and Landau & Lifshitz (1976). The text fully covers the standard themes of an upper-level undergraduate course in analytical mechanics, such as Lagrangian and Hamiltonian dynamics, variational principles, rigid-body dynamics, small oscillations, canonical transformations, and Hamilton-Jacobi theory. The original edition, published in Portuguese in 2004, was well received by both students and colleagues in Brazil. For the present English translation the opportunity has been taken to significantly increase the number of advanced topics addressed. Properly selected, these advanced topics make the book suitable for a graduate course: Bertrand's theorem (Chapter 1); a rigorous proof that action is least if the time interval is short enough (Chapter 2); Frobenius integrability condition and basic theory of differential forms (Appendix B); spontaneous symmetry breakdown (Chapter 5); quartic oscillator and Jacobi elliptic functions (Chapter 5); Van Dam-Wigner no-interaction theorem (Chapter 6); time as a canonical variable and parameterised-time systems (Chapter 7); constrained Hamiltonian systems (Chapter 8); limitations of the Hamilton-Jacobi theory based on the separation of variables technique (Chapter 9); non-integrability criteria (Chapter 9); canonical perturbation theory and introductory KAM theory (Chapter 10); classical field theory (Chapter 11); stability of dynamical systems and Lyapunov functions (Appendix E); geometric phases (Appendix F); and Poisson manifolds (Appendix G).

The exposition is characterised by many fully worked-out examples so as to make it accessible to typical students. At the same time, the text is intended to be stimulating and challenging to the most gifted students. The end-of-chapter problems are aimed at testing the understanding of the theory exposed in the main body of the text as well as enhancing the reader's problem-solving skills, but some are more demanding and introduce new ideas or techniques, sometimes originated in research articles. The straightforward exercises sprinkled through the chapters and appendices are intended to stimulate an active involvement of the serious student with the development of the subject matter and are designed to be taken up on the spot. The References section, which includes very advanced books and recent research papers, aims at exciting the reader's curiosity and showing that classical mechanics is neither a museum piece nor a closed chapter of physics. On the contrary, its power to charm resists the passage of time, and there are many problems worthy of investigation, some monumental, such as the question of the stability of the solar system, which remains without a conclusive answer. Effort has been put forth to make the exposition even-handed, trying to strike the right balance between accessibility and the desire to afford a not too abrupt transition to the highly sophisticated treatments based on differential geometry and topology (Arnold, 1989; Thirring, 1997).

It is hoped that, besides imparting the fundamental notions to fill the needs of most students, the book may prepare and persuade some readers to a deeper immersion in such a beautiful subject.

Lagrangian Dynamics

Lagrange has perhaps done more than any other analyst by showing that the most varied consequences respecting the motion of systems of bodies may be derived from one radical formula; the beauty of the method so suiting the dignity of the results, as to make of his great work a kind of scientific poem.

William Rowan Hamilton, *On a General Method in Dynamics*

Mechanical systems subject to restrictions (constraints) of a geometric or kinematic nature occur very often. In such situations the Newtonian formulation of dynamics turns out to be inconvenient and wasteful, since it not only requires the use of redundant variables but also the explicit appearance of the constraint forces in the equations of motion. Lagrange's powerful and elegant formalism allows one to write down the equations of motion of most physical systems from a single scalar function expressed in terms of arbitrary independent coordinates, with the additional advantage of not involving the constraint forces.

1.1 Principles of Newtonian Mechanics

A fair appreciation of the meaning and breadth of the general formulations of classical mechanics demands a brief overview of Newtonian mechanics, with which the reader is assumed to be familiar. Virtually ever since they first appeared in the *Principia*, Newton's three laws of motion have been controversial regarding their physical content and logical consistency, giving rise to proposals to cast the traditional version in a new form free from criticism (Eisenbud, 1958; Weinstock, 1961). Although the first and second laws are sometimes interpreted as a definition of force (Marion & Thornton, 1995; José & Saletan, 1998; Fasano & Marmi, 2006), we shall adhere to what we believe is the correct viewpoint that regards them as genuine laws and not mere definitions (Feynman, Leighton & Sands, 1963, p. 12–1; Anderson, 1990). A detailed analysis of the physical aspects and logical structure of Newton's laws is beyond the scope of this section, whose main purpose is to serve as a reference for the rest of the exposition.

Laws of Motion

The postulates stated in this section are equivalent to Newton's three laws of motion but seek to avoid certain logical difficulties of the original proposition.

LAW I   *There exist* **inertial reference frames** *with respect to which every isolated particle remains at rest or in uniform motion in a straight line.*

The existence of one inertial reference frame implies the existence of infinitely many, all moving with respect to each other in a straight line with constant speed. Implicit in this postulate is the Newtonian notion of absolute time, which *of itself, and from its own nature, flows equably without relation to anything external*, and is the same in all inertial reference frames. A particle is said to be "isolated" if it is far enough from all material objects.

LAW II   *In any inertial reference frame the motion of a particle is governed by the equation*

$$m\mathbf{a} = \mathbf{F}, \tag{1.1}$$

where **a** *is the particle's acceleration, m is its mass and* **F** *is the total force acting on the particle.*

This postulate implicitly assumes that, associated with each particle, there is a positive constant m, called mass, which is the same in all inertial reference frames. Different types of force can be identified and the intuititive notion of force can be given an operational definition (Taylor, 2005, pp. 11–12). Force is supposed to have inherent properties that can be ascertained independently of Newton's second law (Feynman, Leighton & Sands, 1963, p. 12–1). Furthermore, Eq. (1.1) cannot be a mere definition of force because it is changed to Eq. (6.99) in the special theory of relativity.

LAW III   *To every action there corresponds an equal and opposite reaction, that is, if* $\mathbf{F}_{ij}$ *is the force on particle i exerted by particle j, then*

$$\mathbf{F}_{ij} = -\mathbf{F}_{ji}. \tag{1.2}$$

This is the law of action and reaction in its weak form. The strong version states that, besides being equal and opposite, the forces are directed along the line connecting the particles; in other words, two particles can only attract or repel one another. The third law is not valid in general, since moving electric charges violate it. This is due to the finite propagation speed of the electromagnetic interactions, which requires the introduction of the electromagnetic field as mediator of such interactions.

In the case of a system with several particles, it is assumed that the force on each particle can be decomposed in **external forces**, exerted by sources outside the system, and **internal forces**, due to the other particles of the system.[1] Thus, the equation of motion for the ith particle of a system of N particles is, according to the second law,

$$\frac{d\mathbf{p}_i}{dt} = \sum_{\substack{j=1 \\ j \neq i}}^{N} \mathbf{F}_{ij} + \mathbf{F}_i^{(e)}, \tag{1.3}$$

where

$$\mathbf{p}_i = m_i \, \mathbf{v}_i = m_i \, \frac{d\mathbf{r}_i}{dt} \tag{1.4}$$

---

[1] The possibility that a particle acts on itself is excluded.

is the **linear momentum** or just **momentum** of the ith particle, m_i its mass, $\mathbf{r}_i$ its position vector, $\mathbf{v}_i$ its velocity and $\mathbf{F}_i^{(e)}$ denotes the net external force on it.

Conservation Theorems

By summing Eqs. (1.3) over all particles one finds[2]

$$\sum_i m_i \frac{d^2\mathbf{r}_i}{dt^2} = \sum_{\substack{i,j \\ i\neq j}} \mathbf{F}_{ij} + \sum_i \mathbf{F}_i^{(e)} = \sum_i \mathbf{F}_i^{(e)} \tag{1.5}$$

because

$$\sum_{\substack{i,j \\ i\neq j}} \mathbf{F}_{ij} = \frac{1}{2} \sum_{\substack{i,j \\ i\neq j}} (\mathbf{F}_{ij} + \mathbf{F}_{ji}) = 0, \tag{1.6}$$

where we have used Eq. (A.9) from Appendix A as well as Eq. (1.2). Defining the centre-of-mass position vector by

$$\mathbf{R} = \frac{\sum_i m_i \mathbf{r}_i}{\sum_i m_i} \equiv \frac{\sum_i m_i \mathbf{r}_i}{M} , \tag{1.7}$$

Eq. (1.5) takes the form

$$M\frac{d^2\mathbf{R}}{dt^2} = \sum_i \mathbf{F}_i^{(e)} \stackrel{\text{def}}{=} \mathbf{F}^{(e)} , \tag{1.8}$$

where $\mathbf{F}^{(e)}$ is the total external force. This last equation can be written as

$$\frac{d\mathbf{P}}{dt} = \mathbf{F}^{(e)} \tag{1.9}$$

in terms of the total linear momentum of the system, defined by

$$\mathbf{P} = \sum_i m_i \mathbf{v}_i = \sum_i m_i \frac{d\mathbf{r}_i}{dt} = M\frac{d\mathbf{R}}{dt} . \tag{1.10}$$

Thus, an important conservation law is inferred.

Theorem 1.1.1 (Linear Momentum Conservation)   *If the total external force is zero, the total linear momentum of a system of particles is conserved.*

The system's **total angular momentum** with respect to a point Q with position vector $\mathbf{r}_Q$ is given by[3]

$$\mathbf{L}_Q = \sum_i m_i(\mathbf{r}_i - \mathbf{r}_Q) \times (\dot{\mathbf{r}}_i - \dot{\mathbf{r}}_Q), \tag{1.11}$$

---

[2] From now on $\sum_{i=1}^{N}$ will be abbreviated to $\sum_i$, it being understood that the sum extends over all particles in the system except when otherwise indicated.

[3] An overdot means d/dt, derivative with respect to time.

where $\mathbf{r}_i^{(Q)} = \mathbf{r}_i - \mathbf{r}_Q$ and $\mathbf{v}_i^{(Q)} = \dot{\mathbf{r}}_i - \dot{\mathbf{r}}_Q$ are, respectively, the position vector and velocity vector of the ith particle relative to point Q. Therefore,

$$
\begin{aligned}
\frac{d\mathbf{L}_Q}{dt} &= \sum_i m_i \mathbf{v}_i^{(Q)} \times \mathbf{v}_i^{(Q)} + \sum_i (\mathbf{r}_i - \mathbf{r}_Q) \times \dot{\mathbf{p}}_i - \sum_i m_i (\mathbf{r}_i - \mathbf{r}_Q) \times \ddot{\mathbf{r}}_Q \\
&= \sum_{\substack{i,j \\ i \neq j}} (\mathbf{r}_i - \mathbf{r}_Q) \times \mathbf{F}_{ij} + \sum_i (\mathbf{r}_i - \mathbf{r}_Q) \times \mathbf{F}_i^{(e)} - M(\mathbf{R} - \mathbf{r}_Q) \times \ddot{\mathbf{r}}_Q,
\end{aligned}
\tag{1.12}
$$

where Eqs. (1.3), (1.4) and (1.7) have been used. Now, availing ourselves again of equations (A.9) and (1.2), we can write

$$
\begin{aligned}
\sum_{\substack{i,j \\ i \neq j}} (\mathbf{r}_i - \mathbf{r}_Q) \times \mathbf{F}_{ij} &= \frac{1}{2} \sum_{\substack{i,j \\ i \neq j}} [(\mathbf{r}_i - \mathbf{r}_Q) \times \mathbf{F}_{ij} + (\mathbf{r}_j - \mathbf{r}_Q) \times \mathbf{F}_{ji}] \\
&= \frac{1}{2} \sum_{\substack{i,j \\ i \neq j}} [(\mathbf{r}_i - \mathbf{r}_Q) - (\mathbf{r}_j - \mathbf{r}_Q)] \times \mathbf{F}_{ij} = \frac{1}{2} \sum_{\substack{i,j \\ i \neq j}} (\mathbf{r}_i - \mathbf{r}_j) \times \mathbf{F}_{ij}.
\end{aligned}
\tag{1.13}
$$

If the internal forces obey the strong form of Newton's third law, $\mathbf{F}_{ij}$ is collinear to the vector $\mathbf{r}_{ij} \equiv \mathbf{r}_i - \mathbf{r}_j$ that points from the jth to the ith particle. It follows that $\mathbf{r}_{ij} \times \mathbf{F}_{ij} = 0$ and the sum of internal torques is zero. Thus, Eq. (1.12) can be rewritten in the form

$$
\frac{d\mathbf{L}_Q}{dt} = \mathbf{N}_Q^{(e)} - M(\mathbf{R} - \mathbf{r}_Q) \times \ddot{\mathbf{r}}_Q,
\tag{1.14}
$$

where

$$
\mathbf{N}_Q^{(e)} = \sum_i (\mathbf{r}_i - \mathbf{r}_Q) \times \mathbf{F}_i^{(e)}
\tag{1.15}
$$

is the **total external torque** with respect to point Q. If Q is at rest or is the centre of mass, the second term on the right-hand side of (1.14) vanishes[4] and we are left with

$$
\frac{d\mathbf{L}_Q}{dt} = \mathbf{N}_Q^{(e)}.
\tag{1.16}
$$

This last equation implies an important conservation law, in which both angular momentum and torque are taken with respect to a fixed point or the centre of mass.

Theorem 1.1.2 (Angular Momentum Conservation) *The total angular momentum of a system of particles is conserved if the total external torque is zero.*

Equation (1.10) means that the total linear momentum of a system of particles coincides with the one computed as if its entire mass were concentrated on the centre of mass. In the case of the total angular momentum there is an additional contribution. If $\mathbf{R}$ is the centre-of-mass position vector with respect to the origin O of an inertial reference frame and $\mathbf{r}_i'$ is the ith particle's position vector with respect to the centre of mass, then

$$
\mathbf{r}_i = \mathbf{r}_i' + \mathbf{R}, \qquad \mathbf{v}_i = \mathbf{v}_i' + \mathbf{V},
\tag{1.17}
$$

---

[4] Another possibility, which will not be considered, is a point Q whose acceleraton is parallel to the line connecting the centre of mass to Q itself (Tiersten, 1991).

where $\mathbf{V} \equiv \dot{\mathbf{R}}$ is the centre-of-mass velocity relative to O and $\mathbf{v}'_i \equiv \dot{\mathbf{r}}'_i$ is the ith particle's velocity relative to the centre of mass. The total angular momentum with respect to the origin O is

$$\mathbf{L} = \sum_i m_i \mathbf{r}_i \times \mathbf{v}_i = \sum_i m_i \mathbf{r}'_i \times \mathbf{v}'_i + \left(\sum_i m_i \mathbf{r}'_i\right) \times \mathbf{V} + \mathbf{R} \times \frac{d}{dt}\left(\sum_i m_i \mathbf{r}'_i\right) + M\mathbf{R} \times \mathbf{V}, \quad (1.18)$$

where (1.17) has been used. From the first of equations in (1.17) it follows that

$$\sum_i m_i \mathbf{r}'_i = \sum_i m_i \mathbf{r}_i - \sum_i m_i \mathbf{R} = M\mathbf{R} - M\mathbf{R} = 0. \quad (1.19)$$

Thus, the total angular momentum with respect to the origin admits the decomposition

$$\mathbf{L} = \mathbf{R} \times M\mathbf{V} + \sum_i \mathbf{r}'_i \times \mathbf{p}'_i. \quad (1.20)$$

In words: the total angular momentum with respect to the origin is the angular momentum as if the system were entirely concentrated on the centre of mass plus the angular momentum associated with the motion of the particles about the centre of mass.

The conservation of energy is a bit more involved than the two previous conservation theorems because the internal forces enter the picture. Let us start with a decompositon of the kinetic energy that is quite similar to the above decomposition of the angular momentum of a system of particles. The total kinetic energy is defined by

$$T = \frac{1}{2} \sum_i m_i v_i^2. \quad (1.21)$$

With the use of (1.17) one finds

$$T = \frac{1}{2} \sum_i m_i v_i'^2 + \frac{1}{2} \sum_i m_i V^2 + \mathbf{V} \cdot \frac{d}{dt}\left(\sum_i m_i \mathbf{r}'_i\right), \quad (1.22)$$

whence, owing to (1.19),

$$T = \frac{M}{2} V^2 + \frac{1}{2} \sum_i m_i v_i'^2. \quad (1.23)$$

The total kinetic energy is the kinetic energy of the system as if its entire mass were concentrated on the centre of mass plus the kinetic energy associated with the motion of the particles about the centre of mass. This result is particularly useful in rigid-body dynamics.

The work done by all forces as the system goes from a configuration A to a configuation B is defined by

$$W_{AB} = \sum_i \int_A^B \left(\mathbf{F}_i^{(e)} + \sum_{\substack{j \\ j \neq i}} \mathbf{F}_{ij}\right) \cdot d\mathbf{r}_i = \sum_i \int_A^B \mathbf{F}_i^{(e)} \cdot d\mathbf{r}_i + \sum_{\substack{i,j \\ i \neq j}} \int_A^B \mathbf{F}_{ij} \cdot d\mathbf{r}_i. \quad (1.24)$$

On the other hand, by using the equation of motion (1.3), one deduces

$$W_{AB} = \sum_i \int_A^B m_i \dot{\mathbf{v}}_i \cdot \mathbf{v}_i dt = \sum_i \int_A^B d\left(\frac{1}{2}m_i v_i^2\right) = \int_A^B dT, \quad (1.25)$$

hence

$$W_{AB} = T_B - T_A ,\tag{1.26}$$

that is, the work done by all forces is equal to the variation of the kinetic energy.

In most cases the forces are **conservative**, that is, derive from a scalar potential. Let us assume that the external forces admit a potential energy function $V^{(e)}(\mathbf{r}_1, \ldots, \mathbf{r}_N)$ such that

$$\mathbf{F}_i^{(e)} = -\nabla_i V^{(e)},\tag{1.27}$$

where $\nabla_i = \hat{\mathbf{x}}\,\partial/\partial x_i + \hat{\mathbf{y}}\,\partial/\partial y_i + \hat{\mathbf{z}}\,\partial/\partial z_i$ is the nabla operator with respect to the variable $\mathbf{r}_i$. In this case,

$$\sum_i \int_A^B \mathbf{F}_i^{(e)} \cdot d\mathbf{r}_i = -\int_A^B \sum_i \nabla_i V^{(e)} \cdot d\mathbf{r}_i = -\int_A^B dV^{(e)} = V_A^{(e)} - V_B^{(e)}.\tag{1.28}$$

If $\mathbf{F}_{ij}$ depends only on the relative positions $\mathbf{r}_{ij} = \mathbf{r}_i - \mathbf{r}_j$ and can be derived from a potential energy function $V_{ij}(\mathbf{r}_{ij})$ with $V_{ij} = V_{ji}$, then

$$\mathbf{F}_{ij} = -\nabla_i V_{ij}.\tag{1.29}$$

This form ensures the validity of the weak version of the action-reaction law. Indeed,

$$\mathbf{F}_{ij} = -\nabla_i V_{ij} = +\nabla_j V_{ij} = +\nabla_j V_{ji} = -\mathbf{F}_{ji}.\tag{1.30}$$

If, in addition, V_{ij} depends only on the distance $s_{ij} = |\mathbf{r}_{ij}|$ between the particles (central forces), we have

$$\mathbf{F}_{ij} = -\nabla_i V_{ij}(s_{ij}) = -\frac{\mathbf{r}_{ij}}{s_{ij}} V_{ij}'(s_{ij})\tag{1.31}$$

where V_{ij}' is the derivative of V_{ij} with respect to its argument. Thus, $\mathbf{F}_{ij}$ points along the line connecting the particles and the strong version of Newton's third law holds.

Because of Eq. (1.30), we are allowed to write

$$\sum_{\substack{i,j \\ i\neq j}} \int_A^B \mathbf{F}_{ij} \cdot d\mathbf{r}_i = \frac{1}{2} \sum_{\substack{i,j \\ i\neq j}} \int_A^B (\mathbf{F}_{ij} \cdot d\mathbf{r}_i + \mathbf{F}_{ji} \cdot d\mathbf{r}_j) = \frac{1}{2} \sum_{\substack{i,j \\ i\neq j}} \int_A^B \mathbf{F}_{ij} \cdot d(\mathbf{r}_i - \mathbf{r}_j)$$

$$= -\frac{1}{2} \sum_{\substack{i,j \\ i\neq j}} \int_A^B \nabla_i V_{ij} \cdot d\mathbf{r}_{ij} = -\frac{1}{2} \sum_{\substack{i,j \\ i\neq j}} \int_A^B \nabla_{ij} V_{ij} \cdot d\mathbf{r}_{ij} = -\frac{1}{2} \sum_{\substack{i,j \\ i\neq j}} V_{ij} \bigg|_A^B ,$$

$$\tag{1.32}$$

where ∇_{ij} denotes the gradient with respect to the vector $\mathbf{r}_{ij}$ and use has been made of the evident property $\nabla_i V_{ij} = \nabla_{ij} V_{ij}$. Finally, combining Eqs. (1.24), (1.26), (1.28) and (1.32), one finds

$$(T + V)_A = (T + V)_B ,\tag{1.33}$$

with

$$V = V^{(e)} + \frac{1}{2} \sum_{\substack{i,j \\ i\neq j}} V_{ij} .\tag{1.34}$$

The 2-form Ω defined by $\Omega = \omega^1 \wedge \omega^2$ is given by

$$\Omega = dx \wedge dy - R\sin\theta \, dx \wedge d\phi + R\cos\theta \, dy \wedge d\phi, \qquad (1.47)$$

and we have

$$d\omega^1 = R\sin\theta \, d\theta \wedge d\phi, \quad d\omega^2 = -R\cos\theta \, d\theta \wedge d\phi. \qquad (1.48)$$

It follows that

$$d\omega^1 \wedge \Omega = R\sin\theta \, dx \wedge dy \wedge d\theta \wedge d\phi \neq 0. \qquad (1.49)$$

According to the Frobenius theorem (Appendix B), the constraints (1.44) are not integrable. Since $d\omega^1 \wedge \Omega = 0$ for $\sin\theta = 0$, it seems one can only guarantee non-integrability of the constraints (1.44) in the region of the space of variables x, y, θ, ϕ such that $\sin\theta \neq 0$. However, an additional computation yields

$$d\omega^2 \wedge \Omega = -R\cos\theta \, dx \wedge dy \wedge d\theta \wedge d\phi, \qquad (1.50)$$

which vanishes only if $\cos\theta = 0$. Since $\cos\theta$ and $\sin\theta$ do not vanish simultaneously, $d\omega^1 \wedge \Omega$ and $d\omega^2 \wedge \Omega$ do not vanish together at any point, with the consequence that the constraints (1.44) are not integrable in the full space of the variables x, y, θ, ϕ that describe the configurations of the rolling disc.

In Appendix B, as well as in Lemos (2015), there is a proof that the constraints (1.45) are not integrable either. Knowing whether or not velocity-dependent constraints are integrable not only serves to quench an intellectual curiosity – perfectly legitimate, by the way – but also has practical consequences. The practical importance of determining the nature of the constraints as regards integrability is explained in the Warning right after Exercise 2.4.2 in Chapter 2.

Constraints expressed by non-integrable differential equations represent restrictions on the velocities alone. The centre of mass of the disc in Example 1.6 may travel along a closed curve such that x, y and θ return to their original values. But, at the end of the journey, ϕ may take an arbitrary value, depending on the length of the trajectory described by the centre of mass. This indicates that the variables describing the instantaneous configuration of the disc are functionally independent.

1.3 Virtual Displacements and d'Alembert's Principle

The **configuration** of a mechanical system is defined by the instantaneous position of each of its particles. For a mechanical system subject to constraints, at any instant t there are infinitely many possible configurations, that is, configurations consistent with the constraints.

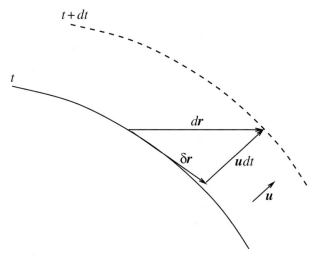

Fig. 1.4 Virtual and real displacements of a particle confined to a moving surface.

Virtual Displacements

Infinitesimal displacements that change a possible configuration into another possible configuration at the *same instant* are called **virtual displacements**. More precisely, given a system of N particles, the virtual displacements $\delta\mathbf{r}_i, i = 1,\ldots,N$, are infinitesimal displacements from positions $\mathbf{r}_1,\ldots,\mathbf{r}_N$ carried out instantaneously[9] and with the property of being compatible with the constraints. In short, the defining attributes of virtual displacements are: (1) they are infinitesimal; (2) they take place at a fixed instant; (3) they do not violate the constraints.

Example 1.9   A particle is confined to a moving surface. Let $f(\mathbf{r},t) = 0$ be the equation of the surface. A virtual displacement must be consistent with the constraint, that is, the point $\mathbf{r}$ and the displaced point $\mathbf{r} + \delta\mathbf{r}$ must belong to the surface at the same time t:

$$f(\mathbf{r} + \delta\mathbf{r}, t) = 0 \implies f(\mathbf{r}, t) + \nabla f \cdot \delta\mathbf{r} = 0 \implies \nabla f \cdot \delta\mathbf{r} = 0. \qquad (1.51)$$

Since ∇f is perpendicular to the surface at time t, the virtual displacement $\delta\mathbf{r}$ is tangent to the surface at this same instant (Fig. 1.4).

Note, however, that a real displacement $d\mathbf{r}$ takes place during a time interval dt. Therefore, in order that the particle remain on the surface, it is necessary that

$$f(\mathbf{r} + d\mathbf{r}, t + dt) = 0 \implies \nabla f \cdot d\mathbf{r} + \frac{\partial f}{\partial t} dt = 0. \qquad (1.52)$$

---

[9] The requirement that the time remain fixed confers a ficticious aspect to the virtual displacements, though in many situations they turn out to be equal to real displacements occurring during a time interval dt, as it will be seen in forthcoming examples. If all constraints are time independent, virtual displacements coincide with real displacements.

It is clear that the real displacement $d\mathbf{r}$ is not tangent to the surface as long as $\partial f/\partial t \neq 0$. Only the virtual displacement $\delta\mathbf{r}$ performed at a fixed time is tangent to the surface even if it is moving. Fig. 1.4 illustrates the difference between virtual and real displacements for a particle confined to a surface that moves with velocity $\mathbf{u}$ at instant t.

Virtual Work

The importance of introducing the notion of virtual displacement stems from the following observation: if the surface to which the particle is confined is ideally smooth, the contact force of the surface on the particle, which is the constraint force, has no tangential component and therefore is normal to the surface. Thus, the work done by the constraint force as the particle undergoes a virtual displacement is zero even if the surface is in motion, differently from the work done during a real displacement, which does not necessarily vanish. In most physically interesting cases, the *total* virtual work of the constraint forces is zero, as the next examples attest.

Example 1.10   Two particles joined by a rigid rod move in space. Let $\mathbf{f}_1$ and $\mathbf{f}_2$ be the constraint forces on the particles. By Newton's third law $\mathbf{f}_1 = -\mathbf{f}_2$ with both $\mathbf{f}_1$ and $\mathbf{f}_2$ parallel to the line connecting the particles. The virtual work done by the constraint forces is

$$\delta W_v = \mathbf{f}_1 \cdot \delta\mathbf{r}_1 + \mathbf{f}_2 \cdot \delta\mathbf{r}_2 = \mathbf{f}_2 \cdot (\delta\mathbf{r}_2 - \delta\mathbf{r}_1).$$

Setting $\mathbf{r} = \mathbf{r}_2 - \mathbf{r}_1$, the constraint equation takes the form (1.38), namely $r^2 - l^2 = 0$. In terms of the variable $\mathbf{r}$, the situation is equivalent to the one discussed in Example 1.9. Taking $f(\mathbf{r}, t) = r^2 - l^2$, Eq. (1.51) reduces to $\mathbf{r} \cdot \delta\mathbf{r} = 0$. Since $\mathbf{f}_2$ and $\mathbf{r}$ are collinear, there exists a scalar λ such that $\mathbf{f}_2 = \lambda\mathbf{r}$, hence $\delta W_v = \lambda\mathbf{r} \cdot \delta\mathbf{r} = 0$. Inasmuch as a rigid body consists of a vast number of particles whose mutual distances are invariable, one concludes that the total virtual work done by the forces responsible for the body's rigidity is zero.

Example 1.11   A rigid body rolls without slipping on a fixed surface. As a rule, in order to prevent slipping, a friction force between the fixed surface and the surface of the body is needed, that is, the surfaces in contact must be rough. Upon rolling without slipping, the body's particles at each instant are rotating about an axis that contains the body's point of contact with the surface. Thus, the friction force acts on a point of the body whose velocity at each instant is zero, because it is on the instantaneous axis of rotation. Virtual displacements are such that the body does not slip on the surface, that is, $\delta\mathbf{r} = 0$ at the point of contact between the body and the fixed surface. Therefore, the virtual work done by the constraint force is $\delta W_v = \mathbf{f} \cdot \delta\mathbf{r} = 0$ because $\delta\mathbf{r} = 0$, even though $\mathbf{f} \neq 0$.

The previous analysis makes it apparent that in a wide range of physically relevant situations the total virtual work of the constraint forces is zero. Situations in which there are sliding friction forces are exceptions, for the relevant virtual displacements no longer vanish and the virtual work done by the constraint forces is not zero. Friction is a strictly macroscopic phenomenon of scant interest to the development of general formulations of mechanics, especially in the light of contemporary physics. Constraints whose associated forces do no work in the course of virtual displacements are called **ideal constraints**. From now on we shall limit ourselves, with no significant loss of generality, to considering mechanical systems subject exclusively to ideal constraints.

Principle of Virtual Work

Newton's formulation of mechanics is characterised by the set of differential equations

$$m_i \ddot{\mathbf{r}}_i = \mathbf{F}_i , \quad i = 1, \ldots, N , \tag{1.53}$$

where $\mathbf{F}_i$ is the *total* or *resultant* force on the ith particle, supposedly a known function of positions, velocities and time. This system of differential equations determines a unique solution for the $\mathbf{r}_i(t)$ once the positions and velocities are specified at an initial instant.[10]

In the presence of constraints, it is patently clear how inconvenient the Newtonian formulation is. First of all, it usually requires the use of more coordinates than are necessary to specify the configuration of the system. When the constraints are holonomic, for instance, the positions $\mathbf{r}_1, \ldots, \mathbf{r}_N$ are not mutually independent, making the Newtonian approach uneconomical by demanding the employment of redundant variables. Furthermore, the total force on the ith particle can be decomposed as

$$\mathbf{F}_i = \mathbf{F}_i^{(a)} + \mathbf{f}_i , \tag{1.54}$$

where $\mathbf{F}_i^{(a)}$ is the **applied force** and $\mathbf{f}_i$ is the **constraint force**. In the case of the double pendulum in Example 1.4, $\mathbf{F}_1^{(a)}$ and $\mathbf{F}_2^{(a)}$ are the weights of the particles, whereas $\mathbf{f}_1$ and $\mathbf{f}_2$ are determined by the tensions on the rods or strings. The difficulty here lies in that one does not a priori know how the constraint forces depend on the positions and velocities. What one knows, in fact, are the *effects* produced by the constraint forces. One may also argue that the applied forces are the true causes of the motion, the constraint forces merely serving to ensure the preservation of the geometric or kinematic restrictions in the course of time. No less important is the fact that Newton's laws – the second law together with the strong version of the third law – turn out to be incapable of correctly describing the motion of certain constrained systems (Stadler, 1982; Casey, 2014).

For all these reasons, it is highly desirable to obtain a formulation of classical mechanics as parsimonious as possible, namely involving only the applied forces and employing only independent coordinates. We shall soon see that this goal is achieved by the Lagrangian formalism when all constraints are holonomic. As an intermediate step towards Lagrange's formulation, we shall discuss d'Alembert's principle, which is a method of writing down the equations of motion in terms of the applied forces alone, the derivation of which explores the fact that the virtual work of the constraint forces is zero.

Let us first consider a static situation, a system of particles in equilibrium. In this case $\mathbf{F}_i = 0$ and, whatever the virtual displacements $\delta \mathbf{r}_i$ may be,

$$\sum_i \mathbf{F}_i \cdot \delta \mathbf{r}_i = 0 . \tag{1.55}$$

---

[10] Uniqueness may fail in the presence of singular forces (Dhar, 1993).

With the help of the decomposition (1.54), one gets

$$\sum_i \mathbf{F}_i^{(a)} \cdot \delta \mathbf{r}_i + \sum_i \mathbf{f}_i \cdot \delta \mathbf{r}_i = 0 \,. \tag{1.56}$$

By limiting ourselves to the sufficiently extensive set of circumstances in which the virtual work of the constraint forces is zero, we arrive at the **principle of virtual work**:

$$\sum_i \mathbf{F}_i^{(a)} \cdot \delta \mathbf{r}_i = 0 \,. \tag{1.57}$$

This principle allows one to express the equilibrium conditions for a constrained system in terms of the applied forces alone.[11]

d'Alembert's Principle

We are interested in dynamics, which can be formally reduced to statics by writing Newton's second law in the form $\mathbf{F}_i - \dot{\mathbf{p}}_i = 0$, with $\mathbf{p}_i = m_i \dot{\mathbf{r}}_i$. According to d'Alembert's interpretation, each particle in the system is in "equilibrium" under a resultant force, which is the real force plus an "effective reversed force" equal to $-\dot{\mathbf{p}}_i$. This fictitious additional force is an inertial force existent in the non-inertial frame that moves along with the particle – that is, in which it remains at rest (Sommerfeld, 1952; Lanczos, 1970). Interpretations aside, the fact is that now, instead of (1.55), the equation

$$\sum_i (\dot{\mathbf{p}}_i - \mathbf{F}_i) \cdot \delta \mathbf{r}_i = 0 \tag{1.58}$$

is obviously true no matter what the virtual displacements $\delta \mathbf{r}_i$ are. Using again the decomposition (1.54) and assuming the virtual work of the constraint forces vanishes, we are led to **d'Alembert's principle**:

$$\sum_i \left(\dot{\mathbf{p}}_i - \mathbf{F}_i^{(a)} \right) \cdot \delta \mathbf{r}_i = 0 \,. \tag{1.59}$$

This principle is an extension of the principle of virtual work to mechanical systems in motion. For constrained systems, d'Alembert's principle is a substantial leap forward with respect to the Newtonian approach because it excludes any reference to the constraint forces. In concrete applications, however, one must take into account that the virtual displacements $\delta \mathbf{r}_i$ are not independent because they have to be in harmony with the constraints.

> **Example 1.12**   Use d'Alembert's principle to find the equations of motion for the mechanical system of Fig. 1.5, known as Atwood's machine.

---

[11] Formulated by Johann Bernoulli in 1717, the principle of virtual work traces back to Jordanus in the 13th century (Dugas, 1988). Illustrations of its use in several interesting cases can be found in Sommerfeld (1952) and Synge and Griffith (1959).

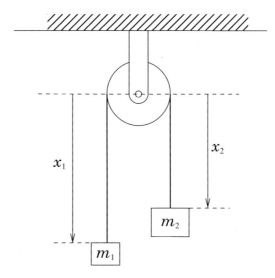

Atwood's machine.

Solution

The pulley in Fig. 1.5 is assumed massless and mounted on a frictionless axle. We also assume that the string does not slip on the pulley. With the coordinate system shown in the figure, we have $\mathbf{r}_1 = x_1\hat{\mathbf{x}}$, $\mathbf{r}_2 = x_2\hat{\mathbf{x}}$ and the holonomic constraint is

$$x_1 + x_2 = l, \tag{1.60}$$

where the constant l is determined by the radius of the pulley and the length of the string, assumed massless and inextensible. Clearly, the virtual displacements δx_1 and δx_2 compatible with the constraint (1.60) are related by

$$\delta x_1 + \delta x_2 = 0 \implies \delta x_2 = -\delta x_1. \tag{1.61}$$

In words, if one of the masses goes down, the other goes up the same distance, and vice versa. In virtue of the last equations, we have $\delta \mathbf{r}_1 = \delta x_1 \hat{\mathbf{x}}$ and $\delta \mathbf{r}_2 = \delta x_2 \hat{\mathbf{x}} = -\delta x_1 \hat{\mathbf{x}} = -\delta \mathbf{r}_1$. Noting that $\ddot{\mathbf{r}}_1 = \ddot{x}_1\hat{\mathbf{x}}$, $\ddot{\mathbf{r}}_2 = \ddot{x}_2\hat{\mathbf{x}}$ and also taking into account that $\ddot{x}_2 = -\ddot{x}_1$, which follows at once from (1.60), d'Alembert's principle

$$m_1\ddot{\mathbf{r}}_1 \cdot \delta\mathbf{r}_1 + m_2\ddot{\mathbf{r}}_2 \cdot \delta\mathbf{r}_2 = \mathbf{F}_1^{(a)} \cdot \delta\mathbf{r}_1 + \mathbf{F}_2^{(a)} \cdot \delta\mathbf{r}_2 = m_1g\hat{\mathbf{x}} \cdot \delta\mathbf{r}_1 + m_2g\hat{\mathbf{x}} \cdot \delta\mathbf{r}_2 \tag{1.62}$$

reduces to

$$m_1\ddot{x}_1\delta x_1 + (-m_2\ddot{x}_1)(-\delta x_1) = m_1g\delta x_1 + m_2g(-\delta x_1), \tag{1.63}$$

whence

$$(m_1 + m_2)\ddot{x}_1\delta x_1 = (m_1 - m_2)g\delta x_1. \tag{1.64}$$

In view of the arbitrariness of δx_1, the equation of motion for mass m_1 follows:

$$(m_1 + m_2)\ddot{x}_1 = (m_1 - m_2)g. \tag{1.65}$$

The acceleration of mass m_1 is

$$\ddot{x}_1 = \frac{m_1 - m_2}{m_1 + m_2} g, \tag{1.66}$$

which coincides with the result given by the elementary Newtonian treatment. The acceleration of mass m_2 is just $\ddot{x}_2 = -\ddot{x}_1$.

For other applications of d'Alembert's principle, the reader is referred to Sommerfeld (1952).

1.4 Generalised Coordinates and Lagrange's Equations

The d'Alembert principle still requires working with more coordinates than necessary, since not only the $\delta\mathbf{r}_i$ are not independent but also, in most cases, the positions $\mathbf{r}_i$ themselves.

Generalised Coordinates

Provided the system is holonomic, it is possible to introduce a certain number n of independent variables, generically denoted by $q_1, \ldots, q_n$ and called **generalised coordinates**, such that: (a) the position vector of each particle is unambiguously determined at any instant by the values of the qs; (b) the constraints, assumed of the form (1.41), are *identically* satisfied if expressed in terms of the qs. Let us see two illustrative cases.

Example 1.13 In the case of the plane double pendulum, defined in Example 1.4, a possible choice of generalised coordinates is $q_1 = \theta_1, q_2 = \theta_2$ (see Fig. 1.1). Then,

$$x_1 = l_1 \sin\theta_1, \quad y_1 = l_1 \cos\theta_1,$$
$$x_2 = l_1 \sin\theta_1 + l_2 \sin\theta_2, \quad y_2 = l_1 \cos\theta_1 + l_2 \cos\theta_2. \tag{1.67}$$

Note that the values of θ_1 and θ_2 completely specify the positions of the particles – that is, the system's configuration. In terms of θ_1 and θ_2, the constraint equations (1.40) reduce to the identities $l_1^2 \sin^2\theta_1 + l_1^2 \cos^2\theta_1 - l_1^2 \equiv 0$ and $l_2^2 \sin^2\theta_2 + l_2^2 \cos^2\theta_2 - l_2^2 \equiv 0$.

Example 1.14 A particle is restricted to the surface of a sphere in uniform motion. Let $\mathbf{u} = (u_x, u_y, u_z)$ be the sphere's constant velocity relative to an inertial reference frame. At instant t the centre of the sphere has coordinates $(u_x t, u_y t, u_z t)$ and the constraint equation takes the form

$$(x - u_x t)^2 + (y - u_y t)^2 + (z - u_z t)^2 - R^2 = 0, \tag{1.68}$$

where R is the radius of the sphere. Introducing the angles θ and ϕ by means of the equations

$$x = u_x t + R\sin\theta \cos\phi, \quad y = u_y t + R\sin\theta \sin\phi, \quad z = u_z t + R\cos\theta, \tag{1.69}$$

the constraint equation is now identically satisfied. Therefore, $q_1 = \theta$ and $q_2 = \phi$ is a possible choice of generalised coordinates.

Consider a mechanical system composed of N particles subject to the p holonomic constraints

$$f_1(\mathbf{r}_1,\ldots,\mathbf{r}_N,t) = 0\,,$$
$$\vdots \tag{1.70}$$
$$f_p(\mathbf{r}_1,\ldots,\mathbf{r}_N,t) = 0\,.$$

Out of the $3N$ coordinates $(x_1,y_1,z_1),\ldots,(x_N,y_N,z_N)$, only $n = 3N - p$ of them can be taken as mutually independent, and the system is said to possess n **degrees of freedom**. It is possible to introduce n generalised coordinates $q_1,\ldots,q_n$ in terms of which

$$\mathbf{r}_i = \mathbf{r}_i(q_1,\ldots,q_n,t)\,, \quad i = 1,\ldots,N \tag{1.71}$$

and Eqs. (1.70) are identically satisfied. A holonomic mechanical system has as many degrees of freedom as generalised coordinates necessary to specify its configuration at any instant. In geometric language, one can say that at each instant Eqs. (1.70) define a surface of dimension n in a space of dimension $3N$, and that (1.71) are the parametric equations of the surface.

Each set of values given to the generalised coordinates defines a configuration of the system – that is, the positions of all particles at any instant. The Cartesian space whose coordinate axes correspond to the generalised coordinates is called the **configuration space** of the system. The representation of the configuration space as a Cartesian space is merely symbolic, however. For example, in the case of a particle restricted to the surface of a sphere and described by the spherical angular coordinates (θ,ϕ), a single configuration corresponds to the infinitely many points $\theta = 0$ with arbitrary ϕ. Strictly speaking, the configuration space has the mathematical structure of a differentiable manifold, the reason why it is also known as the **configuration manifold**.[12]

Once generalised coordinates have been introduced via (1.71), the virtual displacements $\delta\mathbf{r}_i$ can be expressed in terms of the independent virtual displacements δq_k by means of

$$\delta\mathbf{r}_i = \sum_{k=1}^{n} \frac{\partial\mathbf{r}_i}{\partial q_k}\delta q_k\,, \tag{1.72}$$

since time is to be held fixed.[13] On the other hand,

$$\mathbf{v}_i = \frac{d\mathbf{r}_i}{dt} = \sum_{k=1}^{n} \frac{\partial\mathbf{r}_i}{\partial q_k}\dot{q}_k + \frac{\partial\mathbf{r}_i}{\partial t}\,. \tag{1.73}$$

[12] Intuitively, a differentiable manifold of dimension n is an abstract "surface" that locally looks like a neighbourhood of $\mathbb{R}^n$. In some concrete examples of mechanical systems, the pertinent differentiable manifold is $\mathbb{R}^n$ itself. For a rigorous definition of differentiable manifold the reader is referred to the second chapter of Hawking and Ellis (1973) or to the fourth chapter of Arnold (1989).

[13] The independent virtual displacements can be taken as tangent vectors to the configuration space understood as a differentiable manifold, thus avoiding the use of infinitesimal quantities (Arnold, 1989; José & Saletan, 1998; Lemos, 2005).

This is due to the fact that Lagrange's equations (1.100) have been derived from Newton's second law $\dot{\mathbf{p}}_i = \mathbf{F}_i$, which only holds true in inertial reference frames.

Invariance of Lagrange's Equations

Although the invariance of Lagrange's equations under a general coordinate transformation is obvious from the above derivation, a direct proof is instructive. If $Q_1, \ldots, Q_n$ are new generalised coordinates which are differentiable functions of the original coordinates $q_1, \ldots, q_n$, we have[15]

$$Q_k = G_k(q_1, \ldots, q_n, t), \quad k = 1, \ldots, n \tag{1.101}$$

and, conversely,

$$q_k = g_k(Q_1, \ldots, Q_n, t), \quad k = 1, \ldots, n. \tag{1.102}$$

Invertibility requires the following condition on the Jacobian of the transformation:

$$\frac{\partial(q_1, \ldots, q_n)}{\partial(Q_1, \ldots, Q_n)} \equiv \det\left(\frac{\partial q_k}{\partial Q_l}\right) = \left(\frac{\partial(Q_1, \ldots, Q_n)}{\partial(q_1, \ldots, q_n)}\right)^{-1} \neq 0. \tag{1.103}$$

The coordinate change (1.101) is called a **point transformation** because it maps points from the configuration space described by the qs into points of the configuration space described by the Qs. In mathematical terminology, a bijective differentiable application G whose inverse $g = G^{-1}$ is also differentiable is called a **diffeomorphism**, and the configuration space of the Qs is said to be **diffeomorphic** to the configuration space of the qs.

Taking the time derivative of Eq. (1.102), we find

$$\dot{q}_k = \sum_{l=1}^{n} \frac{\partial q_k}{\partial Q_l} \dot{Q}_l + \frac{\partial q_k}{\partial t}, \tag{1.104}$$

whence

$$\frac{\partial \dot{q}_k}{\partial \dot{Q}_l} = \frac{\partial q_k}{\partial Q_l}. \tag{1.105}$$

The transformed Lagrangian $\bar{L}(Q, \dot{Q}, t)$ is just the original Lagrangian $L(q, \dot{q}, t)$ expressed in terms of $(Q, \dot{Q}, t)$:

$$\bar{L}(Q, \dot{Q}, t) = L(q(Q, t), \dot{q}(Q, \dot{Q}, t), t). \tag{1.106}$$

Since $\partial q_k / \partial \dot{Q}_l = 0$, we can write

$$\frac{\partial \bar{L}}{\partial \dot{Q}_i} = \sum_{k=1}^{n} \left(\frac{\partial L}{\partial q_k} \frac{\partial q_k}{\partial \dot{Q}_i} + \frac{\partial L}{\partial \dot{q}_k} \frac{\partial \dot{q}_k}{\partial \dot{Q}_i}\right) = \sum_{k=1}^{n} \frac{\partial L}{\partial \dot{q}_k} \frac{\partial q_k}{\partial Q_i}, \tag{1.107}$$

---

[15] In spite of the coincident notation, these Qs are not to be confused with the components of the generalised force.

hence

$$\frac{d}{dt}\left(\frac{\partial \bar{L}}{\partial \dot{Q}_i}\right) = \sum_{k=1}^{n}\left[\frac{d}{dt}\left(\frac{\partial L}{\partial \dot{q}_k}\right)\frac{\partial q_k}{\partial Q_i} + \frac{\partial L}{\partial \dot{q}_k}\frac{d}{dt}\left(\frac{\partial q_k}{\partial Q_i}\right)\right]$$

$$= \sum_{k=1}^{n}\left[\frac{d}{dt}\left(\frac{\partial L}{\partial \dot{q}_k}\right)\frac{\partial q_k}{\partial Q_i} + \frac{\partial L}{\partial \dot{q}_k}\frac{\partial \dot{q}_k}{\partial Q_i}\right], \tag{1.108}$$

where have we used Eq. (1.78) with $\mathbf{r}_i$ replaced by q_k and q_k replaced by Q_i. This last equation combined with

$$\frac{\partial \bar{L}}{\partial Q_i} = \sum_{k=1}^{n}\left(\frac{\partial L}{\partial q_k}\frac{\partial q_k}{\partial Q_i} + \frac{\partial L}{\partial \dot{q}_k}\frac{\partial \dot{q}_k}{\partial Q_i}\right) \tag{1.109}$$

gives

$$\frac{d}{dt}\left(\frac{\partial \bar{L}}{\partial \dot{Q}_i}\right) - \frac{\partial \bar{L}}{\partial Q_i} = \sum_{k=1}^{n}\left[\frac{d}{dt}\left(\frac{\partial L}{\partial \dot{q}_k}\right) - \frac{\partial L}{\partial q_k}\right]\frac{\partial q_k}{\partial Q_i} = 0 \tag{1.110}$$

in virtue of (1.100), completing the proof. In mathematical language: Lagrange's equations are invariant under diffeomorphisms.

Even at the risk of annoying you, kind reader, we repeat that, although it can be expressed in terms of arbitrary generalised coordinates, the Lagrangian $L = T - V$ must be *initially* written in terms of positions and velocities relative to an inertial reference frame.

1.5 Applications of Lagrange's Equations

In order to write down Lagrange's equations associated with a given mechanical system, the procedure to be followed is rather simple. First, generalised coordinates $q_1, \ldots, q_n$ must be chosen. Next, the kinetic and potential energies relative to an inertial reference frame must be expressed exclusively in terms of the qs and $\dot{q}$s, so that the Lagrangian $L = T - V$ is also expressed only in terms of the generalised coordinates and velocities. Finally, all one has to do is compute the relevant partial derivatives of L, insert them into Eqs. (1.100) and the process of constructing the equations of motion for the system is finished. Let us see some examples of this procedure.

Example 1.17 A bead slides on a smooth straight massless rod which rotates with constant angular velocity on a horizontal plane. Describe its motion by Lagrange's formalism.

Solution

Let xy be the horizontal plane containing the rod and let us use polar coordinates to locate the bead of mass m (see Fig. 1.7). The two variables r, θ cannot be taken as generalised coordinates because θ is restricted to obey $\theta - \omega t = 0$, which is a holonomic constraint (ω is the rod's constant angular velocity, supposed known). The system has

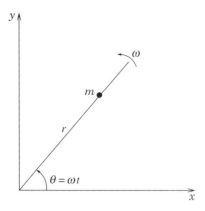

Fig. 1.7 Bead on horizontal rotating rod.

only one degree of freedom associated with the radial motion and we can choose $q_1 = r$ as generalised coordinate. According to Example 1.16, the kinetic energy can be put in the form

$$T = \frac{m}{2}(\dot{r}^2 + r^2\dot{\theta}^2) = \frac{m}{2}(\dot{r}^2 + \omega^2 r^2), \tag{1.111}$$

where $\dot{\theta} = \omega$ has been used. Setting the plane of motion as the zero level of the gravitational potential energy, the Lagrangian for the system reduces to the kinetic energy:

$$L = T - V = \frac{m}{2}(\dot{r}^2 + \omega^2 r^2). \tag{1.112}$$

Now that we have the Lagrangian expressed only in terms of r and $\dot{r}$, the equation of motion for the system follows at once:

$$\frac{d}{dt}\left(\frac{\partial L}{\partial \dot{r}}\right) - \frac{\partial L}{\partial r} = 0 \implies \frac{d}{dt}(m\dot{r}) - m\omega^2 r = 0 \implies \ddot{r} = \omega^2 r. \tag{1.113}$$

One concludes that the bead tends to move away from the rotation axis due to a "centrifugal force", which is the well-known result.

Example 1.18 Find a Lagrangian and the equations of motion for the mechanical system represented in Fig. 1.8 under the following conditions: the spring's natural length is l; the pulley and the inextensible string that connects masses m_1 and m_2 have negligible mass; the string remains taut all the time.

Solution

Let us employ the coordinates shown in Fig. 1.8 and assume that the string remains taut, which requires initial conditions carefully chosen (see Problem 1.2). The constraint $x_1 + x_2 = l_0$, where l_0 is a constant determined by the string's length and the radius of the pulley, shows that out of the three coordinates, only x_3 and either x_1 or x_2 can be

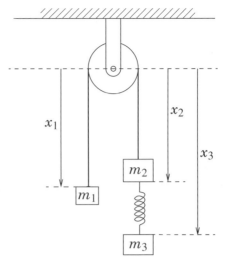

Fig. 1.8 Mechanical system of Example 1.18.

taken as generalised coordinates (the system has only two degrees of freedom). Let us choose x_2 and x_3 as generalised coordinates. The kinetic energy of the system is

$$T = \frac{m_1}{2}\dot{x}_1^2 + \frac{m_2}{2}\dot{x}_2^2 + \frac{m_3}{2}\dot{x}_3^2 = \frac{m_1 + m_2}{2}\dot{x}_2^2 + \frac{m_3}{2}\dot{x}_3^2, \qquad (1.114)$$

since from $x_1 = l_0 - x_2$ one derives $\dot{x}_1 = -\dot{x}_2$. Setting the zero level of the gravitational potential energy on the horizontal plane through the centre of the pulley, we have

$$V = -m_1 g x_1 - m_2 g x_2 - m_3 g x_3 + \frac{k}{2}(x_3 - x_2 - l)^2$$

$$= -(m_2 - m_1)g x_2 - m_1 g l_0 - m_3 g x_3 + \frac{k}{2}(x_3 - x_2 - l)^2, \qquad (1.115)$$

since points below the zero level have negative gravitational potential energy. Dropping an immaterial additive constant, the Lagrangian is

$$L = T - V = \frac{m_1 + m_2}{2}\dot{x}_2^2 + \frac{m_3}{2}\dot{x}_3^2 + (m_2 - m_1)g x_2 + m_3 g x_3 - \frac{k}{2}(x_3 - x_2 - l)^2. \quad (1.116)$$

We have

$$\frac{\partial L}{\partial \dot{x}_2} = (m_1 + m_2)\dot{x}_2, \qquad \frac{\partial L}{\partial x_2} = (m_2 - m_1)g + k(x_3 - x_2 - l), \qquad (1.117)$$

$$\frac{\partial L}{\partial \dot{x}_3} = m_3 \dot{x}_3, \qquad \frac{\partial L}{\partial x_3} = m_3 g - k(x_3 - x_2 - l), \qquad (1.118)$$

and Lagrange's equations are

$$\frac{d}{dt}\left(\frac{\partial L}{\partial \dot{x}_2}\right) - \frac{\partial L}{\partial x_2} = 0 \implies (m_1 + m_2)\ddot{x}_2 - (m_2 - m_1)g - k(x_3 - x_2 - l) = 0,$$

$$(1.119a)$$

$$\frac{d}{dt}\left(\frac{\partial L}{\partial \dot{x}_3}\right) - \frac{\partial L}{\partial x_3} = 0 \implies m_3 \ddot{x}_3 - m_3 g + k(x_3 - x_2 - l) = 0. \qquad (1.119b)$$

If $k = 0$ there is no interaction between m_2 and m_3. In this limiting case Lagrange's equations predict correctly that m_3 falls freely ($\ddot{x}_3 = g$) and that the accceleration of m_2 is $\ddot{x}_2 = (m_2 - m_1)g/(m_1 + m_2)$, in agreement with the result obtained in the treatment of Atwood's machine by d'Alembert's principle in Example 1.12.

Example 1.19 Apply the Lagrangian formalism to obtain the equations of motion for the plane double pendulum.

Solution

Let (x_1, y_1) and (x_2, y_2) be the Cartesian coordinates of masses m_1 and m_2, respectively. Taking the angles θ_1 and θ_2 as generalised coordinates (see Fig. 1.1 in Section 1.2), we have

$$x_1 = l_1 \sin\theta_1, \qquad\qquad y_1 = l_1 \cos\theta_1,$$
$$x_2 = l_1 \sin\theta_1 + l_2 \sin\theta_2, \quad y_2 = l_1 \cos\theta_1 + l_2 \cos\theta_2, \qquad (1.120)$$

whence

$$\dot{x}_1 = l_1\dot{\theta}_1 \cos\theta_1, \qquad \dot{x}_2 = l_1\dot{\theta}_1 \cos\theta_1 + l_2\dot{\theta}_2 \cos\theta_2, \qquad (1.121)$$
$$\dot{y}_1 = -l_1\dot{\theta}_1 \sin\theta_1, \qquad \dot{y}_2 = -l_1\dot{\theta}_1 \sin\theta_1 - l_2\dot{\theta}_2 \sin\theta_2. \qquad (1.122)$$

The kinetic energy relative to the inertial reference frame (x, y) is

$$T = \frac{m_1}{2}(\dot{x}_1^2 + \dot{y}_1^2) + \frac{m_2}{2}(\dot{x}_2^2 + \dot{y}_2^2), \qquad (1.123)$$

which in terms of the generalised coordinates and velocities takes the form

$$T = \frac{m_1 + m_2}{2}l_1^2\dot{\theta}_1^2 + \frac{m_2}{2}l_2^2\dot{\theta}_2^2 + m_2 l_1 l_2 \dot{\theta}_1 \dot{\theta}_2 \cos(\theta_1 - \theta_2). \qquad (1.124)$$

On the other hand, with the zero level of the gravitational potential energy on the horizontal plane through the point of suspension of m_1, we have

$$V = -m_1 g y_1 - m_2 g y_2 = -(m_1 + m_2)g l_1 \cos\theta_1 - m_2 g l_2 \cos\theta_2. \qquad (1.125)$$

Finally, the Lagrangian $L = T - V$ is given by

$$L = \frac{m_1 + m_2}{2}l_1^2\dot{\theta}_1^2 + \frac{m_2 l_2^2}{2}\dot{\theta}_2^2 + m_2 l_1 l_2 \dot{\theta}_1 \dot{\theta}_2 \cos(\theta_1 - \theta_2)$$
$$+ (m_1 + m_2)g l_1 \cos\theta_1 + m_2 g l_2 \cos\theta_2. \qquad (1.126)$$

Making use of

$$\frac{\partial L}{\partial \dot{\theta}_1} = (m_1 + m_2)l_1^2\dot{\theta}_1 + m_2 l_1 l_2 \dot{\theta}_2 \cos(\theta_1 - \theta_2), \qquad (1.127)$$

$$\frac{\partial L}{\partial \theta_1} = -m_2 l_1 l_2 \dot{\theta}_1 \dot{\theta}_2 \sin(\theta_1 - \theta_2) - (m_1 + m_2)g l_1 \sin\theta_1, \qquad (1.128)$$

the Lagrange equation

$$\frac{d}{dt}\left(\frac{\partial L}{\partial \dot{\theta}_1}\right) - \frac{\partial L}{\partial \theta_1} = 0 \qquad (1.129)$$

becomes

$$(m_1+m_2)l_1^2\ddot{\theta}_1+m_2l_1l_2\ddot{\theta}_2\cos(\theta_1-\theta_2)+m_2l_1l_2\dot{\theta}_2^2\sin(\theta_1-\theta_2)+(m_1+m_2)gl_1\sin\theta_1=0.\tag{1.130}$$

In an entirely analogous fashion, one finds

$$m_2l_2^2\ddot{\theta}_2+m_2l_1l_2\ddot{\theta}_1\cos(\theta_1-\theta_2)-m_2l_1l_2\dot{\theta}_1^2\sin(\theta_1-\theta_2)+m_2gl_2\sin\theta_2=0\tag{1.131}$$

for the second of Lagrange's equations.

1.6 Generalised Potentials and Dissipation Function

Lagrange's equations can have their range of validity extended to include a certain class of applied forces that depend on the generalised velocities.

Generalised Potentials

If the generalised forces can be derived from a function $U(q_1,\ldots,q_n,\dot{q}_1,\ldots,\dot{q}_n,t)$ by means of the equations

$$Q_k=-\frac{\partial U}{\partial q_k}+\frac{d}{dt}\left(\frac{\partial U}{\partial \dot{q}_k}\right),\tag{1.132}$$

then Eqs. (1.83) still imply Eqs. (1.100) with the Lagrangian defined by

$$L=T-U.\tag{1.133}$$

The function U is called a **generalised potential** or **velocity-dependent potential**. The set of forces encompassed by Eq. (1.132) is larger than the set of conservative forces, the latter corresponding to the particular case in which U depends neither on the generalised velocities nor on the time. The inclusion of forces derivable from a generalised potential is not fruit of a desire to achieve the greatest mathematical generality without physical consequences, for the electromagnetic force on a moving charge can only be derived from a generalised potential. In virtue of its great importance, this topic deserves a detailed discussion.

Lagrangian for a Charged Particle in an Electromagnetic Field

The force experienced by a particle with electric charge e moving in an external electromagnetic field is the **Lorentz force**[16]

$$\mathbf{F}=e\left(\mathbf{E}+\frac{1}{c}\mathbf{v}\times\mathbf{B}\right).\tag{1.134}$$

---

[16] In CGS Gaussian units.

Maxwell's equations allow us to represent the fields in terms of a scalar potential $\phi(\mathbf{r}, t)$ and a vector potential $\mathbf{A}(\mathbf{r}, t)$ in the following way (Marion & Heald, 1980):

$$\mathbf{E} = -\nabla\phi - \frac{1}{c}\frac{\partial \mathbf{A}}{\partial t}, \qquad \mathbf{B} = \nabla \times \mathbf{A}. \qquad (1.135)$$

Choosing the Cartesian coordinates of the particle themselves as generalised coordinates, the generalised force components are just the Cartesian components of the Lorentz force

$$\mathbf{F} = e\left\{-\nabla\phi - \frac{1}{c}\frac{\partial \mathbf{A}}{\partial t} + \frac{1}{c}\mathbf{v} \times (\nabla \times \mathbf{A})\right\}. \qquad (1.136)$$

We intend to show that $\mathbf{F}$ can be represented in the form (1.132) for some function U. But in (1.132) there is a *total* time derivative, whereas in (1.136) there appears a *partial* derivative with respect to time. We can bring about a total time derivative in (1.136) by noting that

$$\frac{d\mathbf{A}}{dt} = \frac{\partial \mathbf{A}}{\partial x}\dot{x} + \frac{\partial \mathbf{A}}{\partial y}\dot{y} + \frac{\partial \mathbf{A}}{\partial z}\dot{z} + \frac{\partial \mathbf{A}}{\partial t} = (\mathbf{v} \cdot \nabla)\mathbf{A} + \frac{\partial \mathbf{A}}{\partial t}. \qquad (1.137)$$

Using also

$$\mathbf{v} \times (\nabla \times \mathbf{A}) = \nabla(\mathbf{v} \cdot \mathbf{A}) - (\mathbf{v} \cdot \nabla)\mathbf{A}, \qquad (1.138)$$

which follows from the general identity

$$\nabla(\mathbf{a} \cdot \mathbf{b}) = (\mathbf{a} \cdot \nabla)\mathbf{b} + (\mathbf{b} \cdot \nabla)\mathbf{a} + \mathbf{a} \times (\nabla \times \mathbf{b}) + \mathbf{b} \times (\nabla \times \mathbf{a}) \qquad (1.139)$$

since the nabla operator affects only the position variables, Eq. (1.136) becomes

$$\mathbf{F} = e\left\{-\nabla\phi - \frac{1}{c}\frac{d\mathbf{A}}{dt} + \frac{1}{c}\nabla(\mathbf{v} \cdot \mathbf{A})\right\}. \qquad (1.140)$$

With the use of the operator $\nabla_v = \hat{\mathbf{x}}\partial/\partial\dot{x} + \hat{\mathbf{y}}\partial/\partial\dot{y} + \hat{\mathbf{z}}\partial/\partial\dot{z}$ and taking into account that generalised coordinates and velocities are treated as independent variables, we are left with

$$\mathbf{F} = e\left\{-\nabla\left(\phi - \frac{1}{c}\mathbf{v} \cdot \mathbf{A}\right)\right\} - \frac{e}{c}\frac{d\mathbf{A}}{dt} = -\nabla\left(e\phi - \frac{e}{c}\mathbf{v} \cdot \mathbf{A}\right) + \frac{d}{dt}\left[\nabla_v\left(e\phi - \frac{e}{c}\mathbf{v} \cdot \mathbf{A}\right)\right], \qquad (1.141)$$

inasmuch as ϕ and $\mathbf{A}$ do not depend on the velocities.

Recalling that $q_1 = x, q_2 = y, q_3 = z$, the force $\mathbf{F}$ is of the form (1.132) with

$$U = e\phi - \frac{e}{c}\mathbf{v} \cdot \mathbf{A}, \qquad (1.142)$$

and it follows that

$$L = T - U = \frac{mv^2}{2} - e\phi + \frac{e}{c}\mathbf{v} \cdot \mathbf{A} \qquad (1.143)$$

is the Lagrangian for a charged particle in an external electromagnetic field.

Rayleigh's Dissipation Function

Whenever the generalised forces are of the form

$$Q_k = -\frac{\partial U}{\partial q_k} + \frac{d}{dt}\left(\frac{\partial U}{\partial \dot{q}_k}\right) + Q_k', \tag{1.144}$$

where Q_k' denotes the part of the generalised forces that cannot be derived from a generalised potential, the equations of motion (1.83) become

$$\frac{d}{dt}\left(\frac{\partial L}{\partial \dot{q}_k}\right) - \frac{\partial L}{\partial q_k} = Q_k', \tag{1.145}$$

with $L = T - U$. A case of some importance is the one in which the Q_k' represent viscous frictional forces proportional to the velocities of the particles. In Cartesian components,

$$F_{ix}' = -k_{ix}v_{ix}, \quad F_{iy}' = -k_{iy}v_{iy}, \quad F_{iz}' = -k_{iz}v_{iz}, \tag{1.146}$$

where $\mathbf{F}_i'$ is the dissipative force on the ith particle and k_{ix}, k_{iy}, k_{iz} are positive constants. In order to facilitate a more general treatment of such situations, Rayleigh introduced the **dissipation function** defined by

$$\mathcal{F} = \frac{1}{2}\sum_{i=1}^{N}(k_{ix}v_{ix}^2 + k_{iy}v_{iy}^2 + k_{iz}v_{iz}^2), \tag{1.147}$$

with the property that

$$F_{ix}' = -\frac{\partial \mathcal{F}}{\partial v_{ix}}, \quad F_{iy}' = -\frac{\partial \mathcal{F}}{\partial v_{iy}}, \quad F_{iz}' = -\frac{\partial \mathcal{F}}{\partial v_{iz}}. \tag{1.148}$$

An examination of the work per unit time done by the dissipative forces brings to the fore the physical meaning of Rayleigh's dissipation function:

$$\frac{dW'}{dt} = \sum_{i=1}^{N}\mathbf{F}_i' \cdot \frac{d\mathbf{r}_i}{dt} = \sum_{i=1}^{N}\mathbf{F}_i' \cdot \mathbf{v}_i = -\sum_{i=1}^{N}(k_{ix}v_{ix}^2 + k_{iy}v_{iy}^2 + k_{iz}v_{iz}^2) = -2\mathcal{F}, \tag{1.149}$$

and it turns out that $2\mathcal{F}$ is the rate of dissipation of the system's energy.

The dissipative part of the generalised forces can be written as

$$
\begin{aligned}
Q_k' &= \sum_{i=1}^{N}\mathbf{F}_i' \cdot \frac{\partial \mathbf{r}_i}{\partial q_k} = \sum_{i=1}^{N}\mathbf{F}_i' \cdot \frac{\partial \mathbf{v}_i}{\partial \dot{q}_k} \\
&= -\sum_{i=1}^{N}\left(\frac{\partial \mathcal{F}}{\partial v_{ix}}\frac{\partial v_{ix}}{\partial \dot{q}_k} + \frac{\partial \mathcal{F}}{\partial v_{iy}}\frac{\partial v_{iy}}{\partial \dot{q}_k} + \frac{\partial \mathcal{F}}{\partial v_{iz}}\frac{\partial v_{iz}}{\partial \dot{q}_k}\right) = -\frac{\partial \mathcal{F}}{\partial \dot{q}_k},
\end{aligned} \tag{1.150}
$$

where we have used (1.79) and the chain rule. With this last result, the equations of motion (1.145) become

$$\frac{d}{dt}\left(\frac{\partial L}{\partial \dot{q}_k}\right) - \frac{\partial L}{\partial q_k} + \frac{\partial \mathcal{F}}{\partial \dot{q}_k} = 0. \tag{1.151}$$

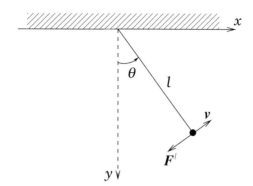

Fig. 1.9   Pendulum with air resistance.

The main (and probably only) advantage derived from resorting to $\mathcal{F}$ lies in that it allows the equations of motion to be written in the same form (1.151) for any choice of generalised coordinates.

Example 1.20   Use Rayleigh's dissipation function to describe the damped oscillations of a simple pendulum with air resistance proportional to velocity.

Solution

Using polar coordinates in the xy-plane of Fig. 1.9, we have $r = l$ and, according to Example 1.16, $T = ml^2\dot{\theta}^2/2$. Since $V = -mgl\cos\theta$, it follows that

$$L = \frac{ml^2}{2}\dot{\theta}^2 + mgl\cos\theta. \tag{1.152}$$

On the assumption that $k_x = k_y = k$, Rayleigh's dissipation function (1.147) is given by

$$\mathcal{F} = \frac{1}{2}kv^2 = \frac{1}{2}kl^2\dot{\theta}^2. \tag{1.153}$$

Equation (1.151) for θ takes the form

$$\ddot{\theta} + \frac{k}{m}\dot{\theta} + \frac{g}{l}\sin\theta = 0. \tag{1.154}$$

In the case of small oscillations, this equation of motion reduces to that of a damped harmonic oscillator.

1.7 Central Forces and Bertrand's Theorem

The problem of the motion of a particle in a central potential, solved by Isaac Newton in the gravitational case in the *Principia*, is certainly the most celebrated of all applications of the laws of motion.

Consider a particle subject to a force that depends only on the particle's distance to a fixed point (the centre of force) and is directed along the line determined by the particle and the centre of force, which will be taken at the origin of the coordinate system chosen for the description of the motion. On account of the symmetry of the problem, it is convenient to employ spherical coordinates r, θ, ϕ defined by $x = r \sin \theta \cos \phi, y = r \sin \theta \sin \phi$, $z = r \cos \theta$. In terms of spherical coordinates, the Cartesian components of the particle's velocity are

$$
\begin{aligned}
\dot{x} &= \dot{r} \sin \theta \cos \phi + r \dot{\theta} \cos \theta \cos \phi - r \dot{\phi} \sin \theta \sin \phi \,, \\
\dot{y} &= \dot{r} \sin \theta \sin \phi + r \dot{\theta} \cos \theta \sin \phi + r \dot{\phi} \sin \theta \cos \phi \,, \\
\dot{z} &= \dot{r} \cos \theta - r \dot{\theta} \sin \theta \,,
\end{aligned}
\tag{1.155}
$$

and the Lagrangian for the system takes the form

$$
L = \frac{m}{2} (\dot{x}^2 + \dot{y}^2 + \dot{z}^2) - V(r) = \frac{m}{2} (\dot{r}^2 + r^2 \dot{\theta}^2 + r^2 \dot{\phi}^2 \sin^2 \theta) - V(r) \,,
\tag{1.156}
$$

where $V(r)$ is the potential energy associated with the central force $\mathbf{F}$. The Lagrangian (1.156) can be more readily obtained by using $\mathbf{v} = \dot{r} \hat{\mathbf{r}} + r \dot{\theta} \hat{\boldsymbol{\theta}} + r \dot{\phi} \sin \theta \hat{\boldsymbol{\phi}}$.

The force is collinear to the position vector $\mathbf{r}$, therefore the torque with respect to the origin is zero and the angular momentum $\boldsymbol{l} = \mathbf{r} \times \mathbf{p}$ is a constant vector. Since $\boldsymbol{l}$ is orthogonal to the plane defined by $\mathbf{r}$ and $\mathbf{p}$, the motion takes place in a plane perpendicular to the vector $\boldsymbol{l}$. Taking the z-axis parallel to $\boldsymbol{l}$, the particle moves in the plane defined by $z = 0$ or, equivalently, $\theta = \pi/2$. Thus, r and ϕ become polar coordinates on the xy-plane and, using $\theta = \pi/2$ and $\dot{\theta} = 0$, the Lagrangian (1.156) takes the simpler form

$$
L = \frac{m}{2} (\dot{r}^2 + r^2 \dot{\phi}^2) - V(r) \,.
\tag{1.157}
$$

Lagrange's equations are

$$
m \ddot{r} - m r \dot{\phi}^2 + \frac{dV}{dr} = 0 \,,
\tag{1.158}
$$

$$
\frac{d}{dt} (m r^2 \dot{\phi}) = 0 \,.
\tag{1.159}
$$

The second of these equations has for solution

$$
m r^2 \dot{\phi} = l = \text{constant} \,,
\tag{1.160}
$$

which corresponds to Kepler's law of areas: the line from the Sun to the planet sweeps out equal areas in equal times (see Exercise 1.7.2). The constant l is the magnitude of the angular momentum.

By energy conservation,

$$
T + V = \frac{m}{2} (\dot{r}^2 + r^2 \dot{\phi}^2) + V(r) = E = \text{constant} \,.
\tag{1.161}
$$

With the use of (1.160) this equation can be recast in the form

$$
E = \frac{m}{2} \dot{r}^2 + \frac{l^2}{2 m r^2} + V(r) \equiv \frac{m}{2} \dot{r}^2 + V_{\text{eff}}(r) \,,
\tag{1.162}
$$

where

$$V_{\text{eff}}(r) = \frac{l^2}{2mr^2} + V(r) \,. \tag{1.163}$$

Equation (1.162) shows that the radial motion is equivalent to the one-dimensional motion of a particle in the **effective potential** $V_{\text{eff}}(r)$.

Exercise 1.7.1 If $r = r_0$ is a circular orbit with energy E_0, prove that $E_0 = V_{\text{eff}}(r_0)$ and $V'_{\text{eff}}(r_0) = 0$ or, equivalently, $l^2/mr_0^3 = V'(r_0)$. Prove further that

$$V''_{\text{eff}}(r_0) = \frac{3}{r_0} V'(r_0) + V''(r_0) \,. \tag{1.164}$$

With the help of

$$\frac{d}{dt} = \frac{d\phi}{dt}\frac{d}{d\phi} = \dot{\phi}\frac{d}{d\phi} = \frac{l}{mr^2}\frac{d}{d\phi} \tag{1.165}$$

time can be eliminated from Eq. (1.162), which takes the new form

$$\frac{l^2}{2mr^4}\left(\frac{dr}{d\phi}\right)^2 + V_{\text{eff}}(r) = E \,, \tag{1.166}$$

from which one derives the equation for the trajectory:

$$\phi = \int \frac{l\,dr}{r^2\,\sqrt{2m[E - V_{\text{eff}}(r)]}} \,. \tag{1.167}$$

The Kepler Problem

The potential energy of a planet in the Sun's gravitational field is $V(r) = -\kappa/r$ with $\kappa = GmM$, where m is the mass of the planet, M is the mass of the Sun and G is the gravitational constant. Introducing the variable $u = 1/r$, the integration in (1.167) becomes elementary:

$$\phi = -\int \frac{l\,du}{\sqrt{2mE + 2m\kappa u - l^2 u^2}} = \cos^{-1}\frac{l/r - m\kappa/l}{\sqrt{2mE + \kappa^2 m^2/l^2}} \,, \tag{1.168}$$

where the origin of the angle was chosen so as to make the constant of integration zero. Defining

$$e = \sqrt{1 + \frac{2El^2}{m\kappa^2}} \,, \tag{1.169}$$

the solution of (1.168) for r is

$$\frac{1}{r} = \frac{m\kappa}{l^2}(1 + e\cos\phi) \,. \tag{1.170}$$

This equation represents a conic section with a focus at the origin. If $E < 0$ then $e < 1$ and the orbit is an ellipse with eccentricity e. If $E = 0$ ($e = 1$) the orbit is a parabola, and if $E > 0$ ($e > 1$) the orbit is a hyperbola. In Goldstein (1980) and Fasano and Marmi (2006) the reader will find an extensive and detailed discussion of the problem of the motion under a central force in general, and of the Kepler problem in particular.

Exercise 1.7.2 (1) For $0 < e < 1$, show that the maximum and minimum values of r are given by $r_{\max} = p(1-e)^{-1}$ and $r_{\min} = p(1+e)^{-1}$, where $p = l^2/m\kappa$. Conclude that the semi-major axis of the ellipse, $a = (r_{\max} + r_{\min})/2$, can be expressed as $a = p(1-e^2)^{-1}$. (2) If the angle ϕ increases by $d\phi$, the area swept out by the radius vector $\mathbf{r}$ is $dA = \frac{1}{2}r^2 d\phi$. Prove Kepler's second law or law of areas: $dA/dt = l/2m$. (3) During a period τ of the motion, the area swept out by the radius vector equals the area of the ellipse, $A = \pi ab$. Using the fact that by definition of eccentricity the semi-minor axis of the ellipse is $b = a\sqrt{1-e^2}$, prove that $\tau = 2\pi a^{3/2}\sqrt{m/\kappa} = 2\pi a^{3/2}/\sqrt{GM}$. Neglecting the motion of the Sun, this is Kepler's third law, according to which the square of the period is proportional to the cube of the semi-major axis of the elliptical orbit with the same proportionality constant for all planets.

Bertrand's Theorem

An orbit is said to be bounded if the distance from the particle to the centre of force remains between two extreme values – that is, $r_{\min} \le r \le r_{\max}$. A bounded orbit does not need to be closed: the particle may go around indefinitely always inside the annulus $r_{\min} \le r \le r_{\max}$ without the trajectory ever closing. As we have just seen, in the case of the gravitational or the attractive Coulomb potential all bounded orbits are closed. A famous result, obtained by J. Bertrand in 1873, establishes that this is a rare phenomenon for central potentials (Bertrand, 1873).

Theorem 1.7.1 (Bertrand) *The only central potentials for which all bounded orbits are closed are $V(r) = -\kappa/r$ and $V(r) = \kappa r^2$, with $\kappa > 0$ – that is, either the force is inversely proportional to the distance squared or obeys Hooke's law.*

Proof Let Φ be the angle of rotation during a complete radial oscillation, from $r = r_{\min}$ to $r = r_{\max}$ and back to $r = r_{\min}$. From Eq. (1.167),

$$\Phi = 2\int_{r_{\min}}^{r_{\max}} \frac{l\,dr}{r^2\sqrt{2m[E - V_{\mathrm{eff}}(r)]}} = 2\int_{u_{\min}}^{u_{\max}} \frac{l\,du}{\sqrt{2m[E - W(u)]}}, \qquad (1.171)$$

where $u = 1/r$ and

$$W(u) = V_{\mathrm{eff}}\left(\frac{1}{u}\right) = \frac{l^2 u^2}{2m} + V\left(\frac{1}{u}\right). \qquad (1.172)$$

A bounded orbit is closed if, after a whole number (n) of complete radial oscillations, the angular displacement is a whole number (m) of 2π radians: $n\Phi = m2\pi$. If all bounded orbits are closed, those near circular orbits are equally closed. As implied by the equivalent one-dimensional problem defined by Eq. (1.162), the effective potential $V_{\mathrm{eff}}(r)$ goes through a minimum at some r_0 between $r_{\min}$ and $r_{\max}$. Let us consider a bounded orbit caused by a slight perturbation of the circular orbit of radius r_0 and energy E_0. Putting $E = E_0 + \Delta E$ and expanding $W(u)$ about $u = u_0 = 1/r_0$ up to second-order terms, the result is

$$E - W(u) = E_0 + \Delta E - W(u_0) - W'(u_0)(u - u_0) - \frac{1}{2}W''(u_0)(u - u_0)^2. \qquad (1.173)$$

According to Exercise 1.7.1 we have $E_0 - W(u_0) = E_0 - V_{\mathrm{eff}}(r_0) = 0$,

$$W'(u_0) = \left[\frac{l^2 u}{m} - \frac{1}{u^2} V'\left(\frac{1}{u}\right) \right]_{u=1/r_0} = \frac{1}{r_0} \left[\frac{l^2}{m} - r_0^3 V'(r_0) \right] = 0, \qquad (1.174)$$

and also

$$W''(u_0) = \left[\frac{l^2}{m} + \frac{2}{u^3} V'\left(\frac{1}{u}\right) + \frac{1}{u^4} V''\left(\frac{1}{u}\right) \right]_{u=1/r_0}$$

$$= 3r_0^3 V'(r_0) + r_0^4 V''(r_0) = r_0^4 V_{\mathrm{eff}}''(r_0). \qquad (1.175)$$

Consequently,

$$E - W(u) = \Delta E - \frac{1}{2} W''(u_0)(u - u_0)^2 \qquad (1.176)$$

and the expression (1.171) for the rotation angle reduces to

$$\Phi = 2l \int_{u_{\min}}^{u_{\max}} \frac{du}{\sqrt{2m\Delta E - mW''(u_0)(u - u_0)^2}}. \qquad (1.177)$$

Setting $a^2 = 2\Delta E / W''(u_0)$ and $u = u_0 + a \sin\theta$, we get

$$\Phi = \frac{2l}{\sqrt{mW''(u_0)}} \int_{-\pi/2}^{\pi/2} \frac{a \cos\theta \, d\theta}{\sqrt{a^2(1 - \sin^2\theta)}}$$

$$= \frac{2\pi l}{r_0^2 \sqrt{m V_{\mathrm{eff}}''(r_0)}} = 2\pi \sqrt{\frac{V'(r_0)}{3V'(r_0) + r_0 V''(r_0)}}, \qquad (1.178)$$

where we have once again made use of the results in Exercise 1.7.1. Varying the angular momentum l the radius r_0 varies continuously, but Φ cannot vary continuously because it is a rational multiple of 2π. The impasse is resolved by demanding that the rightmost term in the chain of equalities (1.178) be a constant independent of r_0, which requires the potential $V(r)$ to satisfy the differential equation $rV''(r) + 3V'(r) = \alpha V'(r)$, where α is a positive constant. Therefore, the admissible potentials are only $V(r) = \kappa r^\beta$ ($\beta > -2$, $\beta \neq 0$) or $V(r) = \kappa \ln r$. For these potentials

$$\Phi = \frac{2\pi}{\sqrt{\beta + 2}}, \qquad (1.179)$$

with $\beta = 0$ corresponding to the logarithmic case. The logarithmic potential is not acceptable because $2\pi/\sqrt{2}$ is not a rational multiple of 2π. The result (1.179) does not depend on the energy, so it must remain valid in the limits $E \to \infty$ or $E \to 0$. The essential idea of the rest of the proof consists in finding expressions for the rotation angle in these limiting cases and requiring that they coincide with (1.179).

Since the force has to be attractive in order that bounded orbits exist, $V'(r) > 0$. Hence, $\kappa > 0$ if $\beta > 0$ and $\kappa < 0$ if $-2 < \beta < 0$. In the case $V(r) = \kappa r^\beta$ with $\kappa > 0$ and $\beta > 0$, let us consider the limit $E \to \infty$. From equation $E = l^2/2mr^2 + \kappa r^\beta$, valid for $r = r_{\min}$ or $r = r_{\max}$, it follows that $r_{\max} \to \infty$ and $r_{\min} \to 0$ as $E \to \infty$, so that $u_{\min} \to 0$ and

$u_{max} \to \infty$. With the change of variable $u = y u_{max}$, Eq. (1.171) for the rotation angle is transformed into

$$\Phi = \frac{2l}{\sqrt{2m}} \int_{y_{min}}^{1} \frac{dy}{\sqrt{\mathcal{U}(1) - \mathcal{U}(y)}}, \tag{1.180}$$

where

$$\mathcal{U}(y) = y^2 \left[\frac{l^2}{2m} + \frac{\kappa}{(y u_{max})^{\beta+2}} \right]. \tag{1.181}$$

In the limit $E \to \infty$ the integral (1.180) reduces to

$$\Phi = \frac{2l}{\sqrt{2m}} \int_0^1 \frac{dy}{\sqrt{\frac{l^2}{2m} - \frac{l^2 y^2}{2m}}} = \frac{2l}{\sqrt{2m}} \frac{\sqrt{2m}}{l} \int_0^1 \frac{dy}{\sqrt{1 - y^2}} = \pi. \tag{1.182}$$

Comparing this result with (1.179), we find $\beta = 2$. For the harmonic potential $V(r) = \kappa r^2$ it is easily shown that all orbits are bounded and closed.

In the case of both κ and β negative, it is better to write $\kappa = -\sigma$ and $\beta = -\lambda$ with $\sigma > 0$ and $0 < \lambda < 2$. In the limit $E \to 0$ the extreme values of u are the solutions to $0 = \frac{l^2}{2mr^2} - \frac{\sigma}{r^\lambda} = \frac{l^2 u^2}{2m} - \sigma u^\lambda$, namely $u_{min} = 0$ and $u_{max} = (2m\sigma/l^2)^{1/(2-\lambda)}$. With $E = 0$ and the same previous substitution $u = y u_{max}$ the expression (1.171) for the rotation angle becomes

$$\Phi = \frac{2l}{\sqrt{2m}} \int_0^1 \frac{dy}{\sqrt{-\frac{l^2 y^2}{2m} + \frac{\sigma y^\lambda}{u_{max}^{2-\lambda}}}} = 2 \int_0^1 \frac{dy}{\sqrt{y^\lambda - y^2}}. \tag{1.183}$$

The introduction of the new variable $x = y^{(2-\lambda)/2}$ leads to

$$\Phi = \frac{4}{2-\lambda} \int_0^1 \frac{dx}{\sqrt{1 - x^2}} = \frac{2\pi}{2-\lambda} = \frac{2\pi}{2+\beta}. \tag{1.184}$$

A comparison of this result with (1.179) yields $\beta = -1$, which completes the proof. $\quad\square$

The present proof, that shows some similarity to Bertrand's original proof, has followed the indications of Arnold (1989). For alternative, but much less elegant, proofs the reader is referred to Goldstein (1980) or Desloge (1982).

Note that $\Phi = 2\pi$ in the case of a force inversely proportional to distance squared, so that the orbit closes in a single radial oscillation. In the case of a Hooke's law force, $\Phi = \pi$ and the orbit closes only after two complete radial oscillations.

At planetary distances, the general theory of relativity corrects the Newtonian potential due to the Sun by the addition of a small term proportional to $1/r^3$. As predicted by Bertrand's theorem, the orbits of the planets are not closed: each planet describes an ellipse whose axis slowly rotates about the Sun. The effect is most pronounced for Mercury. The French astronomer Urbain Le Verrier was the first to report, in 1859, that Newtonian mechanics and perturbations by the other known planets could not completely explain the rate of precession of Mercury's orbit around the Sun. The correct prediction of the precession rate of Mercury's perihelion is one of the greatest triumphs of Einstein's theory of gravitation.

Problems

1.1   In order to decide whether or not a system is holonomic one has to consider all constraints jointly, for the constraints may be integrable when taken together although each separately is non-integrable. For instance, consider the constraints (Neĭmark & Fufaev, 1972)

$$(x^2 + y^2)dx + xzdz = 0, \qquad (x^2 + y^2)dy + yzdz = 0.$$

(a) With the help of the Frobenius theorem in Appendix B, prove that each of these constraints is separately non-integrable. (b) Prove, however, that taken together they constitute an integrable system and are equivalent to

$$xd(x^2 + y^2 + z^2) = 0, \qquad d\ln\frac{y}{x} = 0.$$

(c) Conclude that the system is holonomic with constraints

$$x^2 + y^2 + z^2 = C_1, \qquad y = C_2 x.$$

1.2   Consider the system from Example 1.18 in the case in which all masses are equal ($m_1 = m_2 = m_3 = m$) and the system is released from rest with $x_2 = 0$ and $x_3 = l$. Solve the equations of motion (1.119) to show that

$$x_2(t) = \frac{2mg}{9k}(\cos\omega t - 1) + \frac{g}{6}t^2, \qquad \omega = \sqrt{\frac{3k}{2m}},$$

and deduce $\dot{x}_2(t) = (g/3\omega)(\omega t - \sin\omega t)$. Prove that $\dot{x}_2(t) > 0$ for all $t > 0$ and conclude that the string remains always taut. This justifies *a posteriori* the use of equations (1.119).

1.3   A bead of mass m slides along a frictionless rigid rod of negligible mass which rotates in a vertical plane with constant angular velocity ω. Show that, with an appropriate choice of coordinate r, the Lagrangian for the system is

$$L = \frac{m}{2}\dot{r}^2 + \frac{m\omega^2}{2}r^2 - mgr\sin\omega t.$$

Given the initial conditions $r(0) = r_0$, $\dot{r}(0) = 0$, find the solution to Lagrange's equation for r.

1.4   Obtain a Lagrangian and Lagrange's equations for a spherical pendulum – that is, a mass m suspended by a light and inextensible string of length l whose motion is not restricted to a plane.

1.5   Consider the so-called swinging Atwood's machine shown in Fig. 1.10, in which M moves only vertically (Tufillaro, Abbott & Griffiths, 1984). Using the coordinates indicated in the figure, show that the Lagrangian is given by

$$L = \frac{m + M}{2}\dot{r}^2 + \frac{m}{2}r^2\dot{\theta}^2 - gr(M - m\cos\theta)$$

and write down Lagrange's equations.

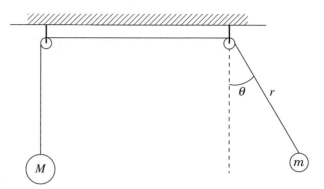

Fig. 1.10   Swinging Atwood's machine of Problem 1.5.

1.6 Huygens' cycloidal pendulum consists of a particle which oscillates under gravity in a vertical plane along a frictionless cycloidal track with parametric equations

$$x = R(\theta - \sin\theta), \quad y = R(1 - \cos\theta),$$

where the vertical y-axis points downward. Show that a Lagrangian for this system is

$$L = 2mR^2 \sin^2\left(\frac{\theta}{2}\right)\dot\theta^2 + mgR(1 - \cos\theta).$$

Making the point transformation $u = \cos(\theta/2)$, find the Lagrangian and Lagrange's equation in terms of the coordinate u. Prove that the period of oscillation is equal to $4\pi(R/g)^{1/2}$, therefore independent of the amplitude θ_0.

1.7 A projectile is fired near the surface of the Earth. Assuming the force of air resistance is proportional to the velocity, obtain the projectile's equations of motion using the dissipation function $\mathcal{F} = \lambda v^2/2$.

1.8 Certain dissipative systems admit a Lagrangian formulation that dispenses with the Rayleigh dissipation function. Consider a projectile in the constant gravitational field $\mathbf{g} = -g\hat{\mathbf{y}}$ and assume the force of air resistance is proportional to the velocity. (a) Show that the equations of motion generated by the Lagrangian

$$L = e^{\lambda t/m}\left[\frac{m}{2}\left(\dot{x}^2 + \dot{y}^2\right) - mgy\right]$$

coincide with those obtained in Problem 1.7. (b) Solve the equations of motion for x and y assuming the projectile is fired from the origin with velocity of magnitude v_0 making the angle θ_0 with the horizontal. (c) Eliminate time to get the equation for the trajectory of the projectile.

1.9 The point of suspension of a pendulum of mass m and length l can move horizontally connected to two identical springs with force constant k (see Fig. 1.11). There is no friction and the point of suspension has negligible mass. (a) Taking as generalised coordinates the displacement x of the point of suspension and the angle θ the pendulum makes with the vertical, find the Lagrangian and the equations of motion. (b) Obtain an approximate form for the equations of motion in the case of small angular oscillations: $\sin\theta \approx \theta$, $\cos\theta \approx 1$ and terms containing θ^n or $\dot\theta^n$ with $n > 1$

The arc length s is a functional of y, since to each continuously differentiable function $y(x)$ there corresponds the only real number $s[y]$ defined by (2.2).

Example 2.2 A light ray propagates in the xy-plane in an optically heterogeneous medium, with variable index of refraction $n = n(x, y)$. The function $n(x, y)$ depends only on the medium and is supposed known. The time dt spent by the light ray to travel the infinitesimal distance ds is

$$dt = \frac{ds}{v} = \frac{ds}{c/n} = \frac{1}{c} n(x, y) \sqrt{1 + y'^2}\, dx, \qquad (2.3)$$

where c is the speed of light in vacuum. Thus, the time T the light ray takes to propagate from the point (x_1, y_1) to the point (x_2, y_2) along the curve $y(x)$ is

$$T[y] = \frac{1}{c} \int_{x_1}^{x_2} n(x, y(x)) \sqrt{1 + y'(x)^2}\, dx. \qquad (2.4)$$

Therefore, T is a functional of y.

The simplest problem of the calculus of variations can be thus stated: given the functional

$$J[y] = \int_{x_1}^{x_2} f(y(x), y'(x), x)\, dx, \qquad (2.5)$$

where $f : \mathbb{R}^3 \to \mathbb{R}$ is a sufficiently differentiable known function, find among all twice continuously differentiable curves $y(x)$ that pass through the fixed points (x_1, y_1) and (x_2, y_2) the one that minimises or, more generally, extremises J.

In order to find the curve that extremises J, we shall make use of a device due to Lagrange that reduces the problem to that of finding extrema of a real function of a real variable. Let $y(x)$ be the sought function that extremises[1] the functional J and consider a neighbouring curve $\bar{y}$ defined by (Fig. 2.1)

$$\bar{y}(x) = y(x) + \epsilon \eta(x), \qquad (2.6)$$

where ϵ is a real parameter and $\eta(x)$ is a continuously differentiable function that vanishes at $x = x_1$ and $x = x_2$:

$$\eta(x_1) = \eta(x_2) = 0. \qquad (2.7)$$

These conditions are necessary in order that the varied curve $\bar{y}$ also pass through the endpoints (x_1, y_1) and (x_2, y_2). Substituting $\bar{y}$ for y in (2.5) the result is a function only of ϵ:

$$\Phi(\epsilon) \equiv J[\bar{y}] = \int_{x_1}^{x_2} f(\bar{y}(x), \bar{y}'(x), x)\, dx. \qquad (2.8)$$

Since, by hypothesis, the curve $y(x)$ furnishes an extremum (minimum, maximum or stationary point) of J, the function $\Phi(\epsilon)$ must have an extremum for $\epsilon = 0$ because, in

---

[1] We shall assume that such a function exists. The issue of existence of solutions to variational problems is a very delicate mathematical question (Courant & Hilbert, 1953).

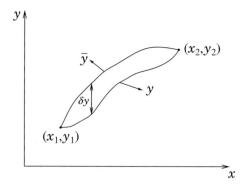

Variation of a curve: $\delta y(x) = \epsilon \eta(x)$.

this case, $\bar{y}$ becomes identical to y. Therefore, a necessary condition for $y(x)$ to extremise J is

$$0 = \left(\frac{d\Phi}{d\epsilon} \right)_{\epsilon=0} = \int_{x_1}^{x_2} \left(\frac{\partial f}{\partial \bar{y}} \frac{\partial \bar{y}}{\partial \epsilon} + \frac{\partial f}{\partial \bar{y}'} \frac{\partial \bar{y}'}{\partial \epsilon} \right)_{\epsilon=0} dx \,, \qquad (2.9)$$

which results from differentiating (2.8) under the integral sign. Using

$$\frac{\partial \bar{y}}{\partial \epsilon} = \eta \,, \qquad \frac{\partial \bar{y}'}{\partial \epsilon} = \eta' \,, \qquad (2.10)$$

equation (2.9) takes the form (inasmuch as $\bar{y} = y$ for $\epsilon = 0$)

$$\left(\frac{d\Phi}{d\epsilon} \right)_{\epsilon=0} = \int_{x_1}^{x_2} \left(\frac{\partial f}{\partial y} \eta + \frac{\partial f}{\partial y'} \eta' \right) dx = 0 \,. \qquad (2.11)$$

This last equation is true for $\eta(x)$ arbitrary except for the requirement that it vanish at $x = x_1$ and $x = x_2$. In order to extract an equation for $y(x)$, one must exploit the arbitrariness of η. But to this end one must first remove η' from (2.11), which is easily achieved thanks to an integration by parts accompanied by the use of (2.7):

$$\int_{x_1}^{x_2} \frac{\partial f}{\partial y'} \eta' \, dx = \frac{\partial f}{\partial y'} \eta \Big|_{x_1}^{x_2} - \int_{x_1}^{x_2} \eta \frac{d}{dx} \left(\frac{\partial f}{\partial y'} \right) dx = - \int_{x_1}^{x_2} \eta \frac{d}{dx} \left(\frac{\partial f}{\partial y'} \right) dx \,. \qquad (2.12)$$

With the help of this last result, Eq. (2.11) reduces to

$$\int_{x_1}^{x_2} \left[\frac{\partial f}{\partial y} - \frac{d}{dx} \left(\frac{\partial f}{\partial y'} \right) \right] \eta \, dx = 0 \,. \qquad (2.13)$$

Now we can infer a differential equation for $y(x)$ by appealing to an important auxiliary result.

Lemma 2.1.1 (Fundamental Lemma of the Calculus of Variations)   *If $M(x)$ is a continuous function such that $\int_{x_1}^{x_2} M(x)\eta(x)dx = 0$ for any continuous function $\eta(x)$ such that $\eta(x_1) = \eta(x_2) = 0$, then $M(x) \equiv 0$ in $[x_1, x_2]$.*

Proof   Let M be a continuous function such that $\int_{x_1}^{x_2} M(x)\eta(x)dx = 0$ for any continuous function η with $\eta(x_1) = \eta(x_2) = 0$. The idea of the proof consists in deriving a contradiction

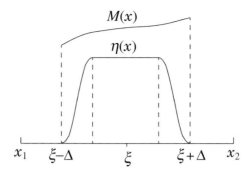

Fundamental lemma of the calculus of variations.

if the function M is not identically zero in the whole interval $[x_1, x_2]$. Suppose $M(\xi) > 0$ for some $x_1 < \xi < x_2$. In virtue of the continuity of M, there exists a subinterval $(\xi - \Delta, \xi + \Delta) \subset (x_1, x_2)$ in which $M(x) > c$ for some real number $c > 0$. Pick a continuous function η (Fig. 2.2) such that $\eta = 1$ in the interval $(\xi - \frac{\Delta}{2}, \xi + \frac{\Delta}{2})$, $\eta = 0$ outside the interval $(\xi - \Delta, \xi + \Delta)$ and $\eta(x) \geq 0$ for all $x \in [x_1, x_2]$. In order to give the proof a purely analytical character, here is a possible choice for $\eta(x)$:

$$
\eta(x) = \begin{cases}
0 & x_1 \leq x \leq \xi - \Delta \\
(4/\Delta^2)(x - \xi + \Delta)^2(1 + x - \xi + \Delta/2)^2 & \xi - \Delta \leq x \leq \xi - \Delta/2 \\
1 & \xi - \Delta/2 \leq x \leq \xi + \Delta/2 \\
(4/\Delta^2)(x - \xi - \Delta)^2(1 + x - \xi - \Delta/2)^2 & \xi + \Delta/2 \leq x \leq \xi + \Delta \\
0 & \xi + \Delta \leq x \leq x_2
\end{cases}
$$

Clearly,

$$
\int_{x_1}^{x_2} M(x)\eta(x)dx = \int_{\xi - \Delta}^{\xi + \Delta} M(x)\eta(x)dx \geq \int_{\xi - \Delta/2}^{\xi + \Delta/2} M(x)\eta(x)dx \geq c\Delta > 0 .
$$

This contradiction shows that $M(\xi)$ cannot be positive, and the same argument applies if $M(\xi) < 0$. Since ξ is arbitrary, it follows that $M(x) \equiv 0$ in $[x_1, x_2]$ and the proof is complete.   □

This lemma applied to Eq. (2.13) leads to **Euler's equation**[2]:

$$
\frac{\partial f}{\partial y} - \frac{d}{dx}\left(\frac{\partial f}{\partial y'}\right) = 0 . \tag{2.14}
$$

Euler's differential equation is but a *necessary*[3] condition for the existence of an extremum. It is a second-order differential equation whose general solution contains two arbitrary

---

[2] First derived by Leonhard Euler in 1744.
[3] Sufficient conditions are known which are relatively easy to apply (Gelfand & Fomin, 1963; Elsgoltz, 1969; Fox, 1987; Kot, 2014).

constants, which normally allows one to satisfy the boundary conditions that the curve must pass through the two prescribed endpoints.

Example 2.3 Find the geodesics of the sphere.

Solution

The infinitesimal arc length in spherical coordinates is

$$ds^2 = dx^2 + dy^2 + dz^2 = dr^2 + r^2 d\theta^2 + r^2 \sin^2 \theta \, d\phi^2 . \tag{2.15}$$

For a spherical surface of radius R with centre at the origin one has $r = R$, so that the distance between two points measured along a curve $\phi(\theta)$ on the surface of the sphere is given by

$$s[\phi] = R \int_{\theta_1}^{\theta_2} \sqrt{1 + \sin^2 \theta \, \phi'(\theta)^2} \, d\theta . \tag{2.16}$$

The geodesic is the curve that gives the shortest distance between two points, the one that minimises $s[\phi]$. Therefore $\phi(\theta)$ has to satisfy Euler's Eq. (2.14) with θ in the place of x, ϕ in the place of y and $f(\phi, \phi', \theta) = \sqrt{1 + \sin^2 \theta \, \phi'^2}$. It follows that

$$\frac{\partial f}{\partial \phi} - \frac{d}{d\theta}\left(\frac{\partial f}{\partial \phi'}\right) = 0 \implies \frac{d}{d\theta}\left(\frac{\partial f}{\partial \phi'}\right) = 0 \implies \frac{\partial f}{\partial \phi'} = \frac{\sin^2 \theta \, \phi'}{[1 + \sin^2 \theta \, \phi'(\theta)^2]^{1/2}} = C_1 , \tag{2.17}$$

where C_1 is an arbitrary constant. Solving the above equation for ϕ' we get

$$\frac{d\phi}{d\theta} = \frac{C_1}{\sin \theta \sqrt{\sin^2 \theta - C_1^2}} = \frac{C_1}{\sin^2 \theta \sqrt{1 - C_1^2/\sin^2 \theta}} = \frac{C_1/\sin^2 \theta}{\sqrt{(1 - C_1^2) - C_1^2 \cot^2 \theta}} . \tag{2.18}$$

An elementary integration yields

$$\phi = -\cos^{-1}\left(\frac{C_1 \cot \theta}{\sqrt{1 - C_1^2}}\right) + C_2 , \tag{2.19}$$

whence

$$C \cot \theta = \cos(\phi - C_2) , \qquad C = \frac{C_1}{\sqrt{1 - C_1^2}} . \tag{2.20}$$

This last equation can be recast in the equivalent form

$$C \cot \theta = A \cos \phi + B \sin \phi , \qquad A = \cos C_2 , \quad B = \sin C_2 . \tag{2.21}$$

For a better understanding of what the above equation means, multiply it by $R \sin \theta$ to obtain

$$CR \cos \theta = AR \sin \theta \cos \phi + BR \sin \theta \sin \phi . \tag{2.22}$$

Thus, each geodesic curve is the intersection of the spherical surface $r = R$ with a plane through the origin characterised by equation

$$Cz = Ax + By. \tag{2.23}$$

Therefore the geodesics of the sphere are arcs of great circles, those circles whose centres coincide with the centre of the sphere.

Example 2.4   Find the surface of revolution with least area.

Solution

Consider a surface generated by the rotation about the y-axis of a plane curve passing through two fixed points (Fig. 2.3). The problem consists in finding the curve for which the area of the surface of revolution is smallest. The area of an infinitesimal ribbon of the surface is $dA = 2\pi x \, ds = 2\pi x \sqrt{1 + y'^2} \, dx$, and so the area of the surface is

$$A[y] = 2\pi \int_{x_1}^{x_2} x\sqrt{1 + y'^2} \, dx. \tag{2.24}$$

With $f(y, y', x) = x\sqrt{1 + y'^2}$, Euler's equation for this problem is

$$\frac{d}{dx}\left(\frac{\partial f}{\partial y'}\right) = 0 \implies \frac{\partial f}{\partial y'} = a \implies \frac{xy'}{\sqrt{1 + y'^2}} = a, \tag{2.25}$$

where a is an arbitrary constant of integration (obviously less than x_1). Solving the last equation for y' we obtain

$$y' = \frac{a}{\sqrt{x^2 - a^2}}, \tag{2.26}$$

whence, by an immediate integration,

$$y = a \cosh^{-1}\left(\frac{x}{a}\right) + b \tag{2.27}$$

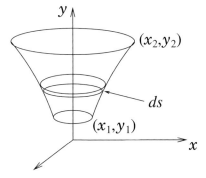

Fig. 2.3   Surface of revolution with least area.

or, equivalently,

$$x = a \cosh \frac{y - b}{a}. \tag{2.28}$$

The sought curve is an arc of a catenary. The constants a and b are to be determined by requiring that the catenary pass through the endpoints. If there is only one such catenary, it yields the minimum area. For certain positions of the endpoints there may be two catenaries, only one of them giving the minimum, or no catenary exists that passes through the endpoints and the minimum is furnished by a curve that does not even have a continuous derivative (Weinstock, 1974; Arfken & Weber, 1995; Kot, 2014).

Example 2.5   Solve the problem of the brachistochrone.

Solution

The inaugural and most celebrated problem[4] in the calculus of variations is that of the line of quickest descent or *brachistochrone*, from the Greek *brachistos* (shortest) and *chronos* (time). It is about determining the curve in a vertical plane, connecting two given points P and Q not belonging to the same vertical line, along which a particle slides without friction in the shortest possible time under the action of gravity. Proposed by Johann Bernoulli as a challenge to European mathematicians in 1696, the problem was solved – long before the appearance of Euler's equation – by the challenger himself and, independently, by l'Hôpital, Jakob Bernoulli, Gottfried Wilhelm Leibniz and Isaac Newton.[5] We choose the origin at point P and orient the y-axis vertically downward. By energy conservation, after falling the vertical distance y along the curve $y(x)$, the speed of the particle is $ds/dt = v = \sqrt{2gy}$. Using $ds = \sqrt{1 + y'^2} dx$, the time spent to slide from $P = (0,0)$ to Q= (x_0, y_0) is

$$T[y] = \int_P^Q \frac{ds}{v} = \frac{1}{\sqrt{2g}} \int_0^{x_0} \left(\frac{1 + y'^2}{y} \right)^{1/2} dx, \quad y(0) = 0, \ y(x_0) = y_0. \tag{2.29}$$

This problem is made easier by picking y as the independent variable and the curve in the form $x(y)$. In this case, with $x' = dx/dy$,

$$T[x] = \frac{1}{\sqrt{2g}} \int_0^{y_0} \left(\frac{1 + x'^2}{y} \right)^{1/2} dy, \quad x(0) = 0, \ x(y_0) = x_0. \tag{2.30}$$

---

[4] Actually, the first problem of physical importance in the calculus of variations was proposed by Newton in the *Principia* (1687): to determine the shape of the solid of revolution that suffers least resistance as it moves through an incompressible perfect fluid. However, it was the brachistochrone problem that had a big impact and aroused interest in the subject.

[5] On the night of 29 January 1697, when Newton received the challenge letter, he could not sleep until he had solved the problem, which happened by four o'clock in the morning. Next the solution was anonymously sent to the challenger. Upon receiving it, Johann Bernoulli famously stated that he immediately recognised the author *tanquam ex ungue leonem*, that is, "as the lion is recognised by his paw" (Westfall, 1983).

For the problem so formulated we have $f(x, x', y) = \sqrt{(1 + x'^2)/y}$ and Euler's equation permits an immediate first integration:

$$\frac{\partial f}{\partial x} - \frac{d}{dy}\left(\frac{\partial f}{\partial x'}\right) = 0 \quad \Longrightarrow \quad \frac{\partial f}{\partial x'} = \frac{x'}{\sqrt{y(1 + x'^2)}} = C_1, \quad (2.31)$$

where C_1 is an arbitrary constant. This last equation is more easily solved by introducing the parameter t by means of $x' = \tan t$. Then, with $C_2 = 1/C_1^2$, we have

$$y = C_2 \frac{\tan^2 t}{1 + \tan^2 t} = C_2 \sin^2 t = \frac{C_2}{2}(1 - \cos 2t) \quad (2.32)$$

and, as a consequence,

$$dx = \tan t \, dy = 2C_2 \sin^2 t \, dt = C_2(1 - \cos 2t) \, dt \quad \Longrightarrow \quad x = \frac{C_2}{2}(2t - \sin 2t) + C_3 . \quad (2.33)$$

Taking into account that $C_3 = 0$ because $y = 0$ for $x = 0$ and putting $\theta = 2t$, we arrive at the parametric equations for a family of cycloids

$$x = C(\theta - \sin \theta), \qquad y = C(1 - \cos \theta). \quad (2.34)$$

The constant $C = C_2/2$ is determined by the condition that the curve pass through the point $Q = (x_0, y_0)$. Thus, the brachistochrone is an arc of a cycloid.

The cycloid is also a tautochrone: the period of oscillation of a particle sliding on a frictionless cycloidal track does not depend on the point from which it starts to fall (see Problem 1.6). However, unlike the brachistochrone, the tautochrone is not unique (Terra, Souza & Farina, 2016).

2.2 Variational Notation

The **variation** δy, depicted in Fig. 2.1, is defined by $\delta y = \epsilon \eta$, so that Eq. (2.6) is written

$$\bar{y} = y + \delta y. \quad (2.35)$$

Similarly, the variation of the functional J is defined by

$$\delta J = \epsilon \Phi'(0) = \int_{x_1}^{x_2} \left(\frac{\partial f}{\partial y}\delta y + \frac{\partial f}{\partial y'}\delta y'\right) dx, \quad (2.36)$$

where we have used the first equality in (2.11) and

$$\delta y' = \epsilon \eta' = (\epsilon \eta)' = (\delta y)' \equiv \frac{d}{dx}(\delta y). \quad (2.37)$$

The traditional notation δJ is analogous to the notation of differential calculus, in which the expression $\epsilon f'(x) \equiv df$ with an arbitrary parameter ϵ is called the differential of the function $f(x)$. An integration by parts, with the use of $\delta y(x_1) = \delta y(x_2) = 0$, reduces Eq. (2.36) to the form

$$\delta J = \int_{x_1}^{x_2} \left[\frac{\partial f}{\partial y} - \frac{d}{dx}\left(\frac{\partial f}{\partial y'}\right) \right] \delta y\, dx. \tag{2.38}$$

Thus, a necessary condition for an extremum of a functional $J[y]$ is that its variation δJ vanish for variations δy that are arbitrary except for the condition that they vanish at the ends of the integration interval. The condition $\delta J = 0$ for a functional is analogous to the condition $df = 0$ or, equivalently, $f'(x) = 0$ for the determination of the critical points of a function. In general, the functions $y(x)$ for which $\delta J = 0$ are called *extremals* or *stationary curves*. This terminology is justified because $\delta J = 0$ means only that a variation δy of the curve y leaves the functional $J[y]$ stationary up to second order terms, there being no guarantee that the curve yields a maximum or a minimum for the functional.

2.3 Hamilton's Principle and Lagrange's Equations

The mere change of notation

$$x \to t, \quad y \to q, \quad y' \equiv \frac{dy}{dx} \to \dot{q} \equiv \frac{dq}{dt}, \quad f \to L, \quad J \to S \tag{2.39}$$

shows that Euler's equation takes the form of Lagrange's equation

$$\frac{d}{dt}\left(\frac{\partial L}{\partial \dot{q}}\right) - \frac{\partial L}{\partial q} = 0, \tag{2.40}$$

which, therefore, follows from the variational principle

$$\delta S \equiv \delta \int_{t_1}^{t_2} L(q, \dot{q}, t)\, dt = 0 \tag{2.41}$$

with $\delta q(t_1) = \delta q(t_2) = 0$. The generalisation for holonomic systems with any number of degrees of freedom is straightforward. Let S be the **action** defined by

$$S = \int_{t_1}^{t_2} L(q_1, \dots, q_n, \dot{q}_1, \dots, \dot{q}_n, t)\, dt \tag{2.42}$$

and consider

$$\bar{q}_1(t) = q_1(t) + \delta q_1(t),$$
$$\vdots \tag{2.43}$$
$$\bar{q}_n(t) = q_n(t) + \delta q_n(t),$$

with the variations $\delta q_1, \dots, \delta q_n$ mutually independent and arbitrary except for the endpoint conditions $\delta q_k(t_1) = \delta q_k(t_2) = 0, k = 1, \dots, n$. It is worth stressing that it is the mutual independence of the generalised coordinates that ensures that each can be varied independently of the others.

The variation of the action $S = \int_{t_1}^{t_2} L \, dt$ is given by

$$\delta S = \int_{t_1}^{t_2} \sum_k \left(\frac{\partial L}{\partial q_k} \delta q_k + \frac{\partial L}{\partial \dot{q}_k} \delta \dot{q}_k \right) dt. \tag{2.44}$$

An integration by parts yields

$$\delta S = \int_{t_1}^{t_2} dt \sum_k \left[\frac{\partial L}{\partial q_k} - \frac{d}{dt} \left(\frac{\partial L}{\partial \dot{q}_k} \right) \right] \delta q_k + \sum_k \frac{\partial L}{\partial \dot{q}_k} \delta q_k \Big|_{t_1}^{t_2}. \tag{2.45}$$

Taking into account that the variations of the generalised coordinates vanish at both limits of integration, this last equation reduces to

$$\delta S = \int_{t_1}^{t_2} dt \sum_k \left[\frac{\partial L}{\partial q_k} - \frac{d}{dt} \left(\frac{\partial L}{\partial \dot{q}_k} \right) \right] \delta q_k. \tag{2.46}$$

Imposing $\delta S = 0$ we are led to

$$\int_{t_1}^{t_2} dt \sum_k \left[\frac{\partial L}{\partial q_k} - \frac{d}{dt} \left(\frac{\partial L}{\partial \dot{q}_k} \right) \right] \delta q_k = 0. \tag{2.47}$$

Because the δqs are mutually independent, we can take all of them equal to zero except for a particular δq_{k_0}. Then, the sum in (2.47) reduces to a single term corresponding to $k = k_0$. But, inasmuch as δq_{k_0} is an arbitrary function, the fundamental lemma of the calculus of variations establishes that the coefficient of δq_{k_0} in (2.47) is identically zero. Finally, since the previous argument is applicable to any k_0, one concludes that (2.47) is equivalent to

$$\frac{d}{dt} \left(\frac{\partial L}{\partial \dot{q}_k} \right) - \frac{\partial L}{\partial q_k} = 0, \quad k = 1, \ldots, n. \tag{2.48}$$

This completes the derivation of Lagrange's equations from the variational principle $\delta S = 0$ for holonomic systems.

The occasion is propitious to introduce a suggestive geometric description of Lagrangian dynamics. As seen in Section 1.4, the n-dimensional Cartesian space whose points are the n-tuples $(q_1, \ldots, q_n)$ is called configuration space.[6] The name is apt, for each point of the configuration space corresponds to a set of specific values $q_1, \ldots, q_n$ defining uniquely the instantaneous configuration of the mechanical system. As time passes, the representative point of the system describes a curve in configuration space whose parametric equations are $q_1 = q_1(t), \ldots, q_n = q_n(t)$ with t as a parameter. This path in configuration space represents the motion of the whole system and, in most cases, has nothing to do with the curve described in three-dimensional space by any of the particles that make up the system. From now on, except when otherwise noted, the n-tuple $(q_1, \ldots, q_n)$ will be denoted just by q – that is, $q \equiv (q_1, \ldots, q_n)$. The time evolution is a path in configuration space and we can state the following fundamental principle which selects, among all conceivable paths, the one actually followed by the mechanical system.

---

[6] As noted in Section 1.4, in general the configuration space is not the same as $\mathbb{R}^n$, but possesses the mathematical structure of a differentiable manifold.

Hamilton's Principle   *Let a holonomic mechanical system be described by the Lagrangian $L(q,\dot{q},t)$. The actual motion of the system from instant t_1 to instant t_2 is such that the action*

$$S = \int_{t_1}^{t_2} L(q,\dot{q},t)\,dt \tag{2.49}$$

is least (more generally, stationary) for fixed initial and final points of the comparison paths in configuration space.

Equivalent Lagrangians

Hamilton's principle, also known as the **principle of least action**,[7] entails that the same equations of motion are generated by two Lagrangians that differ by a total time derivative of an arbitrary function of the generalised coordinates and time.

Definition 2.3.1   *Two Lagrangians $\bar{L}(q,\dot{q},t)$ and $L(q,\dot{q},t)$ are said to be **equivalent** if they differ by the total time derivative of an arbitrary function $f(q,t)$ of the generalised coordinates and time:*

$$\bar{L}(q,\dot{q},t) = L(q,\dot{q},t) + \frac{d}{dt}f(q,t). \tag{2.50}$$

Theorem 2.3.1   *Equivalent Lagrangians give rise to the same equations of motion.*

Proof   The action $\bar{S}$ associated with $\bar{L}$ is

$$\bar{S} = \int_{t_1}^{t_2} \bar{L}(q,\dot{q},t)\,dt = \int_{t_1}^{t_2} L(q,\dot{q},t)\,dt + \int_{t_1}^{t_2} \frac{df}{dt}\,dt = S + f(q(t_2),t_2) - f(q(t_1),t_1). \tag{2.51}$$

Since the variation of the action leaves the endpoints $q(t_1)$ and $q(t_2)$ fixed, $\delta\bar{S} = \delta S$. Therefore, the conditions $\delta\bar{S} = 0$ and $\delta S = 0$ are identical, showing that $\bar{L}$ and L engender exactly the same equations of motion. $\square$

Exercise 2.3.1   Prove, by direct substitution into Lagrange's equations, that equivalent Lagrangians give rise to the same equations of motion.

Example 2.6   Discuss the Lagrangian for a plane pendulum whose suspension point moves on a horizontal straight line with constant speed.

Solution

Using the coordinates shown in Fig. 2.4 we have

$$T = \frac{m}{2}\left[(\dot{x} + l\dot{\theta}\cos\theta)^2 + (-l\dot{\theta}\sin\theta)^2\right], \tag{2.52}$$

---

[7] For a charming and delightful discussion of the principle of least action, see Feynman, Leighton & Sands (1963), Volume II, Chapter 19.

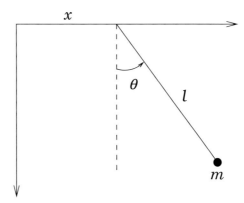

Fig. 2.4   Plane pendulum with moving suspension point.

hence

$$L = T - V = \frac{m}{2}\left(\dot{x}^2 + l^2\dot{\theta}^2 + 2l\dot{x}\dot{\theta}\cos\theta\right) + mgl\cos\theta. \qquad (2.53)$$

If $\dot{x} = v = $ constant we can write

$$L = \frac{ml^2}{2}\dot{\theta}^2 + mgl\cos\theta + \frac{d}{dt}\left(\frac{m}{2}v^2 t + mlv\sin\theta\right). \qquad (2.54)$$

Discarding the total time derivative, the resulting Lagrangian coincides with the one that would be obtained by an observer relative to which the suspension point of the pendulum is at rest. Can you explain why?

It can be shown (Problem 2.6) that two Lagrangians produce exactly the same equations of motion if and only if they differ by the total time derivative of some function of the generalised coordinates and time. However, there are Lagrangians that do not differ by a total time derivative but even so give rise to *equivalent* equations of motion – that is, equations which are not identical but have exactly the same set of solutions (Henneaux & Shepley, 1982; Problem 2.7).

Least Action

Hamilton's principle guarantees only that the action $S = \int_{t_1}^{t_2} L\,dt$ is stationary for the actual path described by the system. In general, one can only ensure that the value of the action for the true motion is a local minimum – for comparison paths sufficiently close to the physical path – and as long as the time interval $[t_1, t_2]$ is sufficiently small. In particular cases its is possible to obtain stronger results by elementary arguments. For a particle in a uniform gravitational field, the action is an absolute minimum without restriction on the time interval (Problem 2.2). In the case of a harmonic oscillator, the action is an absolute minimum provided the time interval is shorter than half the period of the oscillator (Problem 2.3).

In the calculus of variations (Gelfand & Fomin, 1963; Elsgoltz, 1969; Fox, 1987; Kot, 2014) sufficient conditions for a minimum are formulated which can be applied to variational problems such as Hamilton's principle. Here, however, we settle upon a direct approach which, for the sake of simplicity, will be restricted to the case of a particle in one-dimensional motion under a force with potential energy V. The admission of a general potential energy comes with a price: in order to prove a weaker result – compared with the results valid for the projectile and the harmonic oscillator – one has to resort to rather sophisticated mathematical reasonings. Therefore, let the reader be duly warned that the understanding of the proof of the theorem that follows requires a degree of mathematical maturity somewhat above the one presupposed in the rest of this book.

Theorem 2.3.2   *If $V : \mathbb{R} \to \mathbb{R}$ is a twice continuously differentiable function such that $V'' \neq 0$ and the time interval $[t_1, t_2]$ is sufficiently short, the action*

$$S[x] = \int_{t_1}^{t_2} \left[\frac{m}{2} \dot{x}(t)^2 - V(x(t)) \right] dt \tag{2.55}$$

is a local minimum for the physical path as compared to all other sufficiently close neighbouring paths with fixed endpoints $x(t_1)$ and $x(t_2)$.

Proof   We follow Gallavotti (1983). Let $[t_1, t_2]$ be a time interval so short that a single physical path goes through the fixed endpoints $x(t_1)$ and $x(t_2)$. Let $\eta : [t_1, t_2] \to \mathbb{R}$ be an infinitely differentiable function such that $\eta(t_1) = \eta(t_2) = 0$. The physical path shall be denoted simply by $x(t)$. The action for the varied path $\bar{x} = x + \eta$ is

$$S[\bar{x}] = \int_{t_1}^{t_2} \left[\frac{m}{2} (\dot{x} + \dot{\eta})^2 - V(x + \eta) \right] dt. \tag{2.56}$$

By Taylor's theorem with Lagrange's form of the remainder (Spivak, 1994) we can write

$$V(x + \eta) = V(x) + V'(x)\,\eta + \frac{1}{2} V''(x + \xi)\,\eta^2, \qquad |\xi| \leq |\eta|. \tag{2.57}$$

Since we are only interested in establishing that the action is a local minimum for the physical path, let us choose η such that $|\eta(t)| \leq 1$ for all $t \in [t_1, t_2]$, which implies $|\xi| \leq 1$. Substituting (2.57) into (2.56), we are led to

$$S[\bar{x}] = \int_{t_1}^{t_2} \left[\frac{m}{2} \dot{x}^2 + m\dot{x}\dot{\eta} + \frac{m}{2} \dot{\eta}^2 - V(x) - V'(x)\,\eta - \frac{1}{2} V''(x + \xi)\,\eta^2 \right] dt$$

$$= S[x] + \int_{t_1}^{t_2} \left[m\dot{x}\dot{\eta} - V'(x)\,\eta \right] dt + \int_{t_1}^{t_2} \left[\frac{m}{2} \dot{\eta}^2 - \frac{1}{2} V''(x + \xi)\,\eta^2 \right] dt. \tag{2.58}$$

Integrating by parts the term containing $\dot{\eta}$ and taking into account that the physical path obeys the equation of motion $m\ddot{x} + V'(x) = 0$, we obtain

$$S[\bar{x}] - S[x] = \int_{t_1}^{t_2} \left[\frac{m}{2} \dot{\eta}^2 - \frac{1}{2} V''(x + \xi)\,\eta^2 \right] dt. \tag{2.59}$$

We need to prove that the right-hand side of (2.59) is positive. Since V'' is a continuous function, there exists the positive number K defined by

$$K = \max_{\substack{t \in [t_1, t_2] \\ |\xi| \le 1}} |V''(x(t) + \xi)|. \tag{2.60}$$

Thus, $V''(x + \xi)\eta^2 \le |V''(x + \xi)|\eta^2 \le K\eta^2$ and the following inequality holds:

$$S[x + \eta] - S[x] \ge \int_{t_1}^{t_2} \left[\frac{m}{2}\dot{\eta}^2 - \frac{K}{2}\eta^2 \right] dt. \tag{2.61}$$

The final idea is to majorise the integral of η^2 by the integral of $\dot{\eta}^2$, which is accomplished with the help of

$$\left(\int_a^b f(x)g(x)dx \right)^2 \le \int_a^b f(x)^2 dx \int_a^b g(x)^2 dx, \tag{2.62}$$

the Schwarz inequality for integrals.[8] Since $\eta(t_1) = 0$, we have $\eta(t) = \int_{t_1}^t \dot{\eta}(\tau)d\tau$. Therefore, with the use of (2.62),

$$\int_{t_1}^{t_2} \eta(t)^2 dt = \int_{t_1}^{t_2} dt \left(\int_{t_1}^t \dot{\eta}(\tau) \cdot 1 \, d\tau \right)^2 \le \int_{t_1}^{t_2} dt \int_{t_1}^t \dot{\eta}(\tau)^2 d\tau \int_{t_1}^t 1^2 d\tau$$

$$\le \int_{t_1}^{t_2} (t - t_1)dt \int_{t_1}^t \dot{\eta}(\tau)^2 d\tau \le \int_{t_1}^{t_2} (t - t_1)dt \int_{t_1}^{t_2} \dot{\eta}(\tau)^2 d\tau$$

$$= \frac{(t_2 - t_1)^2}{2} \int_{t_1}^{t_2} \dot{\eta}(t)^2 dt. \tag{2.63}$$

Substituting this inequality into (2.61), we finally obtain

$$S[\bar{x}] - S[x] \ge \left[\frac{m}{2} - \frac{K}{4}(t_2 - t_1)^2 \right] \int_{t_1}^{t_2} \dot{\eta}(t)^2 dt. \tag{2.64}$$

Since the integral in (2.64) is positive, it follows that $S[\bar{x}] > S[x]$ if $t_2 - t_1 < \sqrt{2m/K}$, and the proof is complete.   □

   This theorem shows that Hamilton's principle, with due qualifications, deserves to be called the "principle of least action". It is left for the reader to show that, with some simple adaptations, the above proof can be generalised to a system of n particles under conservative forces derived from a sufficiently differentiable potential energy function $V(\mathbf{r}_1, \dots, \mathbf{r}_n)$.

   In Gray and Taylor (2007) there is a nice discussion of circumstances under which the action is not least for several one-dimensional mechanical systems. It turns out that the action may be only stationary but it is never a maximum. This is easy to show for an action

---

[8] This inequality for integrals follows at once from the general Schwarz inequality (D.20) in Appendix D and from the easily proved fact that $(f, g) = \int_a^b f(x)g(x)dx$ is an inner product on the vector space of real continuous functions on $[a, b]$.

of the form (2.55). All one has to do is choose the variation η arbitrarily small with a derivative arbitrarily large, such as[9]

$$\eta(t) = \frac{1}{n} \sin\left[\frac{n^2 \pi (t - t_1)}{t_2 - t_1}\right], \tag{2.65}$$

where n is a very big natural number. Then, as n grows without bound so does the integral of $\dot{\eta}^2$ in (2.61), while the integral of η^2 becomes as small as one pleases. As a result, the action becomes less for the physical path than for some arbitrarily close varied path.

2.4 Hamilton's Principle in the Non-Holonomic Case

In the previous section we assumed that the mechanical system was described by mutually independent coordinates, an independence which was crucial for the derivation of (2.48) from (2.47). Such independent generalised coordinates always exist if all of the constraints imposed on the system are holonomic. When non-holonomic constraints are present it is impossible to introduce generalised coordinates such that the constraint conditions are identically satisfied. It is possible, nevertheless, to deduce the equations of motion from Hamilton's principle in the special case in which the non-holonomic constraints are differential equations of the form

$$\sum_{k=1}^{n} a_{lk} dq_k + a_{lt} dt = 0, \quad l = 1, \ldots, p, \tag{2.66}$$

whose coefficients a_{lk} and a_{lt} are functions solely of $q_1, \ldots, q_n$ and t. Let us suppose, therefore, that the system is described by n coordinates $q_1, \ldots, q_n$ and is subject to the p independent differential constraints (2.66), where the index l distinguishes the constraint equations from one another. In spite of their seemingly restricted form, constraints of the type (2.66) encompass virtually all non-holonomic constraints of physical interest.

Let $L = T - U$ be the Lagrangian for the system *written as if there were no constraints*. The generalised potential in L refers to the applied forces alone, so that the constraint forces responsible for the enforcement of Eqs. (2.66) are not taken into account by the Lagrangian. Hamilton's principle $\delta S = 0$ implies

$$\int_{t_1}^{t_2} dt \sum_k \left[\frac{\partial L}{\partial q_k} - \frac{d}{dt}\left(\frac{\partial L}{\partial \dot{q}_k}\right)\right] \delta q_k = 0, \tag{2.67}$$

but now we *cannot* infer that the coefficient of each δq_k is zero because the δqs *are not mutually independent*. Indeed, Eqs. (2.43) show that each δq_k is a virtual displacement since time remains fixed as one performs the variation that leads from $q_k(t)$ to $\bar{q}_k(t) = q_k(t) + \delta q_k(t)$. But, according to Eqs. (2.66), for there to be compatibility with the constraints the virtual displacements must obey

---

[9] This is an example of a strong variation.

$$\sum_{k=1}^{n} a_{lk}\delta q_k = 0, \quad l = 1, \ldots, p, \tag{2.68}$$

where we have used $dt = 0$ for virtual displacements. Since the n variations $\delta q_1, \ldots, \delta q_n$ have to satisfy the p equations (2.68), only $n-p$ variations of the qs are independent of each other.[10] Thus, we are faced with the problem of determining a conditional extremum for the functional $S = \int_{t_1}^{t_2} L dt$, and the treatment imitates the method of Lagrange multipliers of differential calculus.

Lagrange Multipliers

As a consequence of (2.68), the equation

$$\int_{t_1}^{t_2} dt \sum_{l=1}^{p} \lambda_l \left(\sum_{k=1}^{n} a_{lk}\delta q_k \right) = \int_{t_1}^{t_2} dt \sum_{k=1}^{n} \left(\sum_{l=1}^{p} \lambda_l a_{lk} \right) \delta q_k = 0 \tag{2.69}$$

is obviously valid for all values of the Lagrange multipliers $\lambda_1(t), \ldots, \lambda_p(t)$. Adding (2.69) to (2.67) there results

$$\int_{t_1}^{t_2} dt \sum_{k=1}^{n} \left\{ \frac{\partial L}{\partial q_k} - \frac{d}{dt}\left(\frac{\partial L}{\partial \dot{q}_k} \right) + \sum_{l=1}^{p} \lambda_l a_{lk} \right\} \delta q_k = 0. \tag{2.70}$$

Since the δqs are not independent, nothing can be said about the coefficient of each δq_k in this last equation. With an adequate numbering of the variables, we can take the first $n - p$ variations $\delta q_1, \ldots, \delta q_{n-p}$ as mutually independent, the last p variations being determined in terms of the first ones by solving the p equations (2.68). On the other hand, we have p Lagrange multipliers at our disposal and they can be chosen such that the coefficients of the last p variations in (2.70) vanish – that is,

$$\frac{\partial L}{\partial q_k} - \frac{d}{dt}\left(\frac{\partial L}{\partial \dot{q}_k} \right) + \sum_{l=1}^{p} \lambda_l a_{lk} = 0, \quad k = n - p + 1, \ldots, n. \tag{2.71}$$

With the λs determined by Eqs. (2.71), Eq. (2.70) reduces to

$$\int_{t_1}^{t_2} dt \sum_{k=1}^{n-p} \left\{ \frac{\partial L}{\partial q_k} - \frac{d}{dt}\left(\frac{\partial L}{\partial \dot{q}_k} \right) + \sum_{l=1}^{p} \lambda_l a_{lk} \right\} \delta q_k = 0, \tag{2.72}$$

which involves only the independent variations $\delta q_1, \ldots, \delta q_{n-p}$, implying

$$\frac{\partial L}{\partial q_k} - \frac{d}{dt}\left(\frac{\partial L}{\partial \dot{q}_k} \right) + \sum_{l=1}^{p} \lambda_l a_{lk} = 0, \quad k = 1, \ldots, n - p. \tag{2.73}$$

[10] Holonomic systems distinguish themselves from the non-holonomic ones by the property that the number of independent virtual displacements coincides with the number of independent coordinates necessary to specify the configuration of the system.

In short, equations (2.71) and (2.73) show that the equations of motion for the system are

$$\frac{d}{dt}\left(\frac{\partial L}{\partial \dot{q}_k}\right) - \frac{\partial L}{\partial q_k} = \sum_{l=1}^{p} \lambda_l a_{lk}, \quad k = 1, \ldots, n. \tag{2.74}$$

But these are not enough, for now there are only n equations for $n + p$ unknowns, namely the n coordinates q_k and the p Lagrange multipliers λ_l. The additional p equations are, of course, the constraint equations (2.66), which can be written in the equivalent form

$$\sum_{k=1}^{n} a_{lk}\dot{q}_k + a_{lt} = 0, \quad l = 1, \ldots, p. \tag{2.75}$$

Equations (2.74) and (2.75) make up a set of $n + p$ equations for $n + p$ unknowns which uniquely determine the motion of the system once initial conditions are stipulated.

Constraint Forces

It remains to investigate the physical meaning of the Lagrange multipliers, which are also determined in the process of solving the equations of motion. Let us imagine the constraints removed and generalised forces Q'_k acting without causing any change to the motion of the system. This would only occur if Q'_k were the constraint forces because only then the system would obey the restrictions (2.75). But, taking into account that L includes only the applied forces, the additional forces Q'_k would appear in the equations of motion in the form (1.145) – that is,

$$\frac{d}{dt}\left(\frac{\partial L}{\partial \dot{q}_k}\right) - \frac{\partial L}{\partial q_k} = Q'_k. \tag{2.76}$$

In order for the motions to be identical, the equations of motion (2.74) and (2.76) must be identical. In other words,

$$Q'_k = \sum_{l=1}^{p} \lambda_l a_{lk} \tag{2.77}$$

is the kth component of the generalised constraint force. Thus, in addition to the motion of the system, in the present formulation the constraint forces appear as part of the answer. It must not be forgotten that the components of the generalised constraint force and the constraint forces proper are related by (1.75).

The conclusion of the above analysis is that, suitably modified, Lagrange's equations are valid even in the presence of an important class of non-holonomic constraints. Over and above the motion of the mechanical system, the constraint forces are determined as an important subproduct of the Lagrange multiplier formalism.[11]

Exercise 2.4.1   Prove that the total virtual work of the generalised constraint forces (2.77) is zero. This shows that even in the presence of non-holonomic constraints Hamilton's principle can be in harmony with d'Alembert's principle.

---

[11] It is possible to formulate the equations of motion for the present class of non-holonomic systems without the use of Lagrange multipliers. It is a method based on the introduction of so-called quasi-coordinates that culminates in the Gibbs-Appell equations (Pars, 1965; Gantmacher, 1970; Desloge, 1982).

Remark   In the more general case in which the constraint equations are of the form

$$g_l(q, \dot{q}, t) = 0, \quad l = 1, \ldots, p, \tag{2.78}$$

the result considered correct (Rund, 1966; Saletan & Cromer, 1970) for the equations of motion is

$$\frac{d}{dt}\left(\frac{\partial L}{\partial \dot{q}_k}\right) - \frac{\partial L}{\partial q_k} = \sum_{l=1}^{p} \lambda_l \frac{\partial g_l}{\partial \dot{q}_k}, \quad k = 1, \ldots, n. \tag{2.79}$$

It is clear that these equations reduce to (2.74) when the constraint equations involve the velocities linearly.

Example 2.7   As a first and purely academic illustration, consider a particle moving freely on a plane except for the non-holonomic constraint

$$\dot{x} - \alpha y = 0, \tag{2.80}$$

where α is a constant. In Appendix B it is proved that this constraint is non-holonomic. Employing the notation $q_1 = x$ and $q_2 = y$, we see that (2.80) is of the form (2.75) with $a_{11} = 1$, $a_{12} = 0$, $a_{1t} = -\alpha y$. As there is only one constraint, the index l takes the single value $l = 1$. Using the Lagrangian

$$L = \frac{m}{2}(\dot{x}^2 + \dot{y}^2), \tag{2.81}$$

equations (2.74) take the form

$$m\ddot{x} = \lambda_1 a_{11} = \lambda_1, \quad m\ddot{y} = \lambda_1 a_{12} = 0. \tag{2.82}$$

The general solution to the second of these equations is

$$y = y_0 + v_0 t, \tag{2.83}$$

with y_0 and v_0 arbitrary constants. Inserting (2.83) into (2.80) and integrating one finds

$$x = x_0 + \alpha y_0 t + \frac{1}{2}\alpha v_0 t^2, \tag{2.84}$$

with x_0 a further arbitrary constant. The general solution for the motion of the system is contained in (2.83) and (2.84). Finally, the Lagrange multiplier is determined by

$$\lambda_1 = m\ddot{x} = m\alpha\dot{y} = m\alpha v_0, \tag{2.85}$$

leading to

$$Q_1' \equiv Q_x' \equiv F_x' = \lambda_1 a_{11} = \lambda_1 = m\alpha\dot{y}, \quad F_y' = \lambda_1 a_{12} = 0 \tag{2.86}$$

for the Cartesian components of the constraint force.

Exercise 2.4.2   Solve the Newtonian equations of motion for a particle on a plane subject to the force $\mathbf{F}'$ with components (2.86) – that is, $\mathbf{F}' = m\alpha\dot{y}\hat{\mathbf{x}}$. Does the general solution you obtained automatically satisfy the constraint (2.80)? What is to be required of the initial conditions in order that the constraints be respected during the entire motion of the system?

Warning   Non-integrable constraints **must not** be substituted into the Lagrangian with the intent of symplifying it or decreasing the number of variables. If a variable be so eliminated from the Lagrangian, the accessible configurations will be unduly reduced because non-integrable constraints restrict the velocities alone. For instance, if the variable x is eliminated from the Lagrangian (2.81) by means of (2.80), the new Lagrangian $\bar{L}(y, \dot{y})$ so obtained leads to an equation of motion whose solutions are completely different from (2.83), which is the correct solution (check it yourself). In certain circumstances the elimination of variables by substitution of non-integrable constraints into the Lagrangian is permitted, but in such cases the correct equations of motion are Voronec's equations (Neĭmark & Fufaev, 1972; Lemos, 2003).

Example 2.8   Using the Lagrange multiplier method, describe the motion of a skate on a horizontal plane.

Solution

As an oversimplified model for a skate, let us take a rigid homogeneous rod restricted to move in such a way that the centre-of-mass velocity is always parallel to the rod. Convenient configuration coordinates are the centre-of-mass position vector $\mathbf{r} = x\hat{\mathbf{x}} + y\hat{\mathbf{y}}$ and the angle θ the rod makes with the x-axis (Fig. 2.5). Since $\hat{\mathbf{n}} = \cos\theta\,\hat{\mathbf{x}} + \sin\theta\,\hat{\mathbf{y}}$ is the unit vector parallel to the rod, the constraint is expressed by $\dot{\mathbf{r}} \times \hat{\mathbf{n}} = 0$ or, in components,

$$\dot{x}\sin\theta - \dot{y}\cos\theta = 0. \qquad (2.87)$$

Using the notation $q_1 = x$, $q_2 = y$, $q_3 = \theta$ we identify $a_{11} = \sin\theta$, $a_{12} = -\cos\theta$, $a_{13} = a_{1t} = 0$ by simply comparing (2.87) and (2.75). Inasmuch as $V = 0$, the Lagrangian is just the kinetic energy, which, by Eq. (1.23), is the sum of the centre-of-mass kinetic energy with the energy of rotation about the centre of mass:

$$L = \frac{m}{2}(\dot{x}^2 + \dot{y}^2) + \frac{I}{2}\dot{\theta}^2, \qquad (2.88)$$

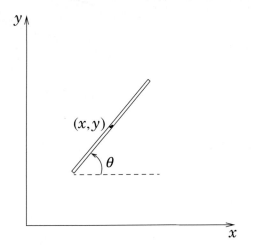

Fig. 2.5   Skate on a horizontal plane.

where I is the moment of inertia of the rod with respect to an axis that goes through the centre of mass and is perpendicular to the rod. The equations of motion are

$$m\ddot{x} = \lambda \sin\theta, \quad m\ddot{y} = -\lambda \cos\theta, \quad I\ddot{\theta} = 0. \tag{2.89}$$

The general solution of the equation for θ is

$$\theta = \theta_0 + \Omega t, \tag{2.90}$$

where θ_0 and Ω are arbitrary constants. Combining the first two equations of motion to eliminate the Lagrange multiplier λ, we get

$$\ddot{x}\cos\theta + \ddot{y}\sin\theta = 0. \tag{2.91}$$

From the constraint Eq. (2.87) it follows that

$$\dot{y} = \dot{x}\tan\theta \implies \ddot{y} = \ddot{x}\tan\theta + \Omega\dot{x}\sec^2\theta. \tag{2.92}$$

Inserting this result into (2.91) we obtain

$$\frac{\ddot{x}}{\cos\theta} + \Omega\dot{x}\frac{\sin\theta}{\cos^2\theta} = 0 \implies \frac{d}{dt}\left(\frac{\dot{x}}{\cos\theta}\right) = 0 \implies \dot{x} = C'\cos(\Omega t + \theta_0). \tag{2.93}$$

An immediate integration yields

$$x = x_0 + C\sin(\Omega t + \theta_0), \tag{2.94}$$

where x_0 and $C(= C'/\Omega)$ are arbitrary constants. Finally, the substitution of (2.90) and (2.94) into the first of equations (2.92), and a subsequent integration, lead to

$$y = y_0 - C\cos(\Omega t + \theta_0). \tag{2.95}$$

The skate turns around its own centre of mass with the constant angular velocity Ω at the same time that the centre of mass describes a circle in the plane of the motion with the same angular velocity Ω. If $\Omega = 0$, the centre of mass of the skate moves in a straight line with constant speed.

Exercise 2.4.3   By means of the Frobenius theorem (Appendix B), show that the constraint (2.87) is non-holonomic.

Holonomic Constraints and Lagrange Multipliers

Although equations of the form (2.75) do not include the most general non-holonomic constraints, they do include the holonomic ones. Indeed, holonomic constraints take the form

$$f_l(q, t) = 0, \quad l = 1, \dots, p, \tag{2.96}$$

hence, by taking the total time derivative,

$$\sum_{k=1}^{n} \frac{\partial f_l}{\partial q_k}\dot{q}_k + \frac{\partial f_l}{\partial t} = 0, \quad l = 1, \dots, p. \tag{2.97}$$

These equations are of the form (2.75) with

$$a_{lk} = \frac{\partial f_l}{\partial q_k}, \qquad a_{lt} = \frac{\partial f_l}{\partial t}. \tag{2.98}$$

Thus, the Lagrange multiplier method can be applied to holonomic systems when it is inconvenient to replace the qs by a smaller set of independent variables or when one wishes to determine the constraint forces. By way of illustration, let us apply the Lagrange multiplier formalism to a holonomic case.

Example 2.9 Making use of the Lagrange multiplier technique, study the motion of a cylinder that rolls without slipping on another fixed cylinder.

Solution

We will use as coordinates the distance r between the axes of the cylinders as well as the angles θ and ϕ depicted in Fig. 2.6, where ϕ is the mobile cylinder's angle of rotation about its centre of mass. Note that r and θ are polar coordinates of the mobile cylinder's centre of mass with origin at the axis of the fixed cylinder. The Lagrangian for the system must be written as if r, θ, ϕ were independent coordinates. The kinetic energy of the mobile cylinder is the centre-of-mass kinetic energy plus the energy of rotation about the centre of mass – that is,

$$T = \frac{m}{2}(\dot{r}^2 + r^2\dot{\theta}^2) + \frac{1}{2}I\dot{\phi}^2, \tag{2.99}$$

where $I = ma^2/2$ is the moment of inertia of the cylinder about its symmetry axis. Therefore, the Lagrangian is

$$L = \frac{m}{2}(\dot{r}^2 + r^2\dot{\theta}^2) + \frac{ma^2}{4}\dot{\phi}^2 - mgr\cos\theta. \tag{2.100}$$

In this problem there are two holonomic constraints, namely

$$r\dot{\theta} = a\dot{\phi} \qquad \text{(rolling constraint)}, \tag{2.101}$$

$$r = a + b \qquad \text{(contact constraint)}. \tag{2.102}$$

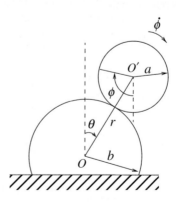

Fig. 2.6   Cylinder rolling on a fixed cylinder.

Note that since point O' (the mobile cylinder's centre of mass) has speed $v_{CM} = r\dot{\theta}$, the rolling constraint is just the usual $v_{CM} = \omega a$, where $\omega = \dot{\phi}$ is the angular velocity associated with the cylinder's rotation about its symmetry axis. In differential form the contact constraint becomes

$$\dot{r} = 0 \,. \tag{2.103}$$

Introducing the notation $q_1 = r$, $q_2 = \theta$, $q_3 = \phi$, we identify

$$a_{11} = 0 \,, \quad a_{12} = r \,, \quad a_{13} = -a \,, \quad a_{21} = 1 \,, \quad a_{22} = a_{23} = 0 \,, \tag{2.104}$$

where $l = 1$ corresponds to (2.101) and $l = 2$ to (2.103).

The equations of motion take the form

$$m\ddot{r} - mr\dot{\theta}^2 + mg\cos\theta = \lambda_1 a_{11} + \lambda_2 a_{21} = \lambda_2 \,, \tag{2.105}$$

$$m\frac{d}{dt}(r^2\dot{\theta}) - mgr\sin\theta = \lambda_1 a_{12} + \lambda_2 a_{22} = r\lambda_1 \,, \tag{2.106}$$

$$\frac{ma^2}{2}\ddot{\phi} = \lambda_1 a_{13} + \lambda_2 a_{23} = -a\lambda_1 \,. \tag{2.107}$$

The use of the constraint equations reduces the equations of motion to

$$-m(a+b)\dot{\theta}^2 + mg\cos\theta = \lambda_2 \,, \tag{2.108}$$

$$m(a+b)^2\ddot{\theta} - mg(a+b)\sin\theta = (a+b)\lambda_1 \,, \tag{2.109}$$

$$\frac{m}{2}(a+b)\ddot{\theta} = -\lambda_1 \,. \tag{2.110}$$

Combining the two last equations so as to eliminate λ_1, we get

$$\frac{3}{2}(a+b)\ddot{\theta} - g\sin\theta = 0 \,. \tag{2.111}$$

The normal contact force exerted by the fixed cylinder on the mobile one is the radial component of the constraint force – that is,

$$N = Q'_r \equiv Q'_1 = \sum_{l=1}^{2} \lambda_l a_{l1} = \lambda_2 = mg\cos\theta - m(a+b)\dot{\theta}^2 \,. \tag{2.112}$$

Now it is possible to answer the following question: at which point does the mobile cylinder detach itself from the fixed cylinder? Of course, this takes place at the instant when $N = 0$. Suppose the cylinder starts at rest ($\dot{\theta} = 0$) from the vertical position ($\theta = 0$). Since there is no sliding friction, the mechanical energy is conserved:

$$T + V = \frac{m}{2}(\dot{r}^2 + r^2\dot{\theta}^2) + \frac{ma^2}{4}\dot{\phi}^2 + mgr\cos\theta$$
$$= \frac{3}{4}m(a+b)^2\dot{\theta}^2 + mg(a+b)\cos\theta = E \,, \tag{2.113}$$

where the constraints have been used. Inasmuch as $E = mg(a+b)$ at the initial instant, we deduce

$$\frac{3}{4}(a+b)\dot{\theta}^2 = g(1 - \cos\theta) \,. \tag{2.114}$$

Because of this last result we have

$$N = \frac{mg}{3}(7\cos\theta - 4).\tag{2.115}$$

The mobile cylinder abandons the fixed cylinder at the angle $\theta_0 = \arccos(4/7)$, that is, $\theta_0 = 55°09'$, which does not depend on the radii of the cylinders.[12] Once the contact is undone, the constraints cease to exist and the mobile cylinder becomes a three-degree-of-freedom system, with coordinates r, θ, ϕ independent of each other.

In the holonomic case it is possible to choose generalised coordinates in terms of which the constraint equations are identically satisfied. Nevertheless, in the discussion of certain general theoretical questions it is sometimes advantageous to use an excessive number of coordinates and treat the holonomic constraints by the Lagrange multiplier technique. Given the holonomic constraints (2.96), consider the new Lagrangian $\mathcal{L}$ defined by

$$\mathcal{L} = L(q, \dot{q}, t) + \sum_{l=1}^{p} \lambda_l f_l(q, t).\tag{2.116}$$

Introducing the $n + p$ variables $\xi_1, \ldots, \xi_{n+p}$ defined by

$$\xi_1 = q_1, \ldots, \xi_n = q_n, \xi_{n+1} = \lambda_1, \ldots, \xi_{n+p} = \lambda_p,\tag{2.117}$$

and treating them as $n + p$ independent generalised coordinates, the equations of motion and the constraint equations can be written in one fell swoop as

$$\frac{d}{dt}\left(\frac{\partial \mathcal{L}}{\partial \dot{\xi}_k}\right) - \frac{\partial \mathcal{L}}{\partial \xi_k} = 0, \qquad k = 1, \ldots, n+p.\tag{2.118}$$

Exercise 2.4.4   Show that the first n equations (2.118) are identical to equations (2.74) with the a_{lk} given by (2.98), whereas the last p equations (2.118) reproduce the constraint equations (2.96).

Equations (2.116) and (2.118) show that the correct variational principle to formulate the equations of motion for a system described by the Lagrangian $L(q, \dot{q}, t)$ and subject to the holonomic constraints (2.96) is

$$\delta \int_{t_1}^{t_2} \left\{ L(q, \dot{q}, t) + \sum_{l=1}^{p} \lambda_l f_l(q, t) \right\} dt = 0,\tag{2.119}$$

where the coordinates $q_1, \ldots, q_n$ and the Lagrange multipliers $\lambda_1, \ldots, \lambda_n$ are varied independently.

It can be checked in particular cases (Problem 2.25) and it can be proved in general (Marsden & Ratiu, 1999) that Eqs. (2.118) are equivalent to Lagrange's equations given by the reduced Lagrangian L_c obtained by inserting the constraints $f_l(q, t) = 0$ into the unconstrained Lagrangian L.

---

[12] This analysis is not fully realistic because the tangential constraint force will become too small to prevent sliding before the point at which $N = 0$ is reached (Problem 2.26).

Remark   Physically, constraints result from stiff elastic forces exerted by springs, strings, rods or other devices that confine the motion. It is to be expected that Lagrange's equations for the constrained system will be reproduced if the potential energy associated with these forces is included in the unconstrained Lagrangian and the limit of infinite stiffness is taken. This is not trivial, however, and the limit must be taken very carefully (Gallavotti, 1983; van Kampen & Lodder, 1984).

2.5 Symmetry Properties and Conservation Laws

A remarkable discovery, which acquired a fundamental importance in the course of the twentieth century, was that of the existence of an intimate relation between conservation laws and the invariance of physical systems under symmetry operations.

Constants of the Motion

A constant of the motion is a conserved physical quantity – that is, a quantity associated with a mechanical system whose value does not change during the dynamical evolution of the system. Mathematically, a **constant of the motion**, an **invariant**, a **first integral** or simply an **integral** of a mechanichal system is a function f of the coordinates, velocities and, possibly, time that remains constant throughout the motion of the system:

$$\frac{df}{dt} = 0 \quad \text{or} \quad f(q_1(t), \ldots, q_n(t), \dot{q}_1(t), \ldots, \dot{q}_n(t), t) = \text{constant} \tag{2.120}$$

where the functions $q_k(t)$ satisfy the equations of motion for the system.

Example 2.10   In the case of a harmonic oscillator with angular frequency ω,

$$f(x, \dot{x}, t) = \arctan\left(\frac{\omega x}{\dot{x}}\right) - \omega t \tag{2.121}$$

is a constant of the motion. In order to check this claim we have to compute the *total* time derivative of f, which is given by

$$\frac{df}{dt} = \frac{\partial f}{\partial x}\frac{dx}{dt} + \frac{\partial f}{\partial \dot{x}}\frac{d\dot{x}}{dt} + \frac{\partial f}{\partial t} = \frac{1}{1 + \omega^2 x^2/\dot{x}^2}\frac{\omega}{\dot{x}}\dot{x} + \frac{1}{1 + \omega^2 x^2/\dot{x}^2}\left[-\frac{\omega x}{\dot{x}^2}\right]\ddot{x} - \omega. \tag{2.122}$$

Using the oscillator equation of motion, $\ddot{x} = -\omega^2 x$, Eq. (2.122) reduces to

$$\frac{df}{dt} = \omega\frac{1 + \omega^2 x^2/\dot{x}^2}{1 + \omega^2 x^2/\dot{x}^2} - \omega = \omega - \omega = 0, \tag{2.123}$$

and f is indeed a constant of the motion.

Constants of the motion supply first-order differential equations that give important information about the time evolution of the system. Even though the exact solution to the equations of motion may be unknown, the knowledge of certain first integrals often reveals physically relevant facts regarding the nature of the motion. Sometimes it is possible to obtain exact answers to certain questions thanks to the constants of the motion, without having to completely solve the dynamical problem. Therefore, investigating general conditions that ensure the existence of constants of the motion is highly relevant. The first and simplest result of such an investigation will be stated after two useful definitions.

Definition 2.5.1 *Let $L(q, \dot{q}, t)$ be the Lagrangian for a system with n degrees of freedom. The quantity p_k defined by*

$$p_k = \frac{\partial L}{\partial \dot{q}_k} \tag{2.124}$$

is called the **generalised momentum**, **conjugate momentum** *or* **canonical conjugate momentum** *associated with the generalised coordinate q_k.*

Example 2.11 If $L = (m/2)(\dot{x}^2 + \dot{y}^2 + \dot{z}^2) - V(x, y, z)$ then $p_x = m\dot{x}$ is the x-component of the particle's linear momentum, with similar results for p_y and p_z.

Example 2.12 Let $L = (m/2)\dot{\mathbf{r}}^2 - e\phi(\mathbf{r}, t) + (e/c)\dot{\mathbf{r}} \cdot \mathbf{A}(\mathbf{r}, t)$ be the Lagrangian for a particle with electric charge e in an external electromagnetic field. Then $p_x = m\dot{x} + (e/c)A_x(\mathbf{r}, t)$, which does not coincide with the x-component of the particle's linear momentum because of the additional term $(e/c)A_x(\mathbf{r}, t)$. Analogous results hold for the components p_y and p_z.

Definition 2.5.2 *If the Lagrangian for a mechanical system does not contain a coordinate q_k (although it does contain $\dot{q}_k$), then the said coordinate is called* **cyclic**[13] *or* **ignorable**.

Theorem 2.5.1 *The momentum conjugate to a cyclic coordinate is a constant of the motion.*

Proof Let q_k be a cyclic coordinate of L. The kth of Lagrange's equations

$$\frac{d}{dt}\left(\frac{\partial L}{\partial \dot{q}_k}\right) - \frac{\partial L}{\partial q_k} = 0 \tag{2.125}$$

reduces to

$$\frac{dp_k}{dt} = 0 \tag{2.126}$$

in virtue of (2.124) and because $\partial L/\partial q_k = 0$ for a cyclic coordinate. It follows that $p_k = $ constant. $\qquad\square$

Example 2.13 In spherical coordinates the Lagrangian for a particle subject to a central force is

$$L = \frac{m}{2}(\dot{r}^2 + r^2\dot{\theta}^2 + r^2 \sin^2\theta\,\dot{\phi}^2) - V(r). \tag{2.127}$$

---

[13] In problems of great physical importance, the coordinates that do not appear in the Lagrangian are usually angles, hence generally calling such coordinates cyclic.

Clearly ϕ is a cyclic coordinate, whence

$$p_\phi = \frac{\partial L}{\partial \dot\phi} = mr^2 \sin^2\theta\,\dot\phi = \text{constant}. \tag{2.128}$$

It is easy to check that p_ϕ is the z-component of the particle's angular momentum.

Example 2.14 A particle oscillates on a vertical plane suspended from a point that can slide horizontally without friction. As seen in Example 2.6, a Lagrangian for the system is

$$L = \frac{m}{2}\left(\dot x^2 + l^2\dot\theta^2 + 2l\dot x\dot\theta\cos\theta\right) + mgl\cos\theta. \tag{2.129}$$

Since x is a cyclic coordinate, we have

$$p_x = \frac{\partial L}{\partial \dot x} = m\dot x + ml\dot\theta\cos\theta = \text{constant}. \tag{2.130}$$

It should be clear that p_x is the x-component of the particle's linear momentum relative to an inertial reference frame.

The absence of a coordinate can be interpreted as a symmetry property of the Lagrangian. Indeed, if q_k is a cyclic coordinate changing the value of q_k does not change the Lagrangian. In other words, the Lagrangian is invariant under a displacement of a cyclic coordinate. But a displacement of a cyclic position coordinate amounts to a translation, whereas a displacement of a cyclic angular coordinate amounts to a rotation. In the previous example, conservation of the x-component of the linear momentum is a consequence of the invariance of the Lagrangian under translations along the x-direction. Likewise, in Example 2.13 conservation of the z-component of the angular momentum ensues from the invariance of the Lagrangian under rotations about the z-axis. These observations suggest the existence of a general connection between symmetries of the Lagrangian under translations and rotations and the linear momentum and angular momentum conservation laws.

Infinitesimal Translations and Rotations

Consider the infinitesimal transformation

$$\mathbf{r}_i \rightarrow \mathbf{r}_i' = \mathbf{r}_i + \delta\mathbf{r}_i, \qquad \mathbf{v}_i \rightarrow \mathbf{v}_i' = \mathbf{v}_i + \delta\mathbf{v}_i, \tag{2.131}$$

where $\delta\mathbf{r}_i$ and $\delta\mathbf{v}_i$ are infinitesimal displacements of the positions and velocities of a system of N particles. Let

$$L(\mathbf{r}_1,\ldots,\mathbf{r}_N,\dot{\mathbf{r}}_1,\ldots,\dot{\mathbf{r}}_N,t) \equiv L(\mathbf{r},\mathbf{v},t) \tag{2.132}$$

be the Lagrangian for the system. The variation of L under transformation (2.131) is defined by

$$\delta L = L(\mathbf{r}',\mathbf{v}',t) - L(\mathbf{r},\mathbf{v},t) = \sum_{i=1}^{N}\left[\frac{\partial L}{\partial \mathbf{r}_i}\cdot\delta\mathbf{r}_i + \frac{\partial L}{\partial \mathbf{v}_i}\cdot\delta\mathbf{v}_i\right], \tag{2.133}$$

where we have used the notation

$$\frac{\partial L}{\partial \mathbf{r}_i} \equiv \frac{\partial L}{\partial x_i}\hat{\mathbf{x}} + \frac{\partial L}{\partial y_i}\hat{\mathbf{y}} + \frac{\partial L}{\partial z_i}\hat{\mathbf{z}}, \qquad (2.134)$$

with a similar definition for $\partial L/\partial \mathbf{v}_i$.

Example 2.15 A rigid translation of the system of particles consists in the same displacement $\boldsymbol{\epsilon} = \epsilon\hat{\mathbf{n}}$ of all particles of the system, so that the velocities, as well as the relative positions of the particles, remain unchanged. In this case, equations (2.131) are valid with

$$\delta\mathbf{r}_i = \epsilon\hat{\mathbf{n}}, \qquad \delta\mathbf{v}_i = 0, \qquad \text{(translation)} \qquad (2.135)$$

where ϵ is an infinitesimal parameter with dimension of length.

Example 2.16 A rigid rotation of the system of particles consists in a rotation of all vectors of the system through the same angle $\delta\theta$ about an axis defined by the unit vector $\hat{\mathbf{n}}$. Inspecting Fig. 2.7 one infers that $|\delta\mathbf{r}_i| = r_i \sin\alpha\, \delta\theta$, which has the appearence of magnitude of a vector product. In fact, defining the vector $\boldsymbol{\delta\theta} = \delta\theta\hat{\mathbf{n}}$, it follows at once that the correct expression for the vector $\delta\mathbf{r}_i$ is $\delta\mathbf{r}_i = \boldsymbol{\delta\theta} \times \mathbf{r}_i$. Since each velocity $\delta\mathbf{v}_i$ undergoes the same rotation, equations (2.131) are valid with

$$\delta\mathbf{r}_i = \boldsymbol{\delta\theta} \times \mathbf{r}_i, \qquad \delta\mathbf{v}_i = \boldsymbol{\delta\theta} \times \mathbf{v}_i, \qquad \text{(rotation)} \qquad (2.136)$$

where $\delta\theta$ is the infinitesimal parameter associated with the transformation.

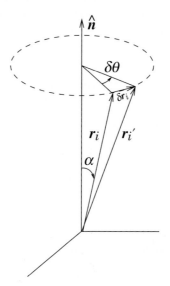

Fig. 2.7   Rigid infinitesimal rotation of a system of particles.

Conservation Theorems

In order to state the results in a sufficiently general form, let us assume that the system described by the Lagrangian (2.132) is subject to the holonomic constraints

$$f_s(\mathbf{r}_1, \ldots, \mathbf{r}_N, t) = 0 , \qquad s = 1, \ldots, p . \tag{2.137}$$

Theorem 2.5.2 *Let a mechanical system be described by the Lagrangian $L = T - V$, where V is a velocity-independent potential. If the Lagrangian and the constraints (2.137) are invariant under an arbitrary rigid translation, then the total linear momentum of the system is conserved.*

Proof   According to the discussion at the end of Section 2.4, the Lagrangian

$$\mathcal{L} = L(\mathbf{r}, \mathbf{v}, t) + \sum_{s=1}^{p} \lambda_s f_s(\mathbf{r}, t) \tag{2.138}$$

yields, in one fell swoop, the equations of motion and the constraint equations. Since L and the constraint equations are invariant under arbitrary translations, so is $\mathcal{L}$. In particular, $\mathcal{L}$ is invariant under any infinitesimal translation and we have

$$0 = \delta \mathcal{L} = \sum_{i=1}^{N} \frac{\partial \mathcal{L}}{\partial \mathbf{r}_i} \cdot (\epsilon \hat{\mathbf{n}}) = \epsilon \hat{\mathbf{n}} \cdot \sum_{i=1}^{N} \frac{\partial \mathcal{L}}{\partial \mathbf{r}_i} , \tag{2.139}$$

where we have used (2.133) and (2.135). Since ϵ is arbitrary, we obtain

$$\hat{\mathbf{n}} \cdot \sum_{i=1}^{N} \frac{\partial \mathcal{L}}{\partial \mathbf{r}_i} = 0 . \tag{2.140}$$

Noting that

$$\frac{\partial \mathcal{L}}{\partial \mathbf{v}_i} \equiv \frac{\partial \mathcal{L}}{\partial \dot{\mathbf{r}}_i} = \frac{\partial L}{\partial \dot{\mathbf{r}}_i} = m_i \dot{\mathbf{r}}_i = \mathbf{p}_i , \tag{2.141}$$

where $\mathbf{p}_i$ is the linear momentum of the ith particle, Lagrange's equations can be written in the form

$$\frac{\partial \mathcal{L}}{\partial \mathbf{r}_i} = \frac{d}{dt} \left(\frac{\partial \mathcal{L}}{\partial \dot{\mathbf{r}}_i} \right) = \frac{d \mathbf{p}_i}{dt} . \tag{2.142}$$

Thus, Eq. (2.140) becomes

$$\frac{d}{dt} \left[\hat{\mathbf{n}} \cdot \sum_{i=1}^{N} \mathbf{p}_i \right] \equiv \frac{d}{dt} \left(\hat{\mathbf{n}} \cdot \mathbf{P} \right) = 0 , \tag{2.143}$$

proving that the component of the total linear momentum $\mathbf{P}$ along direction $\hat{\mathbf{n}}$ is conserved. Choosing in succession $\hat{\mathbf{n}} = \hat{\mathbf{x}}, \hat{\mathbf{y}}, \hat{\mathbf{z}}$, one concludes that P_x, P_y and P_z are constants of the motion – that is, the total linear momentum vector is conserved. $\qquad \square$

Theorem 2.5.3   *Let V be a velocity-independent potential. If the Lagrangian $L = T - V$ and the constraints (2.137) are invariant under an arbitrary rigid rotation then the total angular momentum of the system is conserved.*

Proof   The invariance of $\mathcal{L}$ under any infinitesimal rotation implies

$$0 = \delta\mathcal{L} = \sum_{i=1}^{N} \left[\frac{\partial\mathcal{L}}{\partial\mathbf{r}_i} \cdot (\delta\boldsymbol{\theta} \times \mathbf{r}_i) + \frac{\partial\mathcal{L}}{\partial\mathbf{v}_i} \cdot (\delta\boldsymbol{\theta} \times \mathbf{v}_i) \right], \tag{2.144}$$

where we have used (2.133) and (2.136). The scalar triple product is invariant under cyclic permutations of the vectors involved, which allows us to write

$$\delta\theta\,\hat{\mathbf{n}} \cdot \sum_{i=1}^{N} \left[\mathbf{r}_i \times \frac{\partial\mathcal{L}}{\partial\mathbf{r}_i} + \mathbf{v}_i \times \frac{\partial\mathcal{L}}{\partial\mathbf{v}_i} \right] = 0. \tag{2.145}$$

Taking account of (2.141), Lagrange's equations (2.142) and the arbitrariness of $\delta\theta$, we are led to

$$\hat{\mathbf{n}} \cdot \sum_{i=1}^{N} (\mathbf{r}_i \times \dot{\mathbf{p}}_i + \dot{\mathbf{r}}_i \times \mathbf{p}_i) = \frac{d}{dt}\left[\hat{\mathbf{n}} \cdot \sum_{i=1}^{N} \mathbf{r}_i \times \mathbf{p}_i \right] \equiv \frac{d}{dt}(\hat{\mathbf{n}} \cdot \mathbf{L}) = 0. \tag{2.146}$$

Therefore, the component of the total angular momentum $\mathbf{L}$ about direction $\hat{\mathbf{n}}$ is a constant of the motion. The arbitrariness of $\hat{\mathbf{n}}$ implies the conservation of the total angular momentum vector.   □

The two last results apply to isolated systems of particles and also to isolated systems of rigid bodies because, in the latter case, the rigidity constraints – fixed distances between all pairs of particles – are obviously preserved by rotations and translations. It is also important to stress that a given system may possess a partial symmetry under rotations or translations, which will imply conservation of some components of $\mathbf{P}$ or $\mathbf{L}$, but not all. The kinetic energy is invariant under arbitrary translations and rotations, so the invariance of the Lagrangian is determined by the symmetries of the potential energy. Consider, for instance, a particle moving in the gravitational potential of an infinite uniform plane lamina. The potential is clearly invariant under translations parallel to the plane and under rotations about an axis perpendicular to the plane (which we will take as the z-axis). Therefore, the Lagrangian is invariant under translations along the x and y directions, and under rotations around the z-axis. Thus, P_x, P_y and L_z are conserved quantities.

Incidentally, equations (2.143) and (2.146) show that, for an isolated system therefore invariant under translations and rotations, the sum of internal forces and torques is zero. It is remarkable that the symmetry under translations and rotations guarantees conservation of the total linear and angular momenta without the need to invoke Newton's third law even in its weak form.

2.6 Conservation of Energy

Naturally, the Lagrangian formalism is expected to contain the energy conservation theorem for systems distinguished by forces arising from a potential that depends solely on the positions of the particles. In truth, it is possible to prove a very general conservation theorem which reduces to the energy conservation law in a special case. Although Cartesian coordinates are advantageous to deal with translations and rotations, here it will be more convenient to consider a holonomic system described by the generalised coordinates $q_1, \ldots, q_n$ and the Lagrangian $L(q, \dot{q}, t)$. The generalisation to the non-holonomic case is left as an exercise (Problem 2.24).

Theorem 2.6.1 *Let L be the Lagrangian for a mechanical system expressed in terms of the generalised coordinates $q_1, \ldots, q_n$. If L does not depend explicitly on time, then the quantity h defined by*[14]

$$h = \sum_{k=1}^{n} \dot{q}_k \frac{\partial L}{\partial \dot{q}_k} - L \tag{2.147}$$

is a constant of the motion, often called the **Jacobi integral**.

Proof The total time derivative of h is

$$\frac{dh}{dt} = \sum_{k=1}^{n} \left\{ \ddot{q}_k \frac{\partial L}{\partial \dot{q}_k} + \dot{q}_k \frac{d}{dt} \left(\frac{\partial L}{\partial \dot{q}_k} \right) \right\} - \left\{ \sum_{k=1}^{n} \left(\frac{\partial L}{\partial q_k} \dot{q}_k + \frac{\partial L}{\partial \dot{q}_k} \ddot{q}_k \right) + \frac{\partial L}{\partial t} \right\}, \tag{2.148}$$

from which, by using Lagrange's equations (2.48), it follows that

$$\frac{dh}{dt} = -\frac{\partial L}{\partial t} . \tag{2.149}$$

If L does not depend explicitly on time, $\partial L / \partial t = 0$ and this last equation establishes that h is a constant of the motion. □

Under well-defined circumstances, which we now examine, h is the total energy of the system. The kinetic energy, expressed in terms of the generalised coordinates and velocities, is given by

$$T = \frac{1}{2} \sum_{i} m_i v_i^2 = \frac{1}{2} \sum_{i} m_i \left(\sum_{k} \frac{\partial \mathbf{r}_i}{\partial q_k} \dot{q}_k + \frac{\partial \mathbf{r}_i}{\partial t} \right) \cdot \left(\sum_{l} \frac{\partial \mathbf{r}_i}{\partial q_l} \dot{q}_l + \frac{\partial \mathbf{r}_i}{\partial t} \right), \tag{2.150}$$

where Eq. (1.73) has been used. This last equation can be rewritten in the form

$$T = M_0 + \sum_{k} M_k \dot{q}_k + \frac{1}{2} \sum_{k,l} M_{kl} \dot{q}_k \dot{q}_l, \tag{2.151}$$

---

[14] Some authors call h the "energy function".

where M_0, M_k and M_{kl} are the following functions of the qs and t:

$$M_0 = \frac{1}{2} \sum_i m_i \frac{\partial \mathbf{r}_i}{\partial t} \cdot \frac{\partial \mathbf{r}_i}{\partial t}, \tag{2.152}$$

$$M_k = \sum_i m_i \frac{\partial \mathbf{r}_i}{\partial t} \cdot \frac{\partial \mathbf{r}_i}{\partial q_k}, \tag{2.153}$$

$$M_{kl} = \sum_i m_i \frac{\partial \mathbf{r}_i}{\partial q_k} \cdot \frac{\partial \mathbf{r}_i}{\partial q_l}. \tag{2.154}$$

Exercise 2.6.1   Derive Eqs. (2.152)–(2.154).

If $L = T - V$ with V independent of the velocities, we have

$$h = \sum_k \frac{\partial T}{\partial \dot{q}_k} \dot{q}_k - (T - V). \tag{2.155}$$

If, moreover, Eqs. (1.71) defining the generalised coordinates do not involve the time explicitly, then $\partial \mathbf{r}_i / \partial t = 0$ hence $M_0 = M_k = 0$. In this case, Eq. (2.151) tells us that T is a second-degree homogeneous function of the generalised velocities. Therefore, by Euler's theorem on homogeneous functions (Appendix C),

$$h = 2T - (T - V) = T + V = E, \tag{2.156}$$

and h is the total energy of the system. In the present circumstances, if V does not depend explicitly on time neither does L, and it follows that the total energy – in this case equal to h – will be conserved. It is worth noting that even in the absence of explicitly time-dependent potentials, the presence of time dependent constraints (as in Example 1.17) introduces an explicit time dependence in equations (1.71), preventing h from coinciding with the total energy. It should be further stressed that the conditions that ensure energy conservation are separate from those that guarantee that h equals the total energy. Thus, h may be conserved without being the total energy or be the total energy without being conserved. On the other hand, differently from the Lagrangian, whose value is independent of the choice of generalised coordinates, both the value and the functional form of h depend on the chosen set of generalised coordinates. As a rule, when there are time-dependent constraints the total energy is not conserved, and the reason is simple. As seen in Section 1.3, if the constraints change with time the constraint forces do work during *real* displacements although the work done in the course of virtual displacements is zero. The variation of the total energy during a time interval dt is equal to the work done by the constraint forces during the real displacements $d\mathbf{r}_i = \mathbf{v}_i dt$, which in general does not vanish. Therefore, the total energy cannot remain constant.

The last three theorems shed new light on the origin of the fundamental constants of the motion of mechanics in associating the conservation laws with geometric properties of space and time. Homogeneity and isotropy of space entail invariance of the Lagrangian and the constraints under translations and rotations, leading to conservation of linear and angular momenta. Homogeneity of time entails immutability of the Lagrangian and the constraints under a time displacement (Lagrangian and constraints without explicit time dependence), leading to energy conservation.

In the presence of dissipative forces that can be described by a dissipation function $\mathcal{F}$, Theorem 2.6.1 is no longer true and Eq. (2.149) is replaced by

$$\frac{dh}{dt} = -\sum_k \dot{q}_k \frac{\partial \mathcal{F}}{\partial \dot{q}_k} - \frac{\partial L}{\partial t} . \tag{2.157}$$

Exercise 2.6.2   Derive Eq. (2.157).

Note that $\mathcal{F}$, as concerns its functional dependence on the generalised velocities, has the same structure as the kinetic energy. Thus, if equations (1.71), which define the generalised coordinates, do not depend explicitly on time, $h = E$ and $\mathcal{F}$ is a second-degree homogeneous function of the generalised velocities. In this case (Lemos, 1991), Euler's theorem on homogeneous functions allows us to write (2.157) in the form

$$\frac{dE}{dt} = -2\mathcal{F} - \frac{\partial L}{\partial t} . \tag{2.158}$$

If, in addition to the circumstances already specified, L does not depend explicitly on time, then

$$\frac{dE}{dt} = -2\mathcal{F} , \tag{2.159}$$

a result obtained by other means in Section 1.6.

2.7 Noether's Theorem

In Lagrangian dynamics the general connection between symmetry (invariance) properties and conserved quantities is established by an important theorem due to Emmy Noether.[15] This theorem (Noether, 1918) contains all of the results discussed in Sections 2.5 and 2.6 as particular cases.

Let X and Ψ_i be known functions of $n + 1$ real variables and let ϵ be an arbitrary infinitesimal parameter. Consider the infinitesimal transformation

$$t \quad \longrightarrow \quad t' = t + \epsilon X(q(t), t) ,$$

$$q_i(t) \quad \longrightarrow \quad q_i'(t') = q_i(t) + \epsilon \Psi_i(q(t), t) . \tag{2.160}$$

---

[15] Widely regarded the greatest female mathematician, Amalie Emmy Noether was born on 23 March 1882 in Germany. Her father was a mathematics professor, but she spent her childhood learning the arts, as was then expected from girls. Eventually she moved to mathematics and obtained her doctoral degree in 1907. In 1915 she joined the Mathematical Institute in Göttingen and started working with Felix Klein and David Hilbert on general relativity theory. In 1918 she proved two theorems that were basic for both general relativity and quantum field theory, one of them known to physicists as "Noether's theorem". But she was not allowed to join the faculty at Göttingen University because she was a woman. During the 1920s she did fundamental work in abstract algebra. In 1933 she was forbidden to teach by the Nazi government and decided to accept a professorship at Bryn Mawr College in Pennsylvania. She also accepted an invitation to be a visiting member and to lecture at the Institute for Advanced Study in Princeton. On 14 April 1935, she died of a postsurgical infection in Bryn Mawr, USA.

These equations define a continuous family of transformations parameterised by ϵ. The action integral is said to be invariant under this family of transformations if

$$\Delta S = \int_{t_1'}^{t_2'} L\left(q'(t'), \frac{dq'(t')}{dt'}, t'\right) dt' - \int_{t_1}^{t_2} L\left(q(t), \frac{dq(t)}{dt}, t\right) dt = 0. \qquad (2.161)$$

In this case, transformation (2.160) is said to be a **continuous symmetry** of the action. Noether's theorem states that to each continuous symmetry of the action there corresponds a constant of the motion.

Theorem 2.7.1 (Noether's Theorem) *Given a mechanical system with n degrees of freedom, if the action is invariant under transformation (2.160) then the quantity*

$$C = \sum_{i=1}^{n} \frac{\partial L}{\partial \dot{q}_i}(\dot{q}_i X - \Psi_i) - LX \qquad (2.162)$$

is a constant of the motion, where $L(q, \dot{q}, t)$ is the Lagrangian for the system.

Proof Let us assume that $q_k(t)$ in Eq. (2.160) is the physical path, which satisfies Lagrange's equations. In the calculations below we will retain only the first-order terms in ϵ. Note first that

$$\frac{dt'}{dt} = 1 + \epsilon \dot{X}, \qquad \frac{dt}{dt'} = (1 + \epsilon \dot{X})^{-1} = 1 - \epsilon \dot{X}, \qquad (2.163)$$

where the binomial series $(1 + x)^\lambda = 1 + \lambda x + \cdots$ has been used. We also have

$$\frac{dq_i'(t')}{dt'} = \frac{dt}{dt'}\frac{dq_i'(t')}{dt} = (1 - \epsilon \dot{X})(\dot{q}_i + \epsilon \dot{\Psi}_i) = \dot{q}_i + \epsilon \xi_i, \qquad (2.164)$$

where

$$\xi_i = \dot{\Psi}_i - \dot{q}_i \dot{X}. \qquad (2.165)$$

Therefore, the invariance of the action, that is,

$$\begin{aligned}
\Delta S &= \int_{t_1}^{t_2} L(q + \epsilon \Psi, \dot{q} + \epsilon \xi, t + \epsilon X)(1 + \epsilon \dot{X})dt - \int_{t_1}^{t_2} L(q, \dot{q}, t)dt \\
&= \int_{t_1}^{t_2} \left\{ L + \sum_{i=1}^{n} \left[\frac{\partial L}{\partial q_i}\epsilon \Psi_i + \frac{\partial L}{\partial \dot{q}_i}\epsilon \xi_i \right] + \frac{\partial L}{\partial t}\epsilon X \right\}(1 + \epsilon \dot{X})dt - \int_{t_1}^{t_2} Ldt \\
&= \epsilon \int_{t_1}^{t_2} \left\{ \sum_{i=1}^{n} \left[\frac{\partial L}{\partial q_i}\Psi_i + \frac{\partial L}{\partial \dot{q}_i}\xi_i \right] + \frac{\partial L}{\partial t}X + L\dot{X} \right\}dt = 0
\end{aligned} \qquad (2.166)$$

leads at once to the **Noether condition**

$$\sum_{i=1}^{n} \left[\Psi_i \frac{\partial L}{\partial q_i} + (\dot{\Psi}_i - \dot{q}_i \dot{X})\frac{\partial L}{\partial \dot{q}_i} \right] + L\dot{X} + \frac{\partial L}{\partial t}X = 0 \qquad (2.167)$$

inasmuch as ϵ and the integration interval are arbitrary. *This is the necessary and sufficient condition for the action to be invariant under transformation (2.160).* With the use of

Lagrange's equations as well as equations (2.147) and (2.149), the Noether condition (2.167) can be rewritten in the form

$$\sum_{i=1}^{n} \left[\Psi_i \frac{d}{dt}\left(\frac{\partial L}{\partial \dot{q}_i}\right) + \dot{\Psi}_i \frac{\partial L}{\partial \dot{q}_i} \right] - h\dot{X} - \frac{dh}{dt}X = \frac{d}{dt}\left\{ \sum_{i=1}^{n} \frac{\partial L}{\partial \dot{q}_i}\Psi_i - hX \right\} = 0 \qquad (2.168)$$

or, equivalently,

$$\frac{d}{dt}\left\{ \sum_{i=1}^{n} \frac{\partial L}{\partial \dot{q}_i}(\dot{q}_i X - \Psi_i) - LX \right\} = 0, \qquad (2.169)$$

which completes the proof.   □

Generalisation of Noether's Theorem

According to Theorem 2.3.1, Lagrangians that differ by a total time derivative of an arbitrary function of the generalised coordinates and time yield exactly the same equations of motion. Noether's theorem admits a generalisation that reflects this important property of equivalent Lagrangians. Suppose the action integral is quasi-invariant under the infinitesimal transformation (2.160) – that is, assume there exists a function $G(q,t)$ such that

$$\int_{t'_1}^{t'_2} L\left(q'(t'), \frac{dq'(t')}{dt'}, t'\right) dt' - \int_{t_1}^{t_2} \left\{ L\left(q(t), \frac{dq(t)}{dt}, t\right) + \epsilon \frac{d}{dt}G(q(t), t) \right\} dt = 0. \quad (2.170)$$

Then the Noether condition takes the generalised form

$$\sum_{i=1}^{n} \left[\Psi_i \frac{\partial L}{\partial q_i} + (\dot{\Psi}_i - \dot{q}_i \dot{X}) \frac{\partial L}{\partial \dot{q}_i} \right] + L\dot{X} + \frac{\partial L}{\partial t}X = \dot{G} \qquad (2.171)$$

and the conserved quantity associated with the symmetry is

$$\bar{C} = \sum_{i=1}^{n} \frac{\partial L}{\partial \dot{q}_i}(\dot{q}_i X - \Psi_i) - LX + G. \qquad (2.172)$$

The details of the proof are left to the reader.

Some Simple Applications

The fundamental conservation laws are particular cases of Noether's theorem, revealing its power and generality.

Example 2.17   Prove Theorem 2.6.1 by means of Noether's theorem.

Solution

Consider a displacement of nothing but the time – that is, take transformation (2.160) with $\Psi_i = 0$ and $X = 1$. According to the Noether condition (2.167), the action will be invariant if and only if $\partial L/\partial t = 0$. If this is the case, by (2.162) we conclude that

$$h = \sum_{i=1}^{n} \dot{q}_i \frac{\partial L}{\partial \dot{q}_i} - L = \text{constant}, \tag{2.173}$$

which proves Theorem 2.6.1.

Example 2.18 Prove Theorem 2.5.3 by resorting to Noether's theorem.

Solution

Define $q_1 = x_1, q_2 = y_1, q_3 = z_1, \ldots, q_{3N} = z_N$ and take $\delta\boldsymbol{\theta} = \epsilon\hat{\mathbf{n}}$ with $\epsilon = \delta\theta$. Equations (2.136) are of the form (2.160) with $X = 0$, $\Psi_1 = (\hat{\mathbf{n}} \times \mathbf{r}_1)_x$, $\Psi_2 = (\hat{\mathbf{n}} \times \mathbf{r}_1)_y$, $\Psi_3 = (\hat{\mathbf{n}} \times \mathbf{r}_1)_z, \ldots, \Psi_{3N} = (\hat{\mathbf{n}} \times \mathbf{r}_N)_z$. On the other hand, the invariance of S is due to the invariance of $\mathcal{L}$ since the time variable is not changed. Thus, if $\mathcal{L}$ is invariant Noether's theorem ensures that

$$-\sum_{k=1}^{3N} \frac{\partial \mathcal{L}}{\partial \dot{q}_k} \Psi_k \equiv -\sum_{i=1}^{N} \frac{\partial \mathcal{L}}{\partial \dot{\mathbf{r}}_i} \cdot (\hat{\mathbf{n}} \times \mathbf{r}_i) = \text{constant}, \tag{2.174}$$

that is,

$$\hat{\mathbf{n}} \cdot \sum_{i=1}^{N} \mathbf{r}_i \times \mathbf{p}_i = \text{constant}, \tag{2.175}$$

which reproduces Eq. (2.146).

Exercise 2.7.1 Prove Theorem 2.5.2 by means of Noether's theorem.

We have saved for last an original application of Noether's theorem that involves a transformation of space and time together.

Example 2.19 The equation of motion for a damped harmonic oscillator can be derived from the Lagrangian (see Problem 1.16)

$$L = e^{\lambda t}\left(\frac{m}{2}\dot{x}^2 - \frac{m\omega^2}{2}x^2\right). \tag{2.176}$$

Show that the action is invariant under the *finite* transformation (α is an arbitrary constant)

$$t' = t + \alpha, \qquad x'(t') = e^{-\lambda\alpha/2}x(t), \tag{2.177}$$

and infer that

$$C = e^{\lambda t}\left(\frac{m}{2}\dot{x}^2 + \frac{m\omega^2}{2}x^2 + \frac{m\lambda}{2}x\dot{x}\right) \tag{2.178}$$

is a constant of the motion.

Solution

Note that (2.177) consists of a time translation jointly with a space dilation. We have

$$
\begin{aligned}
S' &= \int_{t_1'}^{t_2'} e^{\lambda t'} \left\{ \frac{m}{2}\left(\frac{dx'}{dt'}\right)^2 - \frac{m\omega^2}{2}x'^2 \right\} dt' \\
&= \int_{t_1'}^{t_2'} e^{\lambda t} e^{\lambda\alpha} e^{-\lambda\alpha} \left(\frac{m}{2}\dot{x}^2 - \frac{m\omega^2}{2}x^2\right) dt' \\
&= \int_{t_1}^{t_2} e^{\lambda t} \left(\frac{m}{2}\dot{x}^2 - \frac{m\omega^2}{2}x^2\right) dt = S,
\end{aligned}
\tag{2.179}
$$

where we have used $dt' = dt$. Therefore, the action is invariant and Noether's theorem applies. The infinitesimal version of (2.177) is obtained by letting $\alpha = \epsilon$. Expanding the exponential up to the first order in ϵ there results

$$
t' = t + \epsilon, \qquad x'(t') = x(t) - \frac{\epsilon\lambda}{2}x(t),
\tag{2.180}
$$

and from a comparison with (2.160) one infers

$$
X = 1, \qquad \Psi = -\frac{\lambda}{2}x.
\tag{2.181}
$$

Substituting (2.176) and (2.181) into (2.162) with $q_1 = x$, $\Psi_1 = -\lambda x/2$ and the sum reduced to a single term, the constant of the motion (2.178) emerges at once. (Lemos, 1993.)

Exercise 2.7.2   (1) Using (2.176) and (2.181), establish the invariance of the action by verifying that the Noether condition (2.167) is satisfied. (2) Taking account of the equation of motion for the damped harmonic oscillator, prove directly that the quantity C defined by (2.178) is a constant of the motion.

Constants of the Motion and Symmetries of the Action

Noether's theorem ensures that to each continuous symmetry of the action there corresponds a conserved quantity, but the converse is not true: there are constants of the motion that are not associated with any continuous symmetry of the action. This is just what happens, for instance, with the constant of the motion in Example 2.10. If that constant of the motion were originated from a continuous symmetry of the action, there would exist functions $X(x,t)$, $\psi(x,t)$ and $G(x,t)$ such that, according to Eq. (2.172),

$$
\frac{\partial L}{\partial \dot{x}}(\dot{x}X - \psi) - LX + G = f(x,\dot{x},t)
\tag{2.182}
$$

or, using $L = m\dot{x}^2/2 - m\omega^2 x^2/2$ and the explicit form of f,

$$
\left(\frac{m}{2}\dot{x}^2 + \frac{m\omega^2}{2}x^2\right)X(x,t) - m\dot{x}\psi(x,t) + G(x,t) = \tan^{-1}\left(\frac{\omega x}{\dot{x}}\right) - \omega t.
\tag{2.183}
$$

It so happens that this equation cannot be satisfied because the left-hand side is a second-degree polynomial in $\dot{x}$ whereas the right-hand side is a transcendental function of $\dot{x}$.

As a general rule, the set of continuous symmetries of the equations of motion is larger than the set of continuous symmetries of the action (Lutzky, 1978), with the consequence that certain constants of the motion are associated with symmetries of the equations of motion that are not symmetries of the action.

Variational Principles: Physics and Metaphysics

As has been seen in this chapter, the formulation of a physical law in the form of a variational principle offers important advantages. First, it makes manifest the invariance of the law under arbitrary coordinate transformations because variational principles do not depend on the particular coordinates used to express them. Second, it allows the establishment of a very general connection between symmetries and conservation laws. As to the aesthetic appeal, variational principles express the laws of physics in an incomparably concise and elegant form. Moreover, it appears that no physical law is fundamental unless it can be expressed as a variational principle. Last, but not least, variational principles serve as basis for highly accurate approximate calculations in problems that resist exact solution (Feynman, Leighton & Sands, 1963, Volume II, Chapter 19; see also Problem 2.4).

On the other hand, on account of making reference to the motion in its totality, variational principles seem to smuggle into physics the notion of purpose or design, endowing mechanical systems with anthropomorphic features. Does a particle somehow "sniff" at every possible path before deciding for the one that gives the least action? Because they arouse questions of this nature, variational principles have exerted a great fascination on scientists and philosophers ever since they were first introduced in mechanics. In the course of the centuries, philosophers and religious people of various hues, sometimes appealing to scientific theories, have dedicated themselves to vain attempts to prove the existence of God. What is surprising is that even a physicist of the stature of Max Planck, succumbing to mysticism, has come to advocate a teleological interpretation of the principle of least action and to invoke it as an argument for the existence of a Supreme Being. The reader inclined to metaphysical speculations is referred to Yourgrau and Mandelstam (1968) for a lucid and stimulating criticism of these ideas as well as of other alleged philosophical implications of variational principles.

Problems

2.1 Determine the geodesics of the cone. Describe the form of the geodesic curve in the case that its initial point is the cone vertex, taken as the origin of the coordinate system. Hint: in terms of spherical coordinates r, θ, ϕ the equation for the surface of a cone is $\theta = \alpha = $ constant.

2.2 A particle moves on a vertical plane in a uniform gravitational field. Show that there is a unique solution $\mathbf{r}_{ph}(t)$ to the equation of motion with the boundary conditions

$\mathbf{r}(0) = 0$, $\mathbf{r}(T) = \mathbf{r}_0$. Let now $\mathbf{r}(t) = \mathbf{r}_{ph}(t) + \boldsymbol{\eta}(t)$, with $\boldsymbol{\eta}(T) = \boldsymbol{\eta}(0) = 0$, be any other trajectory obeying the same boundary conditions. Performing an integration by parts and using the equation of motion satisfied by $\mathbf{r}_{ph}(t)$, show that the action associated with $\mathbf{r}(t)$ is given by $S[\mathbf{r}] = S[\mathbf{r}_{ph}] + (m/2) \int_0^T |\dot{\boldsymbol{\eta}}|^2 dt$ and hence conclude that the action is a minimum for the physical path.

2.3 For a harmonic oscillator of mass m and angular frequency ω, if $x_{ph}(t)$ is the solution to the equation of motion and $x(t) = x_{ph}(t) + \eta(t)$ with $\eta(0) = \eta(T) = 0$, proceeding as in the previous problem show that $S[x] = S[x_{ph}] + (m/2) \int_0^T (\dot{\eta}^2 - \omega^2 \eta^2) dt$. Expanding η in the Fourier series $\eta(t) = \sum_{n=1}^{\infty} C_n \sin(n\pi t/T)$ (why is this possible?), show that $S[x] = S[x_{ph}] + (mT/4) \sum_{n=1}^{\infty} (\frac{n^2\pi^2}{T^2} - \omega^2) C_n^2$ and conclude that the action is a minimum for the physical path if $T < \pi/\omega$. An elementary treatment of this problem can be found in Moriconi (2017), where it is shown that the condition $T < \pi/\omega$ is also necessary for a minimum.

2.4 Variational problems can be approximately solved by searching the minimum or maximum of the functional in a restricted class of functions obeying the stipulated boundary conditions. Consider the problem of finding the minimum of the functional $J[y] = \int_0^1 (y'^2 + 12xy)dx$ with $y(0) = 0$ and $y(1) = 1$. Try an approximate solution of the form $y_\alpha(x) = x + \alpha x(1 - x)$ and determine the value $\tilde{\alpha}$ of the parameter α that yields the best approximation – that is, the least value for the functional J in the chosen class of functions. Prove that the exact solution is $y^{(e)}(x) = x^3$. Compare the approximate value for the minimum $J[y_{\tilde{\alpha}}]$ with the exact value $J[y^{(e)}]$ and compute the percentage error commited. Compare the approximate $y_{\tilde{\alpha}}(x)$ and the exact $y^{(e)}(x)$ by drawing graphs of these functions on a single plot for $x \in [0, 1]$.

2.5 In the calculus of variations the problem of finding extrema of the functional

$$J[y] = \int_{x_1}^{x_2} F(y_1(x), \ldots, y_n(x), y_1'(x), \ldots y_n'(x), x) \, dx$$

with the subsidiary conditions

$$\int_{x_1}^{x_2} f_i(y_1(x), \ldots, y_n(x), y_1'(x), \ldots y_n'(x), x) \, dx = \ell_i, \qquad i = 1, \ldots, m,$$

where the ℓ_i are constants, is called the *isoperimetric problem*. It can be proved that the extrema of $J[y]$ are determined by means of Euler's equations for the functional

$$\bar{J}[y] = \int_{x_1}^{x_2} \left(F + \sum_{i=1}^{m} \lambda_i f_i \right) dx,$$

where the Lagrange multipliers λ_i are constants (Gelfand & Fomin, 1963; Kot, 2014). Use this result to find the shape of a flexible inextensible homogeneous cable of length ℓ that hangs from two vertical poles. Taking into account that the cable's equilibrium position is the one for which its gravitational potential energy is a minimum, first show that the cable assumes the shape of the curve that minimises the functional $E[y] = \int_a^b y\sqrt{1 + y'^2} \, dx$ with the subsidiary condition $\int_a^b \sqrt{1 + y'^2} \, dx = \ell$. Next prove that the solution is a catenary.

2.6 Show that Lagrange's equations can be written in the explicit form

$$\sum_{j=1}^{n} \frac{\partial^2 L}{\partial \dot{q}_j \partial \dot{q}_i} \ddot{q}_j + \sum_{j=1}^{n} \frac{\partial^2 L}{\partial q_j \partial \dot{q}_i} \dot{q}_j + \frac{\partial^2 L}{\partial t \partial \dot{q}_i} - \frac{\partial L}{\partial q_i} = 0, \quad i = 1, \dots, n.$$

Prove that two Lagrangians $\bar{L}(q, \dot{q}, t)$ and $L(q, \dot{q}, t)$ generate *identical* equations of motion if and only if they differ by the total time derivative of a function of the generalised coordinates and time. You will need the result contained in Appendix F.

2.7 Show that the Lagrangian

$$\bar{L} = \frac{1}{12}\dot{x}^4 + \frac{\omega^2}{2}x^2\dot{x}^2 - \frac{\omega^4}{4}x^4$$

generates an equation of motion which is not identical to but has the same set of solutions as the equation of motion for a simple harmonic oscillator of mass $m = 1$. Prove, however, that $\bar{L}$ does not differ from the usual Lagrangian $L = \dot{x}^2/2 - \omega^2 x^2/2$ by the total time derivative of some function $f(x, t)$.

2.8 Consider the swinging Atwood's machine of Problem 1.5 in the special case $M = 3m$. Prove directly that the quantity

$$I(r, \dot{r}, \theta, \dot{\theta}) = r^2\dot{\theta}\left[\dot{r}\cos\frac{\theta}{2} - \frac{r\dot{\theta}}{2}\sin\frac{\theta}{2}\right] + gr^2\sin\frac{\theta}{2}\cos^2\frac{\theta}{2}$$

is a constant of the motion (Tufillaro, 1986).

2.9 Consider a harmonic oscillator with time dependent frequency, whose Lagrangian is $L = m\dot{q}^2/2 - m\omega^2(t)q^2/2$. (a) If the function $\rho(t)$ satisfies the differential equation

$$\ddot{\rho} - \frac{1}{\rho^3} + \omega^2(t)\rho = 0,$$

prove that

$$I(q, \dot{q}, t) = (\rho\dot{q} - \dot{\rho}q)^2 + \frac{q^2}{\rho^2}$$

is a constant of the motion (Lewis Jr. & Riesenfeld, 1969). (b) An example of interest in cosmology is $\omega^2(t) = \omega_0^2 + 2/(9t^2)$, with a constant ω_0. Show that, in this case, a solution for $\rho(t)$ is

$$\rho(t) = \left(\frac{\pi t}{2}\right)^{1/2}\left[J_{1/6}^2(\omega_0 t) + Y_{1/6}^2(\omega_0 t)\right]^{1/2},$$

where J_ν and Y_ν are Bessel functions of the first and second kind, respectively (Lemos & Natividade, 1987; Lemos, 2014a).

2.10 A particle moves in the gravitational potential produced by the following homogeneous mass distributions: (a) a sphere of radius R; (b) an infinitely long rectangular parallelepiped; (c) a rod of length ℓ; (d) a disc of radius R; (e) an infinite circular cylinder; (f) a semi-infinite plane; (g) a wire twisted in an infinite helix of radius R and pitch p. Which components of **P** and **L** are conserved in each case?

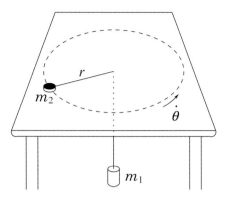

Fig. 2.8   Problem 2.12.

2.11 The electromagnetic field is invariant under a gauge transformation of the potentials defined by

$$\mathbf{A} \;\rightarrow\; \mathbf{A}' = \mathbf{A} + \nabla\Lambda, \qquad \phi \;\rightarrow\; \phi' = \phi - \frac{1}{c}\frac{\partial\Lambda}{\partial t},$$

where $\Lambda(\mathbf{r}, t)$ is an arbitrary differentiable function. How is the Lagrangian (1.143) affected by this transformation? Are the equations of motion altered?

2.12 The system depicted in Fig. 2.8 is such that mass m_2 moves on the frictionless horizontal table and mass m_1 can only move vertically. Use the Lagrange multiplier method to show that the tension on the inextensible string that links the bodies is given by

$$T = \frac{m_1 m_2}{m_1 + m_2}\left[g + \frac{p_\theta^2}{m_2^2 r^3} \right],$$

where p_θ is the constant value of $m_2 r^2 \dot{\theta}$. In the special case that $p_\theta = 0$, determine the fall acceleration of mass m_1.

2.13 In so-called *generalised mechanics* the Lagrangians contain derivatives of order higher than the first. Given a Lagrangian of the form $L(q, \dot{q}, \ddot{q}, t)$, show that Hamilton's principle, with δq_i and $\delta\dot{q}_i$ equal to zero at the endpoints, gives rise to the following generalised form of Lagrange's equations:

$$\frac{d^2}{dt^2}\left(\frac{\partial L}{\partial \ddot{q}_i} \right) - \frac{d}{dt}\left(\frac{\partial L}{\partial \dot{q}_i} \right) + \frac{\partial L}{\partial q_i} = 0.$$

Extend this result to Lagrangians containing derivatives up to order n. Show that the Lagrangian $L = -mq\ddot{q}/2 - kq^2/2$ generates the equation of motion of a harmonic oscillator. Prove that this Lagrangian differs from the usual Lagrangian for the harmonic oscillator by a total time derivative.

2.14 In the case of underdamping, show that with $\Omega = (\omega^2 - \lambda^2/4)^{1/2}$ the Lagrangian

$$L = \frac{2\dot{x} + \lambda x}{2\Omega x}\tan^{-1}\left(\frac{2\dot{x} + \lambda x}{2\Omega x} \right) - \frac{1}{2}\ln(\dot{x}^2 + \lambda x\dot{x} + \omega^2 x^2)$$

yields the equation of motion of the damped harmonic oscillator (Havas, 1957). Prove that the action associated with this Lagrangian is quasi-invariante under the infinitesimal space dilation $x'(t') = (1 + \epsilon)x(t)$, $t' = t$. More precisely, show that Eq. (2.170) holds with $G(x, t) = -t$. Next apply Noether's generalised theorem to conclude that

$$- \tan^{-1} \left(\frac{2\dot{x} + \lambda x}{2\Omega x} \right) - \Omega t = \delta \,,$$

where δ is a constant. Combine this result with the constant of the motion (2.178) to obtain by purely algebraic means (Lemos, 1993)

$$x(t) = A e^{-\lambda t/2} \cos(\Omega t + \delta) \,,$$

which is the general solution to the equation of motion of the damped harmonic oscillator in the underdamped case.

2.15 Consider a system of N interacting particles with Lagrangian $L = T - V$, where $T = \sum_i m_i \dot{\mathbf{r}}_i^2/2$ and V depends only on the interparticle distances $r_{ij} = |\mathbf{r}_i - \mathbf{r}_j|$. Show that, under a Galilean transformation $\mathbf{r}_i' = \mathbf{r}_i - \mathbf{v}t$, $t' = t$, the transformed Lagrangian differs from the original Lagrangian by the total time derivative of a certain function $F(\mathbf{r}_1, \ldots, \mathbf{r}_N, t)$. Letting $\mathbf{v}$ be infinitesimal, show that the action is quasi-invariant and, as a consequence of the generalised version of Noether's theorem, obtain the constant of the motion $\mathbf{C} = t\mathbf{P} - M\mathbf{R}$, where $\mathbf{P}$ is the total linear momentum, M is the total mass and $\mathbf{R}$ is the position vector of the centre of mass of the system.

2.16 A unit mass particle moves in a straight line subject to the non-conservative force $F = -\gamma \dot{x}^2$, with γ a constant. (a) Show that the equation of motion for the particle can be obtained from the Lagrangian $L = e^{2\gamma x} \dot{x}^2/2$. (b) Show that the Jacobi integral h associated with this Lagrangian is equivalent to $\dot{x} e^{\gamma x} = C_1$, where C_1 is a constant. (c) Prove that the action is invariant under the transformation $x' = x + a$, $t' = e^{\beta a} t$ for a certain value of β (determine it). Interpret geometrically the meaning of this transformation. (d) Use Noether's theorem to prove that $(\gamma t \dot{x} - 1)\dot{x} e^{2\gamma x} = C_2 =$ constant. (e) Combining the two constants of the motion to eliminate $\dot{x}$, show that

$$x(t) = A + \frac{1}{\gamma} \ln(B + \gamma t),$$

and check directly that this is the general solution to the equation of motion for the particle.

2.17 The Lagrangian for a two-degree-of-freedom system is

$$L = \frac{m}{2}(\dot{x}^2 + \dot{y}^2) - (\alpha x + \beta y) \,,$$

where α and β are constants such that $\alpha\beta \neq 0$. (a) Prove that the Lagrangian – and, therefore, the action – is invariant under the infinitesimal transformation $x \to x' = x + \epsilon\beta$, $y \to y' = y - \epsilon\alpha$. Using Noether's theorem, show that the quantity A defined by

$$A = \beta\dot{x} - \alpha\dot{y}$$

is a constant of the motion. (b) Introducing the new generalised coordinates $\bar{x}$ and $\bar{y}$ defined by $\bar{x} = \alpha x + \beta y$ and $\bar{y} = \beta x - \alpha y$, show that one of the new coordinates does not appear in the Lagrangian expressed in terms of $\bar{x}, \bar{y}, \dot{\bar{x}}, \dot{\bar{y}}$. Prove that the quantity A in (a) is proportional to the momentum conjugate to the cyclic coordinate and interpret geometrically the whole procedure.

2.18 A particle of mass m and electric charge e moves in a uniform magnetic field $\mathbf{B} = B\hat{\mathbf{z}}$, whose vector potential is $\mathbf{A}(\mathbf{r}) = \mathbf{B} \times \mathbf{r}/2$. (a) Show that Lagrange's equations for the particle are equivalent to $\dot{\mathbf{v}} = -\boldsymbol{\omega} \times \mathbf{v}$ and determine $\boldsymbol{\omega}$ in terms of $\mathbf{B}$. (b) Express the Lagrangian in cylindrical coordinates ρ, ϕ, z and show that, although ϕ is a cyclic coordinate, the z-component of the angular momentum $L_z = m\rho^2\dot{\phi}$ *is not conserved*. Explain.

2.19 Let q_k be a cyclic coordinate of the Lagrangian $L(q, \dot{q}, t)$, so that the conjugate momentum p_k is a constant of the motion. If L is replaced by $\bar{L} = L + dF(q, t)/dt$, in general q_k will not be a cyclic coordinate of $\bar{L}$ and the conjugate momentum will not be conserved. But we know that $\bar{L}$ and L give rise to the same equations of motion. Resolve the apparent paradox: how is it possible that the same equations of motion imply both conservation and nonconservation?

2.20 Reconsider Problem 1.12 by the Lagrange multiplier technique and find the constraint force on the block of mass m.

2.21 Show that the action associated with the Lagrangian of Problem 1.8 is invariant under the translation $x' = x + a$ and quasi-invariant under the translation $y' = y + b$. Find the corresponding constants of the motion via Noether's theorem. Using the equations of motion for the system, check directly that the quantities obtained by Noether's theorem are indeed constants of the motion.

2.22 A particle of mass m and position vector $\mathbf{r}$ moves in the Coulomb or gravitational potential $V = -\kappa/r$, where $r = |\mathbf{r}|$. (a) Show that the Laplace-Runge-Lenz vector

$$\mathbf{A} = \mathbf{p} \times \boldsymbol{l} - m\kappa \frac{\mathbf{r}}{r}$$

is a constant of the motion, where $\boldsymbol{l} = \mathbf{r} \times \mathbf{p}$ is the angular momentum vector. (b) Show that $\mathbf{A}$ lies on the plane of the orbit. (c) Taking the scalar product of $\mathbf{A}$ with $\mathbf{r}$, derive the equation of the orbit. Hint: $\mathbf{r} \cdot (\mathbf{p} \times \boldsymbol{l}) = \boldsymbol{l} \cdot \boldsymbol{l}$. (d) Show that the eccentricity is given by $e = A/m\kappa$ and that $\mathbf{A}$ points along the semi-major axis of the elliptic orbit.

2.23 The Lagrangian for a one-dimensional mechanical system is $L = \dot{x}^2/2 - g/x^2$, where g is a constant. (a) Show that the action is invariant under the finite transformations $x'(t') = e^{\alpha} x(t)$, $t' = e^{2\alpha} t$, where α is an arbitrary constant, and conclude, via Noether's theorem, that $I = x\dot{x} - 2Et$ is a constant of the motion, where E is the total energy. (b) Prove that the action is quasi-invariant under the infinitesimal transformations $t' = t + \epsilon t^2$, $x'(t') = x(t) - \epsilon tx(t)$, and hence infer that $K = Et^2 - tx\dot{x} + x^2/2$ is a constant of the motion. (c) Combine the results in (a) and (b) to obtain the general solution of Lagrange's equation for $x(t)$.

2.24 Show that, in the presence of constraints of the form (2.75), Eq. (2.149) must be replaced by

$$\frac{dh}{dt} = -\frac{\partial L}{\partial t} - \sum_{l=1}^{p} \lambda_l a_{lt}.$$

Thus, time-dependent constraints prevent h from being conserved even though L does not explicitly depend on time.

2.25 Let $L(x, y, z, \dot{x}, \dot{y}, \dot{z})$ be the Lagrangian for a particle subject to a force whose potential energy is $V(x, y, z)$, where x, y, z are Cartesian coordinates. Suppose the particle is constrained to move on a surface with equation $z = \phi(x, y)$. With $q_1 = x, q_2 = y$, $q_3 = z$ and $f(q, t) = z - \phi(x, y)$, show that the equations of motion

$$\frac{d}{dt}\left(\frac{\partial L}{\partial \dot{q}_k}\right) - \frac{\partial L}{\partial q_k} = \lambda \frac{\partial f}{\partial q_k}, \qquad k = 1, 2, 3,$$

together with the constraint equation

$$f(q, t) = 0,$$

are equivalent to the standard Lagrange equations derived from the reduced Lagrangian L_c obtained by inserting the constraint into the unconstrained Lagrangian:

$$L_c(x, y, \dot{x}, \dot{y}) = L(x, y, z, \dot{x}, \dot{y}, \dot{z})\Big|_{z = \phi(x,y)}.$$

2.26 Reconsider the problem of a cylinder rolling on another fixed cylinder, Example 2.9. (a) Show that the tangential constraint force on the mobile cylinder is $f_\theta = -(mg/3)\sin\theta$. This is the friction force necessary to prevent slippage. (b) If μ is the coefficient of static friction between the cylinders, there will be no slippage only as long as $|f_\theta| \leq \mu N$, where N is the normal reaction force given by (2.115). Show that the slipping starts when the angle θ reaches the value θ_s such that

$$\cos\theta_s = \frac{28\mu^2 + \sqrt{1 + 33\mu^2}}{1 + 49\mu^2}.$$

(c) Show that the result in Example 2.9 for the angle at which the mobile cylinder abandons the fixed cylinder is correct only if $\mu \gg 1$.

2.27 A particle of mass m and electric charge e moves in a uniform electromagnetic field: $\mathbf{E}$ and $\mathbf{B}$ are constants. Show that the vector $\mathbf{K} = m\mathbf{v} + \frac{e}{c}\mathbf{B} \times \mathbf{r} - e\mathbf{E}\,t$ is a constant of the motion.

2.28 A particle of mass m, in a uniform gravitational field $\mathbf{g} = -g\hat{\mathbf{z}}$, is restricted to move on the surface of the cone defined by $\theta = \alpha$, with $0 < \alpha < \pi/2$, where θ is the polar angle. (a) Find the Lagrangian in terms of spherical coordinates r, θ, ϕ and write down Lagrange's equations. (b) Identify all constants of the motion and, by means of them, express the solution of the equations of motion for $r(t)$ in the form of an integral of a known function.

2.29 A particle of mass m and electric charge e is subject to the magnetic force $\mathbf{F} = (e/c)\mathbf{v} \times \mathbf{B}$, where $\mathbf{B} = g\mathbf{r}/r^3$ is the field of a hypothetical magnetic monopole fixed at the origin (g is its magnetic charge). Prove that the vector

$$\mathbf{Q} = m\mathbf{r} \times \mathbf{v} - \frac{eg}{c}\hat{\mathbf{r}}$$

is a constant of the motion. Picking the z-axis parallel to $\mathbf{Q}$ and using spherical coordinates, compute $\mathbf{Q} \cdot \hat{\boldsymbol{\phi}}$ and conclude that θ is a constant. Therefore, the charged particle moves on the surface of a cone. This surprising result was obtained by Poincaré in 1896.

2.30 Consider a simple harmonic oscillator of mass m and angular frequency ω. (a) Show that

$$C(x, \dot{x}, t) = \dot{x} \cos \omega t + \omega x \sin \omega t$$

is a constant of the motion. (b) Prove that this constant of the motion is associated with a continuous symmetry of the action characterised by

$$X(x, t) = 0, \quad \psi(x, t) = -\frac{1}{m} \cos \omega t, \quad G(x, t) = \omega x \sin \omega t.$$

2.31 The Neumann problem consists in finding the motion of a collection of n unit-mass harmonic oscillators with positions x_k, $k = 1, \ldots, n$, distinct frequencies ($\omega_k \neq \omega_l$ if $k \neq l$) and restricted to the unit sphere:

$$\sum_{k=1}^{n} x_k^2 = 1.$$

(a) Using the Lagrange multiplier technique, show that the equations of motion are

$$\ddot{x}_k + \omega_k^2 x_k = 2\lambda x_k.$$

(b) Derive the equation $\sum (x_k \ddot{x}_k + \dot{x}_k^2) = 0$ and, availing yourself of it, eliminate the Lagrange multiplier to obtain

$$\ddot{x}_k + \omega_k^2 x_k = x_k \sum_{l=1}^{n} \left(\omega_l^2 x_l^2 - \dot{x}_l^2 \right).$$

(c) Show that

$$F_k(x, \dot{x}) = x_k^2 + \sum_{\substack{l=1 \\ l \neq k}}^{n} \frac{(x_l \dot{x}_k - \dot{x}_l x_k)^2}{\omega_k^2 - \omega_l^2}$$

are n constants of the motion.

> But though the professed aim of all scientific work is to unravel the secrets of nature, it has another effect, not less valuable, on the mind of the worker. It leaves him in possession of methods which nothing but scientific work could have led him to invent.
>
> James Clerk Maxwell, The Theory of Molecules

The dynamics of rigid bodies is a chapter of classical mechanics that deserves to be highlighted not only owing to its intrinsic physical interest but also because it involves important mathematical techniques. Before, however, embarking on the study of dynamics, it is necessary to formulate efficacious methods to describe the motion of rigid bodies. A considerable space will be dedicated to the study of rotational kinematics in the perspective that several of the mathematical tools to be developed are of great generality, finding wide application in other domains of theoretical physics.

3.1 Orthogonal Transformations

A rigid body has, in general, six degrees of freedom. Obviously, three of them correspond to translations of the body as a whole, whereas the other three degrees of freedom describe the orientations of the body relative to a system of axes fixed in space. A simple way to specify the orientation of the rigid body consists in setting up a Cartesian system of axes *fixed in the body*, which move along with it, and consider the angles that these axes make with axes parallel to those that remain fixed in space, represented by dashed lines in Fig. 3.1.

Direction Cosines

Let Σ be a Cartesian coordinate system (x_1, x_2, x_3) with corresponding unit vectors $\hat{\mathbf{e}}_1, \hat{\mathbf{e}}_2, \hat{\mathbf{e}}_3$ representing axes fixed in space, and let Σ' be a Cartesian coordinate system (x'_1, x'_2, x'_3) with unit vectors $\hat{\mathbf{e}}'_1, \hat{\mathbf{e}}'_2, \hat{\mathbf{e}}'_3$ whose axes remain attached to the rigid body, as in Fig. 3.2. An arbitrary vector $\mathbf{g}$ can be expressed in terms of the basis Σ or in terms of the basis Σ':

$$\mathbf{g} = g_1\hat{\mathbf{e}}_1 + g_2\hat{\mathbf{e}}_2 + g_3\hat{\mathbf{e}}_3 = \sum_{j=1}^{3} g_j\hat{\mathbf{e}}_j \,, \tag{3.1}$$

$$\mathbf{g} = g'_1\hat{\mathbf{e}}'_1 + g'_2\hat{\mathbf{e}}'_2 + g'_3\hat{\mathbf{e}}'_3 = \sum_{j=1}^{3} g'_j\hat{\mathbf{e}}'_j. \tag{3.2}$$

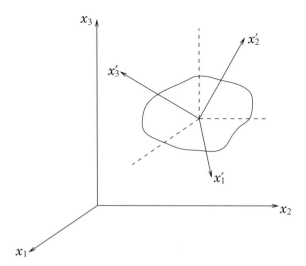

Fig. 3.1   Axes fixed in space and primed axes attached to a rigid body.

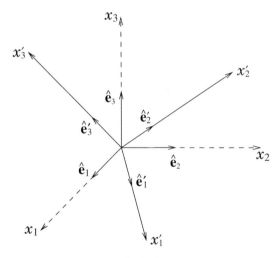

Fig. 3.2   Unit vectors $\hat{\mathbf{e}}_1, \hat{\mathbf{e}}_2, \hat{\mathbf{e}}_3$ fixed in space and unit vectors $\hat{\mathbf{e}}'_1, \hat{\mathbf{e}}'_2, \hat{\mathbf{e}}'_3$ attached to the rigid body.

Of course, the components of $\boldsymbol{g}$ in the bases Σ and Σ' are related to one another. Indeed,

$$g'_i = \hat{\mathbf{e}}'_i \cdot \boldsymbol{g} = \sum_{j=1}^{3} g_j \hat{\mathbf{e}}'_i \cdot \hat{\mathbf{e}}_j \equiv \sum_{j=1}^{3} a_{ij}\, g_j\,, \qquad i = 1, 2, 3\,, \tag{3.3}$$

where

$$a_{ij} = \hat{\mathbf{e}}'_i \cdot \hat{\mathbf{e}}_j = \text{cosine of the angle between axes } x'_i \text{ and } x_j\,. \tag{3.4}$$

The quantities a_{ij} – nine in all – are called **direction cosines** of Σ' relative to Σ. Therefore, the components of $\boldsymbol{g}$ in Σ' are expressed in terms of its components in Σ by means of a *linear transformation* whose coefficients are the direction cosines.

The nine quantities a_{ij} cannot be mutually independent, for three quantities suffice to define the orientation of Σ' relative to Σ. In order to obtain the conditions to which the direction cosines must be subject, all one has to do is take into account that the magnitude of a vector does not depend on the basis used to express it:

$$\boldsymbol{g} \cdot \boldsymbol{g} = \sum_{i=1}^{3} g_i' g_i' = \sum_{i=1}^{3} g_i g_i \,. \tag{3.5}$$

Making use of Eq. (3.3) we find

$$\sum_{i=1}^{3} g_i' g_i' = \sum_{i=1}^{3} \left(\sum_{j=1}^{3} a_{ij} g_j \right) \left(\sum_{k=1}^{3} a_{ik} g_k \right) = \sum_{k,j=1}^{3} \left(\sum_{i=1}^{3} a_{ij} a_{ik} \right) g_j g_k \,. \tag{3.6}$$

With the help of the Kronecker delta symbol δ_{jk} introduced in Appendix A, it is possible to write

$$\sum_{i=1}^{3} g_i g_i \equiv g_1^2 + g_2^2 + g_3^2 = \sum_{k,j=1}^{3} \delta_{jk} g_j g_k \,. \tag{3.7}$$

Comparing equations (3.6) and (3.7), using (3.5) and the arbitrariness of g_1, g_2, g_3, one infers

$$\sum_{i=1}^{3} a_{ij} a_{ik} = \delta_{jk} \,. \tag{3.8}$$

These are the conditions obeyed by the direction cosines, showing that not all of them are independent.

Orthogonal Matrices

Equation (3.3) is a compact representation of the three equations

$$\begin{aligned} g_1' &= a_{11} g_1 + a_{12} g_2 + a_{13} g_3 \,, \\ g_2' &= a_{21} g_1 + a_{22} g_2 + a_{23} g_3 \,, \\ g_3' &= a_{31} g_1 + a_{32} g_2 + a_{33} g_3 \,. \end{aligned} \tag{3.9}$$

Equations of this type can be made much more concise by means of matrix notation. Definining the matrices[1]

$$\boldsymbol{g}_{\Sigma} = \begin{pmatrix} g_1 \\ g_2 \\ g_3 \end{pmatrix}, \quad \boldsymbol{g}_{\Sigma'} = \begin{pmatrix} g_1' \\ g_2' \\ g_3' \end{pmatrix}, \quad \boldsymbol{A} = (a_{ij}) = \begin{pmatrix} a_{11} & a_{12} & a_{13} \\ a_{21} & a_{22} & a_{23} \\ a_{31} & a_{32} & a_{33} \end{pmatrix}, \tag{3.10}$$

Eqs. (3.9) can be written in the form

$$\boldsymbol{g}_{\Sigma'} = \boldsymbol{A} \boldsymbol{g}_{\Sigma} \,. \tag{3.11}$$

---

[1] In this chapter square matrices will be typically denoted by bold capital calligraphic letters such as $\boldsymbol{A}$. A vector written as a column matrix will be denoted by a bold small letter. Whenever necessary, a subscript will be added to indicate the basis in which the vector is expressed, such as in $\boldsymbol{g}_{\Sigma}$.

On the other hand, noting that $a_{ij} = (\mathcal{A}^T)_{ji}$, where $\mathcal{A}^T$ denotes the transpose of $\mathcal{A}$, Eq. (3.8) is equivalent to $\sum_{i=1}^{3}(\mathcal{A}^T)_{ji}(\mathcal{A})_{ik} = \delta_{jk}$. Thus, taking into account the definition of matrix product, Eq. (3.8) can be rewritten in the matrix form

$$\mathcal{A}^T\,\mathcal{A} = I\,, \tag{3.12}$$

where

$$I = (\delta_{jk}) = \begin{pmatrix} 1 & 0 & 0 \\ 0 & 1 & 0 \\ 0 & 0 & 1 \end{pmatrix} \tag{3.13}$$

is the 3×3 identity matrix.

Exercise 3.1.1   Prove that the inverse transformation of (3.3) is

$$g_i = \sum_{j=1}^{3} a_{ji}\, g'_j \tag{3.14}$$

and, using the invariance of $\boldsymbol{g} \cdot \boldsymbol{g}$, prove that

$$\sum_{i=1}^{3} a_{ji}\, a_{ki} = \delta_{jk}\,. \tag{3.15}$$

Show that, in matrix notation, this equations takes the form[2]

$$\mathcal{A}\mathcal{A}^T = I\,. \tag{3.16}$$

Prove, further, the direct and inverse transformation laws

$$\hat{\boldsymbol{e}}'_j = \sum_{l=1}^{3} a_{jl}\,\hat{\boldsymbol{e}}_l\,, \qquad \hat{\boldsymbol{e}}_j = \sum_{l=1}^{3} a_{lj}\,\hat{\boldsymbol{e}}'_l\,. \tag{3.17}$$

Bases $\{\hat{\boldsymbol{e}}_1, \hat{\boldsymbol{e}}_2, \hat{\boldsymbol{e}}_3\}$ and $\{\hat{\boldsymbol{e}}'_1, \hat{\boldsymbol{e}}'_2, \hat{\boldsymbol{e}}'_3\}$ are made up of mutually orthogonal unit vectors, so the following equations hold:

$$\hat{\boldsymbol{e}}_j \cdot \hat{\boldsymbol{e}}_k = \hat{\boldsymbol{e}}'_j \cdot \hat{\boldsymbol{e}}'_k = \delta_{jk}\,. \tag{3.18}$$

Derive (3.8) and (3.15) from (3.17) and (3.18).

The transformation matrix between Cartesian systems obeys equations (3.12) and (3.16), which are equivalent to

$$\mathcal{A}^{-1} = \mathcal{A}^T\,. \tag{3.19}$$

Any matrix that satisfies this equation is said to be an **orthogonal matrix**, and the associated linear transformation (3.11) is called an **orthogonal transformation**.

---

[2] For finite matrices equations (3.12) and (3.16) are equivalent – that is, one is true if and only if the other is true (see Appendix D). This property does not extend to infinite matrices.

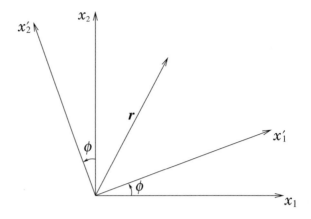

Fig. 3.3 Rotation in the plane.

Example 3.1 In the case of rotations in the plane (Fig. 3.3), we have

$$x_1' = x_1 \cos \phi + x_2 \sin \phi, \tag{3.20a}$$

$$x_2' = -x_1 \sin \phi + x_2 \cos \phi. \tag{3.20b}$$

The matrix associated with this orthogonal transformation is

$$\mathcal{A} = \begin{pmatrix} \cos \phi & \sin \phi \\ -\sin \phi & \cos \phi \end{pmatrix}, \tag{3.21}$$

and a single parameter – the angle ϕ – completely specifies the transformation.

Exercise 3.1.2 Using definition (3.4) and Eq. (3.3) with $g = r$, derive (3.20a) and (3.20b). By computing the product $\mathcal{A}\mathcal{A}^T$, check that the matrix $\mathcal{A}$ given by (3.21) is orthogonal.

The matrix $\mathcal{A}$ is associated with a linear transformation which, according to Eq. (3.11), acts on the components of a vector in Σ and transforms them into the components of the *same vector* in Σ'. This interpretation of Eq. (3.11), in which the vector stays the same and what changes is the coordinate system, is called the *passive point of view*. Alternatively, $\mathcal{A}$ may be thought of as an operator which transforms the vector g into *another vector g'*, both considered expressed in the same coordinate system. This way of interpreting the transformation performed by matrix $\mathcal{A}$ is known as the *active point of view*, and in several occasions, as will be seen, turns out to be more convenient than the passive point of view. Of course, the specific operation represented by $\mathcal{A}$ will depend on the interpretation adopted. A rotation of the coordinate axes keeping the vector fixed has the same effect as keeping the axes fixed and rotating the vector by the same angle in the opposite sense. In Example 3.1, since $\mathcal{A}$ represents a counterclockwise rotation of the coordinate axes by angle ϕ, it will correspond to a clockwise rotation of any vector by the same angle ϕ.

Group of Orthogonal Matrices

It is worthwhile to examine sucessive orthogonal transformations. Let $\mathcal{B} = (b_{ij})$ be the transformation matrix from Σ to Σ', and $\mathcal{A} = (a_{ij})$ be the transformation matrix from Σ' to Σ'', so that

$$g'_i = \sum_j b_{ij} \, g_j \,, \tag{3.22}$$

$$g''_k = \sum_i a_{ki} \, g'_i \,. \tag{3.23}$$

Consequently,

$$g''_k = \sum_{ij} a_{ki} \, b_{ij} \, g_j = \sum_j \left(\sum_i a_{ki} \, b_{ij} \right) g_j \equiv \sum_j c_{kj} \, g_j \,, \tag{3.24}$$

where the c_{kj} defined by

$$c_{kj} = \sum_i a_{ki} \, b_{ij} \tag{3.25}$$

are the elements of the matrix $\mathcal{C}$ which performs the direct transformation from Σ to Σ''. According to the definition of matrix product, Eq. (3.25) is equivalent to

$$\mathcal{C} = \mathcal{A}\mathcal{B} \,. \tag{3.26}$$

It is essential to observe the order of factors in this last equation because matrix multiplication is not commutative. The matrix associated with the first transformation appears at the rightmost position in the product, immediately to the left comes the matrix of the second transformation and so on. It is intuitively clear that two successive rotations are equivalent to a single rotation. So, if $\mathcal{A}$ and $\mathcal{B}$ are orthogonal matrices, their product must also be orthogonal. In order to prove that $\mathcal{C}$ is orthogonal, it suffices to show that $\mathcal{C}$ satisfies (3.16). As the first step, note that

$$\mathcal{C}\mathcal{C}^T = \mathcal{A}\mathcal{B}(\mathcal{A}\mathcal{B})^T = \mathcal{A}\mathcal{B}\mathcal{B}^T\mathcal{A}^T \,, \tag{3.27}$$

where we have used the identity $(\mathcal{A}\mathcal{B})^T = \mathcal{B}^T\mathcal{A}^T$. With the use of the orthogonality of $\mathcal{B}$ and $\mathcal{A}$ we can write

$$\mathcal{C}\mathcal{C}^T = \mathcal{A}\mathcal{I}\mathcal{A}^T = \mathcal{A}\mathcal{A}^T = I \,, \tag{3.28}$$

which verifies the orthogonality of $\mathcal{C}$.

For any orthogonal matrix its unique inverse is just its transpose, and it is immediate that the transpose of an orthogonal matrix is also orthogonal because we can write $(\mathcal{A}^T)^T\mathcal{A}^T = \mathcal{A}\mathcal{A}^T = I$, which shows that (3.12) is satisfied with $\mathcal{A}$ replaced by $\mathcal{A}^T$.

The set of orthogonal matrices enjoys the following properties:

(O1) If $\mathcal{A}$ and $\mathcal{B}$ are elements of the set, so is the product $\mathcal{A}\mathcal{B}$.

(O2) The product is *associative* – that is, $\mathcal{A}(\mathcal{B}\mathcal{C}) = (\mathcal{A}\mathcal{B})\mathcal{C}$ holds true for any elements $\mathcal{A}, \mathcal{B}, \mathcal{C}$ of the set.

(O3) The set contains an element I, called the *identity*, such that for any element $\mathcal{A}$ of the set, $\mathcal{A}I = I\mathcal{A} = \mathcal{A}$.

(O4) For each element $\mathcal{A}$ of the set there exists a unique element $\mathcal{B}$ of the set such that $\mathcal{A}\mathcal{B} = \mathcal{B}\mathcal{A} = I$. The element $\mathcal{B}$ is called the *inverse* of $\mathcal{A}$ and denoted by $\mathcal{A}^{-1}$.

The first and fourth properties have already been proved, the second is satisfied because matrix multiplication is associative and the third is obviously true.

A set of elements equipped with a *composition law* — also called *multiplication* or *product* — having the four properties above make up a **group** (Hamermesh, 1962; Tung, 1985). The group of 3×3 orthogonal matrices is denoted by $O(3)$. Group theory plays an important role in various branches of contemporary physics, most notedly in elementary particle physics.

3.2 Possible Displacements of a Rigid Body

The determinant of an orthogonal matrix can only assume certains restricted values. Indeed, taking the determinant of (3.16) there results

$$(\det \mathcal{A})(\det \mathcal{A}^T) = 1 \quad \implies \quad (\det \mathcal{A})^2 = 1, \tag{3.29}$$

where we have used the following well-known properties of the determinant:

$$\det(\mathcal{A}\mathcal{B}) = (\det \mathcal{A})(\det \mathcal{B}), \qquad \det \mathcal{A}^T = \det \mathcal{A}. \tag{3.30}$$

Therefore, the determinant of an orthogonal matrix can only be $+1$ or -1. A simple orthogonal matrix with determinant -1 is

$$\mathcal{I} = -I = \begin{pmatrix} -1 & 0 & 0 \\ 0 & -1 & 0 \\ 0 & 0 & -1 \end{pmatrix}. \tag{3.31}$$

The transformation effected by $\mathcal{I}$ is

$$g_1' = -g_1, \quad g_2' = -g_2, \quad g_3' = -g_3, \tag{3.32}$$

which corresponds to an inversion of the coordinate axes.

An inversion transforms a right-handed Cartesian system into a left-handed one. It is clear, therefore, that an inversion does not represent a possible displacement of a rigid body: once a right-handed Cartesian system of axes is attached to the body, no physical displacement of the body can change the mutual orientation of those axes unless the body is deformable, which is not the case by hypothesis. On the other hand, any orthogonal matrix with determinant -1 can be written in the form $\mathcal{A} = (-I)(-\mathcal{A}) \equiv \mathcal{I}\mathcal{B}$ where $\det \mathcal{B} \equiv \det(-\mathcal{A}) = +1$ – that is, $\mathcal{A}$ contains an inversion of the axes and does not represent a possible displacement of the rigid body. Orthogonal matrices with determinant $+1$ are said to be *proper*, whereas those with determinant -1 are said to be *improper*. Thus, we conclude that only proper orthogonal transformations correspond to possible displacements of a rigid body. It is easily shown that the set of proper orthogonal transformations is itself a group called the **rotation group**, denoted $SO(3)$.

Example 3.2   The scalar product and the vector product are invariant under rotations. This means that if $\mathcal{R} = (R_{ij})$ is a rotation then

$$(\mathcal{R}\mathbf{a}) \cdot (\mathcal{R}\mathbf{b}) = \mathbf{a} \cdot \mathbf{b} \tag{3.33}$$

and also

$$(\mathcal{R}\mathbf{a}) \times (\mathcal{R}\mathbf{b}) = \mathcal{R}(\mathbf{a} \times \mathbf{b}). \tag{3.34}$$

Proving invariance of the scalar product is very simple with the use of the Einstein convention of sum over repeated indices (Appendix A):

$$(\mathcal{R}\mathbf{a}) \cdot (\mathcal{R}\mathbf{b}) = (\mathcal{R}\mathbf{a})_i (\mathcal{R}\mathbf{b})_i = R_{ij}a_j R_{ik}b_k$$
$$= (\mathcal{R}^T\mathcal{R})_{jk}a_j b_k = \delta_{jk}a_j b_k = a_k b_k = \mathbf{a} \cdot \mathbf{b}. \tag{3.35}$$

As to the vector product, we have (see Appendix A)

$$[(\mathcal{R}\mathbf{a}) \times (\mathcal{R}\mathbf{b})]_i = \epsilon_{ijk}(\mathcal{R}\mathbf{a})_j (\mathcal{R}\mathbf{b})_k$$
$$= \epsilon_{ijk}R_{jm}a_m R_{kn}b_n = \epsilon_{ijk}R_{jm}R_{kn}a_m b_n. \tag{3.36}$$

According to Appendix A,

$$\epsilon_{ijk}R_{il}R_{jm}R_{kn} = (\det \mathcal{R})\epsilon_{lmn} = \epsilon_{lmn}, \tag{3.37}$$

where we used $\det \mathcal{R} = 1$ because $\mathcal{R}$ is a rotation matrix. Multiplying this last equation by R_{rl}, which implies summation over the repeated index l, we find

$$R_{rl}\epsilon_{lmn} = \epsilon_{ijk}R_{rl}R_{il}R_{jm}R_{kn} = \epsilon_{ijk}(\mathcal{R}\mathcal{R}^T)_{ri}R_{jm}R_{kn}$$
$$= \epsilon_{ijk}\delta_{ri}R_{jm}R_{kn} = \epsilon_{rjk}R_{jm}R_{kn}. \tag{3.38}$$

Finally, substituting this result into (3.36) we obtain

$$[(\mathcal{R}\mathbf{a}) \times (\mathcal{R}\mathbf{b})]_i = R_{il}\epsilon_{lmn}a_m b_n = R_{il}(\mathbf{a} \times \mathbf{b})_l = [\mathcal{R}(\mathbf{a} \times \mathbf{b})]_i. \tag{3.39}$$

The proof of Eq. (3.34) is complete.

A classic result, which reveals one of the fundamental features of a rigid body's motion, was established by Euler in 1776.

Theorem 3.2.1 (Euler)   *The most general displacement of a rigid body with one point fixed is a rotation about some axis through the fixed point.*

Proof   Instead of the traditional geometric reasoning, we will resort to the previously developed algebraic techniques, which will confer the proof a rather instructive character. But, first, it is necessary to clarify the content of the theorem. Suppose the rigid body is in an initial configuration and, after moving in any way whatsoever about the fixed point, it reaches a final configuration. As already argued, the final configuration results from the application of a proper orthogonal transformation to the initial configuration. According to the statement of Euler's theorem, the final configuration can be attained by means of a single rotation about some axis through the fixed point – that is, a rotation about a point is always equivalent to a rotation about an axis containing the point. But a vector along the rotation axis remains unaltered, its components before and after the rotation are the same. If $\mathcal{A}$ is the proper orthogonal matrix that takes the body from its initial to its final configuration, Euler's theorem will be proved as soon as one proves the existence of a non-zero vector $\boldsymbol{n}$ such that

$$\mathcal{A}\boldsymbol{n} = \boldsymbol{n} \tag{3.40}$$

or, equivalently,

$$(\mathcal{A} - \boldsymbol{I})\boldsymbol{n} = 0. \tag{3.41}$$

If the matrix $\mathcal{A} - \boldsymbol{I}$ is invertible, multiplying (3.41) on the left by $(\mathcal{A} - \boldsymbol{I})^{-1}$ one infers that $\boldsymbol{n} = 0$. Therefore, there exists a vector $\boldsymbol{n} \neq 0$ satisfying (3.41) if and only if $\mathcal{A} - \boldsymbol{I}$ has no inverse – that is, if and only if

$$\det(\mathcal{A} - \boldsymbol{I}) = 0. \tag{3.42}$$

In short, proving Euler's theorem boils down to showing that (3.42) holds. From the identity

$$\mathcal{A} - \boldsymbol{I} = \mathcal{A} - \mathcal{A}\mathcal{A}^T = \mathcal{A}(\boldsymbol{I} - \mathcal{A}^T) = -\mathcal{A}(\mathcal{A} - \boldsymbol{I})^T = (-\boldsymbol{I})\mathcal{A}(\mathcal{A} - \boldsymbol{I})^T \tag{3.43}$$

one deduces

$$\det(\mathcal{A} - \boldsymbol{I}) = \det(-\boldsymbol{I})\det\mathcal{A}\det(\mathcal{A} - \boldsymbol{I}) = -\det\mathcal{A}\det(\mathcal{A} - \boldsymbol{I}). \tag{3.44}$$

Using $\det\mathcal{A} = 1$, this last equation implies

$$\det(\mathcal{A} - \boldsymbol{I}) = -\det(\mathcal{A} - \boldsymbol{I}), \tag{3.45}$$

from which one infers (3.42), completing the proof of the theorem. □

The traditional geometric proof obscures the role played by the dimensionality of space, which crucially enters the above proof. In a space of even dimension $\det(-\boldsymbol{I}) = 1$ and the previous argument fails. For example, in a two-dimensional space no vector is left unaltered by a rotation. Indeed, for a rotation in the plane the axis of rotation is perpendicular to the plane, with the consequence that any vector along the rotation axis does not belong to the plane. In summary, Euler's theorem does not hold in even-dimensional spaces. Euler's theorem was generalised by Chasles, in 1830, in the terms below.

Theorem 3.2.2 (Chasles) *The most general displacement of a rigid body is a translation together with a rotation. The rotation axis can be so chosen that the translation is parallel to it.*

The first part of Chasles's theorem is obvious, since the removal of the constraint of having one point fixed endows the body with three translational degrees of freedom. A geometric proof of the second part of Chasles's theorem can be found in Whittaker (1944) or Pars (1965). For a modern proof, with the use of purely algebraic methods, see Corben and Stehle (1960).

3.3 Euler Angles

In a Lagrangian formulation of rigid body dynamics, the nine direction cosines a_{ij} are not the most convenient coordinates for describing the instantaneous orientation of a rigid body because they are not mutually independent. The nine equations (3.8) impose only six conditions on the direction cosines, so they can be expressed in terms of three independent parameters. Indeed, there are three distinct conditions corresponding to the diagonal part of Eqs. (3.8), but the six conditions corresponding to the off-diagonal part ($i \neq j$) are pairwise identical. For example, Eq. (3.8) for $j = 1, k = 2$ is the same as the one for $j = 2, k = 1$.

From the practical point of view, a convenient way to parameterise the rotation matrix is by means of the Euler angles. The transformation from the Cartesian system $\Sigma(x, y, z)$ to the $\Sigma'(x', y', z')$ system is accomplished in three successive stages, each serving to define one of the Euler angles (Fig. 3.4).

(a) Rotation of axes (x, y, z) about the z-axis by angle $\phi : (x, y, z) \xrightarrow{\mathcal{D}} (\xi, \eta, \zeta)$.

The transformation equations are the same as equations (3.20) with $x'_1 = \xi$ and $x'_2 = \eta$ supplemented by equation $x'_3 = \zeta = z$. Therefore, the rotation matrix $\mathcal{D}$ is written

$$\mathcal{D} = \begin{pmatrix} \cos\phi & \sin\phi & 0 \\ -\sin\phi & \cos\phi & 0 \\ 0 & 0 & 1 \end{pmatrix}. \tag{3.46}$$

(b) Rotation of axes (ξ, η, ζ) about the ξ-axis by angle $\theta : (\xi, \eta, \zeta) \xrightarrow{\mathcal{C}} (\xi', \eta', \zeta')$.

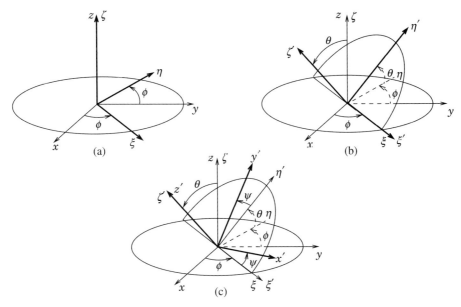

Fig. 3.4 Euler angles.

The ξ'-axis direction is called the *line of nodes* and, in analogy with (3.20), the transformation equations take the form

$$\xi' = \xi\,,$$
$$\eta' = \eta\,\cos\theta + \zeta\,\sin\theta\,,$$
$$\zeta' = -\eta\,\sin\theta + \zeta\,\cos\theta\,,$$

(3.47)

and it follows that

$$\mathcal{C} = \begin{pmatrix} 1 & 0 & 0 \\ 0 & \cos\theta & \sin\theta \\ 0 & -\sin\theta & \cos\theta \end{pmatrix}.$$

(3.48)

(c) Rotation of axes (ξ', η', ζ') about the ζ'-axis by angle $\psi : (\xi', \eta', \zeta') \xrightarrow{\mathcal{B}} (x', y', z')$.
The matrix $\mathcal{B}$ represents a rotation about the third axis, thus has the same form as $\mathcal{D}$:

$$\mathcal{B} = \begin{pmatrix} \cos\psi & \sin\psi & 0 \\ -\sin\psi & \cos\psi & 0 \\ 0 & 0 & 1 \end{pmatrix}.$$

(3.49)

The rotation $(x, y, z) \rightarrow (x', y', z')$ is performed by $\mathcal{A}_E = \mathcal{B}\mathcal{C}\mathcal{D}$. A direct calculation yields

$$\mathcal{A}_E = \begin{pmatrix} \cos\psi\,\cos\phi - \cos\theta\,\sin\phi\,\sin\psi & \cos\psi\,\sin\phi + \cos\theta\,\cos\phi\,\sin\psi & \sin\psi\,\sin\theta \\ -\sin\psi\,\cos\phi - \cos\theta\,\sin\phi\,\cos\psi & -\sin\psi\,\sin\phi + \cos\theta\,\cos\phi\,\cos\psi & \cos\psi\,\sin\theta \\ \sin\theta\,\sin\phi & -\sin\theta\,\cos\phi & \cos\theta \end{pmatrix}.$$

(3.50)

The inverse transformation $(x', y', z') \rightarrow (x, y, z)$ is effected by $\mathcal{A}_E^{-1}$ which is just the transpose of $\mathcal{A}_E$ – that is,

$$\mathcal{A}_E^{-1} = \begin{pmatrix} \cos\psi\,\cos\phi - \cos\theta\,\sin\phi\,\sin\psi & -\sin\psi\,\cos\phi - \cos\theta\,\sin\phi\,\cos\psi & \sin\theta\,\sin\phi \\ \cos\psi\,\sin\phi + \cos\theta\,\cos\phi\,\sin\psi & -\sin\psi\,\sin\phi + \cos\theta\,\cos\phi\,\cos\psi & -\sin\theta\,\cos\phi \\ \sin\psi\,\sin\theta & \cos\psi\,\sin\theta & \cos\theta \end{pmatrix}.$$

(3.51)

In virtue of their definition, the Euler angles take values in the following ranges:

$$0 \le \phi \le 2\pi\,, \quad 0 \le \theta \le \pi\,, \quad 0 \le \psi \le 2\pi\,.$$

(3.52)

3.4 Infinitesimal Rotations and Angular Velocity

The rigid body equations of motion are differential equations, and their formulation requires the study of the body's rotation during an infinitesimal time interval.

Commutativity of Infinitesimal Rotations

Let g be an arbitrary vector submitted to a counterclockwise infinitesimal rotation by an angle $d\Phi$ about an axis defined by the unit vector $\hat{n}$ (active point of view). According to (2.136) we have

$$g' = g + d\boldsymbol{\Omega} \times g, \tag{3.53}$$

where

$$d\boldsymbol{\Omega} = d\Phi\,\hat{\boldsymbol{n}}. \tag{3.54}$$

The symbols $d\Phi$ and $d\boldsymbol{\Omega}$ are to be understood as mere names for infinitesimal quantities, not as differentials of a scalar Φ or of a vector $\boldsymbol{\Omega}$. In fact, generally speaking, there is no vector $\boldsymbol{\Omega}$ whose differential is equal to $d\boldsymbol{\Omega}$ (Problem 3.6).

For successive rotations, with associated vectors $d\boldsymbol{\Omega}_1$ and $d\boldsymbol{\Omega}_2$, we write

$$g' = g + d\boldsymbol{\Omega}_1 \times g, \tag{3.55}$$

$$g'' = g' + d\boldsymbol{\Omega}_2 \times g', \tag{3.56}$$

whence, neglecting second-order infinitesimals,

$$g'' = g + d\boldsymbol{\Omega}_{12} \times g \tag{3.57}$$

with

$$d\boldsymbol{\Omega}_{12} = d\boldsymbol{\Omega}_1 + d\boldsymbol{\Omega}_2. \tag{3.58}$$

This last result shows that successive infinitesimal rotations commute ($d\boldsymbol{\Omega}_{12} = d\boldsymbol{\Omega}_{21}$) as a consequence of the commutativity of vector addition. Furthermore, the vector associated with successive infinitesimal rotations is the sum of the vectors associated with the individual infinitesimal rotations, a property which will be of great value for the forthcoming developments.

It is rewarding to describe infinitesimal rotations in matrix language. Adopting the active point of view, Eq. (3.53) can be written in the form

$$g'_\Sigma = \mathcal{A}g_\Sigma = (I + \varepsilon)g_\Sigma, \tag{3.59}$$

where ε is an infinitesimal matrix. The commutativity of infinitesimal rotations is easily checked, since neglecting second-order infinitesimals and taking into account that matrix sum is commutative,

$$\mathcal{A}_1\mathcal{A}_2 = (I + \varepsilon_1)(I + \varepsilon_2) = I + \varepsilon_1 + \varepsilon_2 = I + \varepsilon_2 + \varepsilon_1 = \mathcal{A}_2\mathcal{A}_1. \tag{3.60}$$

On the other hand, since $\mathcal{A} = I + \varepsilon$ represents a rotation, $\mathcal{A}$ must be an orthogonal matrix. Then,

$$I = \mathcal{A}\mathcal{A}^T = (I + \varepsilon)(I + \varepsilon^T) = I + \varepsilon + \varepsilon^T \implies \varepsilon^T = -\varepsilon. \tag{3.61}$$

A matrix ε which obeys (3.61) is said to be anti-symmetric. In terms of the elements ϵ_{ij} of ε Eq. (3.61) means $\epsilon_{ij} = -\epsilon_{ji}$. Therefore, the most general form of the matrix ε is

$$\varepsilon = \begin{pmatrix} 0 & \alpha & \beta \\ -\alpha & 0 & \gamma \\ -\beta & -\gamma & 0 \end{pmatrix}. \tag{3.62}$$

It follows that the the the sum of the diagonal elements of ε is zero: $\operatorname{tr}\varepsilon = 0$. Requiring that the components of (3.59) be identical to those of (3.53), one easily finds

$$\varepsilon = \begin{pmatrix} 0 & -d\Omega_3 & d\Omega_2 \\ d\Omega_3 & 0 & -d\Omega_1 \\ -d\Omega_2 & d\Omega_1 & 0 \end{pmatrix}. \tag{3.63}$$

Although, as we have just seen, it is possible to associate a vector with each infinitesimal rotation in such a way that the vector associated with two sucessive rotations is the sum of the vectors associated with the individual rotations, such a correspondence is not possible for finite rotations.[3] It should be further noted that Eq. (3.63) establishes the following bijective correspondence between anti-symmetric matrices and vectors in three dimensions:

$$\begin{pmatrix} 0 & -v_3 & v_2 \\ v_3 & 0 & -v_1 \\ -v_2 & v_1 & 0 \end{pmatrix} \longleftrightarrow \begin{pmatrix} v_1 \\ v_2 \\ v_3 \end{pmatrix}. \tag{3.64}$$

Time Rate of Change of a Vector

In general, the time rate of change of a vector measured in a reference frame attached to a rotating rigid body will not coincide with that observed in an inertial reference frame external to the body. For instance, if the vector represents the position of a point fixed in the body, its time rate of change with respect to the system of axes attached to the body will be zero, but its time rate of change will not be zero with respect to the inertial system of axes relative to which the body is rotating. Consider an infinitesimal time interval dt during which an arbitrary vector g changes. Intuitively, we can write

$$(dg)_{inertial} = (dg)_{body} + (dg)_{rot}, \tag{3.65}$$

for the difference between the two systems is due exclusively to the body's rotation. But $(dg)_{rot}$ is a consequence of an infinitesimal rotation of g together with the axes attached to the body – that is,

$$(dg)_{rot} = d\Omega \times g, \tag{3.66}$$

as one infers from Eq. (3.53). The substitution of this last result into (3.65) leads to

$$\left(\frac{dg}{dt}\right)_{inertial} = \left(\frac{dg}{dt}\right)_{body} + \omega \times g, \tag{3.67}$$

where

$$\omega = \frac{d\Omega}{dt} = \frac{d\Phi}{dt}\hat{n} \tag{3.68}$$

is the instantaneous **angular velocity** of the rigid body. Note that the magnitude of ω is the instantaneous rotation speed whereas its direction is along the **instantaneous axis of rotation** characterised by the unit vector $\hat{n}$. We insist that Eq. (3.68) represents

[3] A detailed study of finite rotations can be found in Konopinski (1969). See, also, Example 4.1 in the next chapter.

symbolically the limit of the ratio of two quantities, $d\boldsymbol{\Omega}$ and dt, which jointly tend to zero as $dt \to 0$, and that $\boldsymbol{\omega}$ *is not* the time derivative of a vector $\boldsymbol{\Omega}$ (Problem 3.6). For a purely analytic definition of angular velocity, the reader is referred to Problem 3.7.

The important Eq. (3.67) can be derived in a more rigorous manner, which has the additional virtue of clearing up the meaning of each of its terms. Let $\{\hat{\mathbf{e}}_i\}_{i=1}^3$ be the unit vectors of the inertial Cartesian system and $\{\hat{\mathbf{e}}_i'\}_{i=1}^3$ the unit vectors of the Cartesian system attached to the body. We can write

$$\boldsymbol{g} = \sum_{i=1}^3 g_i\, \hat{\mathbf{e}}_i = \sum_{i=1}^3 g_i'\, \hat{\mathbf{e}}_i'. \tag{3.69}$$

To an inertial observer the $\hat{\mathbf{e}}_i$ are constant in time but the $\hat{\mathbf{e}}_i'$ vary with time because they rotate together with the body. There are, therefore, two distinct ways to calculate $(d\boldsymbol{g}/dt)_{inertial}$, namely

$$\left(\frac{d\boldsymbol{g}}{dt}\right)_{inertial} = \sum_{i=1}^3 \frac{dg_i}{dt}\, \hat{\mathbf{e}}_i \tag{3.70}$$

and

$$\left(\frac{d\boldsymbol{g}}{dt}\right)_{inertial} = \sum_{i=1}^3 \frac{dg_i'}{dt}\, \hat{\mathbf{e}}_i' + \sum_{i=1}^3 g_i'\, \frac{d\hat{\mathbf{e}}_i'}{dt}. \tag{3.71}$$

Equation (3.70) defines $(d\boldsymbol{g}/dt)_{inertial}$ and, similarly, $(d\boldsymbol{g}/dt)_{body}$ is defined by

$$\left(\frac{d\boldsymbol{g}}{dt}\right)_{body} = \sum_{i=1}^3 \frac{dg_i'}{dt}\, \hat{\mathbf{e}}_i'. \tag{3.72}$$

This last equation is perfectly natural since to an observer attached to the body the unit vectors $\hat{\mathbf{e}}_i'$ are constant. But during the time interval dt the mobile system of axes undergoes an infinitesimal rotation, the same occurring to each unit vector $\hat{\mathbf{e}}_i'$ – that is,

$$d\hat{\mathbf{e}}_i' = d\boldsymbol{\Omega} \times \hat{\mathbf{e}}_i', \tag{3.73}$$

where we used (3.66) with $\boldsymbol{g} = \hat{\mathbf{e}}_i'$. Substituting (3.72) and (3.73) into (3.71) there results

$$\left(\frac{d\boldsymbol{g}}{dt}\right)_{inertial} = \left(\frac{d\boldsymbol{g}}{dt}\right)_{body} + \sum_{i=1}^3 g_i'\, \boldsymbol{\omega} \times \hat{\mathbf{e}}_i' = \left(\frac{d\boldsymbol{g}}{dt}\right)_{body} + \boldsymbol{\omega} \times \sum_{i-1}^3 g_i'\, \hat{\mathbf{e}}_i', \tag{3.74}$$

which coincides with (3.67).

Equality (3.67) is valid no matter what the vector $\boldsymbol{g}$ is. This allows us to interpret it not only as a mere equality of vectors, but as a reflection of an equivalence of linear operators that act on vectors. This operator equality can be conveniently expressed in the symbolic form

$$\left(\frac{d}{dt}\right)_{inertial} = \left(\frac{d}{dt}\right)_{body} + \boldsymbol{\omega} \times . \tag{3.75}$$

Exercise 3.4.1 Show that $(d/dt)_{inertial} = (d/dt)_{body}$ when acting on a scalar. Hint: consider the scalar $s = \mathbf{g} \cdot \mathbf{h}$.

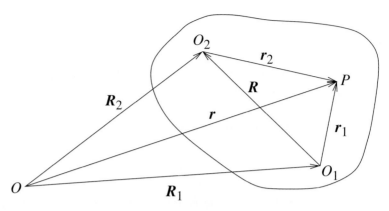

Fig. 3.5 O_1 and O_2 are the origins of two-coordinate systems fixed in the body.

Uniqueness of the Angular Velocity Vector

Intuitively, the angular velocity is expected to be a property of the rigid body as a whole, independent, therefore, of the point chosen as the origin of the coordinate system attached to the body. However, a rigorous proof of this fact is advisable (Lemos, 2000b). In Fig. 3.5, Σ is an inertial coordinate system with origin at point O, while O_1 and O_2 are the origins of two coordinate systems attached to the body, Σ_1' and Σ_2'. Let $\mathbf{r}$ be the vector from O to any point P of the rigid body and let $\boldsymbol{\omega}_1$ and $\boldsymbol{\omega}_2$ be the angular velocities associated with Σ_1' and Σ_2', respectively. Given that $\mathbf{r} = \mathbf{R}_1 + \mathbf{r}_1 = \mathbf{R}_2 + \mathbf{r}_2$, we have

$$\left(\frac{d\mathbf{r}}{dt}\right)_\Sigma = \left(\frac{d\mathbf{R}_1}{dt}\right)_\Sigma + \left(\frac{d\mathbf{r}_1}{dt}\right)_\Sigma = \left(\frac{d\mathbf{R}_1}{dt}\right)_\Sigma + \boldsymbol{\omega}_1 \times \mathbf{r}_1 \qquad (3.76)$$

and, similarly,

$$\left(\frac{d\mathbf{r}}{dt}\right)_\Sigma = \left(\frac{d\mathbf{R}_2}{dt}\right)_\Sigma + \boldsymbol{\omega}_2 \times \mathbf{r}_2\,, \qquad (3.77)$$

because, P being a point of the body, $\mathbf{r}_1$ and $\mathbf{r}_2$ are constant vectors with respect to the coordinate systems Σ_1' and Σ_2'. Putting $\mathbf{R} = \mathbf{R}_2 - \mathbf{R}_1$, from (3.76) and (3.77) one deduces

$$\left(\frac{d\mathbf{R}}{dt}\right)_\Sigma = \boldsymbol{\omega}_1 \times \mathbf{r}_1 - \boldsymbol{\omega}_2 \times \mathbf{r}_2\,. \qquad (3.78)$$

On the other hand,

$$\left(\frac{d\mathbf{R}}{dt}\right)_\Sigma = \left(\frac{d\mathbf{R}}{dt}\right)_{\Sigma_1'} + \boldsymbol{\omega}_1 \times \mathbf{R} = \boldsymbol{\omega}_1 \times \mathbf{R}\,, \qquad (3.79)$$

because $\mathbf{R}$ is a constant vector in the coordinate system Σ_1' attached to the body. Combining (3.78) and (3.79) we get

$$\boldsymbol{\omega}_1 \times \mathbf{R} = \boldsymbol{\omega}_1 \times \mathbf{r}_1 - \boldsymbol{\omega}_2 \times \mathbf{r}_2 \qquad (3.80)$$

or, inasmuch as $\mathbf{R} = \mathbf{r}_1 - \mathbf{r}_2$,

$$(\boldsymbol{\omega}_2 - \boldsymbol{\omega}_1) \times \mathbf{r}_2 = 0 . \tag{3.81}$$

Since P is any point of the body, $\mathbf{r}_2$ is arbitrary and it follows that $\boldsymbol{\omega}_2 = \boldsymbol{\omega}_1$, completing the proof.

Angular Velocity in Terms of the Euler Angles

Some important problems in rigid body dynamics require the angular velocity vector to be expressed in terms of the Euler angles. An infinitesimal rotation of the rigid body can be thought of as the result of three sucessive infinitesimal rotations whose angular velocities have, respectively, magnitudes $\dot{\phi}, \dot{\theta}, \dot{\psi}$. Let $\boldsymbol{\omega}_\phi$, $\boldsymbol{\omega}_\theta$ and $\boldsymbol{\omega}_\psi$ be the corresponding angular velocity vectors. The angular velocity vector $\boldsymbol{\omega}$ is simply given by

$$\boldsymbol{\omega} = \boldsymbol{\omega}_\phi + \boldsymbol{\omega}_\theta + \boldsymbol{\omega}_\psi \tag{3.82}$$

in virtue of Eq. (3.58). One can obtain the components of $\boldsymbol{\omega}$ either along inertial axes (x, y, z) or along axes (x', y', z') attached to the body. Because of its greater utility, we will consider the latter case.

Clearly, the angular velocity $\boldsymbol{\omega}$ depends linearly on $\dot{\phi}, \dot{\theta}, \dot{\psi}$. Therefore, in order to find the general form taken by $\boldsymbol{\omega}$ we can fix a pair of Euler angles at a time, determine the angular velocity associated with the variation of the third angle and then add the results (Epstein, 1982). Fixing θ and ψ, the z-axis, which is fixed in space, also becomes fixed in the body (see Fig. 3.4). So, z is the rotation axis and $\boldsymbol{\omega}_\phi$ is a vector along the z-axis with component $\dot{\phi}$, and from (3.50) we have

$$\begin{pmatrix} (\boldsymbol{\omega}_\phi)_{x'} \\ (\boldsymbol{\omega}_\phi)_{y'} \\ (\boldsymbol{\omega}_\phi)_{z'} \end{pmatrix} = \mathcal{A}_E \begin{pmatrix} 0 \\ 0 \\ \dot{\phi} \end{pmatrix} = \begin{pmatrix} \dot{\phi} \sin\theta \, \sin\psi \\ \dot{\phi} \sin\theta \, \cos\psi \\ \dot{\phi} \cos\theta \end{pmatrix} . \tag{3.83}$$

Fixing now ϕ and ψ, the line of nodes (the ξ'-axis direction in Fig. 3.4) becomes fixed both in space and in the body, so it is the rotation axis. As a consequence, $\boldsymbol{\omega}_\theta$ is a vector with the single component $\dot{\theta}$ along the ξ'-axis and with the help of (3.49) we find

$$\begin{pmatrix} (\boldsymbol{\omega}_\theta)_{x'} \\ (\boldsymbol{\omega}_\theta)_{y'} \\ (\boldsymbol{\omega}_\theta)_{z'} \end{pmatrix} = \mathcal{B} \begin{pmatrix} \dot{\theta} \\ 0 \\ 0 \end{pmatrix} = \begin{pmatrix} \dot{\theta} \cos\psi \\ -\dot{\theta} \sin\psi \\ 0 \end{pmatrix} . \tag{3.84}$$

Finally, with θ and ϕ fixed the z'-axis, which is fixed in the body, becomes fixed in space. Therefore, z' is the rotation axis and $\boldsymbol{\omega}_\psi$ is a vector along the z'-axis with component $\dot{\psi}$, there being no need to apply any transformation matrix to $\boldsymbol{\omega}_\psi$. Collecting the corresponding components the final result is

$$\begin{aligned} \omega_{x'} &= \dot{\phi} \sin\theta \, \sin\psi + \dot{\theta} \, \cos\psi \, , \\ \omega_{y'} &= \dot{\phi} \sin\theta \, \cos\psi - \dot{\theta} \, \sin\psi \, , \\ \omega_{z'} &= \dot{\psi} + \dot{\phi} \, \cos\theta \, . \end{aligned} \tag{3.85}$$

Exercise 3.4.2   Prove that the components of the angular velocity along the axes fixed in
space are

$$
\begin{aligned}
\omega_x &= \dot\theta \, \cos\phi + \dot\psi \, \sin\theta \, \sin\phi \,, \\
\omega_y &= \dot\theta \, \sin\phi - \dot\psi \, \sin\theta \, \cos\phi \,, \\
\omega_z &= \dot\phi + \dot\psi \, \cos\theta \,.
\end{aligned}
\tag{3.86}
$$

3.5 Rotation Group and Infinitesimal Generators

With the help of (3.63) we can write (3.59) in the form

$$
g'_\Sigma = \mathcal{A} g_\Sigma = (I + J_1 \, d\Omega_1 + J_2 \, d\Omega_2 + J_3 \, d\Omega_3) g_\Sigma \,,
\tag{3.87}
$$

where

$$
J_1 = \begin{pmatrix} 0 & 0 & 0 \\ 0 & 0 & -1 \\ 0 & 1 & 0 \end{pmatrix}, \quad
J_2 = \begin{pmatrix} 0 & 0 & 1 \\ 0 & 0 & 0 \\ -1 & 0 & 0 \end{pmatrix}, \quad
J_3 = \begin{pmatrix} 0 & -1 & 0 \\ 1 & 0 & 0 \\ 0 & 0 & 0 \end{pmatrix}.
\tag{3.88}
$$

The matrices J_i are called **infinitesimal generators**[4] of the rotation group $SO(3)$. Note that
$\operatorname{tr} J_i = 0$.

Exercise 3.5.1   Check that the matrices J_i, $i = 1, 2, 3$, obey the algebra

$$
[J_i, J_j] = J_k \,,
\tag{3.89}
$$

where (i, j, k) is a cyclic permutation of $(1, 2, 3)$ and $[\mathcal{A}, \mathcal{B}] = \mathcal{A}\mathcal{B} - \mathcal{B}\mathcal{A}$ is the
commutator of matrices $\mathcal{A}$ and $\mathcal{B}$. Show that, equivalently, one can write

$$
[J_i, J_j] = \epsilon_{ijk} J_k \,, \qquad i, j = 1, 2, 3 \,,
\tag{3.90}
$$

with use of the convention of sum over repeated indices and of the Levi-Civita
symbol ϵ_{ijk} defined in Appendix A.

The matrix associated with a finite rotation by an angle α about the direction $\hat{\mathbf{n}}$ (counterclockwise, active point of view) can be found by applying N successive infinitesimal
rotations by the same angle α/N and taking the limit $N \to \infty$:

$$
g'_\Sigma = \lim_{N \to \infty} \left(I + \frac{\alpha}{N} \hat{\mathbf{n}} \cdot J \right)^N g_\Sigma = e^{\alpha \hat{\mathbf{n}} \cdot J} g_\Sigma \,.
\tag{3.91}
$$

Thus, it turns out that a finite rotation is expressed in terms of the exponential function of
the matrix $\hat{\mathbf{n}} \cdot J = n_1 J_1 + n_2 J_2 + n_3 J_3$. There's nothing mysterious about the exponential
of a matrix. By definition,

$$
e^{\mathcal{A}} = I + \mathcal{A} + \frac{\mathcal{A}^2}{2!} + \frac{\mathcal{A}^3}{3!} + \cdots = \sum_{k=0}^{\infty} \frac{\mathcal{A}^k}{k!} \,,
\tag{3.92}
$$

---

[4] Strictly speaking, the matrices J_i are a particular representation of the generators of $SO(3)$ called the
fundamental representation.

where $\mathcal{A}$ and its powers are $n \times n$ matrices. The series (3.92) is always convergent (Courant & Hilbert, 1953).

Matrix J_k is the generator of infinitesimal rotations about the kth Cartesian axis and all matrices in the three-dimensional rotation group $SO(3)$ can be represented in the form $\mathcal{R}(\boldsymbol{\alpha}) = e^{\boldsymbol{\alpha} \cdot J}$ where $\boldsymbol{\alpha} \equiv \alpha \hat{\mathbf{n}}$. The matrices $\mathcal{R}(\boldsymbol{\alpha})$ form a **Lie group** whose generators obey the **Lie algebra** associated with the group, Eq. (3.90). The quantities ϵ_{ijk} that appear in Eq. (3.90) are called the **structure constants** of the Lie algebra of the three-dimensional rotation group.

Exercise 3.5.2   Making use of the identity

$$\det e^{\mathcal{A}} = e^{\operatorname{tr} \mathcal{A}}, \tag{3.93}$$

verify that all matrices in $SO(3)$ have unit determinant.

3.6 Dynamics in Non-Inertial Reference Frames

Let Σ and Σ' be reference frames *with common origin*, Σ being an inertial frame and Σ' being a rotating frame with angular velocity $\boldsymbol{\omega}$. Let $\mathbf{r}$ be the position vector of a particle of mass m whose equation of motion in Σ is

$$m \left(\frac{d^2 \mathbf{r}}{dt^2} \right)_\Sigma = \mathbf{F}, \tag{3.94}$$

where $\mathbf{F}$ is the total force on the particle. Using Eq. (3.67) we have

$$\mathbf{v}_{in} \equiv \left(\frac{d\mathbf{r}}{dt} \right)_\Sigma = \left(\frac{d\mathbf{r}}{dt} \right)_{\Sigma'} + \boldsymbol{\omega} \times \mathbf{r} \equiv \mathbf{v} + \boldsymbol{\omega} \times \mathbf{r}, \tag{3.95}$$

where $\mathbf{v}_{in}$ is the particle's velocity in the inertial frame Σ and $\mathbf{v}$ denotes the particle's velocity in the rotating frame – owing to the coincidence of origins, $\mathbf{r}$ is also the particle's position vector in Σ'. From (3.95), with further use of (3.67), one derives

$$\left(\frac{d^2 \mathbf{r}}{dt^2} \right)_\Sigma = \left(\frac{d\mathbf{v}_{in}}{dt} \right)_\Sigma = \left(\frac{d\mathbf{v}_{in}}{dt} \right)_{\Sigma'} + \boldsymbol{\omega} \times \mathbf{v}_{in}$$

$$= \left(\frac{d\mathbf{v}}{dt} \right)_{\Sigma'} + \frac{d\boldsymbol{\omega}}{dt} \times \mathbf{r} + 2\boldsymbol{\omega} \times \mathbf{v} + \boldsymbol{\omega} \times (\boldsymbol{\omega} \times \mathbf{r}). \tag{3.96}$$

Equation (3.67) with $\mathbf{g} = \boldsymbol{\omega}$ shows that the angular acceleration is the same in both frames, hence the lack of subscript in $d\boldsymbol{\omega}/dt$ in the last equation. Denoting the particle's acceleration in the rotating frame by $\mathbf{a} = (d\mathbf{v}/dt)_{\Sigma'}$ and using (3.94), the equation of motion for the particle in the rotating frame is finally obtained:

$$m\mathbf{a} = \mathbf{F} + 2m\mathbf{v} \times \boldsymbol{\omega} + m\boldsymbol{\omega} \times (\mathbf{r} \times \boldsymbol{\omega}) + m\mathbf{r} \times \frac{d\boldsymbol{\omega}}{dt}. \tag{3.97}$$

To an observer in the non-inertial frame Σ' it all happens as if the particle were subject to an effective force which is the actual force $\mathbf{F}$ plus: a first term known as the **Coriolis force**;

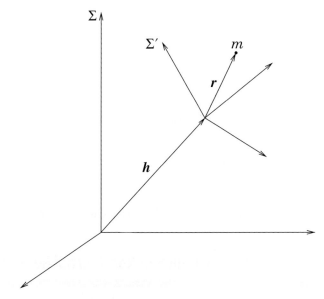

Simultaneous rotation and translation (Exercise 3.6.1).

a second term, quadratic in $\boldsymbol{\omega}$, called the **centrifugal force**; and a third term, proportional to the angular acceleration of the non-inertial frame, sometimes called the **Euler force**.

Exercise 3.6.1 If the origins of Σ and Σ' are not coincident (Fig. 3.6), show that instead of (3.97) we have

$$m\mathbf{a} = \mathbf{F} - m\mathbf{a}_{o'} + 2m\mathbf{v} \times \boldsymbol{\omega} + m\boldsymbol{\omega} \times (\mathbf{r} \times \boldsymbol{\omega}) + m\mathbf{r} \times \frac{d\boldsymbol{\omega}}{dt}, \tag{3.98}$$

where $\mathbf{a}_{o'} = (d^2\mathbf{h}/dt^2)_\Sigma$ is the acceleration of the origin of Σ' relative to Σ.

Dynamics on the Rotating Earth

As one of the most important applications of the previous developments, consider the motion of a particle near the Earth's surface. It is convenient to introduce a Cartesian coordinate system which rotates together with the Earth with origin at the Earth's surface, as shown in Fig. 3.7. The z-axis is vertical, that is, it is perpendicular to the Earth's surface; the x-axis is tangent to the meridian pointing south; the y-axis is tangent to the parallel pointing east. The latitude of the place on the Earth's surface taken as origin of the Cartesian system xyz is λ. Taking the origin of the inertial frame Σ at the centre of the Earth, the position vector of the origin of the rotating frame is $\mathbf{R} = R\hat{\mathbf{z}}$, where R is the radius of the Earth. Therefore, $\mathbf{R}$ is a constant vector in the rotating frame attached to the Earth's surface and the repeated use of Eq. (3.67) immediately gives

$$\mathbf{a}_{o'} = m\left(\frac{d^2\mathbf{R}}{dt^2}\right)_\Sigma = \boldsymbol{\omega} \times (\boldsymbol{\omega} \times \mathbf{R}), \tag{3.99}$$

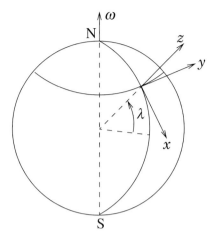

Fig. 3.7 Rotating frame *xyz* fixed on the Earth's surface.

where we have used $d\boldsymbol{\omega}/dt = 0$ because the Earth's angular velocity is practically constant. Substitution of this result into (3.98) yields

$$m\mathbf{a} = \mathbf{T} + m\mathbf{g} + m\boldsymbol{\omega} \times (\mathbf{R} \times \boldsymbol{\omega}) + 2m\mathbf{v} \times \boldsymbol{\omega} + m\boldsymbol{\omega} \times (\mathbf{r} \times \boldsymbol{\omega}), \qquad (3.100)$$

where $\mathbf{F} = m\mathbf{g} + \mathbf{T}$ with $\mathbf{T}$ denoting the sum of all forces on the particle other than its weight. Equation (3.100) can be written in the form

$$m\mathbf{a} = \mathbf{T} + m\mathbf{g}_{\text{eff}} + 2m\mathbf{v} \times \boldsymbol{\omega} + m\boldsymbol{\omega} \times (\mathbf{r} \times \boldsymbol{\omega}), \qquad (3.101)$$

where

$$\mathbf{g}_{\text{eff}} = \mathbf{g} + \boldsymbol{\omega} \times (\mathbf{R} \times \boldsymbol{\omega}) \qquad (3.102)$$

is the effective acceleration of gravity. In order to justify this name, consider a plumb bob, that is, a particle of mass m suspended from a string under tension $\mathbf{T}$ and in equilibrium at the origin ($\mathbf{r} \equiv 0$). Equation (3.101) yields

$$\mathbf{T} + m\mathbf{g}_{\text{eff}} = 0, \qquad (3.103)$$

that is, the direction of the plum line (the measured vertical direction) is determined by the effective acceleration of gravity vector. The vector $\mathbf{g}_{\text{eff}}$ does not point exactly toward the centre of the Earth. In the coordinate system of Fig. 3.7 we have $\mathbf{g} = -g\hat{\mathbf{z}}$ and

$$\boldsymbol{\omega} = -\omega \cos \lambda \, \hat{\mathbf{x}} + \omega \sin \lambda \, \hat{\mathbf{z}}, \qquad (3.104)$$

whence

$$\mathbf{g}_{\text{eff}} = -(g - R\omega^2 \cos^2 \lambda)\,\hat{\mathbf{z}} + R\omega^2 \sin \lambda \, \cos \lambda \, \hat{\mathbf{x}}. \qquad (3.105)$$

Therefore, there is a reduction of the magnitude of the gravitational acceleration in the vertical direction and the appearence of a component in the north-south direction that prevents $\mathbf{g}_{\text{eff}}$ from pointing toward the centre of the Earth – the true vertical direction. The effects are small, however, because

$$\omega \approx \frac{2\pi \text{ rad}}{24 \times 3600 \text{ s}} \approx 7.3 \times 10^{-5} \text{ rad/s} \tag{3.106}$$

and in consequence

$$R\omega^2 \approx (6400 \text{ km}) \times (7.3 \times 10^{-5} \text{ s}^{-1})^2 \approx 3 \times 10^{-2} \text{ m/s}^2, \tag{3.107}$$

which is about 300 times smaller that the acceleration of gravity. This centrifugal effect, although minute, cannot be neglected in studies of the shape of the Earth.

For objects moving near the surface of the Earth, the last term on the right-hand side of Eq. (3.100) is usually much smaller that the Coriolis term. Essentially, the centrifugal effect reduces to converting $\mathbf{g}$ into $\mathbf{g}_{\text{eff}}$, and the main modification to the motion brought about by the rotation of the Earth arises from the Coriolis force. For a supersonic airplane with speed $v \approx 2000$ km/h, the Coriolis acceleration does not exceed $2\omega v \approx 0.08$ m/s^2 or about $0.008g$. Though small, the Coriolis acceleration is important in several circumstances, as will be seen in forthcoming examples. In the northern hemisphere the Coriolis force deflects to the right an object that is moving horizontally – that is, parallelly to the Earth's surface, the deflection being to the left in the southern hemisphere (Problem 3.11). Presumably this effect explains the behaviour of cyclones, which typically rotate counterclockwise in the northern hemisphere but clockwise in the southern hemisphere (Goldstein, 1980).

Example 3.3 An object at height h falls from rest. Find the transversal deflection of the object when it hits the ground caused by the Coriolis force.

Solution

If the object falls vertically, $\mathbf{v} = -v\hat{\mathbf{z}}$ with $v > 0$ and the Coriolis force is

$$\mathbf{F}_{\text{Cor}} = 2m\mathbf{v} \times \boldsymbol{\omega} = -2mv\hat{\mathbf{z}} \times (-\omega \cos \lambda \, \hat{\mathbf{x}} + \omega \sin \lambda \, \hat{\mathbf{z}}) = 2m\omega v \cos \lambda \, \hat{\mathbf{y}}. \tag{3.108}$$

The deviation is in the west-to-east direction in both hemispheres, since $\cos(-\lambda) = \cos \lambda$. In the vertical direction the approximate equation of motion is

$$\ddot{z} = -g \implies \dot{z} = -gt \implies z = h - \frac{1}{2}gt^2, \tag{3.109}$$

where we used the initial conditions $\dot{z}(0) = 0$, $z(0) = h$. In the west-to-east direction we have

$$\ddot{y} = 2\omega v \cos \lambda \approx 2\omega g t \cos \lambda. \tag{3.110}$$

Strictly speaking, only immediately after the beginning of the fall is the velocity vector vertical. As the object falls, it acquires a transversal velocity that remains much smaller that the vertical component because the Coriolis acceleration is negligible compared to the acceleration of gravity. This justifies approximating the speed v by its vertical component alone during the whole fall, as done in (3.110). With the initial conditions $y(0) = \dot{y}(0) = 0$ Eq. (3.110) implies

$$y = \frac{\omega g}{3} t^3 \cos \lambda = \frac{\omega}{3} \sqrt{\frac{8(h-z)^3}{g}} \cos \lambda . \tag{3.111}$$

Upon hitting the ground ($z = 0$) the transversal displacement is

$$\Delta y = \frac{\omega}{3} \sqrt{\frac{8h^3}{g}} \cos \lambda . \tag{3.112}$$

At the equator, where the effect is biggest, for a fall from a height $h = 100$ m the deflection is $\Delta y \approx 2.2$ cm. Although the deviation is measurable, the experiment is hard to do because one has to eliminate the influence of winds, air resistance and other disturbing factors which may mask the Coriolis deflection.

Example 3.4   Determine the effect of the rotation of the Earth on a pendulum whose suspension point can turn freely, known as the Foucault pendulum.

Solution

Disregarding the quadratic term in ω in (3.101) and taking $\mathbf{g}_{\mathrm{eff}} \approx \mathbf{g}$, the equation of motion for a simple pendulum becomes

$$m\mathbf{a} = \mathbf{T} + m\mathbf{g} + 2m\mathbf{v} \times \boldsymbol{\omega} , \tag{3.113}$$

where $\mathbf{T}$ denotes the tension in the string. It is convenient to choose the origin of the coordinate system at the pendulum's equilibrium point (Fig. 3.8). In the case of small oscillations, the pendulum's vertical displacement is much smaller that its horizontal displacement. In fact,

$$z = l(1 - \cos \theta) \approx l\frac{\theta^2}{2} , \tag{3.114}$$

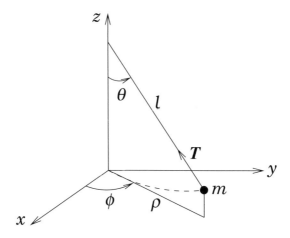

Fig. 3.8   Foucault pendulum.

so, with $\rho = l \sin \theta \approx l\theta$,

$$\frac{z}{\rho} \approx \frac{\theta}{2} \ll 1 \tag{3.115}$$

for small θ. Therefore, in the vertical direction the pendulum remains essentially at rest with the tension in the string balancing its weight:

$$T \cos \theta \approx mg \implies T \approx mg \tag{3.116}$$

since $\cos \theta \approx 1$. The Coriolis force is transversal to the instantaneous plane of oscillation of the pendulum, which prevents it from oscillating in a fixed vertical plane. The horizontal components of (3.113) are

$$\ddot{x} = -\frac{T}{m} \sin \theta \, \cos \phi + 2\omega_z \dot{y}, \qquad \ddot{y} = -\frac{T}{m} \sin \theta \, \sin \phi - 2\omega_z \dot{x}, \tag{3.117}$$

with

$$\omega_z = \omega \sin \lambda. \tag{3.118}$$

Taking into account that

$$\sin \theta = \frac{\rho}{l}, \qquad \rho \cos \phi = x, \qquad \rho \sin \phi = y \tag{3.119}$$

and using (3.116), equations (3.117) take the form

$$\ddot{x} + \omega_0^2 x = 2\omega_z \dot{y}, \qquad \ddot{y} + \omega_0^2 y = -2\omega_z \dot{x}, \tag{3.120}$$

where

$$\omega_0 = \sqrt{\frac{g}{l}}. \tag{3.121}$$

Note that for any pendulum with $l < 1\,\mathrm{km}$ we have $\omega_0 > 0.1\,\mathrm{s}^{-1} \gg \omega_z$. The equations of motion for the pendulum are more easily solved with the help of the complex function $\zeta(t) = x(t) + iy(t)$ because equations (3.120) are the real and imaginary parts of

$$\ddot{\zeta} + 2i\omega_z \dot{\zeta} + \omega_0^2 \zeta = 0. \tag{3.122}$$

This last equation admits solutions in the form $\zeta(t) = \zeta_0 \, e^{-ipt}$ with

$$-p^2 + 2\omega_z p + \omega_0^2 = 0 \implies p = \omega_z \pm \sqrt{\omega_z^2 + \omega_0^2} \approx \omega_z \pm \omega_0. \tag{3.123}$$

The general solution for $\zeta(t)$ is

$$\zeta(t) = A e^{-i(\omega_z - \omega_0)t} + B e^{-i(\omega_z + \omega_0)t}. \tag{3.124}$$

Adopting the initial conditions $x(0) = a$, $\dot{x}(0) = 0$, $y(0) = \dot{y}(0) = 0$, so that the pendulum starts oscillating in the xz-plane, we get

$$A + B = a, \qquad B = \frac{\omega_0 - \omega_z}{\omega_0 + \omega_z} A \approx \frac{\omega_0}{\omega_0} A = A \implies A = B = a/2, \tag{3.125}$$

whence

$$\zeta(t) = a \cos \omega_0 t \, e^{-i\omega_z t}. \tag{3.126}$$

Therefore,

$$x(t) = \operatorname{Re}\zeta(t) = a\cos\omega_z t\,\cos\omega_0 t,$$
$$y(t) = \operatorname{Im}\zeta(t) = -a\sin\omega_z t\,\cos\omega_0 t. \tag{3.127}$$

The angle ϕ that the plane of oscillation makes with the initial plane of oscillation (the xz-plane) is such that

$$\tan\phi = \frac{y}{x} = -\tan\omega_z t \implies \phi = -\omega_z t = -(\omega\sin\lambda)t. \tag{3.128}$$

The plane of oscillation rotates about the vertical with angular velocity $\omega\sin\lambda$, the sense, viewed from above, being clockwise in the northern hemisphere ($\lambda > 0$) and counterclockwise in the southern hemisphere ($\lambda < 0$). Each 24 hours the plane of oscilllation rotates by $360\sin\lambda$ degrees.[5] At the latitude of Rio de Janeiro ($\lambda = -22°54'$), the plane of oscillation turns $140°$ a day counterclockwise. In the North Pole the rate of rotation is $360°$ a day clockwise, which can be understood as follows: The plane of oscillation is fixed in an inertial reference frame, but the Earth rotates under the pendulum by $360°$ a day in the counterclockwise sense. So, to an observer on Earth the sense of rotation of the plane of oscillation is clockwise. As a final remark, note that (3.127) shows that the order of magnitude of $\dot{x}$ and $\dot{y}$ is $\omega_0 a$ since $\omega_0 \gg \omega > \omega_z$. On the other hand, the order of magnitude of $|\boldsymbol{\omega}\times(\mathbf{r}\times\boldsymbol{\omega})|$ is $\omega^2 a$. Thus, the ratio of the centrifugal and Coriolis terms in (3.101) is $\omega^2 a/2\omega_0 a\omega = \omega/2\omega_0 \ll 1$, which justifies *a posteriori* the approximation that led to (3.113).

Problems

3.1   In three dimensions Eq. (3.64) establishes a bijective correspondence between vectors and real antisymmetric matrices. (a) If $\mathcal{A} = (a_{kl})$ is the antisymmetric matrix associated with the vector $\mathbf{v} = (v_1, v_2, v_3)$, show that $a_{kl} = \epsilon_{kml} v_m$ with sum over repeated indices and ϵ_{kml} the Levi-Civita symbol defined in Appendix A. Show that, conversely, $v_m = \frac{1}{2}\epsilon_{kml}a_{kl}$. (b) Prove that the eigenvalues of the antisymmetric matrix associated with $\mathbf{v}$ are zero and $\pm i|\mathbf{v}|$. (c) If $\mathcal{A}$ is a real antisymmetric matrix, show that the matrices $I \pm \mathcal{A}$ are non-singular and that the matrix $\mathcal{B} = (I + \mathcal{A})(I - \mathcal{A})^{-1}$ is orthogonal.

3.2   Construct the vector $\mathbf{r}'$ obtained by reflecting vector $\mathbf{r}$ in the plane whose unit normal vector is $\hat{\mathbf{n}}$. Without any calculations, using only geometric arguments, determine the eigenvalues and eigenvectors of the corresponding transformation matrix $\mathcal{A}$. If $\hat{\mathbf{n}} = (n_1, n_2, n_3)$, show that $\mathcal{A}$ has elements $a_{ij} = \delta_{ij} - 2n_i n_j$ and is an improper orthogonal matrix.

---

[5] Foucault intuited this result in 1851, before the mathematical treatment of the problem taking into account the Coriolis force as the cause of the rotation (Dugas, 1988).

3.3 For the matrix $\mathcal{D}$ in Eq. (3.46), prove that

$$\mathcal{D}^n = \begin{pmatrix} \cos n\phi & \sin n\phi & 0 \\ -\sin n\phi & \cos n\phi & 0 \\ 0 & 0 & 1 \end{pmatrix}$$

for any natural number n. Was this result intuitively expected?

3.4 Any proper orthogonal matrix $\mathcal{A}$ corresponds to a rotation by some angle Φ about some direction. The rotation axis is determined by the eigenvector of $\mathcal{A}$ with eigenvalue 1. Picking coordinates such that the z-axis is along the said eigenvector the matrix $\mathcal{A}$ will have the form (3.46). Taking into account that the trace of a matrix is independent of the coordinate system, show that the rotation angle is given by $\cos \Phi = (\mathrm{tr}\,\mathcal{A} - 1)/2$.

3.5 A sphere of radius R rolls without slipping on a plane surface. If (x, y, R) are Cartesian coordinates of the centre of the sphere, show that, in terms of the Euler angles, the conditions for rolling without slipping are

$$\dot{x} - R(\dot{\theta}\,\sin\phi - \dot{\psi}\,\sin\theta\,\cos\phi) = 0\,, \qquad \dot{y} + R(\dot{\theta}\,\cos\phi + \dot{\psi}\,\sin\theta\,\sin\phi) = 0\,.$$

Prove that these constraints are not holonomic (see Appendix B).

3.6 The translational displacement of a body can be represented by a vector and the linear velocity is the time derivative of the position vector. The angular velocity vector, however, is not, in general, the time derivative of an angular displacement vector. In order to prove this, note that if there exists such a vector $\mathbf{\Omega}$, its components $\Omega_x, \Omega_y, \Omega_z$ can be expressed in terms of the Euler angles and must be such that their time derivatives equal the corresponding components of the angular velocity:

$$\omega_x = \frac{\partial \Omega_x}{\partial \theta}\dot{\theta} + \frac{\partial \Omega_x}{\partial \phi}\dot{\phi} + \frac{\partial \Omega_x}{\partial \psi}\dot{\psi}\,, \qquad \omega_y = \frac{\partial \Omega_y}{\partial \theta}\dot{\theta} + \frac{\partial \Omega_y}{\partial \phi}\dot{\phi} + \frac{\partial \Omega_y}{\partial \psi}\dot{\psi}\,,$$

$$\omega_z = \frac{\partial \Omega_z}{\partial \theta}\dot{\theta} + \frac{\partial \Omega_z}{\partial \phi}\dot{\phi} + \frac{\partial \Omega_z}{\partial \psi}\dot{\psi}\,.$$

Using (3.86), prove that there can be no vector $(\Omega_x, \Omega_y, \Omega_z)$ such that these equations are satisfied.

3.7 Equation (3.67) can be derived in a fully analytic fashion without appealing to the geometric definition of the angular velocity vector. (a) By definition, the derivative of a matrix is the matrix formed by the derivatives of its elements. If $\mathcal{A}$ is a time-dependent orthogonal matrix, by differentiating the identity $\mathcal{A}^T\mathcal{A} = I$ with respect to time prove that the matrix $\dot{\mathcal{A}}^T\mathcal{A}$ is antisymmetric – that is, $\dot{\mathcal{A}}^T\mathcal{A} = -(\dot{\mathcal{A}}^T\mathcal{A})^T$. (b) Let $\mathbf{r}_\Sigma$ and $\mathbf{r}_{\Sigma'}$ be the components of the *same* position vector $\mathbf{r}$ of a particle relative to the inertial frame Σ and the rotating frame Σ', and let $\mathcal{A}$ be the rotation matrix from Σ to Σ'. If $\mathbf{v}$ and $\mathbf{v}'$ are the velocities relative to the respective inertial and rotating frames, we have $\mathbf{v}_\Sigma = d\mathbf{r}_\Sigma/dt$ and $\mathbf{v}'_{\Sigma'} = d\mathbf{r}_{\Sigma'}/dt$. Starting from $\mathbf{r}_\Sigma = \mathcal{A}^T\mathbf{r}_{\Sigma'}$ derive

$$\mathbf{v}_\Sigma = \mathbf{v}'_\Sigma + \dot{\mathcal{A}}^T\mathcal{A}\,\mathbf{r}_\Sigma\,.$$

Taking into account part (a), the correspondence (3.64) between vectors and antisymmetric matrices and the fact that an equality of vectors which is true in one coordinate system is true in all coordinate systems, prove that there exists a vector $\boldsymbol{\omega}$ such that

$$\mathbf{v} = \mathbf{v}' + \boldsymbol{\omega} \times \mathbf{r},$$

which coincides with Eq. (3.67).

3.8 Let $\mathcal{P}$ be the matrix associated with a rotation through $180°$ about an arbitrary axis. (a) Determine $\mathcal{P}^2$ without calculations, reflecting on its meaning. (b) Letting $\mathcal{A} = (\mathbf{I} + \mathcal{P})/2$ and $\mathcal{B} = (\mathbf{I} - \mathcal{P})/2$, prove that $\mathcal{A}^2 = \mathcal{A}$ and $\mathcal{B}^2 = \mathcal{B}$. (c) Show that the matrices $\mathcal{A}$ and $\mathcal{B}$ are singular and compute their product.

3.9 A particle is fired vertically upward from the surface of the Earth with initial speed v_0, reaches its maximum height and returns to the ground. Show that the Coriolis deflection when it hits the ground has the opposite sense and is four times bigger than the deviation for a particle dropped from the same maximum height.

3.10 Show that the equation of motion (3.98) can be derived from the Lagrangian

$$L = \frac{mv^2}{2} - V - m\mathbf{r} \cdot \mathbf{a}_{o'} + \frac{m}{2}(\boldsymbol{\omega} \times \mathbf{r})^2 + m\mathbf{v} \cdot (\boldsymbol{\omega} \times \mathbf{r}),$$

where $\mathbf{v} = \dot{\mathbf{r}}$ and V is the potential energy corresponding to the force $\mathbf{F}$. Obtain the Jacobi integral associated with this Lagrangian in the case in which $\mathbf{a}_{o'}$ and $\boldsymbol{\omega}$ are constant.

3.11 At $t = 0$ a projectile is fired horizontally near the surface of the Earth with speed v. (a) Neglecting gravity, show that, to a good approximation, the horizontal deviation of the projectile due to the Coriolis force is

$$d_H = \omega v |\sin \lambda| t^2,$$

where λ is the latitude. (b) Show that the projectile is deflected to the right in the northern hemisphere and to the left in the southern hemisphere. (c) To the same degree of approximation, express d_H in terms of the projectile's distance D from the firing point at time t. (d) How would the inclusion of the influence of gravity affect the result? (e) What changes if the projectile is fired at an angle above the horizontal? (f) It is told that during a World War I naval battle in the south Atlantic British shells missed German ships by about 90 meters to the left because the British gunsights had been corrected to a latitude $50°$ north (for battles in Europe) instead of $50°$ south where the battle took place. Conclude that the British shells would have missed the targets by twice the deflection calculated above. (g) Assuming the tale is true[6] and $v = 700$ m/s, how distant must the German ships have been?

---

[6] It is a pity that such a savoury story is most likely but an urban legend, as convincingly argued at http://dreadnoughtproject.org/tfs/index.php/Battle_of_the_Falkland_Islands.

4

Dynamics of Rigid Bodies

The wheel is come full circle.

William Shakespeare, *King Lear*, Act 5, Scene 3

Now that we are in possession of the necessary kinematic apparatus, we can proceed to examining some simple but important problems in the dynamics of rigid bodies, which requires us to take into account the causes of the motion: forces and torques. We start by considering two physical quantities which are essential in the study of the motion of a rigid body, the angular momentum and the kinetic energy.

4.1 Angular Momentum and Inertia Tensor

The equations of motion for a rigid body can be written in the form

$$\frac{d\mathbf{P}}{dt} = \mathbf{F}, \tag{4.1}$$

$$\frac{d\mathbf{L}}{dt} = \mathbf{N}, \tag{4.2}$$

where $\mathbf{F}$ is the total force and $\mathbf{N}$ is the total torque, $\mathbf{P}$ and $\mathbf{L}$ being the total linear momentum and total angular momentum, respectively. As discussed in Section 1.1, equations (4.1) and (4.2) are true only if the time rates of change are relative to an inertial reference frame and if certain restrictions are obeyed by the point with respect to which torque and angular momentum are calculated. In particular, (4.2) holds if the reference point for the calculation of $\mathbf{N}$ and $\mathbf{L}$ is at rest in an inertial reference frame – this is convenient if the rigid body is constrained to rotate about a fixed point – or is the centre of mass of the body.

Let O be a fixed point or the centre of mass of a rigid body. From the point of view of the inertial frame Σ depicted in Fig. 4.1, the total angular momentum of the body with respect to point O is

$$\mathbf{L} = \sum_{k=1}^{N} \mathbf{r}_k \times \mathbf{p}_k = \sum_{k=1}^{N} m_k \, \mathbf{r}_k \times \mathbf{v}_k, \tag{4.3}$$

where N is the number of particles in the body and $\mathbf{v}_k = (d\mathbf{r}_k/dt)_\Sigma$ is the velocity of the kth particle relative to point O as seen from the inertial frame Σ. But $\mathbf{r}_k$ is a constant vector in a Cartesian system Σ' attached to the body and we have

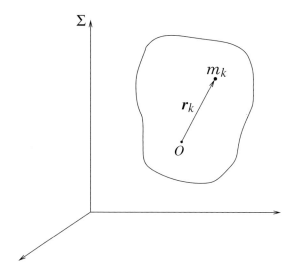

Fig. 4.1 O is a point of the body which remains fixed relative to the inertial frame Σ or is the centre of mass of the body.

$$\mathbf{v}_k = \left(\frac{d\mathbf{r}_k}{dt} \right)_{\Sigma} = \left(\frac{d\mathbf{r}_k}{dt} \right)_{\Sigma'} + \boldsymbol{\omega} \times \mathbf{r}_k = \boldsymbol{\omega} \times \mathbf{r}_k . \tag{4.4}$$

With this result we can write

$$\mathbf{L} = \sum_{k=1}^{N} m_k \, \mathbf{r}_k \times (\boldsymbol{\omega} \times \mathbf{r}_k) = \sum_{k=1}^{N} m_k \left[r_k^2 \, \boldsymbol{\omega} - (\mathbf{r}_k \cdot \boldsymbol{\omega}) \mathbf{r}_k \right], \tag{4.5}$$

where we have used the identity $\mathbf{a} \times (\mathbf{b} \times \mathbf{c}) = (\mathbf{a} \cdot \mathbf{c})\mathbf{b} - (\mathbf{a} \cdot \mathbf{b})\mathbf{c}$. At this stage it is convenient to adopt an indicial notation for the components of $\mathbf{r}_k$, to wit $\mathbf{r}_k = (x_1^{(k)}, x_2^{(k)}, x_3^{(k)})$, which in the traditional notation correspond to $x_1^{(k)} \equiv x_k$, $x_2^{(k)} \equiv y_k$, $x_3^{(k)} \equiv z_k$. Putting $\mathbf{L} = (L_1, L_2, L_3)$, the ith component of (4.5) is given by

$$L_i = \left(\sum_{k=1}^{N} m_k r_k^2 \right) \omega_i - \sum_{k=1}^{N} \sum_{j=1}^{3} m_k x_j^{(k)} \omega_j x_i^{(k)} = \sum_{k=1}^{N} \sum_{j=1}^{3} m_k \left[r_k^2 \delta_{ij} - x_i^{(k)} x_j^{(k)} \right] \omega_j \tag{4.6}$$

inasmuch as

$$\mathbf{r}_k \cdot \boldsymbol{\omega} = \sum_{j=1}^{3} x_j^{(k)} \, \omega_j , \qquad \omega_i = \sum_{j=1}^{3} \delta_{ij} \, \omega_j . \tag{4.7}$$

Consequently,

$$L_i = \sum_{j=1}^{3} I_{ij} \, \omega_j , \tag{4.8}$$

where

$$I_{ij} = \sum_{k=1}^{N} m_k \left[r_k^2 \, \delta_{ij} - x_i^{(k)} x_j^{(k)} \right] . \tag{4.9}$$

The nine quantities I_{ij}, $i, j = 1, 2, 3$, are called components of the **inertia tensor**.

In most problems involving rigid bodies it is more practical to treat them as continuous bodies with mass density $\rho(\mathbf{r})$ rather than discrete systems of particles. The transition from the discrete to the continuous description is accomplished by the correspondence

$$\mathbf{r}_k \longrightarrow \mathbf{r}, \qquad \sum_{k=1}^{N} m_k \longrightarrow \int_V \rho \, dv. \qquad (4.10)$$

In words, the point mass m_k at position $\mathbf{r}_k$ is replaced by the mass element $dm = \rho \, dv$ located at $\mathbf{r}$ and the summation over all particles is replaced by an integration over the volume V occupied by the body. The continuous version of (4.9) is, therefore,

$$I_{ij} = \int_V \left[r^2 \delta_{ij} - x_i x_j \right] \rho \, dv. \qquad (4.11)$$

An intrinsic representation of the inertia tensor which removes any reference to a particular coordinate system is obtained by noting that Eq. (4.5) can be written in the symbolic form

$$\mathbf{L} = \left(\sum_{k=1}^{N} m_k r_k^2 \right) \boldsymbol{\omega} - \left(\sum_{k=1}^{N} m_k \, \mathbf{r}_k \, \mathbf{r}_k \right) \cdot \boldsymbol{\omega}, \qquad (4.12)$$

where the second sum involves a *tensor product* or *dyadic product* of vectors, whose meaning and properties we proceed to discuss.

4.2 Mathematical Interlude: Tensors and Dyadics

The **dyadic product** or **tensor product** of two vectors $\mathbf{A}$ and $\mathbf{B}$, denoted simply by $\mathbf{AB}$, is a linear operator that takes vectors to vectors. The dyadic product is here represented by the mere juxtaposition of the two vectors, without indication of either scalar or vector product, but it is worth noting that another common notation is $\mathbf{A} \otimes \mathbf{B}$. The operator $\mathbf{AB}$ is also called a **dyad**, and the result of its action on an arbitrary vector $\mathbf{C}$ is defined by

$$(\mathbf{AB}) \cdot \mathbf{C} = \mathbf{A}(\mathbf{B} \cdot \mathbf{C}). \qquad (4.13)$$

Since $\mathbf{B} \cdot \mathbf{C}$ is a scalar, the right-hand side of (4.13) is the product of a scalar by a vector, so it is indeed a vector. Linearity is a natural consequence of this definition:

(i) $(\mathbf{AB}) \cdot (\alpha \mathbf{C} + \beta \mathbf{D}) = \alpha(\mathbf{AB}) \cdot \mathbf{C} + \beta(\mathbf{AB}) \cdot \mathbf{D}, \quad \alpha, \beta \in \mathbb{R}. \qquad (4.14)$

The following additional properties hold *by definition*, α and β being any scalars:

(ii) $(\alpha \mathbf{AB} + \beta \mathbf{CD}) \cdot \mathbf{E} = \alpha(\mathbf{AB}) \cdot \mathbf{E} + \beta(\mathbf{CD}) \cdot \mathbf{E}; \qquad (4.15)$

(iii) $(\mathbf{A} + \mathbf{B})\mathbf{C} = \mathbf{AC} + \mathbf{BC}; \qquad (4.16)$

(iv) $\mathbf{A}(\mathbf{B} + \mathbf{C}) = \mathbf{AB} + \mathbf{AC}. \qquad (4.17)$

The tensor product is distributive over vector addition but is not commutative – that is, $\mathbf{AB} \neq \mathbf{BA}$ in general. It is also possible to define the scalar product from the left between a vector and a dyad:

$$\mathbf{C} \cdot (\mathbf{AB}) = (\mathbf{C} \cdot \mathbf{A})\mathbf{B}. \tag{4.18}$$

For this reason, a very suggestive notation for a dyad is a letter topped by a bidirectional arrow, like this: $\overset{\leftrightarrow}{\mathbf{T}} = \mathbf{AB}$. However old-fashioned, this notation comes in handy, as will be seen in the sequel.

The most general **tensor**[1] or **dyadic** is a sum of dyads:

$$\overset{\leftrightarrow}{\mathbf{T}} = \mathbf{AB} + \mathbf{CD} + \cdots. \tag{4.19}$$

By decomposing the vectors $\mathbf{A}, \mathbf{B}, \mathbf{C}, \mathbf{D}, \ldots$ in terms of the unit vectors of an orthonormal basis $\{\hat{\mathbf{e}}_1, \hat{\mathbf{e}}_2, \hat{\mathbf{e}}_3\}$ we find that the most general tensor $\overset{\leftrightarrow}{\mathbf{T}}$ is written as a linear combination of the nine dyads $\hat{\mathbf{e}}_i\, \hat{\mathbf{e}}_j$:

$$\begin{aligned}
\overset{\leftrightarrow}{\mathbf{T}} = {} & T_{11}\,\hat{\mathbf{e}}_1\,\hat{\mathbf{e}}_1 \;+\; T_{12}\,\hat{\mathbf{e}}_1\,\hat{\mathbf{e}}_2 \;+\; T_{13}\,\hat{\mathbf{e}}_1\,\hat{\mathbf{e}}_3 \\
& +\; T_{21}\,\hat{\mathbf{e}}_2\,\hat{\mathbf{e}}_1 \;+\; T_{22}\,\hat{\mathbf{e}}_2\,\hat{\mathbf{e}}_2 \;+\; T_{23}\,\hat{\mathbf{e}}_2\,\hat{\mathbf{e}}_3 \\
& +\; T_{31}\,\hat{\mathbf{e}}_3\,\hat{\mathbf{e}}_1 \;+\; T_{32}\,\hat{\mathbf{e}}_3\,\hat{\mathbf{e}}_2 \;+\; T_{33}\,\hat{\mathbf{e}}_3\,\hat{\mathbf{e}}_3,
\end{aligned} \tag{4.20}$$

where $T_{11}, \ldots, T_{33}$ are nine real numbers, called the components of the tensor $\overset{\leftrightarrow}{\mathbf{T}}$ in the basis in question. Clearly,

$$T_{ij} = \hat{\mathbf{e}}_i \cdot \overset{\leftrightarrow}{\mathbf{T}} \cdot \hat{\mathbf{e}}_j, \tag{4.21}$$

there being no need for parentheses in this last equation because $\mathbf{A} \cdot (\overset{\leftrightarrow}{\mathbf{T}} \cdot \mathbf{B}) = (\mathbf{A} \cdot \overset{\leftrightarrow}{\mathbf{T}}) \cdot \mathbf{B}$ (check this).

The **unit tensor** or **identity tensor**, denoted by $\overset{\leftrightarrow}{\mathbf{1}}$, is defined by

$$\overset{\leftrightarrow}{\mathbf{1}} = \hat{\mathbf{e}}_1\,\hat{\mathbf{e}}_1 \;+\; \hat{\mathbf{e}}_2\,\hat{\mathbf{e}}_2 \;+\; \hat{\mathbf{e}}_3\,\hat{\mathbf{e}}_3. \tag{4.22}$$

Given any vector $\mathbf{A} = A_1\hat{\mathbf{e}}_1 + A_2\hat{\mathbf{e}}_2 + A_3\hat{\mathbf{e}}_3$ we have

$$\overset{\leftrightarrow}{\mathbf{1}} \cdot \mathbf{A} = \hat{\mathbf{e}}_1(\hat{\mathbf{e}}_1 \cdot \mathbf{A}) + \hat{\mathbf{e}}_2(\hat{\mathbf{e}}_2 \cdot \mathbf{A}) + \hat{\mathbf{e}}_3(\hat{\mathbf{e}}_3 \cdot \mathbf{A}) = A_1\,\hat{\mathbf{e}}_1 + A_2\,\hat{\mathbf{e}}_2 + A_3\,\hat{\mathbf{e}}_3 = \mathbf{A}, \tag{4.23}$$

with identical result for $\mathbf{A} \cdot \overset{\leftrightarrow}{\mathbf{1}}$, which justifies the name given to $\overset{\leftrightarrow}{\mathbf{1}}$. Of course, the components of $\overset{\leftrightarrow}{\mathbf{1}}$ in the orthonormal basis $\{\hat{\mathbf{e}}_1, \hat{\mathbf{e}}_2, \hat{\mathbf{e}}_3\}$ are

$$\hat{\mathbf{e}}_i \cdot \overset{\leftrightarrow}{\mathbf{1}} \cdot \hat{\mathbf{e}}_j = \hat{\mathbf{e}}_i \cdot (\overset{\leftrightarrow}{\mathbf{1}} \cdot \hat{\mathbf{e}}_j) = \hat{\mathbf{e}}_i \cdot \hat{\mathbf{e}}_j = \delta_{ij}, \tag{4.24}$$

as expected. It is easy to show that, projected on the basis $\{\hat{\mathbf{e}}_i\}_{i=1}^{3}$, equation

$$\mathbf{A} = \overset{\leftrightarrow}{\mathbf{T}} \cdot \mathbf{B} \tag{4.25}$$

---

[1] Bearing in mind that in this chapter we will not need higher-rank tensors, whenever there is no risk of confusion we will refer to a second-rank tensor simply as a "tensor" (see the antepenultimate paragraph in this section for higher-rank tensors).

is equivalent to

$$A_i = \sum_{j=1}^{3} T_{ij} B_j \,. \tag{4.26}$$

Indeed, owing to linearity,

$$\mathbf{A} = \overset{\leftrightarrow}{\mathbf{T}} \cdot \left(\sum_{j=1}^{3} \hat{\mathbf{e}}_j B_j \right) = \sum_{j=1}^{3} \overset{\leftrightarrow}{\mathbf{T}} \cdot \hat{\mathbf{e}}_j B_j \,, \tag{4.27}$$

whence

$$A_i = \hat{\mathbf{e}}_i \cdot \mathbf{A} = \sum_{j=1}^{3} \hat{\mathbf{e}}_i \cdot \overset{\leftrightarrow}{\mathbf{T}} \cdot \hat{\mathbf{e}}_j B_j \,, \tag{4.28}$$

which corresponds to Eq. (4.26).

Finite Rotation Operator

The dyadic notation is particularly suitable to provide an intrinsic representation for the linear operator which executes a finite rotation of a vector about a given axis. For starters we need a simple definition. The vector product from the left of a vector $\mathbf{A}$ by a tensor $\overset{\leftrightarrow}{\mathbf{T}}$ is another tensor $\mathbf{A} \times \overset{\leftrightarrow}{\mathbf{T}}$ defined in a natural fashion by its action on an arbitrary vector $\mathbf{B}$:

$$(\mathbf{A} \times \overset{\leftrightarrow}{\mathbf{T}}) \cdot \mathbf{B} = \mathbf{A} \times (\overset{\leftrightarrow}{\mathbf{T}} \cdot \mathbf{B}) \,. \tag{4.29}$$

Example 4.1 With the use of dyadics, express the finite rotation operator in an intrinsic, coordinate-independent form.

Solution

Consider the vector $\mathbf{r}'$ produced by a rotation of the vector $\mathbf{r}$ by angle α about the direction defined by the unit vector $\hat{\mathbf{n}}$, as in Fig. 4.2. Let us construct an orthonormal basis $\{\hat{\mathbf{e}}_1, \hat{\mathbf{e}}_2, \hat{\mathbf{e}}_3\}$, shown in Fig. 4.2, by taking $\hat{\mathbf{e}}_3$ along the rotation axis,

$$\hat{\mathbf{e}}_3 = \hat{\mathbf{n}} \,, \tag{4.30}$$

and picking $\hat{\mathbf{e}}_1$ in such a way that the vector $\mathbf{r}$ belongs to the plane defined by $\hat{\mathbf{e}}_1$ and $\hat{\mathbf{e}}_3$, with $\hat{\mathbf{e}}_2$ perpendicular to that plane (Pearlman, 1967). Clearly,

$$\hat{\mathbf{e}}_2 = \frac{\hat{\mathbf{e}}_3 \times \mathbf{r}}{|\hat{\mathbf{e}}_3 \times \mathbf{r}|} = \frac{\hat{\mathbf{n}} \times \mathbf{r}}{r \sin \beta} \,, \tag{4.31}$$

where β is the angle between $\mathbf{r}$ and the rotation axis. Moreover,

$$\hat{\mathbf{e}}_1 = \hat{\mathbf{e}}_2 \times \hat{\mathbf{e}}_3 = \frac{\hat{\mathbf{n}} \times (\mathbf{r} \times \hat{\mathbf{n}})}{r \sin \beta} \,. \tag{4.32}$$

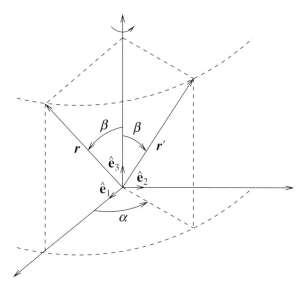

Fig. 4.2   Finite rotation of a vector.

The components x_1', x_2', x_3' of $\mathbf{r}'$ in the basis $\{\hat{\mathbf{e}}_1, \hat{\mathbf{e}}_2, \hat{\mathbf{e}}_3\}$ are

$$\begin{aligned}
x_1' &= r' \sin \beta \, \cos \alpha = r \sin \beta \, \cos \alpha \,, \\
x_2' &= r' \sin \beta \, \sin \alpha = r \sin \beta \, \sin \alpha \,, \\
x_3' &= r' \cos \beta = r \cos \beta \,,
\end{aligned} \qquad (4.33)$$

where we have used $r' = r$. Therefore, from (4.31) and (4.32),

$$\mathbf{r}' = x_1' \, \hat{\mathbf{e}}_1 + x_2' \, \hat{\mathbf{e}}_2 + x_3' \, \hat{\mathbf{e}}_3 = \cos \alpha \, \hat{\mathbf{n}} \times (\mathbf{r} \times \hat{\mathbf{n}}) + \sin \alpha \, \hat{\mathbf{n}} \times \mathbf{r} + (\mathbf{r} \cdot \hat{\mathbf{n}}) \, \hat{\mathbf{n}}. \qquad (4.34)$$

The identity $\hat{\mathbf{n}} \times (\mathbf{r} \times \hat{\mathbf{n}}) = \mathbf{r} - (\mathbf{r} \cdot \hat{\mathbf{n}}) \, \hat{\mathbf{n}}$ allows us to write the *rotation formula*

$$\mathbf{r}' = \cos \alpha \, \mathbf{r} + (1 - \cos \alpha) \, \hat{\mathbf{n}} \, (\hat{\mathbf{n}} \cdot \mathbf{r}) + \sin \alpha \, \hat{\mathbf{n}} \times \mathbf{r}. \qquad (4.35)$$

Introducing the rotation operator $\overset{\leftrightarrow}{\mathbf{R}}_{\hat{\mathbf{n}}}(\alpha)$, defined by

$$\overset{\leftrightarrow}{\mathbf{R}}_{\hat{\mathbf{n}}}(\alpha) = \cos \alpha \, \overset{\leftrightarrow}{\mathbf{1}} + (1 - \cos \alpha) \, \hat{\mathbf{n}} \, \hat{\mathbf{n}} + \sin \alpha \, \hat{\mathbf{n}} \times \overset{\leftrightarrow}{\mathbf{1}} \,, \qquad (4.36)$$

the rotation formula (4.35) can be shortened to

$$\mathbf{r}' = \overset{\leftrightarrow}{\mathbf{R}}_{\hat{\mathbf{n}}}(\alpha) \cdot \mathbf{r} . \qquad (4.37)$$

Expression (4.36) is an intrinsic representation of the rotation operator that mentions only the rotation axis ($\hat{\mathbf{n}}$) and the rotation angle (α), without referring to any particular coordinate system. This representation is advantageous in several situations of physical interest (Leubner, 1979).

Exercise 4.2.1 If $\alpha = \delta\theta$ is an infinitesimal angle, show that (4.35) reduces to Eq. (2.136).

Tensor Transformation Law

It is important to find out how the components of a dyadic change under a rotation of the coordinate axes. From (3.17), we have

$$T'_{rs} = \hat{\mathbf{e}}'_r \cdot \overset{\leftrightarrow}{\mathbf{T}} \cdot \hat{\mathbf{e}}'_s = \left(\sum_k a_{rk}\hat{\mathbf{e}}_k \right) \cdot \overset{\leftrightarrow}{\mathbf{T}} \cdot \left(\sum_l a_{sl}\hat{\mathbf{e}}_l \right) = \sum_{k,l} a_{rk}a_{sl}T_{kl}. \tag{4.38}$$

A tensor of rank n is defined as a set of 3^n quantities $T_{k_1 k_2 \dots k_n}$ which transform under a rotation of the coordinate axes according to the law

$$T'_{k_1 k_2 \dots k_n} = \sum_{l_1, l_2, \dots, l_n} a_{k_1 l_1} a_{k_2 l_2} \dots a_{k_n l_n} T_{l_1 l_2 \dots l_n}. \tag{4.39}$$

Thus, a dyadic is a second-rank tensor. A vector is a first-rank tensor and a scalar can be considered a tensor of rank zero.

In terms of the matrices $\mathcal{T}' = (T'_{rs})$, $\mathcal{T} = (T_{rs})$ and the rotation matrix $\mathcal{A}$, Eq. (4.38) is equivalent to

$$T'_{rs} = \sum_k (\mathcal{A})_{rk} \sum_l T_{kl}(\mathcal{A}^T)_{ls} = \sum_k (\mathcal{A})_{rk}(\mathcal{T}\mathcal{A}^T)_{ks} = (\mathcal{A}\mathcal{T}\mathcal{A}^T)_{rs}, \tag{4.40}$$

that is,

$$\mathcal{T}' = \mathcal{A}\mathcal{T}\mathcal{A}^{-1}. \tag{4.41}$$

Therefore, the matrix associated with the dyadic $\overset{\leftrightarrow}{\mathbf{T}}$ in basis $\{\hat{\mathbf{e}}'_i\}_{i=1}^3$ is obtained from its matrix in basis $\{\hat{\mathbf{e}}_i\}_{i=1}^3$ by means of a similarity transformation executed by the rotation matrix $\mathcal{A}$ that takes $\{\hat{\mathbf{e}}_i\}$ into $\{\hat{\mathbf{e}}'_i\}$.

Tensors in Physics

Second-rank tensors appear naturally whenever there is a linear relation between two vectors or when a scalar depends quadratically on a vector. For instance, the conductivity tensor connects the current density to the electric field in a conductor that obeys Ohm's law, and the magnetic permeability tensor relates **B** to **H** in an magnetically anisotropic medium. In elasticity theory, the variation of the squared distance between two infinitesimally close points is quadratic in the displacement vector and defines the strain tensor. A vector quantity that depends bilinearly on two other vector quantities defines a third rank tensor, and so on. The description of the flow of certain physical quantities also gives rise to tensors. The flow of a scalar quantity is described by a vector. For example, in electrodynamics the current density vector **J** represents the amount of electric charge crossing a unit area perpendicular to the current per unit time. Similarly, the flow of a vector quantity is expressed by a second-rank tensor. Thus, the Maxwell stress tensor T_{ij} describes the flow of linear momentum stored in the electromagnetic field. The need of two indices to characterise the flow of a vector quantity is easy to understand: T_{ij} is the flow of the ith component of the electromagnetic linear momentum in the jth direction. The representation of the flow of a quantity described by a second-rank tensor requires a third-rank tensor, and so forth.

4.3 Moments and Products of Inertia

In dyadic notation Eq. (4.12) is written as

$$\mathbf{L} = \sum_k m_k (r_k^2 \overset{\leftrightarrow}{\mathbf{1}} - \mathbf{r}_k \mathbf{r}_k) \cdot \boldsymbol{\omega} \equiv \overset{\leftrightarrow}{\mathbf{I}} \cdot \boldsymbol{\omega} , \qquad (4.42)$$

where

$$\overset{\leftrightarrow}{\mathbf{I}} = \sum_k m_k (r_k^2 \overset{\leftrightarrow}{\mathbf{1}} - \mathbf{r}_k \mathbf{r}_k) \qquad (4.43)$$

is the **inertia tensor**, whose continuous version is

$$\overset{\leftrightarrow}{\mathbf{I}} = \int_V (r^2 \overset{\leftrightarrow}{\mathbf{1}} - \mathbf{r}\,\mathbf{r}) \rho \, dv . \qquad (4.44)$$

The components of the inertia tensor in a Cartesian basis are given by (4.11), from which it is obvious that the inertia tensor is symmetric – that is,

$$I_{ij} = I_{ji} , \qquad (4.45)$$

this property being valid in any Cartesian basis. The matrix whose elements are the components of the inertia tensor with respect to a given Cartesian coordinate system is called the **inertia matrix**, denoted by I. The diagonal elements of the inertia matrix $\mathrm{I} = (I_{ij})$ are positive quantities called **moments of inertia**:

$$I_{11} \equiv I_{xx} = \int_V (r^2 - x^2) \rho \, dv = \int_V (y^2 + z^2) \rho \, dv ; \qquad (4.46)$$

$$I_{22} \equiv I_{yy} = \int_V (r^2 - y^2) \rho \, dv = \int_V (x^2 + z^2) \rho \, dv ; \qquad (4.47)$$

$$I_{33} \equiv I_{zz} = \int_V (r^2 - z^2) \rho \, dv = \int_V (x^2 + y^2) \rho \, dv . \qquad (4.48)$$

The off-diagonal elements of the inertia matrix are called **products of inertia** and are given by

$$I_{12} = I_{21} \equiv I_{yx} = I_{xy} = - \int_V xy \, \rho \, dv ; \qquad (4.49)$$

$$I_{13} = I_{31} \equiv I_{zx} = I_{xz} = - \int_V xz \, \rho \, dv ; \qquad (4.50)$$

$$I_{23} = I_{32} \equiv I_{zy} = I_{yz} = - \int_V yz \, \rho \, dv . \qquad (4.51)$$

It must be stressed that the moments and products of inertia depend not only on the directions of the chosen coordinate axes but also on their origin.

4.4 Kinetic Energy and Parallel Axis Theorem

The kinetic energy of rotation about point O (see Fig. 4.1) is

$$T = \sum_k \frac{m_k}{2} v_k^2 = \sum_k \frac{m_k}{2} \mathbf{v}_k \cdot (\boldsymbol{\omega} \times \mathbf{r}_k)$$

$$= \sum_k \frac{m_k}{2} \boldsymbol{\omega} \cdot (\mathbf{r}_k \times \mathbf{v}_k) = \frac{1}{2} \boldsymbol{\omega} \cdot \sum_k \mathbf{r}_k \times \mathbf{p}_k = \frac{1}{2} \boldsymbol{\omega} \cdot \mathbf{L} . \tag{4.52}$$

With the use of Eq. (4.42) we can write

$$T = \frac{1}{2} \boldsymbol{\omega} \cdot \overset{\leftrightarrow}{\mathbf{I}} \cdot \boldsymbol{\omega} . \tag{4.53}$$

Let $\hat{\mathbf{n}}$ be the unit vector along the instantaneous rotation axis, so that $\boldsymbol{\omega} = \omega \hat{\mathbf{n}}$. Then,

$$T = \frac{1}{2} \hat{\mathbf{n}} \cdot \overset{\leftrightarrow}{\mathbf{I}} \cdot \hat{\mathbf{n}} \, \omega^2 \equiv \frac{1}{2} I \omega^2 , \tag{4.54}$$

where

$$I = \hat{\mathbf{n}} \cdot \overset{\leftrightarrow}{\mathbf{I}} \cdot \hat{\mathbf{n}} \tag{4.55}$$

is called the moment of inertia with respect to the instantaneous rotation axis. More explicitly, with the use of (4.43),

$$I = \hat{\mathbf{n}} \cdot \overset{\leftrightarrow}{\mathbf{I}} \cdot \hat{\mathbf{n}} = \sum_k m_k [r_k^2 - (\hat{\mathbf{n}} \cdot \mathbf{r}_k)^2] = \sum_k m_k d_k^2 \tag{4.56}$$

where

$$d_k^2 = r_k^2 - (\hat{\mathbf{n}} \cdot \mathbf{r}_k)^2 = r_k^2 (1 - \cos^2 \theta_k) = r_k^2 \sin^2 \theta_k = |\hat{\mathbf{n}} \times \mathbf{r}_k|^2 \tag{4.57}$$

is the square of the distance of the kth particle from the rotation axis (Fig. 4.3). The expression (4.56) coincides with the definition of moment inertia found in elementary

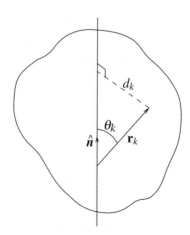

Fig. 4.3   Moment of inertia: d_k is the distance of the kth particle from the rotation axis.

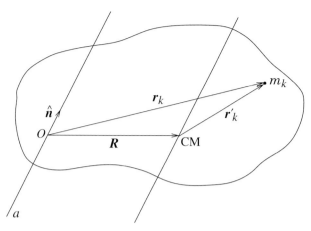

Fig. 4.4 Parallel axis theorem.

physics textbooks. Likewise, (4.54) is the usual formula for the rotational kinetic energy of a rigid body. In general, however, the angular velocity changes direction as time passes, so I varies with time. The moment of inertia I is constant if the body is constrained to rotate about a fixed axis, the case normally treated in elementary physics.

The moment of inertia depends on the choice of the origin of the coordinate system used to calculate it. There is, however, a simple relation between moments of inertia with respect to parallel axes when one of them goes through the centre of mass.

Theorem 4.4.1 (Parallel Axis Theorem) *The moment of inertia with respect to a given axis is equal to the moment of inertia with respect to a parallel axis through the centre of mass plus the moment of inertia of the body with respect to the given axis computed as if the entire mass of the body were concentrated at the centre of mass.[2]*

Proof From Fig. 4.4 it follows that $\mathbf{r}_k = \mathbf{R} + \mathbf{r}'_k$, where $\mathbf{R}$ is the position vector of the centre of mass from origin O on the given axis. The moment of inertia with respect to the given axis is, by (4.56) and (4.57),

$$I_a = \sum_k m_k |\hat{\mathbf{n}} \times \mathbf{r}_k|^2 = \sum_k m_k |\hat{\mathbf{n}} \times \mathbf{R} + \hat{\mathbf{n}} \times \mathbf{r}'_k|^2$$

$$= \left(\sum_k m_k \right) |\hat{\mathbf{n}} \times \mathbf{R}|^2 + \sum_k m_k |\hat{\mathbf{n}} \times \mathbf{r}'_k|^2 + 2(\hat{\mathbf{n}} \times \mathbf{R}) \cdot \left(\hat{\mathbf{n}} \times \sum_k m_k \mathbf{r}'_k \right). \tag{4.58}$$

With the use of (1.19) this last equation reduces to

$$I_a = I_{CM} + M |\hat{\mathbf{n}} \times \mathbf{R}|^2, \tag{4.59}$$

completing the proof, since $|\hat{\mathbf{n}} \times \mathbf{R}|^2$ is the square of the distance of the centre of mass from the given axis. □

---

[2] This result is also known as *Steiner's theorem*.

The inertia tensor itself admits a decomposition analogous to (4.59). Indeed, substituting $\mathbf{r}_k = \mathbf{R} + \mathbf{r}'_k$ into (4.43) and using the properties of the tensor product as well as (1.19) there results

$$\overset{\leftrightarrow}{\mathbf{I}}_O = \sum_k m_k [(r_k'^2 + 2\mathbf{R}\cdot\mathbf{r}'_k + R^2)\overset{\leftrightarrow}{\mathbf{1}} - (\mathbf{R} + \mathbf{r}'_k)(\mathbf{R} + \mathbf{r}'_k)]$$

$$= \sum_k m_k (r_k'^2 \overset{\leftrightarrow}{\mathbf{1}} - \mathbf{r}'_k \mathbf{r}'_k) + (R^2 \overset{\leftrightarrow}{\mathbf{1}} - \mathbf{RR}) \sum_k m_k$$

$$+ 2\mathbf{R}\cdot \left(\sum_k m_k \mathbf{r}'_k \right) \overset{\leftrightarrow}{\mathbf{1}} - \mathbf{R} \left(\sum_k m_k \mathbf{r}'_k \right) - \left(\sum_k m_k \mathbf{r}'_k \right) \mathbf{R}$$

$$= \sum_k m_k (r_k'^2 \overset{\leftrightarrow}{\mathbf{1}} - \mathbf{r}'_k \mathbf{r}'_k) + M(R^2 \overset{\leftrightarrow}{\mathbf{1}} - \mathbf{RR}) , \qquad (4.60)$$

that is,

$$\overset{\leftrightarrow}{\mathbf{I}}_O = \overset{\leftrightarrow}{\mathbf{I}}_{CM} + M(R^2 \overset{\leftrightarrow}{\mathbf{1}} - \mathbf{RR}) . \qquad (4.61)$$

Note that the second term on the right-hand side is the inertia tensor with respect to the origin O as if the whole mass of the body were concentrated at its centre of mass.

Exercise 4.4.1   Derive (4.59) taking (4.61) as the starting point.

Theorem 4.4.2 (Perpendicular Axis Theorem)   *Given a plane lamina with arbitrary form and mass distribution, the sum of its moments of inertia with respect to any two perpendicular axes lying in the plane of the lamina is equal to the moment of inertia with respect to an axis that goes through their point of intersection and is perpendicular to the lamina.*

Exercise 4.4.2   Prove the perpendicular axis theorem.

4.5 Diagonalisation of the Inertia Tensor

The components of the inertia tensor depend both on the origin and the orientation of the coordinate axes. Once the origin is fixed, it would be highly convenient to be able to find coordinate axes ξ_1, ξ_2, ξ_3 with unit vectors $\hat{\boldsymbol{\xi}}_1, \hat{\boldsymbol{\xi}}_2, \hat{\boldsymbol{\xi}}_3$ with respect to which the inertia tensor is diagonal. In this case we would have

$$\overset{\leftrightarrow}{\mathbf{I}} = I_1 \hat{\boldsymbol{\xi}}_1 \hat{\boldsymbol{\xi}}_1 + I_2 \hat{\boldsymbol{\xi}}_2 \hat{\boldsymbol{\xi}}_2 + I_3 \hat{\boldsymbol{\xi}}_3 \hat{\boldsymbol{\xi}}_3 \qquad (4.62)$$

whence

$$\mathbf{L} = \overset{\leftrightarrow}{\mathbf{I}} \cdot \boldsymbol{\omega} = I_1 \omega_1 \hat{\boldsymbol{\xi}}_1 + I_2 \omega_2 \hat{\boldsymbol{\xi}}_2 + I_3 \omega_3 \hat{\boldsymbol{\xi}}_3 , \qquad (4.63)$$

that is,

$$L_1 = I_1 \omega_1 , \quad L_2 = I_2 \omega_2 , \quad L_3 = I_3 \omega_3 . \qquad (4.64)$$

Similarly, the kinetic energy would take the particularly simple form

$$T = \frac{1}{2}\boldsymbol{\omega}\cdot\overset{\leftrightarrow}{\mathbf{I}}\cdot\boldsymbol{\omega} = \frac{1}{2}\boldsymbol{\omega}\cdot\mathbf{L} = \frac{1}{2}(I_1\omega_1^2 + I_2\omega_2^2 + I_3\omega_3^2). \tag{4.65}$$

The inertia tensor is a real symmetric linear operator – therefore self-adjoint – and Theorem D.13 in Appendix D guarantees the existence of an orthonormal basis made up of eigenvectors of $\overset{\leftrightarrow}{\mathbf{I}}$. In other words, there exist three positive numbers I_1, I_2, I_3 and a set of three orthonormal vectors $\hat{\boldsymbol{\xi}}_1, \hat{\boldsymbol{\xi}}_2, \hat{\boldsymbol{\xi}}_3$ such that

$$\overset{\leftrightarrow}{\mathbf{I}}\cdot\hat{\boldsymbol{\xi}}_1 = I_1\hat{\boldsymbol{\xi}}_1, \quad \overset{\leftrightarrow}{\mathbf{I}}\cdot\hat{\boldsymbol{\xi}}_2 = I_2\hat{\boldsymbol{\xi}}_2, \quad \overset{\leftrightarrow}{\mathbf{I}}\cdot\hat{\boldsymbol{\xi}}_3 = I_3\hat{\boldsymbol{\xi}}_3 \tag{4.66}$$

or, without summation over repeated indices,

$$\overset{\leftrightarrow}{\mathbf{I}}\cdot\hat{\boldsymbol{\xi}}_j = I_j\hat{\boldsymbol{\xi}}_j, \qquad j = 1, 2, 3. \tag{4.67}$$

Picking the Cartesian axes along the vectors $\hat{\boldsymbol{\xi}}_1, \hat{\boldsymbol{\xi}}_2, \hat{\boldsymbol{\xi}}_3$, the components of $\overset{\leftrightarrow}{\mathbf{I}}$ are

$$I_{ij} = \hat{\boldsymbol{\xi}}_i\cdot\overset{\leftrightarrow}{\mathbf{I}}\cdot\hat{\boldsymbol{\xi}}_j = I_j\,\hat{\boldsymbol{\xi}}_i\cdot\hat{\boldsymbol{\xi}}_j = I_j\,\delta_{ij}, \tag{4.68}$$

so the inertia matrix in this basis is diagonal:

$$\mathbf{I} = (I_{ij}) = \begin{pmatrix} I_1 & 0 & 0 \\ 0 & I_2 & 0 \\ 0 & 0 & I_3 \end{pmatrix}. \tag{4.69}$$

The diagonal elements, I_1, I_2 and I_3, are called **principal moments of inertia** and the corresponding Cartesian axes are called **principal axes of inertia** or **principal axes**, for short.

The determination of the principal axes amounts to solving the equation

$$\overset{\leftrightarrow}{\mathbf{I}}\cdot\boldsymbol{\xi} = I\,\boldsymbol{\xi} \tag{4.70}$$

for $\boldsymbol{\xi} \neq 0$ and $I \in \mathbb{R}$. The principal axes are determined once three mutually orthogonal solutions have been found, the corresponding values of I being the principal moments of inertia. For an initial choice of coordinate axes, Eq. (4.70), expressed in components, takes the form

$$\sum_{j=1}^{3} I_{ij}\,\xi_j = I\,\xi_i, \qquad i = 1, 2, 3, \tag{4.71}$$

or, in matrix form,

$$\begin{pmatrix} I_{11} - I & I_{12} & I_{13} \\ I_{21} & I_{22} - I & I_{23} \\ I_{31} & I_{32} & I_{33} - I \end{pmatrix}\begin{pmatrix} \xi_1 \\ \xi_2 \\ \xi_3 \end{pmatrix} = 0. \tag{4.72}$$

This system of linear equations for the components of $\boldsymbol{\xi}$ has a non-trivial solution if and only if

$$\det \begin{pmatrix} I_{11} - I & I_{12} & I_{13} \\ I_{21} & I_{22} - I & I_{23} \\ I_{31} & I_{32} & I_{33} - I \end{pmatrix} = 0. \tag{4.73}$$

This equation is cubic in I and its three roots are the principal moments of inertia. For each root found, Eq. (4.72) can be solved for the components of the associated vector $\boldsymbol{\xi}$, which gives the direction of the corresponding principal axis.

Example 4.2 Given a homogeneous plate in the form of an isosceles right triangle, determine the principal axes and principal moments of inertial with respect to the vertex opposite to the hypotenuse.

Solution

Let us start by adopting the system of axes depicted in Fig. 4.5(a). If M is the mass of the plate, its surface mass density is

$$\sigma = \frac{M}{a^2/2} = \frac{2M}{a^2}. \tag{4.74}$$

The inertia tensor has the following components:

$$I_{11} \equiv I_{xx} = \sigma \int (y^2 + z^2)da = \sigma \int_0^a \int_0^{a-y} y^2 dx dy = \frac{Ma^2}{6}; \tag{4.75}$$

$$I_{22} \equiv I_{yy} = \sigma \int (x^2 + z^2)da = I_{11} = \frac{Ma^2}{6} \quad \text{(by symmetry)}; \tag{4.76}$$

$$I_{33} \equiv I_{zz} = I_{xx} + I_{yy} = \frac{Ma^2}{3} \quad \text{(by the perpendicular axis theorem)}; \tag{4.77}$$

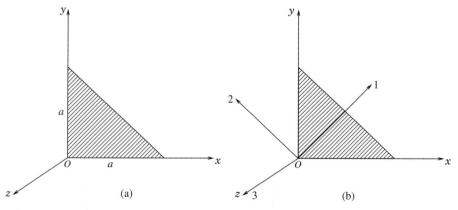

Fig. 4.5 Initially chosen axes and principal axes for a triangular plate with respect to the vertex opposite to the hypotenuse.

$$I_{12} \equiv I_{xy} = -\sigma \int xyda = -\sigma \int_0^a \int_0^{a-y} xydxdy = -\frac{Ma^2}{12} ; \qquad (4.78)$$

$$I_{13} \equiv I_{xz} = -\sigma \int xzda = 0 ; \quad I_{23} \equiv I_{yz} = -\sigma \int yzda = 0 . \qquad (4.79)$$

Therefore,

$$I = (I_{ij}) = \frac{Ma^2}{12} \begin{pmatrix} 2 & -1 & 0 \\ -1 & 2 & 0 \\ 0 & 0 & 4 \end{pmatrix} . \qquad (4.80)$$

Putting $I = Ma^2\lambda/12$ we need to find the roots of

$$\det \begin{pmatrix} 2-\lambda & -1 & 0 \\ -1 & 2-\lambda & 0 \\ 0 & 0 & 4-\lambda \end{pmatrix} = (4-\lambda)[(2-\lambda)^2 - 1] = 0 . \qquad (4.81)$$

The roots are $\lambda_1 = 1, \lambda_2 = 3, \lambda_3 = 4$ and the principal moments of inertia are

$$I_1 = \frac{Ma^2}{12} , \quad I_2 = \frac{Ma^2}{4} , \quad I_3 = \frac{Ma^2}{3} . \qquad (4.82)$$

The eigenvector with eigenvalue I_1 satisfies

$$\begin{pmatrix} 2-\lambda_1 & -1 & 0 \\ -1 & 2-\lambda_1 & 0 \\ 0 & 0 & 4-\lambda_1 \end{pmatrix} \begin{pmatrix} \xi_1 \\ \xi_2 \\ \xi_3 \end{pmatrix} = 0 \implies \begin{cases} \xi_1 - \xi_2 = 0 \\ -\xi_1 + \xi_2 = 0 . \\ 3\xi_3 = 0 \end{cases} \qquad (4.83)$$

Therefore,

$$\begin{pmatrix} \xi_1 \\ \xi_2 \\ \xi_3 \end{pmatrix} = \alpha \begin{pmatrix} 1 \\ 1 \\ 0 \end{pmatrix} \quad \text{or} \quad \boldsymbol{\xi} = \alpha(\hat{\mathbf{x}} + \hat{\mathbf{y}}) , \qquad (4.84)$$

where α is any non-zero real number. Taking $\alpha = 1/\sqrt{2}$ there results a normalised (unit) vector $\hat{\boldsymbol{\xi}}_1$. This vector, which lies on the xy-plane and makes an angle of $45°$ with the x-axis, gives the direction of the first principal axis. Repeating the procedure, the second principal axis is found to lie along the vector $\hat{\boldsymbol{\xi}}_2 = 2^{-1/2}(-\hat{\mathbf{x}} + \hat{\mathbf{y}})$, with the third axis along the vector $\boldsymbol{\xi}_3 = \hat{\mathbf{z}}$, as shown in Fig. 4.5(b).

Example 4.3 Using Eq. (4.61), show that the axes ξ_1, ξ_2, ξ_3 in Fig. 4.5(b) are also principal axes with respect to the centre of mass of the plate and determine the corresponding principal moments of inertia.

Solution

The position vector of the centre of mass from origin O is $\mathbf{R} = (a\sqrt{2}/3)\hat{\boldsymbol{\xi}}_1$. So, by Eq. (4.61),

$$\overset{\leftrightarrow}{\mathbf{I}}_{CM} = \overset{\leftrightarrow}{\mathbf{I}}_O - M(R^2\overset{\leftrightarrow}{\mathbf{1}} - \mathbf{RR})$$

$$= I_1\hat{\boldsymbol{\xi}}_1\hat{\boldsymbol{\xi}}_1 + I_2\hat{\boldsymbol{\xi}}_2\hat{\boldsymbol{\xi}}_2 + I_3\hat{\boldsymbol{\xi}}_3\hat{\boldsymbol{\xi}}_3$$

$$- \frac{2Ma^2}{9}(\hat{\boldsymbol{\xi}}_1\hat{\boldsymbol{\xi}}_1 + \hat{\boldsymbol{\xi}}_2\hat{\boldsymbol{\xi}}_2 + \hat{\boldsymbol{\xi}}_3\hat{\boldsymbol{\xi}}_3 - \hat{\boldsymbol{\xi}}_1\hat{\boldsymbol{\xi}}_1)$$

$$= I_1\hat{\boldsymbol{\xi}}_1\hat{\boldsymbol{\xi}}_1 + \left(I_2 - \frac{2Ma^2}{9}\right)\hat{\boldsymbol{\xi}}_2\hat{\boldsymbol{\xi}}_2 + \left(I_3 - \frac{2Ma^2}{9}\right)\hat{\boldsymbol{\xi}}_3\hat{\boldsymbol{\xi}}_3. \tag{4.85}$$

Therefore, $\overset{\leftrightarrow}{\mathbf{I}}_{CM}$ is also diagonal with respect to axes parallel to ξ_1, ξ_2, ξ_3 but with origin at the centre of mass and

$$I_1^{CM} = \frac{Ma^2}{12}, \qquad I_2^{CM} = \frac{Ma^2}{36}, \qquad I_3^{CM} = \frac{Ma^2}{9} \tag{4.86}$$

are the corresponding principal moments of inertia.

4.6 Symmetries and Principal Axes of Inertia

The identification of principal axes of inertia is greatly facilitated if the body is symmetric. A symmetry plane divides the body into two portions which are the mirror image of each other, and the mass density of the body is the same at points symmetrically situated with respect to the plane.

Lemma 4.6.1 *If a body has a symmetry plane through the point with respect to which the inertia tensor is referred, a principal axis is perpendicular to this plane.*

Proof   Let $z = 0$ be the symmetry plane, so that $\rho(x, y, -z) = \rho(x, y, z)$. Consequently,

$$I_{xz} = -\int\int\int xz\rho(x, y, z)dxdydz = 0 \tag{4.87}$$

because the integrand is an odd function of z and the interval of integration is symmetric about $z = 0$. The same argument applies to I_{yz} and we have

$$\mathbf{I} = \begin{pmatrix} I_{xx} & I_{xy} & 0 \\ I_{yx} & I_{yy} & 0 \\ 0 & 0 & I_{zz} \end{pmatrix}, \tag{4.88}$$

which clearly has the eigenvector

$$\begin{pmatrix} \xi_1 \\ \xi_2 \\ \xi_3 \end{pmatrix} = \begin{pmatrix} 0 \\ 0 \\ 1 \end{pmatrix} \tag{4.89}$$

or $\boldsymbol{\xi} = \hat{\mathbf{z}}$ with eigenvalue I_{zz}. $\square$

An axis is said to be a symmetry axis of a rigid body if the body is a solid of revolution about the axis and its mass density does not depend on the rotation angle about the axis.

Lemma 4.6.2 *A symmetry axis of a rigid body that goes through the point with respect to which the inertia tensor is referred is a principal axis. Any two mutually orthogonal axes in the plane perpendicular to the symmetry axis are principal axes and the corresponding principal moments of inertia are equal to each other.*

Exercise 4.6.1 Prove Lemma 4.6.2. Hint: consider two perpendicular planes whose intersection is the symmetry axis, and apply Lemma 4.6.1 to each of these planes.

A reexamination of Example 4.2 in the light of the previous lemmas reveals that the principal axes for the triangular plate could have been identified without calculations. The z-axis is perpendicular to the plane of the plate (Fig. 4.5), which obviously is a symmetry plane. Therefore, Lemma 4.6.1 ensures that the z-axis is a principal axis. Clearly, a plane perpendicular to the plate that contains O and is orthogonal the the hypotenuse is a symmetry plane too. By using Lemma 4.6.1 again, the 2-axis is a principal axis. Finally, the third axis is simply perpendicular to the other two, and the 1-axis is identified as the last principal axis.

Example 4.4 Determine the principal axes and moments of inertia for a homogeneous cube of mass M and side a, with respect to one of the vertices.

Solution
Let us first choose axes as in Fig. 4.6. By symmetry, the moments of inertia are all equal to each other and the same happens to the products of inertia:

$$I_{zz} = I_{yy} = I_{xx} = \int_0^a \int_0^a \int_0^a (y^2 + z^2)\rho dx dy dz = \frac{2}{3}\rho a^5 = \frac{2}{3}Ma^2 \; ; \qquad (4.90)$$

$$I_{xz} = I_{yz} = I_{xy} = -\int_0^a \int_0^a \int_0^a xy\rho dx dy dz = -\frac{1}{4}\rho a^5 = -\frac{Ma^2}{4} \, . \qquad (4.91)$$

Thus, the inertia matrix is given by

$$I = \frac{2Ma^2}{3} \begin{pmatrix} 1 & -3/8 & -3/8 \\ -3/8 & 1 & -3/8 \\ -3/8 & -3/8 & 1 \end{pmatrix} \, . \qquad (4.92)$$

By inspection of Fig. 4.6 we identify the following planes of symmetry containing the vertex O: $OABCO$ and $ODBEO$. The unit normal vectors to these planes are, respectively,

$$\hat{\mathbf{n}}_1 = \frac{\overrightarrow{OC}}{\sqrt{2}\,a} \times \frac{\overrightarrow{OA}}{a} = \frac{1}{\sqrt{2}}(\hat{\mathbf{x}} + \hat{\mathbf{y}}) \times \hat{\mathbf{z}} = \frac{1}{\sqrt{2}}(\hat{\mathbf{x}} - \hat{\mathbf{y}}) \, ,$$

$$\hat{\mathbf{n}}_2 = \frac{\overrightarrow{OD}}{a} \times \frac{\overrightarrow{OE}}{\sqrt{2}\,a} = \hat{\mathbf{x}} \times \frac{(\hat{\mathbf{y}} + \hat{\mathbf{z}})}{\sqrt{2}} = \frac{1}{\sqrt{2}}(\hat{\mathbf{z}} - \hat{\mathbf{y}}) \, .$$

The two planes of symmetry considered are not mutually perpendicular, so the vectors $\hat{\mathbf{n}}_1$ and $\hat{\mathbf{n}}_2$ are linearly independent but are not orthogonal. This is possible because the

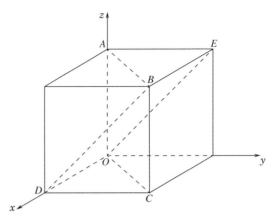

Fig. 4.6 Planes of symmetry of a homogeneous cube with respect to a vertex.

inertia matrix has two degenerate eigenvalues (see Appendix D) – that is, two principal moments of inertia are equal. In fact, from

$$\begin{pmatrix} 1 & \alpha & \alpha \\ \alpha & 1 & \alpha \\ \alpha & \alpha & 1 \end{pmatrix} \begin{pmatrix} 1 \\ -1 \\ 0 \end{pmatrix} = (1-\alpha) \begin{pmatrix} 1 \\ -1 \\ 0 \end{pmatrix},$$

$$\begin{pmatrix} 1 & \alpha & \alpha \\ \alpha & 1 & \alpha \\ \alpha & \alpha & 1 \end{pmatrix} \begin{pmatrix} 0 \\ -1 \\ 1 \end{pmatrix} = (1-\alpha) \begin{pmatrix} 0 \\ -1 \\ 1 \end{pmatrix}$$

with $\alpha = -3/8$ it follows that $I_1 = I_2 = 11Ma^2/12$. Normalised orthogonal eigenvectors can be constructed by, for example, setting $\hat{\boldsymbol{\xi}}_1 = \hat{\mathbf{n}}_1$ and $\hat{\boldsymbol{\xi}}_2 = \gamma(\hat{\mathbf{n}}_1 + \beta\hat{\mathbf{n}}_2)$ with the real numbers β and γ so chosen that $\hat{\boldsymbol{\xi}}_1 \cdot \hat{\boldsymbol{\xi}}_2 = 0$ and $|\hat{\boldsymbol{\xi}}_2| = 1$. An easy calculation yields $\hat{\boldsymbol{\xi}}_2 = (\hat{\mathbf{x}} + \hat{\mathbf{y}} - 2\hat{\mathbf{z}})/\sqrt{6}$. The unit vector along the third principal axis is just

$$\hat{\boldsymbol{\xi}}_3 = \hat{\boldsymbol{\xi}}_1 \times \hat{\boldsymbol{\xi}}_2 = \frac{(\hat{\mathbf{x}} - \hat{\mathbf{y}})}{\sqrt{2}} \times \frac{(\hat{\mathbf{x}} + \hat{\mathbf{y}} - 2\hat{\mathbf{z}})}{\sqrt{6}} = \frac{1}{\sqrt{3}}(\hat{\mathbf{x}} + \hat{\mathbf{y}} + \hat{\mathbf{z}}),$$

which lies along the cube's diagonal $\overline{OB}$. In short, the principal axes are defined by the three orthonormal vectors

$$\hat{\boldsymbol{\xi}}_1 = \frac{\hat{\mathbf{x}} - \hat{\mathbf{y}}}{\sqrt{2}}, \qquad \hat{\boldsymbol{\xi}}_2 = \frac{\hat{\mathbf{x}} + \hat{\mathbf{y}} - 2\hat{\mathbf{z}}}{\sqrt{6}}, \qquad \hat{\boldsymbol{\xi}}_3 = \frac{\hat{\mathbf{x}} + \hat{\mathbf{y}} + \hat{\mathbf{z}}}{\sqrt{3}}. \tag{4.93}$$

Finally, from

$$\begin{pmatrix} 1 & -3/8 & -3/8 \\ -3/8 & 1 & -3/8 \\ -3/8 & -3/8 & 1 \end{pmatrix} \begin{pmatrix} 1 \\ 1 \\ 1 \end{pmatrix} = \frac{1}{4} \begin{pmatrix} 1 \\ 1 \\ 1 \end{pmatrix}$$

one concludes that the third principal moment of inertia is $I_3 = Ma^2/6$.

4.7 Rolling Coin

Even the simplest three-dimensional problems in rigid body dynamics are considerably complicated. In the other extreme are the elementary problems of rotation about a fixed axis. There are interesting problems involving only two rotational degrees of freedom which are characterised by an intermediate degree of difficulty and provide rich and instructive illustrations of the concepts and methods used in investigating the motion of rigid bodies.

Let us consider a coin of mass m and radius R which rolls on a slanting table always remaining upright – its plane is always perpendicular to the table (see Fig. 4.7). Principal axes through the centre of mass are any two orthogonal axes lying in the plane of the coin, with the third axis perpendicular to the plane of the coin. An elementary calculation gives $I_3 = mR^2/2$ and, by symmetry and the perpendicular axis theorem, $I_1 = I_2 = I_3/2 = mR^2/4$. The kinetic energy is the centre-of-mass translational energy plus the kinetic energy of rotation about the centre of mass. Therefore, the Lagrangian is

$$L = T - V = \frac{m}{2}(\dot{x}^2 + \dot{y}^2) + \frac{1}{2}(I_1\omega_1^2 + I_2\omega_2^2 + I_3\omega_3^2) + mgy\sin\alpha, \qquad (4.94)$$

where α is the inclination of the table and x, y are Cartesian coordinates of the centre of mass of the coin. Let ϕ and θ be angles describing the rotations of the coin about its symmetry axis parallel to the table (the third principal axis) and an axis perpendicular to the table, respectively, as shown in Fig. 4.7. Consequently, $\omega_3 = \dot{\phi}$ and, taking account that $\dot{\theta}$ is the projection of the angular velocity vector on the plane of the coin, $\omega_1^2 + \omega_2^2 = \dot{\theta}^2$. Thus, the Lagrangian for the system takes the form

$$L = \frac{m}{2}(\dot{x}^2 + \dot{y}^2) + \frac{mR^2}{4}\dot{\phi}^2 + \frac{mR^2}{8}\dot{\theta}^2 + mgy\sin\alpha. \qquad (4.95)$$

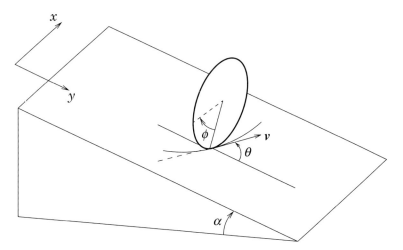

Fig. 4.7   Coin rolling on an inclined table.

As discussed in Section 1.2, the rolling constraints are

$$\dot{x} - R\dot{\phi}\sin\theta = 0, \tag{4.96}$$

$$\dot{y} - R\dot{\phi}\cos\theta = 0. \tag{4.97}$$

Taking advantage of the Lagrange multiplier technique we find the equations of motion

$$m\ddot{x} = \lambda_1, \tag{4.98}$$

$$m\ddot{y} - mg\sin\alpha = \lambda_2, \tag{4.99}$$

$$\frac{mR^2}{4}\ddot{\theta} = 0, \tag{4.100}$$

$$\frac{mR^2}{2}\ddot{\phi} = -\lambda_1 R\sin\theta - \lambda_2 R\cos\theta, \tag{4.101}$$

where λ_1, λ_2 are the Lagrange multipliers associated with the two non-holonomic constraints (4.96) and (4.97), respectively.

Exercise 4.7.1 Derive Eqs. (4.98)–(4.101).

From (4.100) it immediately follows that

$$\theta = \theta_0 + \Omega t \tag{4.102}$$

where θ_0 and Ω are arbitrary constants. Combining equations (4.96) and (4.98) we get

$$\lambda_1 = mR\ddot{\phi}\sin\theta + mR\Omega\dot{\phi}\cos\theta, \tag{4.103}$$

and, proceeding similarly,

$$\lambda_2 = mR\ddot{\phi}\cos\theta - mR\Omega\dot{\phi}\sin\theta - mg\sin\alpha. \tag{4.104}$$

Substitution of these expressions for λ_1 and λ_2 into (4.101) leads to the following differential equation for ϕ:

$$\ddot{\phi} = \frac{2g\sin\alpha}{3R}\cos(\theta_0 + \Omega t). \tag{4.105}$$

The general solution to this equation is

$$\phi = \phi_0 + \omega t - \frac{2g\sin\alpha}{3\Omega^2 R}\cos(\theta_0 + \Omega t). \tag{4.106}$$

Finally, substituting (4.102) and (4.106) into the constraint equations (4.96) and (4.97) and integrating, we find

$$x = x_0 + \frac{g\sin\alpha}{3\Omega}t - \left[\frac{\omega R}{\Omega} + \frac{g\sin\alpha}{3\Omega^2}\sin(\theta_0 + \Omega t)\right]\cos(\theta_0 + \Omega t), \tag{4.107}$$

$$y = y_0 + \left[\frac{\omega R}{\Omega} + \frac{g\sin\alpha}{3\Omega^2}\sin(\theta_0 + \Omega t)\right]\sin(\theta_0 + \Omega t). \tag{4.108}$$

This completes the integration of the equations of motion for the upright rolling coin in terms of the six arbitrary constants $x_0, y_0, \theta_0, \phi_0, \omega, \Omega$. Surprisingly, the coupling between translation and rotation causes the horizontal motion of the centre of mass of the coin to be unbounded but the motion along the slanting y-axis to be bounded and oscillating: if the table is sufficiently long, the coin will never reach the lowest edge of the table.

The solution just found for the motion of the rolling coin applies only if $\Omega \neq 0$. The determination of the motion in the trivial case $\Omega = 0$ is left to the reader.

4.8 Euler's Equations and Free Rotation

Let $\mathbf{N}$ be the torque about point O, the centre of mass of the body or a point of the body which is fixed in an inertial frame. From the point of view of the inertial frame we have

$$\left(\frac{d\mathbf{L}}{dt}\right)_{inertial} = \mathbf{N}, \tag{4.109}$$

whence

$$\left(\frac{d\mathbf{L}}{dt}\right)_{body} + \boldsymbol{\omega} \times \mathbf{L} = \mathbf{N}. \tag{4.110}$$

The inertia tensor is variable in time if it is referred to inertial axes fixed in space. With respect to axes attached to the body, however, the inertia tensor is time independent, which makes the equations of motion for a rigid body considerably simpler. Life is simplified even more by the choice of principal axes for the axes fixed in the body, which will be made from now on. The components of Eq. (4.110) along the principal axes fixed in the body are

$$\frac{dL_1}{dt} + \omega_2 L_3 - \omega_3 L_2 = N_1 \tag{4.111}$$

and the remaining ones obtained from this by cyclic permutations of the subscripts. Making use of $L_1 = I_1\omega_1$, $L_2 = I_2\omega_2$, $L_3 = I_3\omega_3$, we can write

$$I_1\dot{\omega}_1 - (I_2 - I_3)\,\omega_2\,\omega_3 = N_1\,,$$
$$I_2\dot{\omega}_2 - (I_3 - I_1)\,\omega_3\,\omega_1 = N_2\,, \tag{4.112}$$
$$I_3\dot{\omega}_3 - (I_1 - I_2)\,\omega_1\,\omega_2 = N_3\,.$$

These are the celebrated Euler's equations of motion for a rigid body, which describe the motion of the instantaneous axis of rotation (defined by $\boldsymbol{\omega}$) with respect to axes fixed in the body. A complete solution which allows a visualisation of the motion of the body requires determining the instantaneous orientation of the body relative to external inertial axes. Even in the simplest case of free rotation ($\mathbf{N} = 0$), the complete specification of the motion of a rigid body is a problem with a high degree of difficulty. The problem simplifies considerably if the body has a symmetry axis. Instead of treating generically a symmetric body in free rotation, let us consider a special example which exhibits the essential features of the problem.

Example 4.5 A plate (or a frisbee) is thrown almost horizontally into the air, so that it spins and wobbles. Describe the motion of the plate and show that it wobbles twice as fast as it spins about its own symmetry axis.

Solution

Let I_3 be the principal moment of inertia about the symmetry axis of the body (perpendicular to its plane) and $I_1 = I_2$ be the principal moments of inertia with respect to two perpendicular axes lying in the plane of the body. The origin of the principal axes is at the centre of mass of the body. Since the gravitational torque about the centre of mass is zero, Euler's equations (4.112) reduce to

$$I_1 \dot\omega_1 - (I_1 - I_3)\,\omega_2\,\omega_3 = 0\,,$$
$$I_2 \dot\omega_2 - (I_3 - I_1)\,\omega_3\,\omega_1 = 0\,, \qquad\qquad (4.113)$$
$$I_3 \dot\omega_3 = 0\,.$$

The third of the above equations shows that ω_3 stays constant. Defining the constant

$$\Omega = \frac{I_3 - I_1}{I_1}\,\omega_3\,, \qquad\qquad (4.114)$$

we are left with

$$\dot\omega_1 = -\Omega\,\omega_2\,, \qquad \dot\omega_2 = \Omega\,\omega_1\,. \qquad\qquad (4.115)$$

From these equations one immediately derives

$$\ddot\omega_1 + \Omega^2 \omega_1 = 0\,. \qquad\qquad (4.116)$$

With a suitable choice for the origin of time, the solution to the last equation can be put in the form

$$\omega_1 = A \cos \Omega t\,, \qquad\qquad (4.117)$$

whence

$$\omega_2 = A \sin \Omega t\,, \qquad\qquad (4.118)$$

with A an arbitrary constant. The motion performed by the vector $\boldsymbol{\omega}$ is depicted in Fig. 4.8. The vector $\boldsymbol{\omega}_\perp = \omega_1 \hat{\boldsymbol{\xi}}_1 + \omega_2 \hat{\boldsymbol{\xi}}_2$ has constant magnitude and rotates with angular velocity Ω about the the symmetry axis of the body making $\boldsymbol{\omega}$ precess about the same axis. The magnitude of $\boldsymbol{\omega}$ is constant and equal to $(\omega_3^2 + A^2)^{1/2}$, so the instantaneous rotation axis describes a cone around the symmetry axis. The precession of the angular velocity vector around the symmetry axis shows up as a wobbling motion of the body. It is important to stress that the precession of $\boldsymbol{\omega}$ is relative to axes fixed in the body. These, in their turn, rotate in space with angular velocity $\boldsymbol{\omega}$, which makes it difficult to visualise the motion.

The choice of the principal axes in the plane perpendicular to the symmetry axis is arbitrary. Thus, we can pick at any instant the x_1-axis coinciding with the line of nodes (the ξ'-axis direction in Fig. 3.4), with the result that at the instant in question $\psi = 0$ and the components of the angular velocity become simpler:

$$\omega_1 = \dot\theta\,, \qquad \omega_2 = \dot\phi \sin\theta\,, \qquad \omega_3 = \dot\phi \cos\theta + \dot\psi\,. \qquad\qquad (4.119)$$

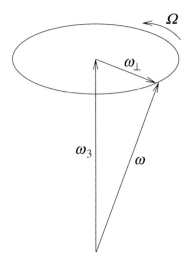

Fig. 4.8 Precession of the instantaneous rotation axis of a symmetric rigid body.

In order to simplify the analysis even more, let us take the z-axis of the inertial frame along the angular momentum vector $\mathbf{L}$, which is constant as seen from the inertial frame. Since the line of nodes is perpendicular to the z-axis, we have (see Fig. 3.4):

$$L_1 = 0, \qquad L_2 = L \sin\theta, \qquad L_3 = L \cos\theta. \qquad (4.120)$$

Thus, with the help of (4.64), we get

$$\dot{\theta} = \frac{L_1}{I_1} = 0, \qquad \dot{\phi} = \frac{\omega_2}{\sin\theta} = \frac{L}{I_1}, \qquad \omega_3 = \frac{L_3}{I_3} = \frac{L\cos\theta}{I_3}. \qquad (4.121)$$

The angle θ that the symmetry axis of the body makes with the angular momentum vector remains constant. Up to this point the results hold true for any symmetric body. Now we specialise to the nearly horizontal plate, for which $I_3 = 2I_1$ in virtue of the perpendicular axis theorem. For a plate thrown almost horizontally, the angular momentum vector is nearly vertical, θ is small and $\dot{\phi} \approx 2\omega_3$ – that is, the plate wobbles twice as fast as it spins around its own symmetry axis.[3]

---

[3] This problem played a singular role in the scientific career of the great American physicist Richard Feynman. In 1947 he was a professor in Cornell University. After having concluded his doctorate at Princeton and having participated in the Manhattan Project for construction of the atom bomb at Los Alamos, he was going through an unproductive phase that made him think he was "burned out" for physics. Feynman preferred to have lunch in the student cafeteria because he liked to ogle the pretty girls. One day a student, fooling around, threw a plate with a medallion of Cornell into the air, and Feynman watched as it wobbled while it rotated about its symmetry axis. With nothing else to do, he wrote and solved the equations of motion for the wobbling plate. Excitedly, Feynman went to Hans Bethe's office and showed him what he had seen at the cafeteria and what he had calculated. "What's the importance of this?", asked Bethe, quite aware that the solution was long known. Feynman said it had no importance but it had been fun. From then on he decided he was going to enjoy physics and do whatever he liked. This episode rekindled Feynman's passion for physics and helped him to get over the depressed state he was in (Feynman, 1985; Mehra, 1994). Just for the record, Feynman's memory betrayed him, and he thought he had seen the plate spin twice as fast as it wobbled.

Stability of Uniform Rotation

In the absence of torques, Euler's equations admit solutions such that $\boldsymbol{\omega}$ has a non-vanishing constant component along a single principal axis. For instance, if $\mathbf{N} = 0$ a solution to (4.112) is $\omega_1 = \omega_1(0)$, $\omega_2 = \omega_3 = 0$. Therefore, uniform rotation about one of the principal axes of inertia represents an equilibrium state for the free rigid body. This is important for mechanical applications, for a wheel remains in uniform rotation about a given axis under no torque only if the axis in question is a principal axis. In order to keep a body rotating uniformly about any other axis it is necessary to apply a torque, causing greater wear of the drive shaft. An interesting issue is whether or not the free uniform rotation motion is stable.

Example 4.6 Show that uniform rotation is stable only if it is about the principal axis with the largest or the smallest moment of inertia.

Solution

Suppose the uniform rotation about a principal axis, say the third axis, is slightly disturbed bringing about the components $\omega_1, \omega_2 \ll \omega_3$. The problem boils down to finding out whether or not ω_1 and ω_2 remain small, in which case the rotational equilibrium will be stable. Euler's equations with $\mathbf{N} = 0$ are, approximately,

$$
\begin{aligned}
I_1 \dot{\omega}_1 - (I_2 - I_3) \omega_2 \omega_3 &= 0, \\
I_2 \dot{\omega}_2 - (I_3 - I_1) \omega_3 \omega_1 &= 0, \\
I_3 \dot{\omega}_3 &= 0,
\end{aligned}
\tag{4.122}
$$

where the term proportional to the product $\omega_1 \omega_2$ has been neglected because it is, by hypothesis, too small. It turns out that ω_3 is practically constant and, differentiating the first and using the second of equations (4.122), one derives

$$
\ddot{\omega}_1 + \Omega_0^2 \omega_1 = 0,
\tag{4.123}
$$

where

$$
\Omega_0^2 = \frac{(I_3 - I_2)(I_3 - I_1)}{I_1 I_2} \omega_3^2.
\tag{4.124}
$$

If $\Omega_0^2 > 0$ we can write

$$
\omega_1(t) = \omega_1(0) \cos \Omega_0 t, \qquad \omega_2(t) = \frac{\Omega_0 I_1}{(I_3 - I_2) \omega_3} \omega_1(0) \sin \Omega_0 t.
\tag{4.125}
$$

Thus, ω_1 and ω_2 oscillate harmonically about the equilibrium state $\omega_1 = \omega_2 = 0$ and the perturbation remains small if it starts small enough. But $\Omega_0^2 > 0$ if: (1) $I_3 > I_2$ and $I_3 > I_1$; (2) $I_3 < I_2$ and $I_3 < I_1$. Therefore, the uniform rotation is stable if it is about the principal axis with largest or smallest moment of inertia. If $\Omega_0^2 < 0$, that is, if $I_3 < I_2$ and $I_3 > I_1$ or $I_3 > I_2$ and $I_3 < I_1$, ω_1 and ω_2 are exponential functions of time and there are initially small perturbations that eventually become large: the rotation is unstable if executed about the principal axis with intermediate moment of inertia. This behaviour can be qualitatively observed with objects in the form of a parallelepiped: a wooden block, a book kept closed by rubber bands or a full matchbox. All one has to do is to hold the object with the chosen principal axis in the vertical direction, spin it vigorously about the said axis and let it fall freely under gravity. It is easy to produce a stable rotation, with at most a slight wobbling, about the axis with the largest or the smallest moment of inertia. But the attempts to give it a steady rotation about the intermediate axis, however careful, are frustrated: the motion quickly degenerates in an very irregular wobble, with the rotation axis wandering at random through the body (Romer, 1978).

Asymmetric Rigid Body: Qualitative Analysis

Let us consider the free rotation of an asymmetric rigid body with the axes so numbered that $I_1 < I_2 < I_3$. Let $\mathbf{L}_b = (L_1, L_2, L_3)$ be the angular momentum as seen in the body frame – that is, $L_1 = I_1\omega_1$, $L_2 = I_2\omega_2$ and $L_3 = I_3\omega_3$ are the components of the angular momentum along the axes attached to the body. In terms of L_1, L_2, L_3 Euler's equations (4.112) take the form

$$\dot{L}_1 = \frac{I_2 - I_3}{I_2 I_3} L_2 L_3 \,,$$

$$\dot{L}_1 = \frac{I_3 - I_1}{I_2 I_3} L_3 L_1 \,, \tag{4.126}$$

$$\dot{L}_1 = \frac{I_1 - I_2}{I_2 I_3} L_1 L_2 \,.$$

These equations can be solved in terms of Jacobi elliptic functions (Landau & Lifshitz, 1976), but we want a qualitative understanding of the motion performed by $\mathbf{L}_b$.

The motion of $\mathbf{L}_b$ can be described geometrically by noting that

$$L_1^2 + L_2^2 + L_3^2 = L^2 \tag{4.127}$$

and also

$$\frac{L_1^2}{2I_1} + \frac{L_2^2}{2I_2} + \frac{L_2^2}{2I_3} = E \,, \tag{4.128}$$

where both the angular momentum squared L^2 and the kinetic energy E are constants (as to the kinetic energy, refer to Problem 4.10). Equation (4.128) can be written in the form

$$\frac{L_1^2}{2EI_1} + \frac{L_2^2}{2EI_2} + \frac{L_3^2}{2EI_3} = 1 \,, \tag{4.129}$$

which represents an ellipsoid with semiaxes $a = \sqrt{2EI_1}$, $b = \sqrt{2EI_2}$, $c = \sqrt{2EI_3}$. Thus, the trajectories described by the tip of the vector $\mathbf{L}_b$ are the intersection curves of the sphere (4.127) and the ellipsoid (4.129). Since

$$\frac{L^2}{I_3} = \frac{L_1^2}{I_3} + \frac{L_2^2}{I_3} + \frac{L_3^2}{I_3} < \frac{L_1^2}{I_1} + \frac{L_2^2}{I_2} + \frac{L_3^2}{I_3} < \frac{L_1^2}{I_1} + \frac{L_2^2}{I_1} + \frac{L_3^2}{I_1} = \frac{L^2}{I_1} \tag{4.130}$$

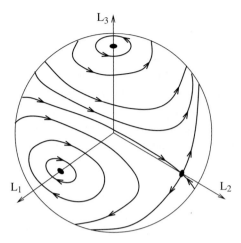

Fig. 4.9 Possible trajectories of the tip of vector $\mathbf{L}_b$.

together with (4.129) imply

$$\frac{L^2}{2EI_3} < 1 < \frac{L^2}{2EI_1} , \tag{4.131}$$

it follows that the radius L of the sphere is comprised between the smallest semiaxis $a = \sqrt{2EI_1}$ and the largest semiaxis $c = \sqrt{2EI_3}$ of the ellipsoid. Therefore, the sphere and the ellipsoid do indeed intersect.

Some trajectories of the tip of vector $\mathbf{L}_b$ on the sphere (4.127) for different values of the energy are shown in Fig. 4.9. The closed curves around the axes with smallest and largest moment of inertia reflect the stability of the rotations about those axes: small perturbations remain small. The point of the sphere pierced by the intermediate axis is a point of unstable equilibrium: there are trajectories that diverge arbitrarily far from it in case of a small perturbation. More precisely, the stable equilibrium points are $(\pm L, 0, 0)$ and $(0, 0, \pm L)$, whereas $(0, \pm L, 0)$ are unstable equilibrium points.

As seen from the external inertial frame, during a period of the motion of the tip of vector $\mathbf{L}_b$, the rigid body rotates by a certain angle about the angular momentum $\mathbf{L}$, which is a constant vector for an observer in the inertial frame. This angle is the sum of a dynamical phase and a phase of geometric origin that has become known as **Berry's phase**. In Appendix G there is a brief discussion of **geometric phases** in classical mechanics.

The Poinsot Construction

In general the angular momentum and the angular velocity are not parallel vectors. Poinsot devised an elegant geometric representation for the motion of the angular velocity vector of a freely rotating asymmetric rigid body by means of the constants of the motion

$$\mathbf{L} = \overset{\leftrightarrow}{\mathbf{I}} \cdot \boldsymbol{\omega} = \textbf{constant} \tag{4.132}$$

and

$$T = \frac{1}{2}\boldsymbol{\omega} \cdot \overset{\leftrightarrow}{\mathbf{I}} \cdot \boldsymbol{\omega} = E = \text{constant}. \tag{4.133}$$

Defining the vector

$$\mathbf{r} = \frac{\boldsymbol{\omega}}{\sqrt{2E}}, \tag{4.134}$$

equation (4.133) takes the form

$$\mathbf{r} \cdot \overset{\leftrightarrow}{\mathbf{I}} \cdot \mathbf{r} = 1. \tag{4.135}$$

Being proportional to the angular velocity, $\mathbf{r}$ is the position vector of a point on the instantaneous rotation axis from the same origin O to which the inertia tensor is referred. Equation (4.135) defines the surface of an elllipsoid because, writtten with respect to principal axes, it reduces to

$$I_1 x_1^2 + I_2 x_2^2 + I_3 x_3^2 = 1, \tag{4.136}$$

which represents an ellipsoid with semiaxes $I_1^{-1/2}$, $I_2^{-1/2}$, $I_3^{-1/2}$, known as the **inertia ellipsoid**. In an arbitrary Cartesian system Eq. (4.135) represents the same ellipsoid with its axes inclined with respect to the coordinate axes. Any vector normal to the surface of the inertia ellipsoid at the point defined by the tip of $\mathbf{r}$ is proportional to

$$\nabla(\mathbf{r} \cdot \overset{\leftrightarrow}{\mathbf{I}} \cdot \mathbf{r}) = \left(\hat{\mathbf{e}}_1 \frac{\partial}{\partial x_1} + \hat{\mathbf{e}}_2 \frac{\partial}{\partial x_2} + \hat{\mathbf{e}}_3 \frac{\partial}{\partial x_3} \right)(I_1 x_1^2 + I_2 x_2^2 + I_3 x_3^2)$$

$$= 2\,(I_1 x_1 \hat{\mathbf{e}}_1 + I_2 x_2 \hat{\mathbf{e}}_2 + I_3 x_3 \hat{\mathbf{e}}_3) = 2\overset{\leftrightarrow}{\mathbf{I}} \cdot \mathbf{r} = \sqrt{\frac{2}{E}}\,\mathbf{L}, \tag{4.137}$$

so the tangent plane to the inertia ellipsoid at the point $\mathbf{r}$ is perpendicular to $\mathbf{L}$, as illustrated in Fig. 4.10. The distance from the origin of the inertia ellipsoid to the tangent plane at $\mathbf{r}$ is

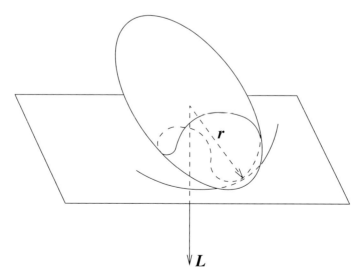

Fig. 4.10   The Poinsot construction.

$$d = \mathbf{r} \cdot \frac{\mathbf{L}}{L} = \frac{\boldsymbol{\omega} \cdot \overset{\leftrightarrow}{\mathbf{I}} \cdot \boldsymbol{\omega}}{L\sqrt{2E}} = \frac{\sqrt{2E}}{L} = \text{constant}. \qquad (4.138)$$

The tangent plane to the inertia ellipsoid is completely fixed in space because not only is the direction normal to the plane fixed, but also the distance from the plane to the origin of the coordinate axes does not change, hence it being called the **invariable plane**. The position of the invariable plane with respect to the centre of the inertia ellipsoid is determined by the initial conditions: $\mathbf{L}$ and E. Taking into consideration that the point of contact of the ellipsoid with the invariable plane belongs to the instantaneous axis of rotation ($\mathbf{r}$ is parallel to $\boldsymbol{\omega}$), the inertia ellipsoid rolls without slipping on the invariable plane. As the ellipsoid rolls without slipping, its point of contact with the invariable plane describes a curve on the ellipsoid called the *polhode*,[4] and a curve on the invariable plane called the *herpolhode*.[5] It would be somewhat of an exaggeration to say that these words add to the beauty of the language.

When the body is symmetric ($I_1 = I_2$), the inertia ellipsoid is a surface of revolution about its symmetry axis and it is not dificult to see that the polhode and the herpolhode are circles. In fact, as seen in Example 4.5, Eqs. (4.113) show that ω_3 is constant and $\boldsymbol{\omega}$ describes a cone around the symmetry axis called the **body cone**. According to Fig. 4.8, the semiangle α_b of the body cone is given by

$$\tan \alpha_b = \frac{\omega_\perp}{\omega_3} = \frac{A}{\omega_3}, \qquad (4.139)$$

where we have used Eqs. (4.117) and (4.118). On the other hand, the angle α_s between $\boldsymbol{\omega}$ and $\mathbf{L}$ is constant since

$$\cos \alpha_s = \frac{\boldsymbol{\omega} \cdot \mathbf{L}}{\omega L} = \frac{\boldsymbol{\omega} \cdot \overset{\leftrightarrow}{\mathbf{I}} \cdot \boldsymbol{\omega}}{\omega L} = \frac{2E}{\omega L}, \qquad (4.140)$$

so the instantaneous rotation axis traces out a cone in space with semiangle α_s, called the **space cone**. The vector $\boldsymbol{\omega}$ belongs simultaneously to the body cone and the space cone, so the instantaneous rotation axis is the line of contact between the two cones. Since the points of the instantaneous rotation axis remain at rest, in order to generate the motion the body cone rolls without slipping on the space cone, which is fixed because $\mathbf{L}$ is fixed in space. Thus, it is clear that the curves traced out by the instantaneous rotation axis on the inertia ellipsoid and the invariable plane are circles. In the general case of an asymmetric body, the analytic expressions for the polhode and herpolhode involve elliptic functions (Whittaker, 1944).

By way of a final remark on the symmetric case, we note that the space cone may be inside or outside the body cone, as illustrated in Fig. 4.11. Let $\boldsymbol{\omega} = \omega \hat{\mathbf{n}}$ and notice that

$$\overset{\leftrightarrow}{\mathbf{I}} = I_1 \hat{\mathbf{e}}_1 \hat{\mathbf{e}}_1 + I_1 \hat{\mathbf{e}}_2 \hat{\mathbf{e}}_2 + I_3 \hat{\mathbf{e}}_3 \hat{\mathbf{e}}_3 = I_1 \overset{\leftrightarrow}{\mathbf{1}} + (I_3 - I_1) \hat{\mathbf{e}}_3 \hat{\mathbf{e}}_3, \qquad (4.141)$$

---

[4] From the Greek *polos* (pole) and *hodos* (path).
[5] The serpentine path.

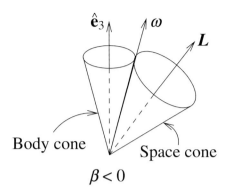
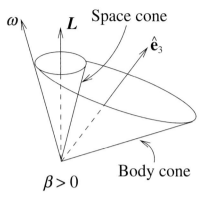

The body cone and the space cone, where $\beta = (I_3 - I_1)/I_1$.

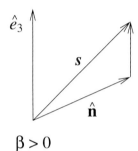
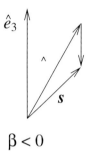

Auxiliary vectors to elucidate the relative positions of the body and space cones.

whence

$$s \equiv \frac{\mathbf{L}}{\omega I_1} = \frac{\overset{\leftrightarrow}{\mathbf{I}} \cdot \boldsymbol{\omega}}{\omega I_1} = \frac{\omega \overset{\leftrightarrow}{\mathbf{I}} \cdot \hat{\mathbf{n}}}{\omega I_1} = \hat{\mathbf{n}} + \beta \cos \alpha_b \, \hat{\mathbf{e}}_3 \,, \tag{4.142}$$

where we have taken into account that $\cos \alpha_b = \hat{\mathbf{n}} \cdot \hat{\mathbf{e}}_3$ and have defined

$$\beta = \frac{I_3 - I_1}{I_1} \,. \tag{4.143}$$

Observing that s is parallel to $\mathbf{L}$ and considering Fig. 4.12, one concludes that the space cone is inside the body cone if $\beta > 0$ and outside the body cone if $\beta < 0$. The latter case corresponds to a body elongated ("cigar-shaped") in the direction of the symmetry axis, whereas $\beta > 0$ corresponds to a body flattened in the direction of the symmetry axis.

4.9 Symmetric Top with One Point Fixed

Generically, a symmetric top is any rigid body with $I_1 = I_2$. Let us assume it is constrained to rotate about a fixed point O on the symmetry axis, located a distance l from the centre of mass in a uniform gravitational field (Fig. 4.13). The motion of the top will be described

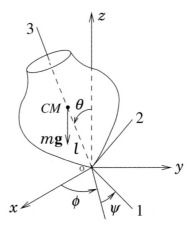

Fig. 4.13   Symmetric top with one point fixed.

by the Euler angles ϕ, θ, ψ. The angle θ measures the inclination of the symmetry axis of the top with respect to the vertical, while ϕ describes the precession of the symmetry axis around the vertical direction. Finally, ψ describes the rotation of the top about its own symmetry axis (see the description of the Euler angles in Section 3.3).

Lagrangian and Constants of the Motion

Instead of using the Euler equations (4.112), it turns out to be more advantageous to resort to the Lagrangian formalism for the analysis of the motion of the top. Since the only applied force is gravity, the Lagrangian is written

$$L = T - V = \frac{1}{2}I_1(\omega_1^2 + \omega_2^2) + \frac{1}{2}I_3\omega_3^2 - mgl\cos\theta, \tag{4.144}$$

because $I_1 = I_2$. Note that the axes x', y', z' in Fig. 3.4 are the same as the axes $1, 2, 3$ in Fig. 4.13. Thus, making use of equations (3.85), the Lagrangian takes the form

$$L = \frac{1}{2}I_1\dot{\theta}^2 + \frac{1}{2}I_1\dot{\phi}^2\sin^2\theta + \frac{1}{2}I_3(\dot{\psi} + \dot{\phi}\cos\theta)^2 - mgl\cos\theta. \tag{4.145}$$

The variables ψ and ϕ are cyclic, so the corresponding conjugate momenta are the constants of the motion p_ψ and p_ϕ:

$$\frac{\partial L}{\partial \dot{\psi}} = I_3(\dot{\psi} + \dot{\phi}\cos\theta) = I_3\omega_3 = p_\psi, \tag{4.146}$$

$$\frac{\partial L}{\partial \dot{\phi}} = I_1\dot{\phi}\sin^2\theta + I_3\cos\theta\,(\dot{\psi} + \dot{\phi}\cos\theta) = I_1\dot{\phi}\sin^2\theta + p_\psi\cos\theta = p_\phi. \tag{4.147}$$

Exercise 4.9.1   Show that the gravitational torque about the fixed point O has no component along the direction of the 3-axis. Next, use the Euler equations (4.112) to derive (4.146) with $p_\psi = $ constant.

Since a variation of the angle ϕ implies a rotation about the vertical z-axis, the constant p_ϕ is the z-component of the angular momentum, L_z. The conservation of L_z, in its turn, is easily understood by noting that the gravitational torque about the fixed point O has no vertical component. Another constant of the motion of crucial importance is the energy:

$$E = T + V = \frac{1}{2}I_1\dot{\theta}^2 + \frac{1}{2}I_1\dot{\phi}^2\sin^2\theta + \frac{1}{2}I_3(\dot{\psi} + \dot{\phi}\cos\theta)^2 + mgl\cos\theta. \quad (4.148)$$

Solution by Quadratures

Solving a system of differential equation by quadratures means expressing the solution in terms of integrals[6] of known functions and, if needed, performing additional "algebraic" operations, including the inversion of functions.

The constants of the motion (4.146) to (4.148) allow us to reduce the solution for the motion of the top to quadratures. Indeed, writing

$$\dot{\psi} + \dot{\phi}\cos\theta = p_\psi/I_3, \quad (4.149)$$

$$\dot{\phi} = \frac{p_\phi - p_\psi\cos\theta}{I_1\sin^2\theta} \quad (4.150)$$

and introducing the new constant

$$E' = E - \frac{p_\psi^2}{2I_3}, \quad (4.151)$$

we derive

$$E' = \frac{1}{2}I_1\dot{\theta}^2 + V_{\mathrm{eff}}(\theta), \quad (4.152)$$

where

$$V_{\mathrm{eff}}(\theta) = \frac{(p_\phi - p_\psi\cos\theta)^2}{2I_1\sin^2\theta} + mgl\cos\theta. \quad (4.153)$$

Separating variables in (4.152) and integrating we obtain

$$\int_{\theta_0}^{\theta} \frac{d\theta}{[E' - V_{\mathrm{eff}}(\theta)]^{1/2}} = \sqrt{\frac{2}{I_1}}\, t, \quad (4.154)$$

where θ_0 is the value of θ at the initial instant $t = 0$.[7] In principle, therefore, a single quadrature (integration) determines t as a function of θ, an additional inversion yielding $\theta(t)$. Once θ as a function of time has been found, one finds $\phi(t)$ by an integration of Eq. (4.150). Finally, with $\theta(t)$ and $\phi(t)$ at one's disposal, one obtains $\psi(t)$ by a last integration using Eq. (4.149). The integral in (4.154), however, cannot be expressed in terms of elementary functions, save for very particular initial conditions. So, in general, θ is expressed as a function of t in terms of the Weierstrass or Jacobi elliptic functions.[8] It is

---

[6] Called "quadratures" in Newton's time.
[7] If $\dot{\theta} < 0$ at $t = 0$ one must take the negative square root on the left-hand side of (4.154).
[8] In Section 5.1, in the context of solving for the motion of the quartic oscillator, a brief introduction is given to the Jacobi elliptic functions.

remarkable that the main features of the motion of the top can be understood without the need to deal with elliptic functions.

Effective Potential

The behaviour of $\theta(t)$ can be studied by the effective potential method with the help of Eq. (4.152). Let us consider a general situation such that $|p_\psi| \neq |p_\phi|$ (an example of the special case $p_\psi = p_\phi$ will be treated later). The physically acceptable interval for θ is $[0, \pi]$, and from (4.153) one deduces that $V_{\text{eff}}(\theta) \longrightarrow \infty$ as $\theta \longrightarrow 0$ or π. On the other hand, the derivative of the effective potential vanishes for θ_0 such that

$$V'_{\text{eff}}(\theta_0) = -mgl \sin\theta_0 + \frac{(p_\phi - p_\psi \cos\theta_0)(p_\psi - p_\phi \cos\theta_0)}{I_1 \sin^3\theta_0} = 0 \,. \qquad (4.155)$$

Putting $u = \cos\theta_0$, this last equation is equivalent to

$$mglI_1(1 - u^2)^2 = (p_\phi - p_\psi u)(p_\psi - p_\phi u) \,. \qquad (4.156)$$

Denoting the left-hand side of (4.156) by $f(u)$ and the right-hand side by $g(u)$, we can graphically find the number of intersections of the curves defined by $f(u)$ and $g(u)$ for $-1 \leq u \leq 1$. The roots of $f(u)$ are ± 1, the tangent to the graph of $f(u)$ is horizontal at $u = 0, \pm 1$ and $f(u) \longrightarrow \infty$ as $u \longrightarrow \pm\infty$. As to $g(u)$, its roots are $u_1 = p_\phi/p_\psi$ and $1/u_1$, so if u_1 belongs to the interval $(-1, 1)$ then $1/u_1$ does not belong to $(-1, 1)$, and vice versa. In other words, $g(u)$ has only one root in the physically acceptable interval. These considerations suffice to draw Fig. 4.14, which makes it clear that Eq. (4.156) possesses a single solution u_0 in the physically significant range, or, equivalently, there is only one $\theta_0 \in (0, \pi)$ for which $V'_{\text{eff}}(\theta_0) = 0$. Thus, we infer that the shape of the graph of $V_{\text{eff}}(\theta)$ is the one sketched in Fig. 4.15.

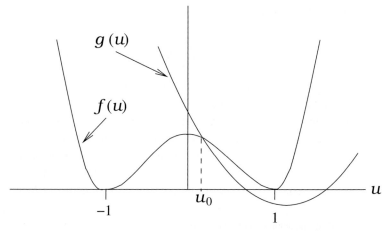

Fig. 4.14 The intersection of the curves defined in the main text by $f(u)$ and $g(u)$ gives the single value $\theta_0 = \cos^{-1} u_0$, for which $V'_{\text{eff}}(\theta)$ vanishes in the physically acceptable interval. The case shown corresponds to $p_\psi p_\phi > 0$.

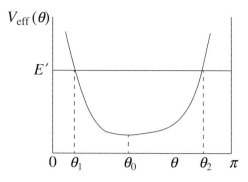

The effective potential $V_{\text{eff}}(\theta)$.

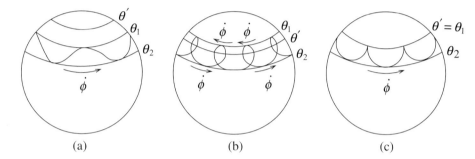

The possible combinations of nutation and precession.

General Features of the Motion

In general, for any given E' the angle θ periodically oscillates between the values θ_1 and θ_2 shown in Fig. 4.15. The symmetry axis of the top is said to perform a **nutation** about the vertical. During the nutation the instantaneous angular velocity associated with the **precession** of the symmetry axis of the top around the vertical is

$$\dot{\phi} = \frac{p_\phi - p_\psi \cos\theta}{I_1 \sin^2\theta} = p_\psi \frac{\frac{p_\phi}{p_\psi} - \cos\theta}{I_1 \sin^2\theta} . \tag{4.157}$$

There are several possibilities, to wit:

(a) $|p_\phi| > |p_\psi|$. The precession rate $\dot{\phi}$ has the same sign all the time.

(b) $|p_\phi| < |p_\psi|$. Introducing the ancillary angle $\theta' \in (0, \pi)$ defined by $\cos\theta' = \frac{p_\phi}{p_\psi}$, we have $\dot{\phi} = p_\psi(\cos\theta' - \cos\theta)/I_1 \sin^2\theta$. In order to determine the position of θ', note that Eq. (4.155) gives $V'_{\text{eff}}(\theta') = -mgl \sin\theta' < 0$, and it follows that $\theta' < \theta_0$. If $\theta' < \theta_1$ then $\dot{\phi}$ has the same sign during the entire motion. In this case, as well as in the previous one, the symmetry axis of the top traces out on the unit sphere with centre at O a curve similar to the one shown in Fig. 4.16(a). If $\theta' > \theta_1$, the precession rate $\dot{\phi}$ changes sign periodically during the nutation and the motion resembles the one depicted in Fig. 4.16(b). The situation portrayed in Fig. 4.16(c) corresponds to $\theta' = \theta_1$. The last case is not so rare as it seems. Suppose, for instance, that the initial conditions are $\dot{\theta} = 0$, $\dot{\phi} = 0$, $\dot{\psi} = \omega_3$, so that the only initial motion of the top is a spin about its own symmetry axis. With these initial conditions one has

$$p_\psi = I_3\omega_3 , \qquad p_\phi = p_\psi \cos\theta_1 , \tag{4.158}$$

whence

$$\cos\theta' = \frac{p_\phi}{p_\psi} = \cos\theta_1 \implies \theta' = \theta_1 . \tag{4.159}$$

If $E' = V_{\text{eff}}(\theta_0)$ the angle θ stays constant and equal to θ_0, there is no nutation and the symmetry axis of the top executes a **regular precession** around the vertical with the constant angular velocity

$$\dot\phi = \frac{p_\phi - p_\psi \cos\theta_0}{I_1 \sin^2\theta_0} . \tag{4.160}$$

There are, in general, two possible values for $\dot\phi$ because Eq. (4.155) is quadratic in

$$\sigma = p_\phi - p_\psi \cos\theta_0 . \tag{4.161}$$

Indeed,

$$p_\psi - p_\phi \cos\theta_0 = p_\psi \sin^2\theta_0 - \sigma\cos\theta_0 , \tag{4.162}$$

so (4.155) becomes

$$\sigma^2 \cos\theta_0 - \sigma p_\psi \sin^2\theta_0 + mglI_1 \sin^4\theta_0 = 0 , \tag{4.163}$$

whose solutions are

$$\sigma = \frac{I_3\omega_3 \sin^2\theta_0}{2\cos\theta_0} \left[1 \pm \sqrt{1 - \frac{4mglI_1 \cos\theta_0}{I_3^2\omega_3^2}} \right] , \tag{4.164}$$

where we have used $p_\psi = I_3\omega_3$. If $\theta_0 < \pi/2$, so that $\cos\theta_0 > 0$, the spin angular velocity ω_3 cannot be smaller than a minimum value[9] in order that the top can regularly precess with inclination angle θ_0:

$$\omega_{\min} = \left(\frac{4mglI_1 \cos\theta_0}{I_3^2} \right)^{1/2} . \tag{4.165}$$

If $\omega_3 > \omega_{\min}$ there are two solutions for $\sigma = p_\phi - p_\psi \cos\theta_0$, corresponding to a slow and a fast precession. For $\omega_3 \gg \omega_{\min}$ we have the following approximate values:

$$\dot\phi_{\text{fast}} \approx \frac{I_3\omega_3}{I_1 \cos\theta_0} , \qquad \dot\phi_{\text{slow}} \approx \frac{mgl}{I_3\omega_3} . \tag{4.166}$$

Exercise 4.9.2   Derive these two last results applying, if necessary, a binomial expansion to Eq. (4.164).

---

[9] In the case of a giroscope suspended from the top, $\theta_0 > \pi/2$ and the regular precession is possible whatever the value of ω_3.

Stability of a Vertical Top

The last important problem to be investigated is that of the stability of a top set to spin in the upright position, also known as the sleeping top.[10] The initial conditions $\theta(0) = 0$, $\dot{\psi}(0) = \omega_3$ imply, by Eq. (4.147), $p_\phi = p_\psi$. In such circumstances, the effective potential becomes

$$V_{\text{eff}}(\theta) = \frac{p_\psi^2}{2I_1} \frac{(1 - \cos\theta)^2}{\sin^2\theta} + mgl\cos\theta = \frac{p_\psi^2}{2I_1}\tan^2\frac{\theta}{2} + mgl\cos\theta \,. \tag{4.167}$$

For small nutations relative to the vertical,

$$V_{\text{eff}}(\theta) \approx \frac{p_\psi^2}{2I_1}\left(\frac{\theta}{2}\right)^2 + mgl\left(1 - \frac{\theta^2}{2}\right) = mgl + \frac{1}{2}k\theta^2, \tag{4.168}$$

where

$$k = \frac{I_3^2\omega_3^2}{4I_1} - mgl\,. \tag{4.169}$$

For small θ the graph of $V_{\text{eff}}(\theta)$ is a parabola with a minimum at $\theta = 0$, if $k > 0$, and a maximum at $\theta = 0$ if $k < 0$. As a consequence, the motion of the upright top will be stable if $k > 0$, that is, if

$$\omega_3 > \omega_c = \left(\frac{4mglI_1}{I_3^2}\right)^{1/2}\,. \tag{4.170}$$

In other words, the spin angular velocity has a critical value ω_c below which the rotational motion of the upright top becomes unstable. If $\omega_3 > \omega_c$ small oscillations of the symmetry axis of the top about the vertical direction stay small, the symmetry axis of the top remains nearly vertical and the top spins serenely, as if asleep. In practice, if a top is set spinning rapidly and almost upright, it remains "asleep" until, owing to friction, ω_3 becomes smaller than ω_c, when the top "wakes up" and starts to wobble, its symmetry axis moving farther and farther away from the vertical until it finally falls. Anyone who has ever played with tops has certainly observed this behaviour.

Problems

4.1 The cube in Example 4.6 is set spinning about the side lying on the z-axis. Find the angular momentum of the cube and the angle it makes with the angular velocity.

4.2 Use Eq. (4.61) to determine the inertia matrix of the cube in Example 4.6 with respect to axes through the centre of mass and perpendicular to the faces of the cube. Prove that any three perpendicular axes through the centre of mass are principal axes. Conclude that, as far as rotations about the centre of mass are concerned, a homogeneous cube is indistinguishable from a homogeneous sphere.

---

[10] The reader is referred to Whittaker (1944) for other aspects of the motion of the top as well as for the exact solution in terms of elliptic functions.

4.3 Define the vector product from the right of a vector $\mathbf{A}$ by a dyad. By linearity, using either (4.19) or (4.20), extend the definition to a general dyadic $\overset{\leftrightarrow}{\mathbf{T}}$. Show that the formula $(\overset{\leftrightarrow}{\mathbf{T}} \times \mathbf{A}) \cdot \mathbf{B} = \overset{\leftrightarrow}{\mathbf{T}} \cdot (\mathbf{A} \times \mathbf{B})$ holds where $\mathbf{B}$ is an arbitrary vector. This formula may be used to define $\overset{\leftrightarrow}{\mathbf{T}} \times \mathbf{A}$ by its action on any vector $\mathbf{B}$.

4.4 Suppose the principal moments of inertia I_1, I_2, I_3 of a rigid body with respect to the centre of mass are all distinct. If now the inertia tensor is referred to a point O displaced by $\mathbf{s}$ from the centre of mass, show that only if $\mathbf{s}$ is along one of the axes $1, 2, 3$ will the principal axes with origin at O be parallel to the axes $1, 2, 3$.

4.5 Show that each moment of inertia is not larger than the sum of the other two.

4.6 The centrifugal acceleration $\mathbf{a}_c = \boldsymbol{\omega} \times (\mathbf{r} \times \boldsymbol{\omega})$ depends linearly on $\mathbf{r}$, so it can be written in the form $\mathbf{a}_c = \overset{\leftrightarrow}{\mathbf{T}}_c \cdot \mathbf{r}$. Express the tensor $\overset{\leftrightarrow}{\mathbf{T}}_c$ in terms of $\boldsymbol{\omega}$ and obtain the matrix of $\overset{\leftrightarrow}{\mathbf{T}}_c$ in an arbitrary Cartesian system.

4.7 Find the principal axes and principal moments of inertia with respect to the vertex of a homogeneous cone with height h and radius of the base R. Given that the centre of mass is located at a distance $3h/4$ from the vertex, obtain the principal axes and principal moments of inertia with respect to the centre of mass.

4.8 (a) Given an arbitrary tensor $\overset{\leftrightarrow}{\mathbf{T}}$, prove that

$$\left(\frac{d\overset{\leftrightarrow}{\mathbf{T}}}{dt}\right)_{inertial} = \left(\frac{d\overset{\leftrightarrow}{\mathbf{T}}}{dt}\right)_{body} + \boldsymbol{\omega} \times \overset{\leftrightarrow}{\mathbf{T}} - \overset{\leftrightarrow}{\mathbf{T}} \times \boldsymbol{\omega}.$$

(b) Using $\mathbf{L} = \overset{\leftrightarrow}{\mathbf{I}} \cdot \boldsymbol{\omega}$, $(d\mathbf{L}/dt)_{inertial} = \mathbf{N}$ and the result in part (a), derive Euler's equations of motion for a rigid body.

4.9 A symmetric spaceship moves in outer space. Symmetrically placed engines apply a constant torque N_3 along the symmetry axis. (a) Assuming that $\omega_3(0) = 0$, find $\omega_3(t)$. (b) Prove that $\omega_1^2 + \omega_2^2$ is a constant of the motion. (c) Given the initial conditions $\omega_1(0) = 0$ and $\omega_2(0) = \Omega$, determine $\omega_1(t)$ and $\omega_2(t)$. Describe the motion executed by the angular velocity vector relative to the principal axes of inertia.

4.10 Show that the kinetic energy relative to the centre of mass or to a fixed point O obeys the equation $dT/dt = \mathbf{N} \cdot \boldsymbol{\omega}$, known as the *kinetic energy theorem*. It follows that the kinetic energy is constant if the torque is zero.

4.11 A homogeneous rod of length $2l$ and mass m is connected to the free end of a light spring with force constant k which hangs from a fixed support. The rod can oscillate in a vertical plane but the string is constrained to move vertically only. Write down a Lagrangian and the equations of motion for this system.

4.12 A homogeneous door of mass m, width l, height h and negligible thickness is closing with a constant angular velocity ω (see Fig. 4.17). (a) Choosing the rotation axis as the z-axis with origin at point P, find the components of the angular momentum along the axes x_1, x_2, x_3 attached to the door. Hint: determine the components of the inertia tensor referred to point P with respect to the Cartesian axes attached to the door. (b) Find the torque about point P. Hint: transform the components of $\mathbf{L}$ to axes x, y, z fixed in space and use $\mathbf{N} = d\mathbf{L}/dt$.

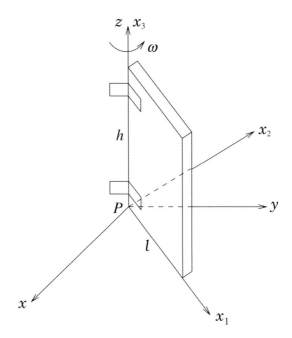

Fig. 4.17 Problem 4.12.

4.13 An automobile starts from rest with constant acceleration a and one of its doors completely open (making $90°$ with the body of the vehicle). If the door is a homogeneous rectangle of width l and height h, prove that the time it takes until the door closes is

$$t = \sqrt{\frac{l}{3a}} \int_0^{\pi/2} \frac{d\theta}{\sqrt{\sin\theta}} .$$

If $l = 1,2m$ and $a = 0,3m/s^2$, show that $t \approx 3s$. Hints: (a) in the automobile reference frame the problem is formally identical to that of a door rotating about a horizontal axis in a uniform gravitational field; (b) compute the integral numerically or look it up in a table.

4.14 A symmetric top with one point fixed is set in motion as follows:

$$\theta = 60^o , \quad \dot\theta = 0 , \quad \dot\phi = 2\left(\frac{mgl}{3I_1}\right)^{1/2} , \quad \dot\psi = (3I_1 - I_3)\left(\frac{mgl}{3I_1 I_3^2}\right)^{1/2} .$$

(a) Find the values of p_ϕ and p_ψ, and determine $V_{\text{eff}}(\theta)$. Sketch the graph of $V_{\text{eff}}(\theta)$ for $0 \le \theta \le \pi$ and make a qualitative analysis of the nutation motion. (b) Show that Eq. (4.152) for θ can be put in the form

$$\dot{u}^2 = \frac{mgl}{I_1}(1 - u)^2(2u - 1),$$

where $u = \cos\theta$. Prove that the solution for $\theta(t)$ is

$$\sec\theta = 1 + \text{sech}\left[\left(\frac{mgl}{I_1}\right)^{1/2} t\right].$$

Is the behaviour of θ as a function of t in agreement with the previous qualitative analysis? Study also the behaviour of $\dot\phi$ and $\dot\psi$.

4.15 Construct, in terms of the Euler angles, the Lagrangian for a symmetric top constrained to move in a uniform gravitational field with its bottom tip always in contact with a smooth horizontal plane. Explain your choice of axes fixed in the body. Identify all cyclic coordinates and the corresponding constants of the motion. Describe how the solution of the equations of motion for the Euler angles can be reduced to quadratures.

5

Small Oscillations

> Time travels in divers paces with divers persons. I'll tell you who Time ambles withal,
> who Time trots withal, who Time gallops withal, and who he stands still withal.
>
> William Shakespeare, *As You Like It*, Act 3, Scene 2

The theory of small oscillations about stable equilibrium positions finds application in molecular spectroscopy, in structural dynamics, in acoustics and in the analysis of coupled electric circuits, among other areas. The concept of normal mode of vibration, which plays a crucial role in the theory, is of enormous importance in almost all branches of physics, from quantum field theory and optics to condensed matter physics and the theory of earthquakes. If the deviations from the equilibrium position are sufficiently small, the general motion can be described as that of a collection of coupled harmonic oscillators.

5.1 One-Dimensional Case

Let us consider a conservative system with a single degree of freedom described by the generalised coordinate q. Without significant loss of generality, let us assume that the relation between the Cartesian coordinates and the generalised coordinate q does not involve the time, so that the Lagrangian has the form[1]

$$L = \frac{1}{2}\alpha(q)\,\dot{q}^2 - V(q), \qquad (5.1)$$

with $\alpha(q)$ a positive function.

Example 5.1 A bead of mass m slides without friction along a rigid wire in a vertical plane. The wire is bent in the form of a parabola with equation $y = x^2/R$. The system is in a uniform gravitational field $\mathbf{g} = -g\hat{\mathbf{y}}$. Using $q = x$ as generalised coordinate we have

$$L = \frac{m}{2}(\dot{x}^2 + \dot{y}^2) - mgy = \frac{m}{2}\left(1 + \frac{4x^2}{R^2}\right)\dot{x}^2 - \frac{mg}{R}x^2, \qquad (5.2)$$

which is of the form (5.1) with $\alpha(x) = m(1 + 4x^2/R^2)$.

---

[1] Consider the one-dimensional version of Eqs. (2.151)–(2.154).

Equilibrium Configurations

If

$$\left(\frac{dV}{dq}\right)_{q=q^{(0)}} = 0 \tag{5.3}$$

then $q = q^{(0)}$ is said to be an **equilibrium configuration**. In fact, the equation of motion generated by the Lagrangian (5.1) is

$$\alpha(q)\ddot{q} + \frac{\alpha'(q)}{2}\dot{q}^2 + \frac{dV}{dq} = 0, \tag{5.4}$$

which possesses the solution $q = q^{(0)} = $ constant if (5.3) holds. In other words, if at a given instant the system is in the configuration $q = q^{(0)}$ with velocity $\dot{q} = 0$, it will remain in this state forever. For example, a pendulum initially at rest in the equilibrium position ($\theta = 0$) will so remain indefinitely.

Stability Criterion

The equilibrium is said to be **stable** if a sufficiently small perturbation does not drive the system very far away from the equilibrium, but gives rise to small oscillations about the equilibrium configuration. This occurs if the potential has a minimum at the equilibrium point. If V has a maximum or a point of inflexion, the equilibrium is unstable. As is clear from Fig. 5.1, if a small velocity is imparted to the particle in the equilibrium configuration, slightly increasing its energy, the displacement will remain small if the potential V has a minimum, but will not be small and may grow indefinitely if V has a maximum or a point of inflexion. A stability criterion is

$$\left(\frac{d^2V}{dq^2}\right)_{q=q^{(0)}} > 0, \tag{5.5}$$

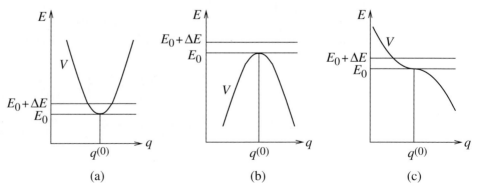

Fig. 5.1 In (a) the equilibrium is stable, but in (b) and (c) it is unstable.

It turns out that the expansion of V about $x = 0$ is dominated by the fourth-order term. Indeed, using the binomial series expansion

$$\left(1 + \frac{x^2}{l^2}\right)^{1/2} = 1 + \frac{x^2}{2l^2} + O(x^4), \qquad |x| < l, \tag{5.21}$$

the potential energy

$$V(x) = \frac{k}{2}\left(\sqrt{l^2 + x^2} - l\right)^2 = \frac{kl^2}{2}\left(\sqrt{1 + \frac{x^2}{l^2}} - 1\right)^2 \tag{5.22}$$

reduces to

$$V(x) = \frac{k}{8\,l^2}x^4 + O(x^6). \tag{5.23}$$

The potential energy has a unique minimum at $x = 0$ but, obviously, $V''(0) = 0$, so the small oscillations about the equilibrium position are not simple harmonic. The oscillations will be small if $|x| \ll l$.

Thus, for small oscillations about the stable equilibrium point, the Lagrangian (5.14) takes the form of the Lagrangian for a **quartic oscillator** (Lemos, 2017a)

$$L = \frac{m}{2}\dot{x}^2 - \frac{\lambda}{2}x^4, \tag{5.24}$$

where

$$\lambda = \frac{1}{12}\frac{d^4V}{dx^4}\bigg|_{x=0} = \frac{k}{4l^2} > 0. \tag{5.25}$$

The graph of $V(x) = \lambda x^4/2$ is qualitatively the same as that of the harmonic oscillator potential energy. So for a given energy $E > 0$ the particle oscillates periodically between the extreme points $x = -A$ and $x = A$. The problem of finding $x(t)$ is reduced to a quadrature by means of energy conservation:

$$\frac{m}{2}\dot{x}^2 + \frac{\lambda}{2}x^4 = E. \tag{5.26}$$

The amplitude of the oscillation is found by putting $x = A$ and $\dot{x} = 0$ – that is,

$$A = \left(\frac{2E}{\lambda}\right)^{1/4}. \tag{5.27}$$

In terms of the amplitude, Eq. (5.26) takes the following form:

$$\dot{x}^2 = \frac{\lambda}{m}(A^4 - x^4). \tag{5.28}$$

With the new dimensionless variables y and τ defined by

$$y = \frac{x}{A}, \qquad \tau = \alpha t, \qquad \alpha = \left(\frac{2\lambda E}{m^2}\right)^{1/4}, \tag{5.29}$$

Eq. (5.28) becomes

$$y'^2 = 1 - y^4 = (1 - y^2)(1 + y^2), \qquad y' = \frac{dy}{d\tau}. \tag{5.30}$$

With the initial conditions $x(0) = 0$ and $\dot{x}(0) > 0$, the solution to this last equation is

$$\tau = \int_0^y \frac{d\xi}{\sqrt{(1 - \xi^2)(1 + \xi^2)}}. \tag{5.31}$$

It is worth remarking that the chosen initial conditions do not imply any loss of generality, since they result from a mere choice of the origin of time.

The integral (5.31) cannot be expressed in terms of elementary functions. In order to put it in the standard form of an elliptic integral, it is necessary to make a change of integration variable that is conveniently expressed in the form of two successive changes of variables suggested by the form of the integrand at each stage. Let us start with the change of variable

$$\xi = \tan \psi, \qquad y = \tan \psi_*, \tag{5.32}$$

which leads to

$$\tau = \int_0^{\psi_*} \frac{\sec^2 \psi \, d\psi}{\sqrt{1 - \tan^2 \psi \, \sec \psi}} = \int_0^{\psi_*} \frac{d\psi}{\sqrt{\cos^2 \psi - \sin^2 \psi}} = \int_0^{\psi_*} \frac{d\psi}{\sqrt{1 - 2\sin^2 \psi}}. \tag{5.33}$$

Next, putting

$$\sqrt{2} \sin \psi = \sin \theta, \qquad \sqrt{2} \sin \psi_* = \sin \theta_*, \tag{5.34}$$

we get

$$\tau = \int_0^{\theta_*} \frac{\cos \theta \, d\theta}{\sqrt{2} \sqrt{1 - \frac{1}{2}\sin^2 \theta} \, \cos \theta} = \frac{1}{\sqrt{2}} \int_0^{\theta_*} \frac{d\theta}{\sqrt{1 - \frac{1}{2}\sin^2 \theta}}. \tag{5.35}$$

The **elliptic integral of the first kind** is defined by (Spiegel, 1963)

$$u = F(\kappa, \phi) = \int_0^\phi \frac{d\theta}{\sqrt{1 - \kappa^2 \sin^2 \theta}}, \qquad 0 < \kappa < 1. \tag{5.36}$$

Since

$$\frac{du}{d\phi} = (1 - \kappa^2 \sin^2 \phi)^{-1/2} > 0, \tag{5.37}$$

for each fixed κ there exists the inverse function

$$\phi = \mathrm{am}(u) \tag{5.38}$$

called the **amplitude function**, κ being its **modulus**. The Jacobi **elliptic sine** function of modulus κ is defined by

$$\mathrm{sn}\, u = \sin(\mathrm{am}(u)). \tag{5.39}$$

Similarly, the **elliptic cosine** is defined by

$$\operatorname{cn} u = \cos\left(\operatorname{am}(u)\right) . \tag{5.40}$$

Other Jacobi elliptic functions are

$$\operatorname{dn} u = \sqrt{1 - \kappa^2 \operatorname{sn}^2 u}, \qquad \operatorname{tn} u = \frac{\operatorname{sn} u}{\operatorname{cn} u} . \tag{5.41}$$

Strictly, we should write $\operatorname{sn}(u,\kappa)$, $\operatorname{cn}(u,\kappa)$, $\operatorname{dn}(u,\kappa)$ and $\operatorname{tn}(u,\kappa)$, but it is customary to leave the modulus implicit.

Therefore, with modulus $\kappa = 1/\sqrt{2}$, Eq. (5.35) gives

$$\theta_* = \operatorname{am}\left(\sqrt{2}\,\tau\right). \tag{5.42}$$

Consequently, by (5.34),

$$\sin\psi_* = \frac{1}{\sqrt{2}} \sin\left(\operatorname{am}\left(\sqrt{2}\,\tau\right)\right) = \frac{\operatorname{sn}\left(\sqrt{2}\,\tau\right)}{\sqrt{2}} . \tag{5.43}$$

Finally, (5.32) yields

$$y = \tan\psi_* = \frac{\sin\psi_*}{\cos\psi_*} = \frac{\sin\psi_*}{\sqrt{1 - \sin^2\psi_*}} = \frac{1}{\sqrt{2}} \frac{\operatorname{sn}\left(\sqrt{2}\,\tau\right)}{\sqrt{1 - \frac{1}{2}\operatorname{sn}^2(\sqrt{2}\,\tau)}} . \tag{5.44}$$

Taking account of (5.41), we can also write

$$y(\tau) = \frac{1}{\sqrt{2}} \frac{\operatorname{sn}\left(\sqrt{2}\,\tau\right)}{\operatorname{dn}\left(\sqrt{2}\,\tau\right)} . \tag{5.45}$$

Exercise 5.2.1 Show that

$$\operatorname{sn}' u = \operatorname{cn} u \operatorname{dn} u, \qquad \operatorname{dn}' u = -\kappa^2 \operatorname{sn} u \operatorname{cn} u . \tag{5.46}$$

From these results, prove that the function $y(\tau)$ given by Eq. (5.45) satisfies the differential Eq. (5.30). Show also that the function $y = \operatorname{sn} u$ obeys the differential equation $y'^2 = (1 - y^2)(1 - k^2 y^2)$.

Expressing (5.45) in terms of the original variables we have, at last,

$$x(t) = \frac{A}{\sqrt{2}} \frac{\operatorname{sn}\omega_0 t}{\operatorname{dn}\omega_0 t}, \qquad \omega_0 = \left(\frac{8\lambda E}{m^2}\right)^{1/4} . \tag{5.47}$$

A graphic representation of this solution to the equation of motion for the quartic oscillator is given in Fig. 5.5. Except for a numerical factor, expression (5.47) for ω_0 can be obtained by simple dimensional analysis (see Exercise 5.2.2).

Exercise 5.2.2 Using dimensional analysis, show that ω_0 is proportional to the fourth root of the energy.

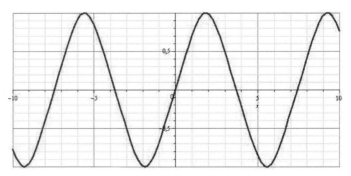

Fig. 5.5 Graph of the function (5.47) in which the elliptic functions have modulus $\kappa = 1/\sqrt{2}$. The values of $\omega_0 t$ are represented on the horizontal axis, while those of x/A are represented on the vertical axis. The period is given by Eq. (5.51).

The oscillation period T is four times the time spent by the particle to travel from $x = 0$ to $x = A$. According to Eq. (5.31),

$$\alpha T = 4 \int_0^1 \frac{d\xi}{\sqrt{(1 - \xi^2)(1 + \xi^2)}} . \qquad (5.48)$$

This integral is a particular case of the integral (5.31) with $y = 1$, and the changes of variables that led to (5.35) are still valid with $\theta_* = \pi/2$. Therefore,

$$T = \frac{4}{\omega_0} \int_0^{\pi/2} \frac{d\theta}{\sqrt{1 - \frac{1}{2} \sin^2 \theta}} . \qquad (5.49)$$

Since the **complete elliptic integral of the first kind** is defined by

$$K(\kappa) = \int_0^{\pi/2} \frac{d\theta}{\sqrt{1 - \kappa^2 \sin^2 \theta}} , \qquad (5.50)$$

it follows that

$$T = \frac{4}{\omega_0} K\left(\frac{1}{\sqrt{2}}\right) \approx \frac{7.416}{\omega_0} . \qquad (5.51)$$

Note that the period is a decreasing function of the amplitude (and of the energy). The motion is periodic but anharmonic since a simple harmonic motion is distinguished by a period independent of the amplitude. It is worth noting that the angular frequency ω of the quartic oscillator is not ω_0, but

$$\omega = \frac{2\pi}{T} = \frac{\pi \omega_0}{2K(1/\sqrt{2})} . \qquad (5.52)$$

Formula (5.51) for the period may be directly inferred from definitions (5.39)–(5.41) of the Jacobi elliptic functions. If $\phi = \operatorname{am} u$ is increased by 2π, the values of $\operatorname{sn} u$ and $\operatorname{cn} u$ do not change. Therefore, $\operatorname{sn} u$ and $\operatorname{cn} u$ are periodic with period U given by $2\pi = \operatorname{am}(U) -$ that is,

$$U = \int_0^{2\pi} \frac{d\theta}{\sqrt{1 - \kappa^2 \sin^2 \theta}} = 4 \int_0^{\pi/2} \frac{d\theta}{\sqrt{1 - \kappa^2 \sin^2 \theta}} = 4K(\kappa). \qquad (5.53)$$

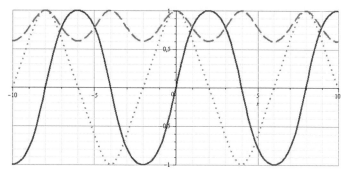

Fig. 5.6 Graphs of the elliptic functions sn (full line), cn (dotted line) and dn (dashed line) with modulus $\kappa = 0.8$.

Thus, with the modulus κ understood, we have

$$\operatorname{sn}(u + 4K) = \operatorname{sn} u, \quad \operatorname{cn}(u + 4K) = \operatorname{cn} u, \tag{5.54}$$

$$\operatorname{dn}(u + 4K) = \operatorname{dn} u, \quad \operatorname{tn}(u + 4K) = \operatorname{tn} u. \tag{5.55}$$

The elliptic sine and the elliptic cosine oscillate periodically between -1 and 1 and somewhat resemble their trigonometric counterparts: $\operatorname{sn}(-u) = \operatorname{sn} u$; $\operatorname{cn}(-u) = \operatorname{cn} u$; $\operatorname{sn}(0) = 0$; $\operatorname{cn}(0) = 1$; $\operatorname{sn}^2 u + \operatorname{cn}^2 u = 1$. The function $\operatorname{dn} u$ is always positive and oscillates between $\sqrt{1 - \kappa^2}$ and 1 (see Fig. 5.6).

The elliptic integrals and the Jacobi elliptic functions are implemented in algebraic, numerical and graphical computing softwares such as Mathematica and Maple (the latter was used to make Figs. 5.5 and 5.6). Graphs of the Jacobi elliptic functions are easily found for free on the Internet. For example, at http://keisan.casio.com/has10/SpecExec.cgi?id=system/2006/1180573437, it is possible to construct graphs with free choice of the modulus and the interval of interest.

It can be shown (Whittaker & Watson, 1927; Lawden, 1989; Armitage & Eberlein, 2006) that the Jacobi elliptic functions possess a second imaginary period: $\operatorname{sn} u$ has period $2iK'$ and $\operatorname{cn} u$ has period $4iK'$ where

$$K' = \int_0^{\pi/2} \frac{d\theta}{\sqrt{1 - \kappa'^2 \sin^2 \theta}}, \qquad \kappa' = \sqrt{1 - \kappa^2}, \tag{5.56}$$

that is,

$$\operatorname{sn}(u + 2iK') = \operatorname{sn} u, \quad \operatorname{cn}(u + 4iK') = \operatorname{cn} u. \tag{5.57}$$

The number $\kappa' = \sqrt{1 - \kappa^2}$ is said to be the **complementary modulus** of the elliptic functions of modulus κ. Therefore, the Jacobi elliptic functions belong to the class of doubly periodic functions (Whittaker & Watson, 1927). The double periodicity of the function sn can be inferred from dynamical considerations (Whittaker, 1944, §44).

Historical note   For 40 years Adrien Marie Legendre (1752–1833) published articles and books on elliptic integrals, so named because they appear in the problem of calculating the arc length of the ellipse (Bottazzini & Gray, 2013). Legendre showed that any integral of the form $\int R(x, w)\, dx$ where $R(x, w)$ is a rational function[3] of x and w, with w the square root of a polinomial of the third or fourth degree in x, can be expressed as a combination of the **elliptic integral of the first kind**

$$F(\kappa, \phi) = \int_0^\phi \frac{d\theta}{\sqrt{1 - \kappa^2 \sin^2 \theta}} , \qquad (5.58)$$

the **elliptic integral of the second kind**

$$E(\kappa, \phi) = \int_0^\phi \sqrt{1 - \kappa^2 \sin^2 \theta}\, d\theta \qquad (5.59)$$

and the **elliptic integral of the third kind**

$$\Pi(\kappa, n, \phi) = \int_0^\phi \frac{d\theta}{(1 + n \sin^2 \theta)\sqrt{1 - \kappa^2 \sin^2 \theta}} . \qquad (5.60)$$

If $\phi = \pi/2$ the elliptic integrals are said to be complete.

Legendre studied the elliptic integrals – called "elliptic functions" by him – as functions of the upper integration limit ϕ and seems to have been disappointed by the little interest aroused by the three volumes of his *Traité des Fonctions Elliptiques*.

It was Carl Friedrich Gauss (1777–1855) who first had the idea of inverting elliptic integrals to obtain elliptic functions in 1797, but with no impact because he did not publish his results. Things changed radically when the young Norwegian mathematician Niels Henrik Abel (1802–1829) studied the inverses of the elliptic integrals – now called elliptic functions – and extended them as analytic functions of a complex variable. In 1827 Abel published a theory of elliptic functions in close analogy to the theory of trigonometric functions. Also in 1827, the still younger German mathematician Carl Gustav Jakob Jacobi (1804–1851) had ideas similar to those of Abel. Jacobi introduced the composite function sine of the amplitude, which he denoted by sinam and was eventually abbreviated to sn. Jacobi also proved the double periodicity of the elliptic functions. Jacobi's initial work on elliptic functions may have been inspired by Abel's ideas. A friendly rivalry developed between the two mathematicians, who competed for priority in the publication of new discoveries. The competition ended abruptly in 1829 with Abel's untimely death of tuberculosis when he was just 26 years old. New ideas erupted from Jacobi at an explosive rate (Stillwell, 2010) and he published in 1829 the first book on elliptic functions, titled *Fundamenta Nova Theoriae Functionum Ellipticarum*. Elliptic functions eventually became one of the most important topics of mathematical research in the nineteenth century.

The aged Legendre expressed his satisfaction with the work of Jacobi and Abel in a letter sent to the former in 1828:

> It is a great satisfaction to me to see two young mathematicians, like you and him, cultivate successfully a branch of analysis which has for so long been my favourite object of study and which has not attracted the attention in my country that it deserves.[4]

---

[3] $R(x, w) = P(x, w)/Q(x, w)$ where $P(x, w)$ and $Q(x, w)$ are polynomials in the two variables x and w.
[4] Translation taken from Bottazzini and Gray (2013).

5.3 Stationary Motion and Small Oscillations

The considerations in Section 5.1 can be extended to certain multidimensional systems capable of performing stationary motions, which may be seen as states of dynamic equilibrium. Such motions typically occur in systems with cyclic coordinates. We shall adopt the following definition: a motion is said to be **stationary** if, in some coordinate system, the **noncyclic coordinates**, that is, those which appear in the Lagrangian, are **constant**.

Example 5.3 In polar coordinates the Lagrangian for a particle moving in a plane under a central force is

$$L = \frac{m}{2}(\dot{r}^2 + r^2\dot{\theta}^2) - V(r).\tag{5.61}$$

The coordinate θ is cyclic, so $mr^2\dot{\theta} = p_\theta = $ constant. The radial equation of motion

$$m\ddot{r} - mr\dot{\theta}^2 + \frac{dV}{dr} = 0\tag{5.62}$$

possesses the stationary solution $r = r_0$ on the condition that $(dV/dr)_{r=r_0} = p_\theta^2/(mr_0^3)$. For example, if $V(r) = -A/r$ with $A > 0$ we have $r_0 = p_\theta^2/(mA)$. Therefore, there are circular orbits in the case of the gravitational or attractive Coulomb potential.

Example 5.4 The Lagrangian for a spherical pendulun (that is, a pendulum not restricted to oscillate in a fixed plane) is given by

$$L = \frac{ml^2}{2}(\dot{\theta}^2 + \dot{\phi}^2\sin^2\theta) - mgl(1 - \cos\theta).\tag{5.63}$$

The coordinate ϕ is cyclic, so $ml^2\dot{\phi}\sin^2\theta = p_\phi = $ constant. The equation of motion for θ is

$$ml^2\ddot{\theta} - ml^2\dot{\phi}^2\sin\theta\cos\theta + mgl\sin\theta = 0,\tag{5.64}$$

which admits the stationary solution $\theta = \theta_0$ with $\cos\theta_0 = g/(l\dot{\phi}^2)$ if $\dot{\phi}^2 > g/l$.

In the two previous examples the Lagrangian does not explicitly depend on time and the kinetic energy is a second-degree homogeneous function of the velocities, so that $h = E$ and E is a constant of the motion. In both cases the conservation of energy is expressed in the form

$$\frac{\alpha(q)}{2}\dot{q}^2 + V_{\text{eff}}(q) = E,\tag{5.65}$$

where q denotes the noncyclic coordinate. Indeed, in Example 5.3 with $mr^2\dot{\theta} = p_\theta$,

$$E = T + V = \frac{m}{2}\dot{r}^2 + \frac{m}{2}r^2\dot{\theta}^2 + V(r) = \frac{m}{2}\dot{r}^2 + \frac{p_\theta^2}{2mr^2} + V(r),\tag{5.66}$$

while in Example 5.4, making use of $ml^2\dot{\phi}^2 \sin^2\theta = p_\phi$,

$$E = T + V = \frac{ml^2}{2}\dot{\theta}^2 + \frac{p_\phi^2}{2ml^2 \sin^2\theta} + mgl(1 - \cos\theta). \qquad (5.67)$$

The equilibrium configurations are the solutions of

$$\frac{dV_{\text{eff}}(q)}{dq} = 0, \qquad (5.68)$$

while stability depends on the sign of the second derivative of the **effective potential** V_{eff} at the equilibrium configuration, the positive sign implying stable equilibrium. It is important to stress that the effective potential is identified from the equation that expresses the conservation of energy after the elimination of the velocity associated with the cyclic coordinate by expressing it in terms of its conjugate momentum, which is a constant of the motion.

Example 5.5 Investigate the stability of the circular orbit of a celestial body about the Sun.

Solution

According to Eq. (5.66),

$$V_{\text{eff}}(r) = \frac{p_\theta^2}{2mr^2} - \frac{A}{r} \qquad (5.69)$$

with $A = GmM > 0$ where M is the mass of the Sun and m is the mass of the celestial body in circular orbit. The radius of the orbit is found from

$$0 = \frac{dV_{\text{eff}}}{dr} = -\frac{p_\theta^2}{mr^3} + \frac{A}{r^2} \implies r = r_0 = \frac{p_\theta^2}{mA}, \qquad (5.70)$$

a result already obtained in a different way in Example 5.3. On the other hand,

$$k^{(0)} = \left(\frac{d^2 V_{\text{eff}}}{dr^2}\right)_{r=r_0} = \frac{3p_\theta^2}{mr_0^4} - \frac{2A}{r_0^3} = \frac{A}{r_0^3} > 0, \qquad (5.71)$$

and the circular orbits are stable under small perturbations,[5] with

$$\omega = \left(\frac{k^{(0)}}{\alpha^{(0)}}\right)^{1/2} = \left(\frac{GM}{r_0^3}\right)^{1/2} \qquad (5.72)$$

being the frequency of the small oscillations around the circular orbit.

Example 5.6 Study the stability of the stationary precession ($\theta = \theta_0$) of a spherical pendulum.

Solution

From (5.67) one identifies

$$V_{\text{eff}}(\theta) = mgl(1 - \cos\theta) + \frac{p_\phi^2}{2ml^2 \sin^2\theta}. \qquad (5.73)$$

[5] The present analysis does not apply to perturbations perpendicular to the plane of the orbit.

The stationary motion $\theta = \theta_0$ requires

$$\left(\frac{dV_{\text{eff}}}{d\theta}\right)_{\theta=\theta_0} = mgl \sin \theta_0 - \frac{p_\phi^2 \cos \theta_0}{ml^2 \sin^3 \theta_0} = 0 \implies p_\phi^2 = m^2 gl^3 \frac{\sin^4 \theta_0}{\cos \theta_0}. \quad (5.74)$$

Making use of $p_\phi = ml^2 \sin^2 \theta_0 \, \dot\phi$ there results $\cos \theta_0 = g/(l\dot\phi^2)$ as long as $|\dot\phi| > (g/l)^{1/2}$, the same restriction found in Example 5.4. On the other hand,

$$k^{(0)} = \left(\frac{d^2 V_{\text{eff}}}{d\theta^2}\right)_{\theta=\theta_0} = mgl \cos \theta_0 + \frac{p_\phi^2}{ml^2 \sin^2 \theta_0} + 3\frac{p_\phi^2 \cos^2 \theta_0}{ml^2 \sin^4 \theta_0}$$

$$= \frac{mgl}{\cos \theta_0}(1 + 3\cos^2 \theta_0) > 0, \quad (5.75)$$

where we have used (5.74). The precession is stable and

$$\omega = \left(\frac{k^{(0)}}{\alpha^{(0)}}\right)^{1/2} = \left(\frac{k^{(0)}}{ml^2}\right)^{1/2} = \left(\frac{g}{l}\frac{1 + 3\cos^2 \theta_0}{\cos \theta_0}\right)^{1/2} \quad (5.76)$$

is the frequency of the small oscillations about the stationary motion.

5.4 Small Oscillations: General Case

With the same hypotheses as in the one-dimensional case, an n-degree-of-freedom conservative system described by the generalised coordinates $q_1, \ldots, q_n$ has a Lagrangian

$$L = \frac{1}{2} \sum_{k,l=1}^{n} M_{kl} \, \dot q_k \, \dot q_l - V(q_1, \ldots, q_n) \quad (5.77)$$

where the $M_{kl} = M_{kl}(q_1, \ldots, q_n)$ are quantities which, according to Eq. (2.154), are symmetric in their indices: $M_{kl} = M_{lk}$. The configuration $q^{(0)} \equiv (q_1^{(0)}, \ldots, q_n^{(0)})$ is an equilibrium configuration if

$$\left(\frac{\partial V}{\partial q_k}\right)_{q^{(0)}} = 0, \quad k = 1, \ldots, n. \quad (5.78)$$

Defining

$$q_k = q_k^{(0)} + \eta_k, \quad k = 1, \ldots, n, \quad (5.79)$$

and expanding $V(q)$ up to the second order in powers of $\eta_1, \ldots, \eta_n$, one has

$$V(q_1, \ldots, q_n) = V(q^{(0)}) + \sum_k \left(\frac{\partial V}{\partial q_k}\right)_{q^{(0)}} \eta_k + \frac{1}{2} \sum_{kl} \left(\frac{\partial^2 V}{\partial q_k \partial q_l}\right)_{q^{(0)}} \eta_k \, \eta_l + \cdots. \quad (5.80)$$

The term linear in the small displacements vanishes in virtue of (5.78), and the dominant approximation to V takes the form

$$V = V_0 + \frac{1}{2} \sum_{kl} V_{kl}\, \eta_k\, \eta_l \,, \tag{5.81}$$

where $V_0 = V(q^{(0)})$ is a constant and

$$V_{kl} = \left(\frac{\partial^2 V}{\partial q_k \partial q_l} \right)_{q^{(0)}} . \tag{5.82}$$

The V_{kl} are constant and symmetric: $V_{kl} = V_{lk}$. The stability criterion is now a little more complicated. If V has a minimum at $q = q^{(0)}$ its value can only increase in consequence of a small displacement from the configuration $q = q^{(0)}$. By Eq. (5.81), therefore, $q = q^{(0)}$ is a stable equilibrium configuration if

$$\sum_{kl} V_{kl}\, \eta_k\, \eta_l > 0 \qquad \forall\, (\eta_1, \ldots, \eta_n) \neq (0, \ldots, 0) \,. \tag{5.83}$$

In short, the equilibrium is stable if the **quadratic form** $\sum_{kl} V_{kl}\, \eta_k\, \eta_l$ constructed from the second derivatives of the potential is **positive**, which will be assumed from now on.

As in the one-dimensional problem, the expansion of the Lagrangian (5.77), correct up to the second order in η and $\dot{\eta}$, is written

$$L = \frac{1}{2} \sum_{kl} T_{kl}\, \dot{\eta}_k\, \dot{\eta}_l - \frac{1}{2} \sum_{kl} V_{kl}\, \eta_k\, \eta_l \,, \tag{5.84}$$

where

$$T_{kl} = M_{kl}(q^{(0)}) \tag{5.85}$$

and the irrelevant additive constant V_0 has been discarded. The equations of motion for the small displacements from equilibrium are

$$\frac{d}{dt} \left(\frac{\partial L}{\partial \dot{\eta}_j} \right) - \frac{\partial L}{\partial \eta_j} = 0, \qquad j = 1, \ldots, n \,. \tag{5.86}$$

We have

$$\frac{\partial L}{\partial \dot{\eta}_j} = \frac{1}{2} \sum_{kl} T_{kl}\, (\delta_{kj}\, \dot{\eta}_l + \dot{\eta}_k\, \delta_{lj})$$

$$= \frac{1}{2} \left\{ \sum_l T_{jl}\, \dot{\eta}_l + \sum_k T_{kj}\, \dot{\eta}_k \right\} = \frac{1}{2} \left\{ \sum_l T_{jl}\, \dot{\eta}_l + \sum_k T_{jk}\, \dot{\eta}_k \right\}, \tag{5.87}$$

because $T_{kj} = T_{jk}$. Taking into account that the summation index is dummy, we may replace l by k in the penultimate sum in equation Eq. (5.87) to obtain

$$\frac{\partial L}{\partial \dot{\eta}_j} = \sum_k T_{jk}\, \dot{\eta}_k \,. \tag{5.88}$$

Similarly,

$$-\frac{\partial L}{\partial \eta_j} = \sum_k V_{jk}\, \eta_k\,, \tag{5.89}$$

and the Lagrange equations (5.86) take the form

$$\sum_k T_{jk}\, \ddot{\eta}_k + \sum_k V_{jk}\, \eta_k = 0\,, \qquad j = 1,\ldots,n\,, \tag{5.90}$$

which are equations of motion for coupled oscillators.

5.5 Normal Modes of Vibration

As we shall soon see, the general solution to the coupled equations of motion (5.90) can be expressed in terms of particular solutions such that all displacements $\eta_1,\ldots,\eta_n$ oscillate harmonically with the same frequency. It is convenient to replace $\eta_k(t)$ by the complex function $z_k(t)$ and define

$$\eta_k(t) = \operatorname{Re} z_k(t)\,, \qquad k = 1,\ldots,n\,. \tag{5.91}$$

Since the coefficients of the system of Eqs. (5.90) are real, the real and imaginary parts of a complex solution are also solutions, which justifies (5.91).

Characteristic Frequencies

Let us search for complex solutions to (5.90) – with η_k replaced by z_k – in the form

$$z_k(t) = z_k^{(0)} e^{i\omega t}\,, \qquad k = 1,\ldots,n\,, \tag{5.92}$$

such that all displacements vibrate with the same frequency ω. Inserting the above expression in the place of $\eta_k(t)$ in (5.90) there results

$$\sum_k (V_{jk} - \omega^2 T_{jk})\, z_k^{(0)} = 0\,. \tag{5.93}$$

Defining the matrices

$$\mathbf{V} = (V_{jk})\,, \qquad \mathbf{T} = (T_{jk})\,, \qquad \mathbf{z}^{(0)} = \begin{pmatrix} z_1^{(0)} \\ \vdots \\ z_n^{(0)} \end{pmatrix}\,, \tag{5.94}$$

equations (5.93) are equivalent to the matrix equation

$$(\mathbf{V} - \omega^2 \mathbf{T})\, \mathbf{z}^{(0)} = 0\,. \tag{5.95}$$

This equation has a non-trivial solution $\mathbf{z}^{(0)} \neq 0$ if and only if

$$\det(\mathbf{V} - \omega^2 \mathbf{T}) = \det \begin{pmatrix} V_{11} - \omega^2 T_{11} & V_{12} - \omega^2 T_{12} & \cdots \\ V_{21} - \omega^2 T_{21} & V_{22} - \omega^2 T_{22} & \cdots \\ \vdots & \vdots & \vdots \end{pmatrix} = 0, \qquad (5.96)$$

which is an algebraic equation of degree n for ω^2, possessing n positive roots

$$\omega = \omega_s, \qquad s = 1, \ldots, n. \qquad (5.97)$$

The positive quantities $\omega_1, \ldots, \omega_n$ are called the **characteristic frequencies** of the system.

Normal Modes of Vibration

It is possible to find n linearly independent vectors $\varrho^{(1)}, \ldots, \varrho^{(n)}$, called the **characteristic vectors**, which satisfy (5.95) – that is,

$$(\mathbf{V} - \omega_{(s)}^2 \mathbf{T}) \, \varrho^{(s)} = 0, \qquad s = 1, \ldots, n. \qquad (5.98)$$

Since $\mathbf{V}$ and $\mathbf{T}$ are real matrices, it is always possible to choose each $\varrho^{(s)}$ so that all of its components are real. Assuming such a choice has been made, the complex solution associated with frequency ω_s is written

$$\mathbf{z}^{(s)}(t) = \varrho^{(s)} \, e^{i\omega_s t}. \qquad (5.99)$$

Each particular solution of the form (5.99) is called a **normal mode of vibration** of the system.

General Solution to the Equations of Motion

Equations (5.90) are linear, so any linear combination of solutions is also a solution. Thus, the most general solution for $\mathbf{z}(t)$ is expressed as a superposition of the solutions associated with the characteristic frequencies, namely

$$\mathbf{z}(t) = \sum_{s=1}^{n} A^{(s)} \varrho^{(s)} e^{i\omega_s t} \qquad (5.100)$$

where the complex coefficients $A^{(s)}$ are arbitrary. Writing $A^{(s)}$ in the polar form

$$A^{(s)} = C^{(s)} e^{i\phi_s} \qquad (5.101)$$

with $C^{(s)}$ and ϕ_s real, we are left with

$$\mathbf{z}(t) = \sum_{s=1}^{n} C^{(s)} \varrho^{(s)} e^{i(\omega_s t + \phi_s)}. \qquad (5.102)$$

Since the $\varrho^{(s)}$ are real vectors by hypothesis, the physical solution is

$$\eta(t) = \mathrm{Re}\,\mathbf{z}(t) = \sum_{s=1}^{n} C^{(s)} \varrho^{(s)} \cos(\omega_s t + \phi_s) \qquad (5.103)$$

or, in components,

$$\eta_k(t) = \sum_{s=1}^{n} C^{(s)} \varrho_k^{(s)} \cos\left(\omega_s t + \phi_s\right), \qquad k = 1, \ldots, n. \tag{5.104}$$

The general solution to the equations of motion (5.90), expressed by the last equation above, contains the $2n$ arbitrary constants $C^{(1)}, \ldots, C^{(n)}, \phi_1, \ldots, \phi_n$ which are determined by the initial conditions.

Example 5.7 Two equal pendulums perform small oscillations coupled by a spring of negligible mass. Determine the normal modes of vibration and the solution to the equations of motion if, when $t = 0$, only one of the pendulums is slightly displaced from its equilibrium position and the motion starts with both pendulums at rest.

Solution

Let us assume that the equilibrium configuration is $\theta_1 = \theta_2 = 0$, in which the spring has its natural length d_0. We shall use as coordinates the small horizontal displacements η_1 and η_2 shown in Fig. 5.7, which are related to the small angles θ_1 and θ_2 by

$$\eta_1 = l \sin\theta_1 \approx l\theta_1, \qquad \eta_2 = l \sin\theta_2 \approx l\theta_2. \tag{5.105}$$

The kinetic energy is

$$T = \frac{ml^2}{2}\dot{\theta}_1^{\,2} + \frac{ml^2}{2}\dot{\theta}_2^{\,2} \approx \frac{m}{2}\dot{\eta}_1^{\,2} + \frac{m}{2}\dot{\eta}_2^{\,2}, \tag{5.106}$$

while the potential energy is written

$$V = \frac{k}{2}[(d_0 + \eta_2 - \eta_1) - d_0]^2 + mg[l(1 - \cos\theta_1) + l(1 - \cos\theta_2)]. \tag{5.107}$$

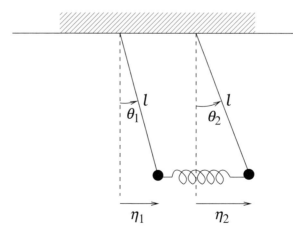

Fig. 5.7 Two equal pendulums coupled by a spring (Example 5.7).

Using $\cos\theta \approx 1 - \theta^2/2$ and (5.105) we get

$$V = \frac{k}{2}(\eta_2 - \eta_1)^2 + \frac{mg}{2l}(\eta_1{}^2 + \eta_2{}^2),\qquad (5.108)$$

so the Lagrangian appropriate to the description of small oscillations is

$$L = \frac{m}{2}(\dot{\eta}_1^2 + \dot{\eta}_2^2) - \frac{1}{2}\left\{\frac{mg}{l}(\eta_1{}^2 + \eta_2{}^2) + k(\eta_2 - \eta_1)^2\right\}.\qquad (5.109)$$

Comparing this with the standard form (5.84) we have

$$\mathbf{T} = \begin{pmatrix} m & 0 \\ 0 & m \end{pmatrix},\qquad \mathbf{V} = \begin{pmatrix} \frac{mg}{l} + k & -k \\ -k & \frac{mg}{l} + k \end{pmatrix}.\qquad (5.110)$$

Warning In the standard form (5.84), the off-diagonal terms appear in duplicate. For example, in the second double sum over k and l there is one term for $k = 1, l = 2$ and another for $k = 2, l = 1$, which are equal to each other because $V_{12} = V_{21}$. Thus, the coefficient of $\eta_1\eta_2$ is $V_{12} + V_{21} = 2V_{12}$. On the other hand, the coefficient of $\eta_1\eta_2$ in the expression within brackets in (5.109) is $-2k$, hence the identification $V_{12} = -k$. This kind of attention is needed when comparing the standard form (5.84) with a specific Lagrangian such as (5.109), otherwise errors by a factor of two will be made, which will damage all subsequent results.

The frequencies of the normal modes are the solutions of

$$\det(\mathbf{V} - \omega^2\mathbf{T}) = \det\begin{pmatrix} \frac{mg}{l} + k - m\omega^2 & -k \\ -k & \frac{mg}{l} + k - m\omega^2 \end{pmatrix} = 0,\qquad (5.111)$$

that is,

$$\left(\frac{mg}{l} + k - m\omega^2\right)^2 - k^2 = 0 \quad\Longrightarrow\quad \frac{mg}{l} + k - m\omega^2 = \pm k.\qquad (5.112)$$

The characteristic frequencies are

$$\omega_1 = \left(\frac{g}{l}\right)^{1/2},\qquad \omega_2 = \left(\frac{g}{l} + \frac{2k}{m}\right)^{1/2},\qquad (5.113)$$

and the corresponding eigenvectors $\varrho^{(s)}$ are such that

$$(\mathbf{V} - \omega_1^2\mathbf{T})\varrho^{(1)} = 0 \implies \begin{pmatrix} k & -k \\ -k & k \end{pmatrix}\begin{pmatrix} \varrho_1^{(1)} \\ \varrho_2^{(1)} \end{pmatrix} = 0 \implies \varrho^{(1)} = \begin{pmatrix} \alpha \\ \alpha \end{pmatrix},$$

$$(\mathbf{V} - \omega_2^2\mathbf{T})\varrho^{(2)} = 0 \implies \begin{pmatrix} -k & -k \\ -k & -k \end{pmatrix}\begin{pmatrix} \varrho_1^{(2)} \\ \varrho_2^{(2)} \end{pmatrix} = 0 \implies \varrho^{(2)} = \begin{pmatrix} \beta \\ -\beta \end{pmatrix},$$

where α and β are non-zero real numbers. The mode with frequency ω_1 corresponds to an oscillatory motion of the pendulums with $\eta_1 = \eta_2$. In this case, it all happens as if the spring did not exist since it remains with its natural length all the time, which explains the value of ω_1 given by (5.113): the pendulums oscillate in unison with the frequency of an individual simple pendulum. The second normal mode consists in an oscillatory motion of the pendulums with frequency ω_2, but now $\eta_1 = -\eta_2$ and the pendulums oscillate in opposite directions, as depicted in Fig. 5.8.

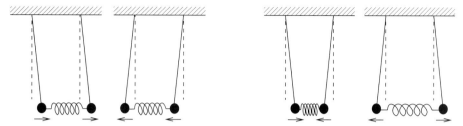

Fig. 5.8   Normal modes of vibration of two equal pendulums coupled by a spring.

The two linearly independent real solutions can be chosen as

$$\varrho^{(1)} = \begin{pmatrix} 1 \\ 1 \end{pmatrix}, \qquad \varrho^{(2)} = \begin{pmatrix} 1 \\ -1 \end{pmatrix} \tag{5.114}$$

and the general solution (5.104) takes the form

$$\eta_1(t) = \sum_{s=1}^{2} C^{(s)} \varrho_1^{(s)} \cos(\omega_s t + \phi_s) = C^{(1)} \cos(\omega_1 t + \phi_1) + C^{(2)} \cos(\omega_2 t + \phi_2),$$

$$\tag{5.115}$$

$$\eta_2(t) = \sum_{s=1}^{2} C^{(s)} \varrho_2^{(s)} \cos(\omega_s t + \phi_s) = C^{(1)} \cos(\omega_1 t + \phi_1) - C^{(2)} \cos(\omega_2 t + \phi_2).$$

$$\tag{5.116}$$

From the initial conditions $\eta_1(0) = 0, \eta_2(0) = a, \dot{\eta}_1(0) = \dot{\eta}_2(0) = 0$, we infer

$$C^{(1)} \cos \phi_1 + C^{(2)} \cos \phi_2 = 0, \tag{5.117a}$$

$$C^{(1)} \cos \phi_1 - C^{(2)} \cos \phi_2 = a, \tag{5.117b}$$

$$-\omega_1 C^{(1)} \sin \phi_1 - \omega_2 C^{(2)} \sin \phi_2 = 0, \tag{5.117c}$$

$$-\omega_1 C^{(1)} \sin \phi_1 + \omega_2 C^{(2)} \sin \phi_2 = 0. \tag{5.117d}$$

By adding the first two and the last two of Eqs. (5.117) we deduce

$$C^{(1)} \cos \phi_1 = -C^{(2)} \cos \phi_2 = a/2, \tag{5.118a}$$

$$C^{(1)} \sin \phi_1 = C^{(2)} \sin \phi_2 = 0. \tag{5.118b}$$

Neither $C^{(1)} = 0$ nor $C^{(2)} = 0$ is admissible because this would give rise to a contradiction with (5.118a). Therefore, $\sin \phi_1 = \sin \phi_2 = 0$, whose simplest solution is $\phi_1 = \phi_2 = 0$, by reason of which $C^{(1)} = -C^{(2)} = a/2$. Finally, the solution to the equations of motion is written

$$\eta_1(t) = \frac{a}{2}(\cos \omega_1 t - \cos \omega_2 t), \tag{5.119a}$$

$$\eta_2(t) = \frac{a}{2}(\cos \omega_1 t + \cos \omega_2 t), \tag{5.119b}$$

with ω_1 and ω_2 given by (5.113).

5.6 Normal Coordinates

The existence of n normal modes of vibration with frequencies $\omega_1, \ldots, \omega_n$ suggests that the Lagrangian (5.84) is equivalent to the Lagrangian for a set of independent harmonic oscillators, each with a frequency equal to one of the characteristic frequencies. This, however, is not manifest in the Lagrangian (5.84) since the coordinates $\eta_1, \ldots, \eta_n$ are mixed with one another, leading to the coupled equations of motion (5.90). It must be possible to introduce new coordinates instead of $\eta_1, \ldots, \eta_n$ in such a way as to eliminate the off-diagonal terms in the sums occurring in the Lagrangian (5.84), reducing it to a sum of Lagrangians for independent harmonic oscillators. Such coordinates always exist, and we proceed to explain how they can generally be found. In order for the exposition to flow smoothly, without frequent interruptions, we defer to the next section the proofs of the mathematical hypotheses or statements made in Section 5.4 and those that will necessarily come forth in the course of the text that follows.

Introducing the matrix notation

$$\boldsymbol{\eta} = \begin{pmatrix} \eta_1 \\ \vdots \\ \eta_n \end{pmatrix}, \qquad \boldsymbol{\eta}^T = (\eta_1 \ \ldots \ \eta_n), \tag{5.120}$$

the Lagrangian (5.84) is written

$$L = \frac{1}{2} \dot{\boldsymbol{\eta}}^T \mathbf{T} \dot{\boldsymbol{\eta}} - \frac{1}{2} \boldsymbol{\eta}^T \mathbf{V} \boldsymbol{\eta} . \tag{5.121}$$

Define new coordinates $\zeta_1, \ldots, \zeta_n$ related to $\eta_1, \ldots, \eta_n$ by the linear transformation

$$\boldsymbol{\eta} = \mathbf{A} \boldsymbol{\zeta} , \tag{5.122}$$

where $\mathbf{A}$ is a matrix to be determined. Substituting (5.122) into (5.121) there results

$$L = \frac{1}{2} \dot{\boldsymbol{\zeta}}^T \mathbf{A}^T \mathbf{T} \mathbf{A} \dot{\boldsymbol{\zeta}} - \frac{1}{2} \boldsymbol{\zeta}^T \mathbf{A}^T \mathbf{V} \mathbf{A} \boldsymbol{\zeta} . \tag{5.123}$$

In order for the Lagrangian not to contain couplings between the new coordinates $\zeta_1, \ldots, \zeta_n$, both matrices $\mathbf{A}^T \mathbf{T} \mathbf{A}$ and $\mathbf{A}^T \mathbf{V} \mathbf{A}$ must be diagonal. The problem, therefore, consists in finding a matrix $\mathbf{A}$ capable of simultaneously diagonalising matrices $\mathbf{T}$ and $\mathbf{V}$.

The introduction of an inner product in the vector space of the n-component column real matrices facilitates the construction of the matrix $\mathbf{A}$ with the required properties. Define, therefore, the inner product $(\boldsymbol{\eta}, \boldsymbol{\xi})$ by

$$(\boldsymbol{\eta}, \boldsymbol{\xi}) = \boldsymbol{\eta}^T \mathbf{T} \boldsymbol{\xi} \equiv \sum_{kl} T_{kl} \eta_k \xi_l . \tag{5.124}$$

If $\boldsymbol{\varrho}^{(1)}, \ldots, \boldsymbol{\varrho}^{(n)}$ are linearly independent solutions to (5.98), let us impose on these vectors the additional condition that they be chosen as mutually orthogonal and normalised in the inner product that has just been defined:

$$(\boldsymbol{\varrho}^{(r)}, \boldsymbol{\varrho}^{(s)}) = \delta_{rs} , \qquad r, s = 1, \ldots, n . \tag{5.125}$$

Define, now, the **modal matrix** $\mathbf{A} = (A_{kr})$ by

$$A_{kr} = \varrho_k^{(r)}, \tag{5.126}$$

whence[6]

$$\mathbf{A} = \begin{pmatrix} \varrho^{(1)} & \varrho^{(2)} & \cdots & \varrho^{(n)} \\ \downarrow & \downarrow & \cdots & \downarrow \end{pmatrix}, \tag{5.127}$$

that is, the pth column of $\mathbf{A}$ is the vector $\varrho^{(p)}$. Equations (5.125) are equivalent to

$$\sum_{kl} \varrho_k^{(r)} T_{kl} \varrho_l^{(s)} = \delta_{rs}, \tag{5.128}$$

which can also be rewritten in the form

$$\sum_k (\mathbf{A}^T)_{rk} \left(\sum_l T_{kl} A_{ls} \right) = \sum_k (\mathbf{A}^T)_{rk} (\mathbf{TA})_{ks} = \delta_{rs}. \tag{5.129}$$

Using once again the definition of matrix product we are led to

$$\mathbf{A}^T \mathbf{TA} = \mathbf{I}. \tag{5.130}$$

The modal matrix (5.127) reduces the positive matrix $\mathbf{T}$ to the simplest diagonal form possible – the identity matrix. It remains to verify if matrix $\mathbf{V}$ is also reduced to the diagonal form.

In components, Eqs. (5.98) are written

$$\sum_l V_{kl} \varrho_l^{(s)} = \omega_s^2 \sum_l T_{kl} \varrho_l^{(s)} \tag{5.131}$$

or, with the help of (5.126),

$$(\mathbf{VA})_{ks} = \omega_s^2 (\mathbf{TA})_{ks}. \tag{5.132}$$

Define the diagonal matrix $\mathbf{W}^D = (W_{rs})$ by

$$W_{rs} = \omega_s^2 \delta_{rs}, \tag{5.133}$$

so that

$$\mathbf{W}^D = \begin{pmatrix} \omega_1^2 & 0 & \cdots & 0 \\ 0 & \omega_2^2 & \cdots & 0 \\ \vdots & \vdots & \vdots & \vdots \\ 0 & 0 & \cdots & \omega_n^2 \end{pmatrix}. \tag{5.134}$$

Equation (5.132) can be rewritten in the equivalent form

$$(\mathbf{VA})_{ks} = \sum_r (\mathbf{TA})_{kr} \omega_s^2 \delta_{rs} = \sum_r (\mathbf{TA})_{kr} (\mathbf{W}^D)_{rs}, \tag{5.135}$$

that is,

$$\mathbf{VA} = \mathbf{TAW}^D. \tag{5.136}$$

---

[6] Here we adopt the notation and terminology of Fetter and Walecka (1980).

Multiplying this equation from the left by $\mathbf{A}^T$ and using (5.130), we finally obtain

$$\mathbf{A}^T \mathbf{V} \mathbf{A} = \mathbf{W}^D, \tag{5.137}$$

completing the proof that $\mathbf{A}$ also reduces $\mathbf{V}$ to the diagonal form.

In terms of the **normal coordinates** $\zeta_1, \dots \zeta_n$ defined by (5.122) and where $\mathbf{A}$ is the modal matrix (5.127), the Lagrangian (5.123) takes the desired form

$$L = \frac{1}{2} \dot{\boldsymbol{\zeta}}^T \dot{\boldsymbol{\zeta}} - \frac{1}{2} \boldsymbol{\zeta}^T \mathbf{W}^D \boldsymbol{\zeta} = \sum_{k=1}^{n} \frac{1}{2} (\dot{\zeta}_k^2 - \omega_k^2 \zeta_k^2), \tag{5.138}$$

which describes a set of n independent harmonic oscillators with angular frequencies $\omega_1, \dots, \omega_n$, respectively. The equation of motion for each normal coordinate is just

$$\frac{d}{dt} \left(\frac{\partial L}{\partial \dot{\zeta}_j} \right) - \frac{\partial L}{\partial \zeta_j} = 0 \quad \Longrightarrow \quad \ddot{\zeta}_j + \omega_j^2 \zeta_j = 0, \quad j = 1, \dots, n, \tag{5.139}$$

whose general solution is

$$\zeta_j(t) = C^{(j)} \cos(\omega_j t + \phi_j), \quad j = 1, \dots, n. \tag{5.140}$$

Each normal mode consists in a motion of the system in which only one of the normal coordinates oscillates harmonically, all of the remaining ones being identically zero. Lastly, the expression of the normal coordinates as functions of the original coordinates $\eta_1, \dots, \eta_n$ can be immediately found by inverting Eq. (5.122): multiply it from the left by $\mathbf{A}^T \mathbf{T}$ and use (5.130) to get

$$\boldsymbol{\zeta} = \mathbf{A}^T \mathbf{T} \boldsymbol{\eta}. \tag{5.141}$$

Example 5.8 Determine the normal coordinates for the coupled pendulums of Example 5.7.

Solution

The two linearly independent solutions to Eqs. (5.98) in the case of the coupled pendulums are

$$\boldsymbol{\varrho}^{(1)} = \alpha \begin{pmatrix} 1 \\ 1 \end{pmatrix}, \quad \boldsymbol{\varrho}^{(2)} = \beta \begin{pmatrix} 1 \\ -1 \end{pmatrix}, \quad \alpha, \beta \in \mathbb{R}, \tag{5.142}$$

which were obtained in Example 5.7. In order to construct the modal matrix, it is necessary that $\boldsymbol{\varrho}^{(1)}$ and $\boldsymbol{\varrho}^{(2)}$ make up an orthonormal set in the inner product defined by the kinetic energy matrix $\mathbf{T} = m\mathbf{I}$ – that is,

$$\boldsymbol{\varrho}^{(1)T} \mathbf{T} \boldsymbol{\varrho}^{(1)} = 1 \implies m\alpha^2 \begin{pmatrix} 1 & 1 \end{pmatrix} \begin{pmatrix} 1 \\ 1 \end{pmatrix} = 1 \implies 2m\alpha^2 = 1, \tag{5.143a}$$

$$\boldsymbol{\varrho}^{(2)T} \mathbf{T} \boldsymbol{\varrho}^{(2)} = 1 \implies m\beta^2 \begin{pmatrix} 1 & -1 \end{pmatrix} \begin{pmatrix} 1 \\ -1 \end{pmatrix} = 1 \implies 2m\beta^2 = 1, \tag{5.143b}$$

$$\boldsymbol{\varrho}^{(1)T} \mathbf{T} \boldsymbol{\varrho}^{(2)} = 0 \implies m\alpha\beta \begin{pmatrix} 1 & 1 \end{pmatrix} \begin{pmatrix} 1 \\ -1 \end{pmatrix} = 0 \implies m\alpha\beta(1 - 1) = 0. \tag{5.143c}$$

The orthogonality condition is identically satisfied and the choice $\alpha = \beta = (2m)^{-1/2}$ ensures normalisation of the vectors. An orthonormal set of solutions of (5.98) is

$$\varrho^{(1)} = \frac{1}{\sqrt{2m}} \begin{pmatrix} 1 \\ 1 \end{pmatrix}, \qquad \varrho^{(2)} = \frac{1}{\sqrt{2m}} \begin{pmatrix} 1 \\ -1 \end{pmatrix}, \tag{5.144}$$

which give rise to the modal matrix

$$\mathbf{A} = \frac{1}{\sqrt{2m}} \begin{pmatrix} 1 & 1 \\ 1 & -1 \end{pmatrix}. \tag{5.145}$$

The normal coordinates are

$$\boldsymbol{\zeta} = \mathbf{A}^T \mathbf{T}\boldsymbol{\eta} = \frac{m}{\sqrt{2m}} \begin{pmatrix} 1 & 1 \\ 1 & -1 \end{pmatrix} \begin{pmatrix} \eta_1 \\ \eta_2 \end{pmatrix}, \tag{5.146}$$

that is,

$$\zeta_1 = \sqrt{\frac{m}{2}}\,(\eta_1 + \eta_2), \qquad \zeta_2 = \sqrt{\frac{m}{2}}\,(\eta_1 - \eta_2). \tag{5.147}$$

The normal mode of frequency ω_1 has $\eta_1 = \eta_2$, so only ζ_1 oscillates while ζ_2 is identically zero; the opposite occurs with the normal mode of frequency ω_2.

As the last illustration, let us consider a system of interest in molecular spectroscopy.

Example 5.9   Determine the normal modes of vibration of the symmetric linear triatomic molecule (CO_2, for instance) represented by the model in Fig. 5.9.

Solution
The potential energy is given by

$$V = \frac{k}{2}(x_2 - x_1 - \ell)^2 + \frac{k}{2}(x_3 - x_2 - \ell)^2, \tag{5.148}$$

where ℓ denotes the natural length of each spring. The equilibrium configuration is such that

$$0 = \frac{\partial V}{\partial x_1} = -k(x_2 - x_1 - \ell), \qquad 0 = \frac{\partial V}{\partial x_2} = k(x_2 - x_1 - \ell) - k(x_3 - x_2 - \ell), \tag{5.149}$$

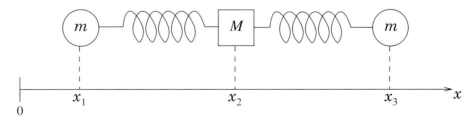

Fig. 5.9   Model for the study of vibrations of a symmetric linear triatomic molecule.

whence

$$x_2 - x_1 = \ell, \qquad x_3 - x_2 = \ell, \tag{5.150}$$

as expected. In order to make our formalism applicable to the present problem, it is necessary to eliminate from the potential energy the terms linear in the generalised coordinates. This can be achieved by defining

$$\eta_1 = x_1 + \ell, \qquad \eta_2 = x_2, \qquad \eta_3 = x_3 - \ell, \tag{5.151}$$

leading to

$$L = \frac{m}{2}(\dot{x}_1^2 + \dot{x}_3^2) + \frac{M}{2}\dot{x}_2^2 - V = \frac{m}{2}(\dot{\eta}_1^2 + \dot{\eta}_3^2) + \frac{M}{2}\dot{\eta}_2^2 - \frac{k}{2}[(\eta_2 - \eta_1)^2 + (\eta_3 - \eta_2)^2]. \tag{5.152}$$

By inspection we identify

$$\mathbf{T} = \begin{pmatrix} m & 0 & 0 \\ 0 & M & 0 \\ 0 & 0 & m \end{pmatrix}, \qquad \mathbf{V} = \begin{pmatrix} k & -k & 0 \\ -k & 2k & -k \\ 0 & -k & k \end{pmatrix}. \tag{5.153}$$

The characteristic equation is written (with $\omega^2 = \lambda$)

$$\det(\mathbf{V} - \lambda\mathbf{T}) = \det \begin{pmatrix} k - m\lambda & -k & 0 \\ -k & 2k - M\lambda & -k \\ 0 & -k & k - m\lambda \end{pmatrix} = 0, \tag{5.154}$$

whence

$$(k - m\lambda)[(2k - M\lambda)(k - m\lambda) - k^2] - k^2(k - m\lambda) = 0. \tag{5.155}$$

One immediate root is $\lambda = k/m$ and the other two are the solutions of

$$mM\lambda^2 - k(2m + M)\lambda = 0. \tag{5.156}$$

Therefore, the frequencies of the normal modes are

$$\omega_1 = 0, \qquad \omega_2 = \sqrt{\frac{k}{m}}, \qquad \omega_3 = \sqrt{\frac{k}{m}\left(1 + \frac{2m}{M}\right)}. \tag{5.157}$$

The first characteristic frequency may look strange at first sight, inasmuch as the associated normal coordinate obeys the equation $\ddot{\zeta}_1 = 0$ which does not correspond to oscillatory motion. This zero frequency mode appears because the molecule can move rigidly with unchanging potential energy. Since the restoring force against such a displacement vanishes, the associated frequency is zero. Mathematically this is due to the fact that the potential energy quadratic form is not strictly positive. Indeed, $V = 0$ for $\eta_1 = \eta_2 = \eta_3 \neq 0$ and this equilibrium configuration is not stable but indifferent. The amplitude $\varrho^{(1)}$ associated with the frequency $\omega_1 = 0$ satisfies

$$(\mathbf{V} - \lambda\mathbf{T})\varrho^{(1)} = 0 \implies \begin{pmatrix} k & -k & 0 \\ -k & 2k & -k \\ 0 & -k & k \end{pmatrix} \begin{pmatrix} \varrho_1^{(1)} \\ \varrho_2^{(1)} \\ \varrho_3^{(1)} \end{pmatrix} = 0 \implies \varrho^{(1)} = A \begin{pmatrix} 1 \\ 1 \\ 1 \end{pmatrix}. \tag{5.158}$$

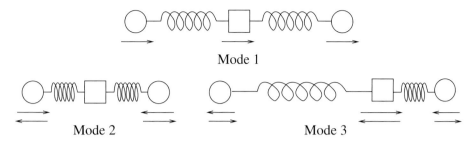

Fig. 5.10 Normal modes of the symmetric linear triatomic molecule.

Proceeding similarly we find

$$\varrho^{(2)} = B \begin{pmatrix} 1 \\ 0 \\ -1 \end{pmatrix}, \qquad \varrho^{(3)} = C \begin{pmatrix} 1 \\ -2m/M \\ 1 \end{pmatrix}. \tag{5.159}$$

The normal modes of vibration are sketched in Fig. 5.10. Only the modes 2 and 3 correspond to vibrations of the molecule, the mode number 1 being a rigid uniform translatory motion. The vectors $\varrho^{(k)}$ are associated with distinct frequencies, so they are automatically orthogonal in the inner product (5.124), as will be generally proved in the next section. The normalisation condition is

$$\varrho^{(s)^T} \mathbf{T} \varrho^{(s)} = \sum_{k\ell} T_{k\ell} \varrho_k^{(s)} \varrho_\ell^{(s)} = 1 \Rightarrow m(\varrho_1^{(s)^2} + \varrho_3^{(s)^2}) + M \varrho_2^{(s)^2} = 1, \; s = 1, 2, 3 \tag{5.160}$$

and the normalised vectors can be chosen as

$$\varrho^{(1)} = \frac{1}{\sqrt{2m+M}} \begin{pmatrix} 1 \\ 1 \\ 1 \end{pmatrix}, \tag{5.161a}$$

$$\varrho^{(2)} = \frac{1}{\sqrt{2m}} \begin{pmatrix} 1 \\ 0 \\ -1 \end{pmatrix}, \tag{5.161b}$$

$$\varrho^{(3)} = \frac{1}{\sqrt{2m(1 + 2m/M)}} \begin{pmatrix} 1 \\ -2m/M \\ 1 \end{pmatrix}. \tag{5.161c}$$

With the help of the modal matrix

$$\mathbf{A} = \begin{pmatrix} \varrho^{(1)} & \varrho^{(2)} & \varrho^{(3)} \\ \downarrow & \downarrow & \downarrow \end{pmatrix}, \tag{5.162}$$

the normal coordinates are given by $\boldsymbol{\zeta} = \mathbf{A}^T \mathbf{T} \boldsymbol{\eta}$, whence, for example,

$$\zeta_1 = \frac{m\eta_1 + M\eta_2 + m\eta_3}{\sqrt{2m+M}} = \frac{mx_1 + Mx_2 + mx_3}{\sqrt{2m+M}} = (2m+M)^{1/2} x_{\mathrm{CM}}. \tag{5.163}$$

There being no external forces, the centre of mass of the molecule moves with constant velocity ($\ddot{x}_{\mathrm{CM}} = 0$), justifying the frequency $\omega_1 = 0$ associated with ζ_1.

More elaborate examples of molecular vibrations can be found in Landau and Lifshitz (1976). For a detailed and general discussion of the problem of driven and damped oscillations, which will not be treated here, an excellent reference is Gantmacher (1970).

5.7 Mathematical Supplement

In this section we present the proofs which, to avoid breaking the continuity of the exposition, have been omitted in the two previous sections. We also take the opportunity to call attention to the general validity of certain properties found in the examples chosen to illustrate the theory. Whenever necessary, reference will be made to Appendix D, with which the reader is assumed to be familiar.

Kinetic Energy and Inner Product

The kinetic energy of any physical system is always positive and vanishes only if all velocities are zero, property that is mathematically expressed by

$$\sum_{kl} T_{kl}\xi_k\xi_l > 0 \qquad \forall\,(\xi_1,\ldots,\xi_n) \neq (0,\ldots,0). \tag{5.164}$$

This positivity property and the symmetry of matrix $\mathbf{T}$, that is, $T_{kl} = T_{lk}$, allow us to prove that (5.124) has all the properties of an inner product as defined in Appendix D, namely

$$(i)\ (\boldsymbol{\eta},\alpha\boldsymbol{\xi}+\beta\boldsymbol{\xi}') = \alpha(\boldsymbol{\eta},\boldsymbol{\xi}) + \beta(\boldsymbol{\eta},\boldsymbol{\xi}'),\quad \alpha,\beta\in\mathbb{R} \qquad \text{(linearity)}; \tag{5.165a}$$

$$(ii)\ (\boldsymbol{\eta},\boldsymbol{\xi}) = (\boldsymbol{\xi},\boldsymbol{\eta}) \qquad \text{(symmetry)}; \tag{5.165b}$$

$$(iii)\ (\boldsymbol{\eta},\boldsymbol{\eta}) \geq 0 \ \text{ and }\ (\boldsymbol{\eta},\boldsymbol{\eta}) = 0 \iff \boldsymbol{\eta} = 0 \qquad \text{(positivity)}. \tag{5.165c}$$

Exercise 5.7.1 Prove that $(\boldsymbol{\eta},\boldsymbol{\xi})$ defined on $\mathbb{R}^n$ by (5.124) enjoys the properties (i)–(iii) that characterise an inner product.

Positivity of the Characteristic Frequencies

We have tacitly assumed that all solutions for ω^2 of the characteristic Eq. (5.96) are positive, implying that the frequencies $\omega_1,\ldots,\omega_n$ are real and positive. This can be easily proved from the positivity of the matrix $\mathbf{V}$. Let $\lambda = \omega^2$ and assume that

$$\mathbf{V}\boldsymbol{\xi} = \lambda\,\mathbf{T}\boldsymbol{\xi} \tag{5.166}$$

with $\boldsymbol{\xi} \neq 0$. Multiplying this last equation from the left by $\boldsymbol{\xi}^T$ and solving for λ we obtain

$$\lambda = \frac{\boldsymbol{\xi}^T\mathbf{V}\boldsymbol{\xi}}{\boldsymbol{\xi}^T\mathbf{T}\boldsymbol{\xi}} > 0 \tag{5.167}$$

because both the numerator and the denominator are positive real numbers in virtue of equations (5.83) and (5.164). If the matrix $\mathbf{V}$ is not strictly positive, there may occur the eigenvalue $\lambda = 0$.

Orthogonality of the Characteristic Vectors

In Example 5.7 the modes of frequencies ω_1 and ω_2 are automatically orthogonal in the inner product (5.124). This is not fortuitous, but takes place whenever the frequencies are *distinct*. In fact, let $\varrho^{(1)}$ and $\varrho^{(2)}$ be solutions of (5.166) for different values of λ:

$$\mathbf{V}\varrho^{(1)} = \lambda_1 \mathbf{T}\varrho^{(1)} ; \tag{5.168a}$$

$$\mathbf{V}\varrho^{(2)} = \lambda_2 \mathbf{T}\varrho^{(2)} . \tag{5.168b}$$

Multiplying (5.168a) from the left by $\varrho^{(2)^T}$, (5.168b) by $\varrho^{(1)^T}$ and subtracting the resulting equations, we get

$$(\lambda_1 - \lambda_2)(\varrho^{(1)}, \varrho^{(2)}) = \varrho^{(2)^T}\mathbf{V}\varrho^{(1)} - \varrho^{(1)^T}\mathbf{V}\varrho^{(2)} , \tag{5.169}$$

where we have used the symmetry of the inner product (5.124). Since $\mathbf{V}$ is also a symmetric matrix, Exercise 5.7.1 establishes that

$$\varrho^{(2)^T}\mathbf{V}\varrho^{(1)} = \varrho^{(1)^T}\mathbf{V}\varrho^{(2)} . \tag{5.170}$$

Consequently,

$$(\lambda_1 - \lambda_2)(\varrho^{(1)}, \varrho^{(2)}) = 0 \quad \Longrightarrow \quad (\varrho^{(1)}, \varrho^{(2)}) = 0 \tag{5.171}$$

if $\lambda_1 \neq \lambda_2$. If all frequencies are distinct, the linearly independent solutions of (5.95) will automatically be orthogonal to each other, and it suffices to normalise them in the inner product (5.124) in order to construct the modal matrix. If the frequency spectrum is degenerate – that is, if the characteristic Eq. (5.96) possesses multiple roots – it is necessary to pick suitable linear combinations of the linearly independent solutions of (5.95) associated with the same frequency so as to obtain a set of mutually orthogonal vectors.

Completeness of the Normal Modes of Vibration

Finally, another assumption made to allow writing the general solution to the equations of motion in the form (5.103) was that of the existence of exactly n linearly independent solutions $\varrho^{(1)}, \ldots, \varrho^{(n)}$ of Eq. (5.95). In order to prove this, note first that the positivity of matrix $\mathbf{T}$ ensures the existence of its inverse $\mathbf{T}^{-1}$. Indeed, if $\mathbf{T}$ had no inverse there would exist a vector $\xi \neq 0$ such that $\mathbf{T}\xi = 0$, and multiplying this last equation from the left by ξ^T we would be led to conclude that $(\xi, \xi) = 0$ whence $\xi = 0$, which is a contradiction. Let us, therefore, write Eq. (5.95) in the form

$$\mathbf{T}^{-1}\mathbf{V}\xi = \lambda \xi , \tag{5.172}$$

which is the eigenvalue equation for the operator $\mathbf{K} = \mathbf{T}^{-1}\mathbf{V}$. According to Theorem D.13 in Appendix D, if we prove that $\mathbf{K}$ is symmetric it will be established that $\mathbf{K}$ is self-adjoint and guaranteed the existence of an orthonormal basis of $\mathbb{R}^n$ (and, therefore, of a collection of n linearly independent vectors) made up of eigenvectors of $\mathbf{K}$. On the other hand, in order to prove that an operator is symmetric, it suffices to check that its matrix in some orthonormal basis $\boldsymbol{\xi}^{(1)}, \ldots, \boldsymbol{\xi}^{(n)}$ is symmetric. But, by definition, the matrix elements of $\mathbf{K}$ are

$$K_{ij} = (\boldsymbol{\xi}^{(i)}, \mathbf{K}\boldsymbol{\xi}^{(j)}) = \boldsymbol{\xi}^{(i)^T}\mathbf{T}^{-1}\mathbf{V}\boldsymbol{\xi}^{(j)} = \boldsymbol{\xi}^{(i)^T}\mathbf{V}\boldsymbol{\xi}^{(j)}, \qquad (5.173)$$

whence, upon using Eq. (5.170), we infer at once that $K_{ij} = K_{ji}$. Thus, we have proved the assertion made in Section 5.4 that it is possible to find n linearly independent solutions to Eq. (5.95).

As a last comment, note that the positivity of $\mathbf{V}$ was not used in the proof that the modal matrix simultaneously diagonalises $\mathbf{T}$ and $\mathbf{V}$. Therefore, from a strictly mathematical point of view, we have proved that given two real quadratic forms the positivity of one of them suffices for the existence of a matrix that simultaneously diagonalises them. This simultaneous diagonalisation admits a geometric interpretation which is easily visualisable in the case of systems with only two degrees of freedom (Aravind, 1989).

Problems

5.1 A particle of mass m moves on the positive half-line $x > 0$ with potential energy

$$V(x) = V_0 \left(e^{-kx} + Cx \right),$$

where the constants V_0, k, C are positive. Find the values of the constants for which there exist equilibrium positions and obtain the frequency of the small oscillations about the stable equilibrium positions.

5.2 With the zero level for the gravitational potential energy containing the lowest position, the energy of a simple pendulum is

$$E = \frac{ml^2}{2}\dot{\theta}^2 + mgl(1 - \cos\theta) = \frac{ml^2}{2}\dot{\theta}^2 + 2mgl \sin^2\left(\frac{\theta}{2}\right).$$

Let $\theta_0 < \pi$ be the amplitude of the pendulum's oscillation. With the help of the variable $y = \sin(\theta/2)$, show that the solution of the equation of motion for the angle θ is

$$\theta(t) = 2 \arcsin\left[\kappa \, \text{sn}\left(\omega_0 t + \delta\right)\right],$$

where δ is an arbitrary constant, $\omega_0 = \sqrt{g/l}$ and the modulus of the elliptic sine is

$$\kappa = \sin\left(\frac{\theta_0}{2}\right) = \sqrt{\frac{E}{2mgl}}.$$

Show that the pendulum's period of oscillation is

$$T = \frac{4}{\omega_0} K(\kappa),$$

where $K(\kappa)$ is the complete elliptic integral of the first kind defined by (5.50).

5.3 A bead of mass m slides without friction on a hoop of radius R that is rotating about its vertical diameter with constant angular velocity ω. With the origin at the centre of the hoop and θ the angle relative to the vertical defined by the bead's position, the cartesian coordinates of the bead are $x = R \sin\theta \cos\phi$, $y = R \sin\theta \sin\phi$, $z = -R \cos\theta$, where $\phi = \omega t$ is the angle of rotation about the vertical z-axis. (a) If the zero level for the gravitational potential energy contains the lowest point of the hoop, show that the Lagrangian for the system is given by

$$L = \frac{m}{2} R^2 \dot{\theta}^2 - mgR \left[(1 - \cos\theta) - \frac{1}{2} \frac{\omega^2}{\omega_c^2} \sin^2\theta \right],$$

where $\omega_c = \sqrt{g/R}$. (b) Find the equilibrium positions and classify them as to stability. (c) Show that this system also exhibits spontaneous symmetry breaking (Sivardière, 1983).

5.4 The Lagrangian for a generic double pendulum is given in Example 1.19. Consider the small oscillations in the particular case $l_1 = l_2 = l$, $m_1 = m$, $m_2 = m/3$. (a) Determine the normal modes of vibration and represent them pictorially. (b) Find the position of each particle at time t if when $t = 0$ both were at rest displaced by the same angle θ_0 relative to the vertical. (c) Obtain the normal coordinates of the system.

5.5 Assume that the vectors $\varrho^{(s)}$ obey (5.125) and write the general solution to the equations of motion (5.90) in the form

$$\boldsymbol{\eta}(t) = \frac{1}{2} \sum_{s=1}^{n} \varrho^{(s)} \left[A^{(s)} e^{i\omega_s t} + \overline{A}^{(s)} e^{-i\omega_s t} \right],$$

where $\overline{A}^{(s)}$ is the complex conjugate of $A^{(s)}$. Prove that the complex coefficients $A^{(s)}$ are determined by the initial conditions as

$$A^{(s)} = \left(\boldsymbol{\eta}(0), \varrho^{(s)} \right) - \frac{i}{\omega_s} \left(\dot{\boldsymbol{\eta}}(0), \varrho^{(s)} \right).$$

Conclude that the unique solution to the equations of motion (5.90) can be expressed in terms of the initial conditions as follows:

$$\boldsymbol{\eta}(t) = \sum_{s=1}^{n} \left[\left(\boldsymbol{\eta}(0), \varrho^{(s)} \right) \cos\omega_s t + \frac{\left(\dot{\boldsymbol{\eta}}(0), \varrho^{(s)} \right)}{\omega_s} \sin\omega_s t \right] \varrho^{(s)}.$$

5.6 A particle moves without friction on the concave side of a surface of revolution $z = f(\rho)$, where ρ, ϕ, z are cylindrical coordinates, in a uniform gravitational field $\mathbf{g} = -g\hat{\mathbf{z}}$. (a) Find the condition for the existence of a circular orbit of radius ρ_0.

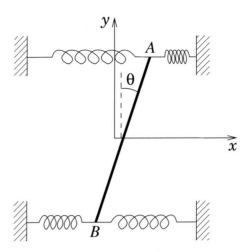

Fig. 5.11 Problem 5.7.

(b) Show that a circular orbit is stable or unstable according to whether $3f'(\rho_0) +$ $\rho_0 f''(\rho_0)$ is positive or negative. (c) Find the frequency of the small oscillations about the stable circular orbits.

5.7 The system depicted in Fig. 5.11 is on a smooth horizontal table. All springs are equal, with force constant k, and the homogeneous rod AB has mass m and length $2l$. The centre of mass of the rod does not move in the y-direction. Using as generalised coordinates the abscissa x of the centre of mass of the rod and the small angle θ the rod makes with the y-axis, find: (a) the Lagrangian, and (b) the characteristic frequencies and the normal modes of vibration. (c) Represent the normal modes pictorially.

5.8 Two equal blocks of mass m on an air track are connected by a spring of force constant k'. From their opposite ends, the blocks are connected to two fixed supports by means of springs with the same force constant k. In the equilibrium configuration all springs have their natural length. (a) Determine the characteristic frequencies and the normal modes of vibration. (b) Justify physically the value of the smallest characteristic frequency. (c) Represent pictorially the normal modes of vibration. (d) Construct the modal matrix and the normal coordinates. Check directly that the normal coordinates diagonalise the Lagrangian. (e) Find the position of each block at time t given the following initial conditions when $t = 0$: (e1) both blocks in the equilibrium position, the first with velocity v_0 and the second at rest; (e2) both blocks at rest, the first in the equilibrium position and the second displaced by a. In this case, describe qualitatively the motion of each block if $k' \ll k$.

5.9 A thin homogeneous ring of mass m and radius R oscillates in a vertical plane with point O fixed (see Fig. 5.12). A bead of the same mass m slides without friction along the ring. (a) Show that a Lagrangian for the system is

$$L = \frac{3mR^2}{2}\dot{\theta}_1^2 + \frac{mR^2}{2}\dot{\theta}_2^2 + mR^2\,\dot{\theta}_1\,\dot{\theta}_2\cos(\theta_1 - \theta_2) + 2mgR\,\cos\theta_1 + mgR\cos\theta_2\,.$$

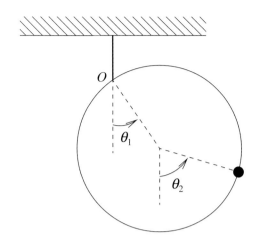

Fig. 5.12   Problem 5.9.

(b) Considering small oscillations, find the characteristic frequencies and the normal modes of vibration. Represent pictorially the normal modes. (c) Obtain the solution of the equations of motion with the initial conditions $\theta_1(0) = 0$, $\theta_2(0) = \theta_0$, $\dot{\theta}_1(0) = \dot{\theta}_2(0) = 0$. (d) Construct the modal matrix and the normal coordinates for the system.

5.10 Given the Lagrangian $L(q, \dot{q}, t)$ consider the new Lagrangian $\lambda(q, \dot{q}, \eta, \dot{\eta}, t)$ defined by Núñes-Yepes, Delgado & Salas-Brito (2001) as

$$\lambda(q, \dot{q}, \eta, \dot{\eta}, t) = \sum_{l=1}^{n} \left(\frac{\partial L}{\partial q_l} \eta_l + \frac{\partial L}{\partial \dot{q}_l} \dot{\eta}_l \right).$$

(a) Show that the variational principle $\delta \int \lambda(q, \dot{q}, \eta, \dot{\eta}, t) dt = 0$ yields in a fell swoop Lagrange's equations for $q_k(t)$ and the equations of motion for the small difference $\eta_k(t)$ between two dynamical trajectories $q_k(t)$ and $\bar{q}_k(t) = q_k(t) + \eta_k(t)$ – that is, prove that the equations

$$\frac{d}{dt}\left(\frac{\partial \lambda}{\partial \dot{\eta}_k} \right) - \frac{\partial \lambda}{\partial \eta_k} = 0, \qquad \frac{d}{dt}\left(\frac{\partial \lambda}{\partial \dot{q}_k} \right) - \frac{\partial \lambda}{\partial q_k} = 0$$

are equivalent to

$$\frac{d}{dt}\left(\frac{\partial L}{\partial \dot{q}_k} \right) - \frac{\partial L}{\partial q_k} = 0, \qquad \sum_{l=1}^{n} \left(M_{kl}\ddot{\eta}_l + C_{kl}\dot{\eta}_l + K_{kl}\eta_l \right) = 0,$$

where

$$M_{kl} = \frac{\partial^2 L}{\partial \dot{q}_k \partial \dot{q}_l}, \quad C_{kl} = \frac{d}{dt}\left(\frac{\partial^2 L}{\partial \dot{q}_k \partial \dot{q}_l} \right) + \frac{\partial^2 L}{\partial \dot{q}_k \partial q_l} - \frac{\partial^2 L}{\partial q_k \partial \dot{q}_l},$$

$$K_{kl} = \frac{d}{dt}\left(\frac{\partial^2 L}{\partial \dot{q}_k \partial q_l} \right) - \frac{\partial^2 L}{\partial q_k \partial q_l}.$$

(b) For the Lagrangian $L = \frac{1}{2}\sum_{k,l=1}^{n} m_{kl}\dot{q}_k\dot{q}_l - V(q)$ with the constant coefficients $m_{kl} = m_{lk}$, show that the equations in (a) for the $\eta_k(t)$ reduce to

$$\sum_{l=1}^{n}\left(m_{kl}\ddot{\eta}_l + \frac{\partial^2 V}{\partial q_k \partial q_l}\eta_l\right) = 0,$$

and verify directly that these equations coincide with the ones satisfied by the small deviations between two solutions of

$$\sum_{l=1}^{n} m_{kl}\ddot{q}_l = -\frac{\partial V}{\partial q_k}.$$

5.11 The Lagrangian

$$L = \frac{m}{2}\sum_{l=1}^{N-1}\dot{\eta}_l^2 - \frac{k}{2}\sum_{l=0}^{N-1}(\eta_{l+1} - \eta_l)^2$$

describes a linear chain of $N - 1$ particles of the same mass m connected by equal springs with force constant k. The particles at both extremities of the chain are fixed:

$$\eta_0 = 0, \quad \eta_N = 0.$$

(a) Find the equations of motion for $\eta_1, \cdots, \eta_{N-1}$ and show that they admit the normal modes of vibration

$$\eta_l^{(s)} = C^{(s)}\sin\frac{sl\pi}{N}\cos(\omega_s t + \phi_s) \equiv \varrho_l^{(s)}\cos(\omega_s t + \phi_s), \quad s,l = 1,\cdots,N-1$$

with frequencies

$$\omega_s = \sqrt{\frac{4k}{m}}\sin\frac{s\pi}{2N}.$$

(b) Show that the frequencies are all distinct and conclude that the corresponding eigenvectors are mutually orthogonal. Prove that each vector $\varrho^{(s)}$ is normalised by the choice $C^{(s)} = \sqrt{2/Nm}$, the same for all vectors. (c) If when $t = 0$ only the first particle is displaced from its equilibrium position and all particles are at rest, find $\eta_l(t)$. Hint: refer to Problem 5.5. (d) Show that the orthonormalisation condition for the vectors $\varrho^{(s)}$ leads to the following non-trivial trigonometric identity:

$$\sum_{j=1}^{N-1}\sin\left(j\frac{r\pi}{N}\right)\sin\left(j\frac{s\pi}{N}\right) = \frac{N}{2}\delta_{rs}, \quad r,s = 1,\ldots,N-1.$$

Multiplying this last equation from the right by $\mathbf{G}$ and noting that $\mathbf{G}^2 = \mathbf{I}$ there results $\mathbf{\Lambda}^T \mathbf{G} \mathbf{\Lambda} \mathbf{G} = \mathbf{I}$, which shows that $\mathbf{\Lambda}^T \mathbf{G}$ is the left inverse of $\mathbf{\Lambda} \mathbf{G}$. Since the left inverse of a matrix coincides with the right inverse, it is also the case that $\mathbf{\Lambda} \mathbf{G} \mathbf{\Lambda}^T \mathbf{G} = \mathbf{I}$, whence

$$\mathbf{\Lambda} \mathbf{G} \mathbf{\Lambda}^T = \mathbf{G}, \tag{6.21}$$

which is a condition equivalent to (6.20). A Lorentz transformation is a kind of rotation in spacetime and the associated matrix is orthogonal with respect to the metric $\mathbf{G}$ of Minkowski space. In the case of three-dimensional rotations, the metric is the identity matrix and the condition corresponding to (6.20) is Eq. (3.12).

Exercise 6.1.2   Using (6.20) and (6.21), prove that the set of Lorentz transformations form a group, called the **Lorentz group**. Hints: (1) the definition of a group is near the end of Section 3.1; (2) if $\mathbf{G}$ and $\mathbf{H}$ are invertible matrices then $(\mathbf{G}\mathbf{H})^{-1} = \mathbf{H}^{-1}\mathbf{G}^{-1}$; (3) the transpose of the inverse is the inverse of the transpose.

It should be stressed that the definition (6.20) comprises a much larger class of linear transformations than the pure Lorentz transformations of the form (6.6). For example, a rotation of the spatial axes is a Lorentz transformation according to our general definition, since it does not affect the time and leaves invariant the spatial distance between two points. The space inversion $t' = t$, $\mathbf{r}' = -\mathbf{r}$ and the time reversal $t' = -t$, $\mathbf{r}' = \mathbf{r}$ are also examples of Lorentz transformations, since both preserve ds^2.

The main properties of the Lorentz transformations can be inferred from (6.20) or (6.21). Taking the determinant of (6.20) we deduce

$$-1 = \det \mathbf{G} = \det(\mathbf{\Lambda}^T \mathbf{G} \mathbf{\Lambda}) = \det \mathbf{\Lambda}^T \det \mathbf{G} \det \mathbf{\Lambda} = (\det \mathbf{\Lambda})(-1)(\det \mathbf{\Lambda}), \tag{6.22}$$

which amounts to

$$(\det \mathbf{\Lambda})^2 = 1 \quad \Longrightarrow \quad \det \mathbf{\Lambda} = \pm 1. \tag{6.23}$$

A Lorentz transformation with determinant $+1$ is said to be **proper**, and it is said to be **improper** if the determinant is -1. Any pure Lorentz transformation such as (6.1) is proper (check it).

Taking $\alpha = \beta = 0$ in (6.19) we obtain

$$1 = g_{00} = \Lambda^\mu{}_0 \, g_{\mu\nu} \, \Lambda^\nu{}_0 = (\Lambda^0{}_0)^2 - (\Lambda^1{}_0)^2 - (\Lambda^2{}_0)^2 - (\Lambda^3{}_0)^2, \tag{6.24}$$

that is,

$$(\Lambda^0{}_0)^2 = 1 + (\Lambda^1{}_0)^2 + (\Lambda^2{}_0)^2 + (\Lambda^3{}_0)^2 \geq 1. \tag{6.25}$$

Therefore,

$$\Lambda^0{}_0 \geq +1 \quad \text{or} \quad \Lambda^0{}_0 \leq -1. \tag{6.26}$$

If $\Lambda^0{}_0 \geq +1$ the Lorentz transformation is said to be **orthocronous** (it preserves the direction of time); if $\Lambda^0{}_0 \leq -1$ the Lorentz transformation is said to be **non-orthocronous** (it reverses the direction of time). Any **pure** Lorentz transformation connecting reference frames with parallel axes in relative motion with constant velocity is orthocronous.[7]

---

[7] Pure Lorentz transformations are also known as *boosts*.

Time reversal is a non-orthocronous Lorentz transformation. The proper orthocronous Lorentz transformations are called **restricted** Lorentz transformations. The infinitesimal Lorentz transformations play a very important role in the study of the properties of the Lorentz group. An infinitesimal Lorentz transformation is necessarily restricted because it differs only infinitesimally from the identity transformation, which is proper and orthocronous, and an infinitesimal variation cannot cause a finite jump either to $\Lambda^0{}_0$ or to the determinant of Λ. Therefore, only the restricted Lorentz transformations can be constructed by the successive application of infinitesimal Lorentz transformations. The present discussion establishes that the set of Lorentz transformations decomposes itself into four disjoint subsets, called the sectors of the Lorentz group and denoted by $L_+^\uparrow, L_+^\downarrow, L_-^\uparrow$ and $L_-^\downarrow$.

Sectors of the Lorentz Group

$L_+^\uparrow :$ $\det \Lambda = +1$ and $\Lambda^0{}_0 \geq +1$

$L_+^\downarrow :$ $\det \Lambda = +1$ and $\Lambda^0{}_0 \leq -1$

$L_-^\uparrow :$ $\det \Lambda = -1$ and $\Lambda^0{}_0 \geq +1$

$L_-^\downarrow :$ $\det \Lambda = -1$ and $\Lambda^0{}_0 \leq -1$

The set of restricted Lorentz transformations $L_+^\uparrow$ is the only sector that forms a subgroup of the Lorentz group. This subgroup is called the **restricted Lorentz group**. The only non-trivial part of the proof that $L_+^\uparrow$ constitutes a group is the verification that the product of two orthocronous Lorentz matrices is also an orthocronous matrix. Let Λ and $\bar{\Lambda}$ be two consecutive orthocronous Lorentz transformations and let $\bar{\bar{\Lambda}} = \bar{\Lambda}\Lambda$. Then $\bar{\bar{\Lambda}}^\mu{}_\nu = \bar{\Lambda}^\mu{}_\alpha \Lambda^\alpha{}_\nu$, so that

$$\bar{\bar{\Lambda}}^0{}_0 = \bar{\Lambda}^0{}_0 \Lambda^0{}_0 + \bar{\Lambda}^0{}_1 \Lambda^1{}_0 + \bar{\Lambda}^0{}_2 \Lambda^2{}_0 + \bar{\Lambda}^0{}_3 \Lambda^3{}_0 . \tag{6.27}$$

Setting $\mathbf{a} = (\bar{\Lambda}^0{}_1, \bar{\Lambda}^0{}_2, \bar{\Lambda}^0{}_3)$, $\mathbf{b} = (\Lambda^1{}_0, \Lambda^2{}_0, \Lambda^3{}_0)$ and using (6.25) as well as its analogue from (6.21), we are left with

$$\bar{\bar{\Lambda}}^0{}_0 = \sqrt{1 + a^2} \sqrt{1 + b^2} + \mathbf{a} \cdot \mathbf{b} , \tag{6.28}$$

where $a = |\mathbf{a}|$ and $b = |\mathbf{b}|$. Since

$$\sqrt{1 + a^2}\sqrt{1 + b^2} = \sqrt{1 + a^2 + b^2 + a^2 b^2} = \sqrt{(1 + ab)^2 + (a - b)^2} \geq 1 + ab , \tag{6.29}$$

it follows that

$$\bar{\bar{\Lambda}}^0{}_0 \geq 1 + ab + \mathbf{a} \cdot \mathbf{b} . \tag{6.30}$$

Taking into account that $\mathbf{a} \cdot \mathbf{b} \geq -ab$ we obtain $\bar{\bar{\Lambda}}^0{}_0 \geq 1$, as we wanted to show.

6.2 Light Cone and Causality

An interval $\Delta s^2 = c^2 \Delta t^2 - \Delta x^2 - \Delta y^2 - \Delta z^2$ between two events is said to be **space-like** if the space part is larger that the time part ($\Delta s^2 < 0$), and **time-like** if the time part is larger than the space part ($\Delta s^2 > 0$). If $\Delta s^2 = 0$ the interval is said to be **light-like**. Note that

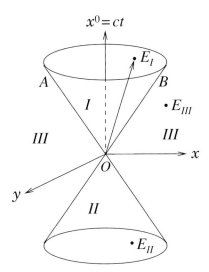

Fig. 6.2 Light cone of event O (one spatial dimension has been suppressed).

this classification of intervals is absolute – in other words, it is valid in all inertial reference frames because Δs^2 is a scalar, an invariant.

Consider, now, an event O and take it as the origin of an inertial frame $\mathcal{K}$. In Fig. 6.2 the vertical axis represents $x^0 = ct$ and the other axes represent space (the z-direction is suppressed). The straight lines A and B, inclined by $45°$ with respect to the time axis, represent the **worldlines** (trajectories in spacetime) of light rays travelling in opposite directions but whose emission is event O. The worldlines of light rays propagating in all possible directions intersecting at event O form a cone with apex at O, called the **light cone** of O, which separates spacetime in three regions. Region I is called the **absolute future** of O because all inertial observers agree that the event O precedes any event in that region. Also, any event E_I in I is separated from O by a time-like interval. The magnitude of the horizontal projection of a vector such as $\overrightarrow{OE_I}$ is the spatial distance between the events, whereas the vertical projection is $c\Delta t$, where Δt is the time interval between the events. Note that $c\Delta t$ is the distance travelled by a light ray during the time interval between the events. Since the spatial distance between events O and E_I is smaller than c times the time lapse between them, there may be a cause and effect relationship between O and E_I. For example, event O might represent an explosion and event E_I represent the arrival of the boom created by the explosion. Region II is the **absolute past** of O: all inertial observers agree that the event O is later than any event in region II. Besides, there may be a cause-and-effect relationship between E_{II} and O because the interval between these events is time-like. Region III, the exterior of the light cone of event O, consists of those events that cannot have any causal connexion with O. For instance, the distance travelled by light during the time interval between events O and E_{III} is smaller than their spatial separation, so the existence of a cause-and-effect relationship between O and E_{III} would require a physical process capable of propagating faster than light, which is forbidden by the special theory of relativity.

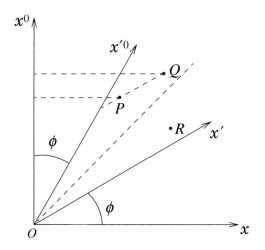

Representation of a Lorentz transformation by a Minkowski diagram.

Graphical representations of spacetime such as Fig. 6.2, known as **Minkowski diagrams**, are useful to illustrate some of the main properties of Lorentz transformations. Let us represent by orthogonal axes the coordinates $x^0 = ct$ and x of the inertial frame $\mathcal{K}$, as in Fig. 6.3. Let us consider now the frame $\mathcal{K}'$ that moves relative to $\mathcal{K}$ with velocity v along the x-direction. The straight line $x' = 0$, which corresponds to the x'^0-axis of $\mathcal{K}'$, satisfies the equation $x = \beta x^0$ with $\beta = v/c$, as one infers from (6.1). Therefore, the angle ϕ between the axes x^0 and x'^0 is smaller than $45°$ because $\tan \phi = \beta < 1$. Similarly, the x'-axis is defined by $x'^0 = 0$ and corresponds to the straight line $x^0 = \beta x$, which makes the same angle ϕ with the x-axis. A Lorentz transformation amounts to a transition from the perpendicular axes (x^0, x) to the inclined axes (x'^0, x'). The coordinates of an event are obtained by drawing from the event lines parallel to each axis and finding where they intersect the other axis. By means of these diagrams it is easy to see the relativity of simultaneity, one of the hallmarks of special relativity. Given two spatially separated events P and Q, which are simultaneous in $\mathcal{K}'$, they are not simultaneous in $\mathcal{K}$, as Fig. 6.3 makes clear (the segment $\overline{PQ}$ is parallel to the x'-axis). On the other hand, if R is an event separated from O by a space-like interval, there exists an inertial frame in which the two events O and R occur at the same time.

The worldline of a particle is represented in a Minkowski diagram by a curve as in Fig. 6.4(a). The curve is not arbitrary, however, because from $dx/dx^0 = \beta < 1$ it follows that the tangent to the worldline of the particle always makes an angle with the time axis less that $45°$. If the particle's spacetime trajectory contains event O, its worldline will be entirely contained inside the light cone of event O. The worldline of a light ray is a straight line inclined by $45°$ with respect to the coordinate axes. As a further example of the utility of spacetime diagrams, let us show in another way the impossibility of causal connexion between events separated by a space-like interval. In Fig. 6.4(b), suppose that O represents the emission and P the reception of a hypothetical signal – which would have to be capable of propagating in $\mathcal{K}$ with a speed bigger than the speed of light in vacuum. Then, as seen in

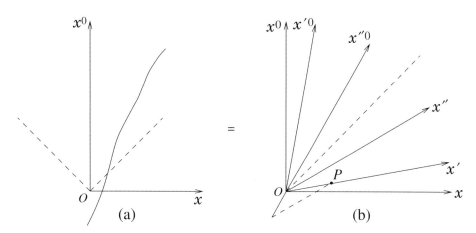

Fig. 6.4   Minkowski diagrams.

the previous paragraph, there would exist an inertial frame $\mathcal{K}'$ in which the propagation of this signal would be instantaneous – that is, its speed would be infinite. Even worse, there would be an inertial frame $\mathcal{K}''$ in which the signal would be received at instant $x_P''^0 < 0$ – that is, before it was emitted! In spite of its undeniable utility, we will make use of these diagrams only occasionally because, in general, algebraic and analytic arguments are much more powerful.

6.3 Vectors and Tensors

The principle of relativity states that the laws of physics take the same form in all inertial frames. In order to enforce this principle, the mathematical expression of physical laws must involve only quantities with definite transformation rules under a change of inertial frames.

Definition 6.3.1   *A **scalar** is a quantity which is invariant under Lorentz transformations – that is, whose value is the same in all inertial frames.*

The electric charge of a particle, the spacetime interval between two events and the speed of light in vacuum are examples of scalars.

Definition 6.3.2   *A **contravariant four-vector** (or just a **contravariant vector**) is a set ot four quantities (V^0, V^1, V^2, V^3) which under a Lorentz transformation change in the same way as the displacement (dx^0, dx^1, dx^2, dx^3), namely*

$$V'^{\mu} = \Lambda^{\mu}{}_{\nu} V^{\nu}. \tag{6.31}$$

In short, the prototypical contravariant vector is the displacement vector dx^{μ}.

Theorem 6.3.1   *If V^μ and W^μ are contravariant vectors then*

$$g_{\mu\nu} V^\mu W^\nu = V^0 W^0 - V^1 W^1 - V^2 W^2 - V^3 W^3 = scalar. \qquad (6.32)$$

This theorem is an immediate consequence of (6.19), and its validity suggests introducing an operation of lowering indices defined by

$$V_\mu = g_{\mu\nu} V^\nu, \qquad (6.33)$$

since with this operation Eq. (6.32) takes the usual form $V^\mu W_\mu$ of a scalar product.

Definition 6.3.3   *The* **scalar product** *of two contravariant vectors V^μ and W^μ is defined by*[8]

$$\mathbf{V} \cdot \mathbf{W} = V_\mu W^\mu = V_0 W^0 + V_1 W^1 + V_2 W^2 + V_3 W^3, \qquad (6.34)$$

with V_μ given by (6.33). In particular,

$$\mathbf{V}^2 \equiv \mathbf{V} \cdot \mathbf{V} = (V^0)^2 - (V^1)^2 - (V^2)^2 - (V^3)^2. \qquad (6.35)$$

Exercise 6.3.1   Prove that

$$V_\mu W^\mu = V^\mu W_\mu. \qquad (6.36)$$

Show further that if $V^\mu = (V^0, V^1, V^2, V^3)$ then

$$V_\mu = (V^0, -V^1, -V^2, -V^3). \qquad (6.37)$$

Thus, the operation of lowering an index merely changes the sign of the spatial components of the four-vector.

It is convenient to introduce the elements $g^{\mu\nu}$ of the inverse of the matrix $\mathbf{G}$ defined by (6.14). But, inasmuch as $\mathbf{G}^{-1} = \mathbf{G}$, it is obvious that $g^{\mu\nu}$ and $g_{\mu\nu}$ take the same numerical values. Therefore

$$\mathbf{G}^{-1} = (g^{\mu\nu}) = \begin{pmatrix} 1 & 0 & 0 & 0 \\ 0 & -1 & 0 & 0 \\ 0 & 0 & -1 & 0 \\ 0 & 0 & 0 & -1 \end{pmatrix} \qquad (6.38)$$

and, evidently,

$$g^{\mu\alpha} g_{\alpha\nu} = \delta^\mu_\nu. \qquad (6.39)$$

Note further that both $g^{\mu\nu}$ and $g_{\mu\nu}$ are symmetric in their indices:

$$g^{\mu\nu} = g^{\nu\mu}, \qquad g_{\mu\nu} = g_{\nu\mu}. \qquad (6.40)$$

With the help of $g^{\mu\nu}$ we define the operation of raising an index by

$$V^\mu = g^{\mu\nu} V_\nu. \qquad (6.41)$$

---

[8] In this chapter, boldface capital letters from the second half of the alphabet will be occasionally employed to denote four-vectors.

It is clear that lowering the index of a contravariant vector V^μ by (6.33) followed by raising the index by (6.41) reproduces the original vector V^μ.

If ϕ is a scalar function, its differential

$$d\phi = \frac{\partial \phi}{\partial x^0}\,dx^0 + \frac{\partial \phi}{\partial x^1}\,dx^1 + \frac{\partial \phi}{\partial x^2}\,dx^2 + \frac{\partial \phi}{\partial x^3}\,dx^3 = \frac{\partial \phi}{\partial x^\mu}\,dx^\mu \qquad (6.42)$$

is also a scalar. Comparing this equation with (6.34) we conclude that the gradient of a scalar function is a vector distinct from the contravariant sort, and its transformation law is identical to that of V_μ as defined by (6.33) from a contravariant vector V^μ. Let us, therefore, see how V_μ transforms:

$$V'_\mu = g_{\mu\nu}\,V'^\nu = g_{\mu\nu}\,\Lambda^\nu{}_\alpha\,V^\alpha = g_{\mu\nu}\,\Lambda^\nu{}_\alpha\,g^{\alpha\beta}\,V_\beta \equiv \Lambda_\mu{}^\beta\,V_\beta\,, \qquad (6.43)$$

with the definition (pay attention to the placement of indices)

$$\Lambda_\mu{}^\beta = g_{\mu\nu}\,g^{\beta\alpha}\,\Lambda^\nu{}_\alpha\,. \qquad (6.44)$$

Definition 6.3.4   *A **covariant four-vector** (or just **covariant vector**) V_μ is defined by the transformation law*

$$V'_\mu = \Lambda_\mu{}^\nu\,V_\nu \qquad (6.45)$$

where the coefficients $\Lambda_\mu{}^\nu$ are determined in terms of the elements of the Lorentz transformation matrix by (6.44). In short, the prototypical covariant vector is the gradient of a scalar function $\partial_\mu \phi \equiv \partial \phi / \partial x^\mu$.

Exercise 6.3.2   Making use of Eqs. (6.19) and (6.39) show that

$$\Lambda_\mu{}^\alpha\,\Lambda^\mu{}_\beta = \delta^\alpha_\beta\,, \qquad (6.46)$$

and from this conclude that the matrix $(\Lambda_\mu{}^\nu)$ is the transpose of the inverse of the Lorentz transformation matrix $\mathbf{\Lambda} = (\Lambda^\mu{}_\nu)$.

Definition 6.3.5   *A **contravariant tensor** of rank r is an object $T^{\alpha_1 \alpha_2 \cdots \alpha_r}$ with 4^r components which under a Lorentz transformation change according to the rule*

$$T'^{\alpha_1 \alpha_2 \cdots \alpha_r} = \Lambda^{\alpha_1}{}_{\beta_1}\,\Lambda^{\alpha_2}{}_{\beta_2}\,\cdots\,\Lambda^{\alpha_r}{}_{\beta_r}\,T^{\beta_1 \beta_2 \cdots \beta_r}\,. \qquad (6.47)$$

*A **covariant tensor** $Z_{\alpha_1 \alpha_2 \cdots \alpha_r}$ of rank r is defined by the transformation law*

$$Z'_{\alpha_1 \alpha_2 \cdots \alpha_r} = \Lambda_{\alpha_1}{}^{\beta_1}\,\Lambda_{\alpha_2}{}^{\beta_2}\,\cdots\,\Lambda_{\alpha_r}{}^{\beta_r}\,Z_{\beta_1 \beta_2 \cdots \beta_r}\,. \qquad (6.48)$$

Mixed tensors, with covariant and contravariant indices, are similarly defined.

Examples of second-rank contravariant tensors are

$$V^\mu W^\nu\,, \qquad V^\mu W^\nu + V^\nu W^\mu\,, \qquad V^\mu W^\nu - V^\nu W^\mu\,, \qquad (6.49)$$

where V^μ and W^μ are contravariant vectors.

The operations of lowering and raising an index can be applied to tensors of any rank. For example:

$$T_\mu{}^\nu = g_{\mu\alpha}\,T^{\alpha\nu}\,, \qquad\qquad T^\mu{}_\nu = g_{\nu\alpha}\,T^{\mu\alpha}\,, \qquad (6.50\text{a})$$

$$T_{\mu\nu} = g_{\mu\alpha}\,g_{\nu\beta}\,T^{\alpha\beta}\,, \qquad\qquad T^{\mu\nu} = g^{\mu\alpha}\,g^{\nu\beta}\,T_{\alpha\beta}\,. \qquad (6.50\text{b})$$

The scalar product of a vector and a tensor reduces by one the rank of the tensor. For example:

$$V_\mu \, T^{\alpha\beta\mu} = \text{second-rank tensor}; \tag{6.51a}$$

$$V_\mu \, Y^{\mu\nu} = \text{first-rank tensor} = \text{vector}. \tag{6.51b}$$

Let us, for instance, verify this last statement. Defining $W^\nu = V_\mu \, Y^{\mu\nu}$ and using equations (6.45)–(6.47) we have

$$W'^\nu = V'_\mu \, Y'^{\mu\nu} = \Lambda_\mu{}^\alpha \, V_\alpha \, \Lambda^\mu{}_\beta \, \Lambda^\nu{}_\gamma \, Y^{\beta\gamma} = \Lambda_\mu{}^\alpha \, \Lambda^\mu{}_\beta \, V_\alpha \, \Lambda^\nu{}_\gamma \, Y^{\beta\gamma}$$

$$= \delta^\alpha_\beta \, V_\alpha \, \Lambda^\nu{}_\gamma \, Y^{\beta\gamma} = V_\beta \, \Lambda^\nu{}_\gamma \, Y^{\beta\gamma} = \Lambda^\nu{}_\gamma \, (V_\beta \, Y^{\beta\gamma}) = \Lambda^\nu{}_\gamma \, W^\gamma, \tag{6.52}$$

which establishes that W^ν is a contravariant vector.

An operation such as

$$T^\mu{}_\mu{}^\alpha, \qquad T^{\mu\nu}{}_\nu, \qquad T^{\mu\nu}{}_\mu, \qquad Z^{\alpha\mu\nu}{}_\alpha, \tag{6.53}$$

is called a **contraction**. Thus, for instance, $T^\mu{}_\mu{}^\alpha$ is the contraction of the first two indices of $T^{\mu\nu\alpha}$. Clearly, a contraction reduces the rank of a tensor by two. In the case of a second-rank tensor $T^{\mu\nu}$ the scalar[9]

$$T^\mu{}_\mu = T_\mu{}^\mu \tag{6.54}$$

is called the trace of the tensor.

Note, finally, that the classification of intervals applies to vectors:

$$V_\mu \, V^\mu > 0 \quad \Longleftrightarrow \quad V^\mu \text{ is a time-like vector};$$
$$V_\mu \, V^\mu < 0 \quad \Longleftrightarrow \quad V^\mu \text{ is a space-like vector}; \tag{6.55}$$
$$V_\mu \, V^\mu = 0 \quad \Longleftrightarrow \quad V^\mu \text{ is a light-like vector}.$$

A light-like vector is also called a **null vector**. This classification is absolute because $V_\mu \, V^\mu$ is invariant under Lorentz transformations.

6.4 Tensor Fields

Fundamental physical theories such as classical electrodynamics, general relativity and the standard model of particle physics require tensor fields for their description.

Definition 6.4.1   *A **scalar field** is a function $\phi(P)$ whose value at each point P of space-time is invariant under Lorentz transformations – that is, $\phi'(P) = \phi(P)$. If the point P has coordinates $x \equiv (x^0, x^1, x^2, x^3)$ in the inertial reference frame $\mathcal{K}$ and $x' \equiv (x'^0, x'^1, x'^2, x'^3)$ in the inertial reference frame $\mathcal{K}'$, we have*

$$\phi'(x') = \phi(x). \tag{6.56}$$

---

[9] A scalar may be thought of as a zero-rank tensor.

Definition 6.4.2   *A* **tensor field** $T^{\mu\nu\cdots\rho}(P)$ *is a tensor associated with each point P of space-time and characterised by the transformation law*

$$T'^{\,\mu\nu\cdots\rho}(x') = \Lambda^{\mu}{}_{\alpha}\,\Lambda^{\nu}{}_{\beta}\cdots\Lambda^{\rho}{}_{\sigma}\,T^{\alpha\beta\cdots\sigma}(x)\,. \tag{6.57}$$

The four-dimensional nabla operator

$$\partial_{\mu} \equiv \frac{\partial}{\partial x^{\mu}} = \left(\frac{\partial}{\partial x^0}, \frac{\partial}{\partial x^1}, \frac{\partial}{\partial x^2}, \frac{\partial}{\partial x^3}\right) = \left(\frac{1}{c}\frac{\partial}{\partial t}, \nabla\right) \tag{6.58}$$

is a covariant vector. This is inferred from the fact that $\partial_{\mu}\phi$ is a covariant vector, as seen in Section 6.3, and from the definition of a scalar field. Indeed, combining

$$(\partial_{\mu}\phi)' = \Lambda_{\mu}{}^{\nu}\,(\partial_{\nu}\phi) \qquad \text{or} \qquad \partial'_{\mu}\phi' = \Lambda_{\mu}{}^{\nu}\,\partial_{\nu}\phi \tag{6.59}$$

with $\phi' = \phi$ we conclude that, as an operator,

$$\partial'_{\mu} = \Lambda_{\mu}{}^{\nu}\,\partial_{\nu}\,, \tag{6.60}$$

confirming the above assertion. Similarly,

$$\partial^{\mu} = g^{\mu\nu}\partial_{\nu} = \frac{\partial}{\partial x_{\mu}} = \left(\frac{\partial}{\partial x_0}, \frac{\partial}{\partial x_1}, \frac{\partial}{\partial x_2}, \frac{\partial}{\partial x_3}\right) = \left(\frac{1}{c}\frac{\partial}{\partial t}, -\nabla\right) \tag{6.61}$$

is a contravariant vector.

Tensor fields may have their rank increased by differentiation. For example, if ϕ is a scalar field then

$$\partial_{\mu}\,\partial_{\nu}\,\phi \equiv \frac{\partial^2\phi}{\partial x^{\mu}\partial x^{\nu}} = \text{second-rank covariant tensor field}\,. \tag{6.62}$$

If $T^{\mu\nu}$ is a second-rank contravariant tensor field it follows that

$$\partial^{\rho}T^{\mu\nu} = \text{third-rank contravariant tensor field}\,. \tag{6.63}$$

By means of a contraction the rank of a tensor field can be reduced by differentiation. For instance, if $F^{\mu\nu}$ is a second-rank contravariant tensor field it is easily checked that

$$\partial_{\nu}F^{\mu\nu} = \text{contravariant vector field}\,. \tag{6.64}$$

Divergence, Curl and d'Alembertian

Differential operations usually defined in three-dimensional space have their counterparts in Minkowski space.

Definition 6.4.3   *The* **divergence** *of a vector field* V^{μ} *is a scalar field denoted by div V and defined by*

$$div\,V = \partial_{\mu}V^{\mu}\,. \tag{6.65}$$

Only in the particular case of three-dimensional space is the curl of a vector field also a vector field. In all higher-dimensional cases the curl of a vector field is naturally defined as a second-rank antisymmetric tensor field. The exceptionality of three-dimensional space is easy to understand because only in three dimensions is there a bijective correspondence between vectors and antisymmetric matrices, as seen in Section 3.4.

Definition 6.4.4   *The **curl** of a vector field V^μ is a second-rank tensor field denoted by curl V and defined by*

$$(curl\,V)^{\mu\nu} = \partial^\mu V^\nu - \partial^\nu V^\mu\,. \tag{6.66}$$

The divergence of the gradient is known as the **d'Alembert operator** or just the **d'Alembertian**, which is the four-dimensional version of the Laplacian:

$$\Box \equiv \partial_\mu \partial^\mu = \frac{1}{c^2}\frac{\partial^2}{\partial t^2} - \frac{\partial^2}{\partial x^2} - \frac{\partial^2}{\partial y^2} - \frac{\partial^2}{\partial z^2} = \frac{1}{c^2}\frac{\partial^2}{\partial t^2} - \nabla^2\,. \tag{6.67}$$

This operator is a scalar, so its application does not change the rank of a tensor field.

6.5 Physical Laws in Covariant Form

The tensor apparatus developed in the previous sections aims at allowing the mathematical expression of the laws of physics in **covariant** form – that is, in a form **valid in all inertial frames**. Suppose that in a given inertial frame $\mathcal{K}$ a physical law can be expressed in the form

$$T^{\mu\nu\cdots\rho} = 0\,, \tag{6.68}$$

where $T^{\mu\nu\cdots\rho}$ is a tensor or a tensor field. In virtue of the homogeneous transformation laws of its components, (6.47) or (6.57), if a tensor or tensor field vanishes in one inertial frame it vanishes in all inertial frames. This means that in another inertial frame $\mathcal{K}'$ it is the case that

$$T'^{\mu\nu\cdots\rho} = 0 \tag{6.69}$$

if (6.68) holds in $\mathcal{K}$. In consequence, the physical law described by (6.68) in $\mathcal{K}$ has the same form in all inertial frames. We say that a physical law represented by an equation such as (6.68) is expressed in a **manifestly covariant** form, a mere glance at the equation being enough to recognise its validity in all inertial reference frames.

Given a set of physical laws known to be valid in all inertial frames, it is important to express them in a manifestly covariant form.[10] This, among other advantages, facilitates finding out how the relevant physical quantities transform under a change of inertial reference frame. A highly important example is provided by classical electrodynamics.

Electrodynamics in Covariant Form

In CGS Gaussian units Maxwell's equations in vacuum are

$$\nabla \cdot \mathbf{E} = 4\pi\,\rho\,, \quad \nabla \cdot \mathbf{B} = 0\,, \quad \nabla \times \mathbf{E} + \frac{1}{c}\frac{\partial \mathbf{B}}{\partial t} = 0\,, \quad \nabla \times \mathbf{B} - \frac{1}{c}\frac{\partial \mathbf{E}}{\partial t} = \frac{4\pi}{c}\,\mathbf{J}\,. \tag{6.70}$$

[10] When referring to an equation or a physical law, the word "covariant" means "valid with the same form in all inertial frames".

These equations predict that light propagates in vacuum with speed c. According to a fundamental postulate of the special theory of relativity, this is true in all inertial reference frames, which is compatible with Lorentz transformations but contradicts Galilean transformations. Therefore, Eqs. (6.70) must be valid in all inertial reference frames, so it must be possible to rewrite them in a manifestly covariant form *without altering their content*.

Let us start with the local conservation law of electric charge

$$\frac{\partial \rho}{\partial t} + \nabla \cdot \mathbf{J} = 0, \tag{6.71}$$

which follows easily from (6.70). Let us introduce the *tentative* four-vector

$$J^\mu = (c\,\rho\,, J_x\,, J_y\,, J_z) \equiv (c\,\rho\,, \mathbf{J})\,, \tag{6.72}$$

which allows us to rewrite (6.71) in the form

$$\partial_\mu J^\mu = 0\,. \tag{6.73}$$

Since experiment is consistent with the validity of the local conservation of electric charge in all inertial frames, we conclude that J^μ is a four-vector. According to Definition 6.4.3, Eq. (6.73) states that a scalar equals zero, which holds true in all inertial frames. Thus, (6.73) is the manifestly covariant form of the continuity equation (6.71). The four-vector J^μ defined by (6.72) is called the **charge-current density four-vector**, or just **four-current**.

Exercise 6.5.1   Prove that the charge and current densities transform as

$$\rho' = \gamma \left(\rho - \frac{v}{c^2} j_x \right)\,, \quad j'_x = \gamma\,(j_x - v\,\rho)\,, \quad j'_y = j_y\,, \quad j'_z = j_z \tag{6.74}$$

in the case of the pure Lorentz transformation (6.1).

It is very rewarding to introduce the potentials ϕ and $\mathbf{A}$ such that

$$\mathbf{B} = \nabla \times \mathbf{A}, \quad \mathbf{E} = -\nabla\phi - \frac{1}{c}\frac{\partial \mathbf{A}}{\partial t}\,. \tag{6.75}$$

Let us consider, therefore,

$$A^\mu = (\phi\,, \mathbf{A}) \tag{6.76}$$

as another tentative four-vector. The first of Eqs. (6.75) suggests introducing a tentative second-rank tensor $F^{\mu\nu}$ defined as the curl of A^μ:

$$F^{\mu\nu} = \partial^\mu A^\nu - \partial^\nu A^\mu\,. \tag{6.77}$$

The components of $F^{\mu\nu}$ are

$$(F^{\mu\nu}) = \begin{pmatrix} 0 & -E_x & -E_y & -E_z \\ E_x & 0 & -B_z & B_y \\ E_y & B_z & 0 & -B_x \\ E_z & -B_y & B_x & 0 \end{pmatrix}\,. \tag{6.78}$$

As a check let us consider, for example, F^{01}:

$$F^{01} = \partial^0 A^1 - \partial^1 A^0 = \partial_0 A_x + \partial_1 \phi = \frac{1}{c}\frac{\partial A_x}{\partial t} + \frac{\partial \phi}{\partial x} = -E_x\,.$$

Exercise 6.5.2 Verify that the remaining components of $F^{\mu\nu}$ are correctly given by Eq. (6.78).

Since its components are expressed in terms of the components of $\mathbf{E}$ and $\mathbf{B}$, $F^{\mu\nu}$ is called the **electromagnetic tensor** or **electromagnetic field tensor**. It is important to stress that $F^{\mu\nu}$ is an antisymmetric tensor:

$$F^{\mu\nu} = -F^{\nu\mu}.\tag{6.79}$$

The first and fourth of Maxwell's equations (6.70) suggest an equality with the divergence of $F^{\mu\nu}$ on the left-hand side and J^μ on the right-hand side. A more careful analysis is needed to uncover the appropriate covariant form for the other two equations. It turns out that Maxwell's equations take the covariant form

$$\partial_\mu F^{\mu\nu} = \frac{4\pi}{c} J^\nu,\tag{6.80}$$

$$\partial^\rho F^{\mu\nu} + \partial^\mu F^{\nu\rho} + \partial^\nu F^{\rho\mu} = 0.\tag{6.81}$$

Since J^μ is a four-vector and Eq. (6.80) is valid in all inertial frames, we conclude that $F^{\mu\nu}$ is indeed a second-rank tensor. Equations (6.80) coincide with the non-homogeneous Maxwell's equations (those that contain ρ and $\mathbf{J}$), whereas equations (6.81) – only four of them are independent – are identical to the homogeneous Maxwell's equations. For example, take $\nu = 0$ in (6.80):

$$\partial_\mu F^{\mu 0} = \frac{4\pi}{c} J^0 \implies \frac{\partial F^{00}}{\partial x^0} + \frac{\partial F^{10}}{\partial x^1} + \frac{\partial F^{20}}{\partial x^2} + \frac{\partial F^{30}}{\partial x^3} = \frac{4\pi}{c} c\rho \implies \nabla \cdot \mathbf{E} = 4\pi\,\rho.$$

Similarly, picking $\rho = 0$, $\mu = 1$, $\nu = 2$ in (6.81) we find

$$\partial^0 F^{12} + \partial^1 F^{20} + \partial^2 F^{01} = 0 \implies \frac{1}{c}\frac{\partial(-B_z)}{\partial t} - \frac{\partial E_y}{\partial x} - \frac{\partial(-E_x)}{\partial y} = 0,$$

which is the same as

$$\left(\nabla \times \mathbf{E} + \frac{1}{c}\frac{\partial \mathbf{B}}{\partial t}\right)_z = 0.$$

In summary, (6.80) and (6.81) are the expression of Maxwell's equations in a manifestly covariant form. It is worth stressing that it is the empirically corroborated validity of Maxwell's equations in all inertial frames that leads us to the conclusion that A^μ is a four-vector and $F^{\mu\nu}$ is a second-rank tensor.

Exercise 6.5.3 Using the transformation law (6.57) of a contravariant tensor field, show that in the case of the pure Lorentz transformation (6.12) the electromagnetic field transforms as

$$E'_x = E_x,\qquad\qquad B'_x = B_x,\tag{6.82a}$$
$$E'_y = \gamma\,(E_y - \beta B_z),\qquad\qquad B'_y = \gamma\,(B_y + \beta E_z),\tag{6.82b}$$
$$E'_z = \gamma\,(E_z + \beta B_y),\qquad\qquad B'_z = \gamma\,(B_z - \beta E_y).\tag{6.82c}$$

Hint: denoting matrix (6.78) by $\mathcal{F}$, show that the transformation rule for $F^{\mu\nu}$ can be written in the matrix form $\mathcal{F}' = \Lambda \mathcal{F} \Lambda^T$.

6.6 Relativistic Dynamics

Newtonian mechanics is invariant under Galilean transformations. In particular, Newton's second law for a particle,

$$m\frac{d^2\mathbf{r}}{dt^2} = \mathbf{F}, \tag{6.83}$$

preserves its form, assuming the invariance of $\mathbf{F}$, inasmuch as the acceleration is invariant under a Galilean transformation. Unlike electromagnetic theory, whose relativistic covariance was already an integral part of the equations as originally proposed by Maxwell, in order to make Newton's second law covariant under Lorentz transformations we expect to have to modify its content.

Four-Velocity, Four-Acceleration and Four-Momentum

In order to construct a four-dimensional version of (6.83), let us start by trying to define a four-velocity. The most immediate idea would be to consider dx^μ/dt, but this object is not a four-vector because dt is not a scalar. We must, therefore, replace dt by a scalar time interval. The time registered by a hypothetical clock carried by the particle – that is, the time τ measured in the reference frame in which the particle is instantaneously at rest – is called the particle's **proper time**.[11] Suppose at instant t in inertial frame $\mathcal{K}$ the particle has velocity $\mathbf{v}$ as seen from $\mathcal{K}$. Let $\mathcal{K}'$ be an inertial frame moving with velocity $\mathbf{v}$ relative to $\mathcal{K}$, so that from the point of view of $\mathcal{K}'$ the particle is at rest at the instant t of $\mathcal{K}$. Let ds^2 be the invariant interval between the events defined by the particle's passing by the points (x, y, z) at instant t and $(x + dx, y + dy, z + dz)$ at instant $t + dt$. Of course $d\mathbf{r} = \mathbf{v}dt$, so in $\mathcal{K}$

$$ds^2 = c^2\,dt^2 - d\mathbf{r}\cdot d\mathbf{r} = c^2\,dt^2\left(1 - v^2/c^2\right). \tag{6.84}$$

But, from the point of view of $\mathcal{K}'$,

$$ds'^2 = c^2\,d\tau^2 \tag{6.85}$$

for during the time interval $d\tau$ measured in $\mathcal{K}'$ the particle remained at rest because the frame $\mathcal{K}'$ moved along with the particle. But $ds^2 = ds'^2$ owing to the invariance of the interval, and we conclude that

$$d\tau = \sqrt{1 - v^2/c^2}\,dt \equiv dt/\gamma\,, \tag{6.86}$$

with γ defined by (6.4). Note that (6.85) establishes that $d\tau$ is a scalar because ds'^2 and the speed of light in vacuum are both scalars.

The **four-velocity** U^μ is the four-vector defined by

$$U^\mu = \frac{dx^\mu}{d\tau}\,, \tag{6.87}$$

---

[11] Probably a mistranslation of the French *temps propre*, which means "own time".

whose components are

$$U^{\mu} = (\gamma c, \gamma \mathbf{v}), \tag{6.88}$$

where $\mathbf{v} = d\mathbf{r}/dt$ is the three-dimensional velocity of the particle. An immediate consequence of this definition is that the square of U^{μ} is always the same constant no matter how the particle is moving:

$$U^{\mu} U_{\mu} = c^2. \tag{6.89}$$

This shows that the four-velocity U^{μ} is a time-like vector that never vanishes.

Similarly, the **four-acceleration** is the four-vector defined by

$$\mathcal{A}^{\mu} = \frac{dU^{\mu}}{d\tau} = \frac{d^2 x^{\mu}}{d\tau^2}, \tag{6.90}$$

whose components are

$$\mathcal{A}^{\mu} = \left(\gamma^4 \frac{\mathbf{v} \cdot \mathbf{a}}{c}, \gamma^4 \frac{\mathbf{v} \cdot \mathbf{a}}{c^2}\mathbf{v} + \gamma^2 \mathbf{a}\right), \tag{6.91}$$

where $\mathbf{a} = d\mathbf{v}/dt$ is the three-dimensional acceleration. In the particle's instantaneous rest frame $\mathcal{K}'$ obviously $\mathbf{v}' = 0$ and the components of the four-acceleration reduce to $\mathcal{A}'^{\mu} = (0, \mathbf{a}')$, showing that the four-acceleration is a space-like vector that vanishes only if the **proper acceleration** – the acceleration measured in the particle's instantaneous rest frame – is zero.

Exercise 6.6.1   Verify (6.89). Prove directly or using (6.89) that

$$\mathcal{A}^{\mu} U_{\mu} = 0, \tag{6.92}$$

that is, the four-velocity and the four-acceleration are orthogonal vectors.

Finally, the **four-momentum** of a particle is defined by

$$P^{\mu} = mU^{\mu} = (\gamma mc, \gamma m\mathbf{v}), \tag{6.93}$$

where m is a scalar called **rest mass** or just **mass** of the particle. From (6.89) it immediately follows that

$$P^{\mu} P_{\mu} = m^2 c^2. \tag{6.94}$$

We can write

$$P^{\mu} = (\gamma mc, \mathbf{p}), \tag{6.95}$$

where

$$\mathbf{p} = \gamma m\mathbf{v} = \frac{m\mathbf{v}}{\sqrt{1 - v^2/c^2}} \tag{6.96}$$

is the **relativistic momentum**. The relativistic momentum reduces to the Newtonian expression $m\mathbf{v}$ in the limit $v \ll c$ and a thought experiment conceived by G. N. Lewis and R. C. Tolman shows that it is exactly the form (6.96) that ensures the validity of the law of conservation of momentum in all inertial frames (Pauli, 1958; Bergmann, 1976).

Covariant Equation of Motion

We are now ready to propose a covariant version of Newton's second law in the form

$$\frac{dP^\mu}{d\tau} = \mathcal{F}^\mu, \tag{6.97}$$

where $\mathcal{F}^\mu$ is the **four-force**, also known as **Minkowski force**. For consistency reasons the four-force $\mathcal{F}^\mu$ has to be orthogonal to P^μ:

$$\mathcal{F}^\mu P_\mu = 0. \tag{6.98}$$

Exercise 6.6.2   Prove (6.98) from (6.94).

Equation (6.97) lacks any content unless we know how to relate $\mathcal{F}^\mu$ to the three-dimensional force **F**. The simplest approach consists in assuming that the Newtonian connexion between force and time rate of change of momentum remains valid:

$$\frac{d\mathbf{p}}{dt} \equiv \frac{d}{dt}\left(\frac{m\,\mathbf{v}}{\sqrt{1-v^2/c^2}}\right) = \mathbf{F}. \tag{6.99}$$

Comparing this equation with the spatial components of (6.97) and using (6.86) we find $\mathcal{F}^\mu = (\mathcal{F}^0, \gamma\,\mathbf{F})$. The time component $\mathcal{F}^0$ is easily obtained from (6.98):

$$\gamma m c \mathcal{F}^0 - \gamma^2 m\mathbf{F}\cdot\mathbf{v} = 0 \quad\Longrightarrow\quad \mathcal{F}^0 = \frac{\gamma}{c}\mathbf{F}\cdot\mathbf{v}. \tag{6.100}$$

Thus, the Minkowski force $\mathcal{F}^\mu$ is given by

$$\mathcal{F}^\mu = \left(\frac{\gamma}{c}\mathbf{F}\cdot\mathbf{v}, \gamma\mathbf{F}\right), \tag{6.101}$$

which shows that it has only three independent components.

The consistency of (6.101) can be put to the test in the electromagnetic case. The Lorentz force

$$\mathbf{F} = e\left(\mathbf{E} + \frac{\mathbf{v}}{c}\times\mathbf{B}\right) \tag{6.102}$$

depends linearly on the electromagnetic field and the velocity of the particle. Therefore, the Minkowski force $\mathcal{F}^\mu$ must be a four-vector built exclusively with $F^{\mu\nu}$ and U^μ, being linear in both. The only possible form for $\mathcal{F}^\mu$ is

$$\mathcal{F}^\mu = \kappa\, F^{\mu\nu}\, U_\nu, \qquad \kappa = \text{scalar}, \tag{6.103}$$

and the covariant equation of motion (6.97) takes the form

$$m\mathcal{A}^\mu = \kappa\, F^{\mu\nu}\, U_\nu. \tag{6.104}$$

In the instantaneous rest frame $\mathcal{K}'$ the particle's velocity is zero, there are no relativistic corrections and the Newtonian equation of motion (6.83) is exact. But in $\mathcal{K}'$ we have $\mathcal{A}^\mu = (0, \mathbf{a}')$ and $U_\mu = (c, 0, 0, 0)$, which reduces the time component of (6.104) to the identity $0 = 0$, whereas the spatial part, obtained by sucessively making $\mu = 1, 2, 3$, coincides with

$$m\mathbf{a}' = \kappa\, c\,\mathbf{E}', \tag{6.105}$$

where Eq. (6.78) has been used. Since the Lorentz force in the instantaneous rest frame is $\mathbf{F}' = e\mathbf{E}'$, the scalar κ is determined: $\kappa = e/c$. Therefore, the covariant equation of motion for a charged particle in an external eletromagnetic field is

$$\frac{dP^\mu}{d\tau} = \frac{e}{c} F^{\mu\nu} U_\nu .$$ (6.106)

Exercise 6.6.3   Show that the components of the eletromagnetic Minkowski force

$$\mathcal{F}^\mu = \frac{e}{c} F^{\mu\nu} U_\nu$$ (6.107)

coincide with (6.101), where $\mathbf{F}$ is the Lorentz force (6.102).

Relativistic Energy

By construction, the spatial part of the covariant equation of motion (6.97) coincides with (6.99), which is what one usually calls the relativistic equation of motion for the particle. A question naturally arises: what is the physical meaning of the time component of the covariant equation of motion? With the help of (6.86) and (6.101) we can write

$$\frac{dP^0}{d\tau} \equiv \gamma \frac{d}{dt}(mc\gamma) = \mathcal{F}^0 = \frac{\gamma}{c}\mathbf{F}\cdot\mathbf{v},$$ (6.108)

that is,

$$\frac{d}{dt}\left(\frac{mc^2}{\sqrt{1 - v^2/c^2}}\right) = \mathbf{F}\cdot\mathbf{v}.$$ (6.109)

Taking into account that $\mathbf{F}\cdot\mathbf{v}$ is the power delivered to the particle by the force $\mathbf{F}$, we must take

$$E = \frac{mc^2}{\sqrt{1 - v^2/c^2}}$$ (6.110)

as the **relativistic energy** of the particle: the time component of the covariant equation of motion represents the law of conservation of energy.

In view of definition (6.110) we can write

$$P^\mu = \left(\frac{E}{c}, \mathbf{p}\right).$$ (6.111)

Thus, energy and linear momentum make up a four-vector, with the consequence that the laws of conservation of energy and linear momentum are no longer independent but have become aspects of a covariant law of conservation of four-momentum.

In the non-relativistic limit, the binomial series expansion applied to (6.110) yields

$$E = mc^2 + \frac{1}{2}mv^2 + \frac{3}{8}m\frac{v^4}{c^2} + \cdots .$$ (6.112)

On the right-hand side of this equation the second term is the Newtonian kinetic energy and the next terms are relativistic corrections to the non-relativistic expression. The first term has no Newtonian analogue and implies that a particle at rest contains an amount of energy (known as **rest energy**) given by Einstein's celebrated equation

$$E = mc^2, \tag{6.113}$$

which establishes an equivalence between mass and energy: a mass m at rest can be converted into an amount of energy equal to mc^2.

Example 6.1   Study the one-dimensional motion of a relativistic particle subject to a constant force, also known as hyperbolic motion.

Solution

This example might describe a charged particle being accelerated by a uniform electric field. In the one-dimensional case, Eq. (6.99) takes the form

$$\frac{d}{dt}\left(\frac{m\dot{x}}{\sqrt{1 - \dot{x}^2/c^2}}\right) = F. \tag{6.114}$$

An immediate integration yields

$$\frac{\dot{x}}{\sqrt{1 - \dot{x}^2/c^2}} = at + b, \qquad a = \frac{F}{m}, \tag{6.115}$$

where b is an arbitrary constant. Solving this equation for $\dot{x}$ we are led to

$$\dot{x} = \frac{at + b}{\sqrt{1 + (at + b)^2/c^2}}, \tag{6.116}$$

which shows that, differently from the Newtonian problem, the particle's acceleration is not constant.[12] After another elementary integration, we find

$$x - d = \frac{c^2}{a}\sqrt{1 + \left(\frac{at + b}{c}\right)^2}. \tag{6.117}$$

Let us consider the particularly simple case in which the particle starts from the origin at rest – that is, $x(0) = 0$ and $\dot{x}(0) = 0$. In such circumstances $b = 0$ and $d = -c^2/a$, whence

$$x = \frac{c^2}{a}\left[-1 + \sqrt{1 + \left(\frac{at}{c}\right)^2}\right]. \tag{6.118}$$

This last equation can be cast in the more elegant form

$$\left(x + \frac{c^2}{a}\right)^2 - c^2 t^2 = \frac{c^4}{a^2}, \tag{6.119}$$

which is the equation of a hyperbola in the $x^0 x$ plane (see Fig. 6.5).

---

[12] It is the *proper acceleration* of the particle that is constant and equal to F/m (Problem 6.13).

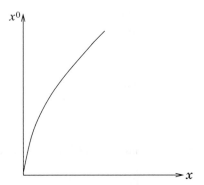

Fig. 6.5　Worldline of a relativistic particle subject to a constant force (hyperbolic motion).

Exercise 6.6.4　Inspecting Eq. (6.116), show that $|\dot{x}| < c$ for all t and that $\dot{x} \to c$ as $t \to \infty$. Find the non-relativistic limit of Eq. (6.118) by formally taking the limit $c \to \infty$ and show that the expected result is obtained.

Zero Mass Particles

An important result is the connexion between the energy and the momentum of a particle, namely

$$E = \sqrt{p^2c^2 + m^2c^4}\,, \qquad p^2 = \mathbf{p}\cdot\mathbf{p}\,. \tag{6.120}$$

Another characteristic of relativistic mechanics not shared by Newtonian mechanics is that of allowing the existence of zero mass particles (photons, gravitons). Putting $m = 0$ in (6.120) there results

$$E = |\mathbf{p}|c\,, \tag{6.121}$$

which is the relativistic link between E and $\mathbf{p}$ for zero mass particles. A fundamental feature of zero mass particles is that their speed is always c in any inertial frame, as the reader is invited to show.

Exercise 6.6.5　From equations (6.94) and (6.111) derive (6.120). Show also that the particle's velocity can be expressed as

$$\mathbf{v} = \frac{c^2\mathbf{p}}{E}\,, \tag{6.122}$$

which does not depend on the mass. Conclude that $|\mathbf{v}| = c$ for a zero mass particle.

6.7 Relativistic Collisions

The laws of conservation of energy and momentum in a collision involving any number of particles can be summarised in the covariant equation

$$\mathbf{P}^{(I)}_{(1)} + \cdots \mathbf{P}^{(I)}_{(N)} = \mathbf{P}^{(F)}_{(1)} + \cdots + \mathbf{P}^{(F)}_{(M)}\,. \tag{6.123}$$

The initial state (before the collision) contains N particles and the final state (after the collision) contains M particles; $\mathbf{P}_{(k)}^{(I)}$ and $\mathbf{P}_{(k)}^{(F)}$ denote the initial and final four-momenta of the kth particle, respectively. Being a tensor equation, (6.123) ensures the validity of the conservaton of energy and momentum in all inertial reference frames. In the following we illustrate the striking effectiveness of Eq. (6.123), allied with the invariance of the scalar product (6.34), to extract in an elegant and concise way valuable information regarding high energy collisions.

Example 6.2   Antiprotons can be produced in a proton-proton collision according to the reaction

$$p + p \rightarrow p + p + p + \bar{p}, \tag{6.124}$$

where p denotes a proton and $\bar{p}$ an antiproton. If the target proton is at rest in the laboratory frame, what is the threshold energy (minimum energy) of the incoming proton in order that the reaction be possible?

Solution

According to (6.110) a free particle has the lowest energy possible when at rest. Therefore, the minimum energy of the incoming proton is the one that allows the formation of the reaction products at rest. However, the reaction products cannot be formed at rest in the laboratory frame because this would violate the conservation of momentum. As a consequence, the reaction products can only be formed at rest in the centre-of-mass frame, for it is in this frame that the total momentum is zero. Denoting by $\mathbf{P}^{(F)}$ the total four-momentum of the reaction products, its components in the centre-of-mass frame are

$$\mathbf{P}^{(F)} = (4mc, 0, 0, 0) \quad \text{(in the centre-of-mass frame)} \tag{6.125}$$

since each of the reaction products has four-momentum with components $(mc, 0, 0, 0)$, where m is mass of the proton (equal to that of the antiproton). The conservation of four-momentum is written

$$\mathbf{P} + \mathbf{Q} = \mathbf{P}^{(F)}, \tag{6.126}$$

where $\mathbf{P}$ and $\mathbf{Q}$ are, respectively, the four-momenta of the incoming proton and of the target proton. Squaring (6.126) there results

$$\mathbf{P}^2 + 2\mathbf{P} \cdot \mathbf{Q} + \mathbf{Q}^2 = \mathbf{P}^{(F)2} \quad \Longrightarrow \quad m^2c^2 + 2\mathbf{P} \cdot \mathbf{Q} + m^2c^2 = \mathbf{P}^{(F)2}, \tag{6.127}$$

where we have used (6.94). The other scalar quantities involved in this last equation may be calculated in the inertial frame that is found the most convenient. Thus, calculating the right-hand side of (6.127) in the centre-of-mass frame with the help of (6.125), we get

$$\mathbf{P}^{(F)2} = 16m^2c^2, \tag{6.128}$$

where (6.35) has been used. Finally, in the laboratory frame we have

$$\mathbf{P} = \left(\frac{E}{c}, \mathbf{p}\right), \quad \mathbf{Q} = (mc, 0, 0, 0) \quad \text{(in the laboratory frame)}, \tag{6.129}$$

whence, by (6.34),

$$\mathbf{P} \cdot \mathbf{Q} = \frac{E}{c} mc = mE\,.$$ (6.130)

Substituting (6.128) and (6.130) into (6.127) we immediately obtain

$$E = 7mc^2$$ (6.131)

as the threshold energy for the reaction (6.124). In order to have this energy, the incoming proton must have a speed of $0.99c$.

In the special case of a photon, its energy is $h\nu$ and its linear momentum has magnitude $h\nu/c$, where h is Planck's constant and ν is the frequency. Thus, the photon four-momentum is

$$\mathbf{P} = \left(\frac{E}{c}, \mathbf{p}\right) = \frac{h\nu}{c}(1, \hat{\mathbf{n}})$$ (6.132)

where $\hat{\mathbf{n}}$ is the unit vector along the propagation direction. Clearly

$$\mathbf{P}^2 = 0\,,$$ (6.133)

so that $\mathbf{P}$ is a light-like or null vector.

Example 6.3 Interpreting the scattering of X-rays by a graphite sheet as caused by the collision of a photon with an electron, A. H. Compton showed in 1923 that

$$\lambda' = \lambda + \frac{2h}{mc}\sin^2\frac{\theta}{2}\,.$$ (6.134)

The energy and momentum transfer from the X-ray to the electron is known as the Compton effect. Derive Compton's formula, where λ is the wavelength of the incoming X-ray, λ' is the wavelength of the outgoing X-ray, m is the mass of the electron and θ is the angle between the outgoing and the incoming X-rays.

Solution

In Fig. 6.6 a photon of frequency ν collides with an electron at rest and is deflected by an angle θ while its frequency ν diminishes to ν', whereas the electron's recoil angle is θ'. Let $\mathbf{P}$ and $\mathbf{P}'$ be the initial and final four-momenta of the photon, with $\mathbf{Q}$ and $\mathbf{Q}'$ being those of the electron. Conservation of four-momentum requires

$$\mathbf{P} + \mathbf{Q} = \mathbf{P}' + \mathbf{Q}'.$$ (6.135)

Inasmuch as we are not interested in the recoiling electron, it turns out to be convenient to eliminate the undesirable four-vector $\mathbf{Q}'$ by isolating it in (6.135) and squaring the resulting equation:

$$\mathbf{Q}'^2 = (\mathbf{P} + \mathbf{Q} - \mathbf{P}')^2 = \mathbf{P}^2 + \mathbf{Q}^2 + \mathbf{P}'^2 + 2\mathbf{Q}\cdot\mathbf{P} - 2\mathbf{P}\cdot\mathbf{P}' - 2\mathbf{Q}\cdot\mathbf{P}'.$$ (6.136)

Since $\mathbf{Q}^2 = \mathbf{Q}'^2 = m^2c^2$ and $\mathbf{P}^2 = \mathbf{P}'^2 = 0$, we are left with

$$\mathbf{P}\cdot\mathbf{P}' = \mathbf{Q}\cdot\mathbf{P} - \mathbf{Q}\cdot\mathbf{P}'.$$ (6.137)

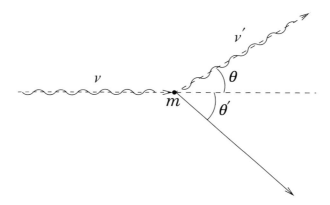

Fig. 6.6
Fig. 6.6   Compton effect.

Calculating the right-hand side of this equation in the laboratory frame, in which $\mathbf{Q} = (mc, 0, 0, 0)$, we find $\mathbf{Q} \cdot \mathbf{P} = mh\nu$ and $\mathbf{Q} \cdot \mathbf{P}' = mh\nu'$, while

$$\mathbf{P} \cdot \mathbf{P}' = \frac{h\nu}{c}\frac{h\nu'}{c}(1 - \hat{\mathbf{n}} \cdot \hat{\mathbf{n}}') = \frac{h^2\nu\nu'}{c^2}(1 - \cos\theta). \qquad (6.138)$$

Inserting these results into (6.137), we finally get

$$\frac{h\nu\nu'}{c^2}(1 - \cos\theta) = m(\nu - \nu'). \qquad (6.139)$$

It is easy to check that in terms of the semi-angle $\theta/2$ and of the wavelength $\lambda = c/\nu$ this last equation becomes identical to Compton's formula (6.134).

6.8 Relativistic Dynamics in Lagrangian Form

In the case of a particle subject to a conservative force

$$\mathbf{F} = -\nabla V(\mathbf{r}) \qquad (6.140)$$

the relativistic equation of motion

$$\frac{d}{dt}\left(\frac{m\,\mathbf{v}}{\sqrt{1 - v^2/c^2}}\right) = -\nabla V(\mathbf{r}) \qquad (6.141)$$

can be put in Lagrangian form. In order to do so it suffices to note that

$$\frac{m\,\dot{x}}{\sqrt{1 - (\dot{x}^2 + \dot{y}^2 + \dot{z}^2)/c^2}} = \frac{\partial}{\partial\dot{x}}\left[-mc^2\sqrt{1 - (\dot{x}^2 + \dot{y}^2 + \dot{z}^2)/c^2}\right]. \qquad (6.142)$$

This shows that the Lagrangian

$$L = -mc^2\sqrt{1 - v^2/c^2} - V(\mathbf{r}) \tag{6.143}$$

allows us to write the components of (6.141) in the form

$$\frac{d}{dt}\left(\frac{\partial L}{\partial \dot{q}_k}\right) - \frac{\partial L}{\partial q_k} = 0, \qquad k = 1,2,3, \tag{6.144}$$

with $q_1 = x, q_2 = y, q_3 = z$. Note that the Lagrangian (6.143) *is not* of the form $T - V$. However, if V does not explicitly depend on time, a constant of the motion of fundamental importance is the Jacobi integral

$$h = \dot{x}\frac{\partial L}{\partial \dot{x}} + \dot{y}\frac{\partial L}{\partial \dot{y}} + \dot{z}\frac{\partial L}{\partial \dot{z}} - L = \frac{mc^2}{\sqrt{1 - v^2/c^2}} + V, \tag{6.145}$$

which turns out to be the particle's total energy.

The Lagrangian (6.143) differs from its non-relativistic counterpart only in the kinetic term, which prompts us to write the following relativistic Lagrangian for a charged particle in an external electromagnetic field:

$$L = -mc^2\sqrt{1 - v^2/c^2} - e\phi + \frac{e}{c}\mathbf{v}\cdot\mathbf{A}, \tag{6.146}$$

where ϕ and $\mathbf{A}$ are the eletromagnetic potentials. If ϕ and $\mathbf{A}$ do not explicitly depend on time, the Jacobi integral coincides with the total energy:

$$h = E_{\text{tot}} = \frac{mc^2}{\sqrt{1 - v^2/c^2}} + e\phi. \tag{6.147}$$

This Lagrangian formulation is not covariant, since it yields only the relativistic equation of motion (6.99), which refers to a particular inertial frame. A covariant Lagrangian formulation should allow us to write the covariant equation of motion

$$\frac{d}{d\tau}\left(mU^\mu\right) = \mathcal{F}^\mu \tag{6.148}$$

in the form

$$\frac{d}{d\tau}\left(\frac{\partial L}{\partial U_\mu}\right) - \frac{\partial L}{\partial x_\mu} = 0, \tag{6.149}$$

where x^μ and U^μ are treated as generalised coordinates and velocities, with τ playing the role of time. The Lagrangian L must be a scalar in order that $\partial L/\partial x_\mu$ and $\partial L/\partial U_\mu$ be four-vectors. In the electromagnetic case a simple Lagrangian that does the job is

$$L = \frac{m}{2}U^\alpha U_\alpha + \frac{e}{c}U_\alpha A^\alpha. \tag{6.150}$$

Let us first note that

$$\frac{\partial}{\partial U_\mu}(U^\alpha U_\alpha) = \frac{\partial}{\partial U_\mu}(g^{\alpha\beta}U_\alpha U_\beta) = g^{\alpha\beta}(\delta^\mu_\alpha U_\beta + U_\alpha \delta^\mu_\beta) = \delta^\mu_\alpha U^\alpha + \delta^\mu_\beta U^\beta = 2U^\mu,$$
$$\tag{6.151}$$

whence

$$\frac{\partial L}{\partial U_\mu} = mU^\mu + \frac{e}{c}A^\mu \, . \tag{6.152}$$

On the other hand,

$$\frac{d}{d\tau}A^\mu = \frac{\partial A^\mu}{\partial x_\alpha}\frac{dx_\alpha}{d\tau} = U_\alpha \partial^\alpha A^\mu \, . \tag{6.153}$$

Thus, the Lagrange equations (6.149) obtained from the Lagrangian (6.150) are

$$m\frac{dU^\mu}{d\tau} + \frac{e}{c}U_\alpha \partial^\alpha A^\mu - \frac{e}{c}U_\alpha \partial^\mu A^\alpha = 0 \, , \tag{6.154}$$

which reproduce (6.106).

A possible objection to the Lagrangian (6.150) is that it gives the correct covariant equation of motion only if the four components of the four-velocity are treated as independent, which is patently false because of the constraint (6.89). However, one may counterargue that the Lagrangian (6.150) yields an equation of motion that is not only correct but also contains in itself the constraint (6.89). Indeed, from (6.154) one derives $U_\mu dU^\mu/d\tau = (e/mc)F^{\mu\nu}U_\mu U_\nu = 0$ because $F^{\mu\nu}$ is antisymmetric (see Problem 6.6). Therefore, $U_\mu U^\mu = $ constant and the constant can be chosen to equal c^2. This justifies the use of the Lagrangian (6.150) treating the components of the four-velocity as if they were all mutually independent.

6.9 Action at a Distance in Special Relativity

In Section 6.6 we described how to construct a covariant formulation of single particle dynamics. The next step would be to develop manifestly covariant equations of motion for two or more interacting particles. One feature of Newtonian mechanics is action at a distance, such as expressed by the law of universal gravitation. Conservation of linear momentum is a direct consequence of Newton's third law. If action and reaction occur at distant points, this law cannot be straightforwardly transplanted to special relativity because of the relativity of simultaneity: a covariant Newton's third law is only possible for contact interactions. It is evident that action at a distance in a strictly Newtonian fashion is incompatible with the special theory of relativity because it involves the instantaneous (infinitely fast) propagation of the interaction between particles. In order to circumvent this difficulty, one might concoct a retarded action at a distance such that any alteration in the state of a particle would be felt by another particle not before the time spent by light to travel the distance between them. In this case, the change of the momentum of a particle would not cause an immediate change of the momentum of the other particles, but there would be a momentum in transit from each particle to the others to ensure conservation of the total linear momentum. This indicates that in a relativistic theory all particle interactions must be mediated by local fields endowed with mechanical attributes such as energy and momentum. Although this is intuitively clear, a formal proof of the impossibility of action at a distance in the special theory of relativity has a considerable instructive value.

Consider a system composed of two particles with worldlines described by the four-vectors $x_{(1)}^{\mu}(\tau_1)$ and $x_{(2)}^{\mu}(\tau_2)$, where τ_1 and τ_2 are their respective proper times. The total four-momentum of the system at instant t of an inertial frame $\mathcal{K}$ is

$$\mathbf{P}(t) = \mathbf{P}_{(1)}(\tau_1) + \mathbf{P}_{(2)}(\tau_2),\qquad(6.155)$$

where the proper times τ_1 and τ_2 on the worldlines of the particles correspond to the same instant t. A reasonable relativistic theory of two particles interacting by action at a distance must possess the following properties: (a) the total four-momentum is conserved in all inertial frames; (b) asymptotically, that is, for $\tau_i \rightarrow \infty$ and $\tau_i \rightarrow -\infty$ the worldlines are straight, the particles are free. The latter requirement aims to enable the description of a scattering process, in which the particles approach each other from far away, interact during a short time then move indefinitely away from each other following straight worldlines deflected from their original directions. This asymptotic condition also guarantees that the total four-momentum transforms as a four-vector. Now you might ask, "Is this not obvious?" It seems so, but it is not because in (6.155) the events corresponding to the proper times τ_1 and τ_2 are simultaneous in frame $\mathcal{K}$ but are not in general simultaneous in other inertial frames moving relative to $\mathcal{K}$. However, asymptotically these four-momenta become constants and can be evaluated at any instant. Thus we can consider them simultaneously calculated in any inertial frame, so the asymptotic total four-momentum transforms as a four-vector. But since the transformation law between inertial frames has to be the same at any time, if the total four-momentum transforms as a four-vector in the asymptotic region, the same transformation law applies at any instant, namely

$$P'^{\mu} = \Lambda^{\mu}{}_{\nu}\, P^{\nu}.\qquad(6.156)$$

Let us consider two inertial frames $\mathcal{K}$ and $\mathcal{K}'$ with common origin on the worldline of the first particle, as illustrated in Fig. 6.7. By the hypothesis of conservation of the total

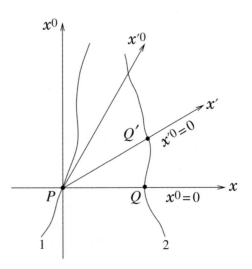

Fig. 6.7 Worldlines of two interacting particles.

four-momentum, each side of Eq. (6.156) is constant and can be evaluated at any instant t or t' one wishes. Taking advantage of this freedom, let us evaluate P'^{μ} at instant $t' = 0$ and P^{μ} at instant $t = 0$ to obtain

$$P'^{\mu}_{(1)}\big|_P + P'^{\mu}_{(2)}\big|_{Q'} = \Lambda^{\mu}_{\ \nu} \left(P^{\nu}_{(1)}\big|_P + P^{\nu}_{(2)}\big|_Q\right). \tag{6.157}$$

But, at points P and Q, the four-momentum transformation law gives

$$P'^{\mu}_{(1)}\big|_P = \Lambda^{\mu}_{\ \nu} P^{\nu}_{(1)}\big|_P, \qquad P'^{\mu}_{(2)}\big|_{Q'} = \Lambda^{\mu}_{\ \nu} P^{\nu}_{(2)}\big|_{Q'}. \tag{6.158}$$

Combining (6.157) and (6.158) we are led to

$$\Lambda^{\mu}_{\ \nu} P^{\nu}_{(2)}\big|_{Q'} = \Lambda^{\mu}_{\ \nu} P^{\nu}_{(2)}\big|_Q. \tag{6.159}$$

Since $\boldsymbol{\Lambda}$ is an invertible matrix, it follows that

$$P^{\mu}_{(2)}\big|_{Q'} = P^{\mu}_{(2)}\big|_Q, \tag{6.160}$$

which means that the four-momentum of the second particle is the same at the points Q and Q' of its worldline. Since Q and Q' are arbitrary events and the roles of the particles can be exchanged, we have shown that the total four-momentum will be conserved if and only if the four-momentum of each particle is constant – that is, if and only if the particles do not interact.

This no-interaction theorem (Van Dam & Wigner, 1966; Mann, 1974; Ohanian, 1976) establishes that the relativistic implementation of the law of conservation of energy and momentum for a system of interacting particles requires the addition of another entity carrying energy and momentum, to wit a field that acts as mediator of the interactions. The interaction of two distant particles can be interpreted as a contact action of one particle on the field, followed by the propagation of the field and, finally, a contact action of the field on the other particle. What is conserved is the the sum of the four-momenta of the particles plus the four-momentum of the field.

Problems

6.1   Two photons travel along the x-axis of the inertial frame $\mathcal{K}$ keeping a distance L from one another. Let $\mathcal{K}'$ be an inertial frame moving relative to $\mathcal{K}$ with velocity v along the x-axis. Prove that in $\mathcal{K}'$, whose axes are parallel to those of $\mathcal{K}$, the distance between the photons is $L(c + v)^{1/2}/(c - v)^{1/2}$.

6.2   (a) Show that the standard Lorentz transformation (6.1) can be written in the form

$$x' = x \cosh \phi - ct \sinh \phi, \qquad ct' = -x \sinh \phi + ct \cosh \phi,$$

where $\tanh \phi = v/c$. Noting that the relations $\cosh \phi = \cos i\phi$ and $i \sinh \phi = \sin i\phi$ allow us to convert any trigonometric identity into an identity for hyperbolic functions, show that formally this is a "rotation" in x and ict by an "angle" $i\phi$; as such, it preserves $x^2 + (ict)^2$. (b) Derive the following useful form for the standard Lorentz transformation:

$$ct' + x' = e^{-\phi} (ct + x), \qquad ct' - x' = e^{\phi} (ct - x).$$

If $\mathcal{K}''$ is an inertial frame that moves with velocity u along the x'-axis of $\mathcal{K}'$, use the Lorentz transformation in this form to show that the velocity of $\mathcal{K}''$ relative to $\mathcal{K}$ is $(u + v)/(1 + uv/c^2)$.

6.3 If $\mathbf{v}'$ is the velocity of a particle as seen from the inertial frame $\mathcal{K}'$ that moves relative to the inertial frame $\mathcal{K}$ with velocity $\mathbf{u}$, prove that the particle's velocity relative to $\mathcal{K}$ is

$$\mathbf{v} = \frac{\mathbf{v}' + (\gamma - 1)\left(\mathbf{u} \cdot \mathbf{v}'/u^2\right)\mathbf{u} + \gamma\mathbf{u}}{\gamma\left(1 + \mathbf{u} \cdot \mathbf{v}'/c^2\right)},$$

where $\gamma = (1 - u^2/c^2)^{-1/2}$. Note that in general this expression is not symmetric in $\mathbf{u}$ and $\mathbf{v}'$. To what does this equation reduce when $\mathbf{v}'$ and $\mathbf{u}$ are parallel?

6.4 Two events A and B are time-like separated. Consider a straight worldline and all curved time-like worldlines connecting A and B. Show that the proper time between A and B,

$$\Delta\tau = \frac{1}{c}\int_A^B ds = \frac{1}{c}\int_A^B \sqrt{dx^\mu dx_\mu}\,,$$

is a maximum when calculated along the straight worldline. Thus, the free particle worldline can be characterised as the time-like space-time curve of maximum proper time. Hint: use a reference frame such that the particle following the straight worldline remains spatially at rest.

6.5 An infinitesimal Lorentz transformation is of the form $\Lambda^\mu{}_\nu = \delta^\mu_\nu + \omega^\mu{}_\nu$ where the $\omega^\mu{}_\nu$ are infinitesimal quantities. Using (6.19) and neglecting higher-order infinitesimals, prove that $\omega_{\mu\nu} = -\omega_{\nu\mu}$.

6.6 (a) Prove that a symmetric (antisymmetric) tensor $A^{\mu\nu}$ is symmetric (antisymmetric) in the contravariant form if and only if it is symmetric (antisymmetric) in the covariant form $A_{\mu\nu}$. (b) If $A^{\mu\nu}$ is a symmetric tensor and $B^{\mu\nu}$ is an antisymmetric tensor, show that $A^{\mu\nu}B_{\mu\nu} = 0$. Hint: $A^{\mu\nu}B_{\mu\nu} = A^{\nu\mu}B_{\nu\mu}$ because the summation indices are dummy (Appendix A).

6.7 Show that a photon cannot spontaneously decay into an electron-positron pair.

6.8 (a) Prove that in relativistic mechanics the equation $d\mathbf{L}/dt = \mathbf{N}$ still holds, where $\mathbf{L} = \mathbf{r} \times \mathbf{p}$ is the angular momentum and $\mathbf{N} = \mathbf{r} \times \mathbf{F}$ is the torque, $\mathbf{p}$ being the relativistic momentum. (b) If $L^{\mu\nu} = x^\mu P^\nu - x^\nu P^\mu$ is the angular momentum tensor and $N^{\mu\nu} = x^\mu \mathcal{F}^\nu - x^\nu \mathcal{F}^\mu$ is the torque tensor, show that $dL^{\mu\nu}/d\tau = N^{\mu\nu}$ and check that the spatial components of this equation corroborate the result in part (a).

6.9 (a) If A^μ is a time-like vector, prove that there exists an inertial frame in which only its time component is non-zero. Hint: a rotation of the spatial axes allows us to eliminate two spatial components of the vector; show that there is a pure Lorentz transformation that eliminates the remaining spatial component. (b) Prove that if a vector is orthogonal to a time-like vector then it is space-like.

6.10 A particle of mass M decays from rest into a particle of smaller mass m and a photon. Find the the energy of each of the decay products.

6.11 Prove that a massive particle cannot decay into a single photon. Which conservation law would necessarily be violated?

6.12 In the laboratory frame a photon collides with a proton at rest. What is the minimum photon frequency if the reaction products are a neutron and a pion? Denote by m the masses of both proton and neutron, assumed equal, and by m_π the pion mass.

6.13 The *proper acceleration* $\mathbf{a}'$ of a particle is the three-dimensional acceleration in the particle's instantaneous rest frame $\mathcal{K}'$. By appropriately transforming the four-acceleration $\mathcal{A}^\mu$, show that the components of the proper acceleration are

$$a'_{x'} = \frac{a_x}{(1 - v^2/c^2)^{3/2}}, \qquad a'_{y'} = \frac{a_y}{1 - v^2/c^2}, \qquad a'_{z'} = \frac{a_z}{1 - v^2/c^2},$$

where the axes of $\mathcal{K}$ and $\mathcal{K}'$ are parallel and have been so chosen that the particle's velocity $\mathbf{v}$ relative to $\mathcal{K}$ is along the x-axis. Show that the proper acceleration in the case of hyperbolic motion (Example 6.1 in Section 6.6) is constant and its only non-zero component is equal to F/m.

6.14 A pion with energy 1200 MeV collides with a proton at rest producing several pions according to the reaction $\pi + p \to p + n\pi$. What is the maximum number n of pions that can be created in the reaction? The rest energy of the proton is 938 MeV and that of the pion is 140 MeV.

6.15 A particle of mass m with speed $v = 4c/5$ collides inelastically with an identical particle at rest, forming a single particle. (a) What is the speed of the particle that results from the collision? (b) What is its mass?

6.16 A particle of mass M decays into two particles of masses m_1 and m_2. Show that the energy of the first particle in the rest frame of the decaying particle is $E_1 = (M^2 + m_1^2 - m_2^2)c^2/2M$.

6.17 The dual electromagnetic field tensor is the antisymmetric tensor $\tilde{F}^{\mu\nu}$ defined by

$$\tilde{F}^{\mu\nu} = \frac{1}{2}\,\epsilon^{\mu\nu\rho\lambda} F_{\rho\lambda}$$

where $\epsilon^{\mu\nu\rho\lambda}$ is the four-dimensional generalisation of the Levi-Civita symbol introduced in Appendix A: $\epsilon^{\mu\nu\rho\lambda} = \pm 1$ according to whether $(\mu, \nu, \rho, \lambda)$ is an even or odd permutation of $(0, 1, 2, 3)$, and $\epsilon^{\mu\nu\rho\lambda} = 0$ whenever repeated values of the indices occur. (a) Show that the components of $\tilde{F}^{\mu\nu}$ can be obtained from those of $F^{\mu\nu}$ by simply replacing $\mathbf{E}$ by $\mathbf{B}$ and $\mathbf{B}$ by $-\mathbf{E}$. (b) Prove that the homogeneous Maxwell's equations can be written as

$$\partial_\nu \tilde{F}^{\mu\nu} = 0.$$

(c) Express the Lorentz scalars $F^{\mu\nu} F_{\mu\nu}$, $\tilde{F}^{\mu\nu} \tilde{F}_{\mu\nu}$ and $\tilde{F}^{\mu\nu} F_{\mu\nu}$ in terms of $\mathbf{E}$ and $\mathbf{B}$. (d) Prove that if $\mathbf{E}$ and $\mathbf{B}$ are perpendicular in some inertial frame then they are perpendicular in all inertial frames.

6.18 The energy-momentum tensor of the eletromagnetic field is defined by

$$T^{\mu\nu} = \frac{1}{4\pi}\left[F^{\alpha\mu} F^\nu{}_\alpha + \frac{1}{4} g^{\mu\nu} F^{\alpha\beta} F_{\alpha\beta} \right].$$

Show that:

$$T^{00} = \frac{1}{8\pi}(E^2 + B^2) = \text{energy density};$$

for $k = 1, 2, 3,$

$$c\,T^{0k} = \frac{c}{4\pi}(\mathbf{E} \times \mathbf{B})_k = \text{Poynting vector}.$$

Show also that the trace of $T^{\mu\nu}$ is zero, that is, $T^{\mu}{}_{\mu} = 0$.

6.19 Consider a possible modification of the theory of the gravitational field with the purpose of making it relativistically covariant: the usual expression $\mathbf{F} = -m\nabla\phi$ for the gravitational force is replaced by

$$\mathcal{F}_{\mu} = m\partial_{\mu}\phi,$$

where m is the mass of the particle and ϕ is the gravitational potential; in the covariant equation of motion

$$d(mU^{\mu})/d\tau = \mathcal{F}^{\mu}$$

the mass of the particle is supposed variable. Derive the equation

$$c^2 dm/d\tau = m\,d\phi/d\tau$$

and, as a consequence,

$$m = m_0 \exp(\phi/c^2)$$

with m_0 a constant. According to such a theory, the mass of each particle would depend on the gravitational potential at the particle's position. This theory was proposed by G. Nordström in 1912 but was soon abandoned (Pais, 1982).

6.20 The constraint $U^{\mu}U_{\mu} = c^2$ can be incorporated into the Lagrangian dynamics of the free particle by means of a Lagrange multiplier treated as an additional degree of freedom of the system. Consider the free particle Lagrangian

$$L(x^{\mu}, U^{\mu}, \lambda, \dot{\lambda}) = \frac{\lambda}{2}\left(U^{\mu}U_{\mu} - c^2\right).$$

Show that

$$\frac{d}{d\tau}\left(\frac{\partial L}{\partial U^{\alpha}}\right) - \frac{\partial L}{\partial U^{\alpha}} = 0 \implies \lambda\frac{dU_{\alpha}}{d\tau} + \dot{\lambda}\,U_{\alpha} = 0$$

and

$$\frac{d}{d\tau}\left(\frac{\partial L}{\partial \dot{\lambda}}\right) - \frac{\partial L}{\partial \lambda} = 0 \implies U^{\mu}U_{\mu} - c^2 = 0.$$

From these equations infer that $\dot{\lambda} = 0$ – that is, $\lambda = $ constant. Conclude, finally, that the above equations of motion are equivalent to the pair

$$\frac{dU^{\alpha}}{d\tau} = 0, \qquad U^{\alpha}U_{\alpha} = c^2,$$

which are the correct equations for the free particle, including the four-velocity constraint.

7 Hamiltonian Dynamics

Le savant n'étudie pas la nature parce que cela est utile; il l'étudie parce qu'il y prend plaisir et il y prend plaisir parce qu'elle est belle. Si la nature n'était pas belle, elle ne vaudrait pas la peine d'être connue, la vie ne vaudrait pas la peine d'être vécue.

Henri Poincaré, *Science et Méthode*

In the Lagrangian formulation the motion of an n-degree-of-freedom mechanical system is governed by n ordinary differential equations of the second order in time. Hamilton's dynamics consist in replacing the n Lagrange's equations by a certain equivalent set of $2n$ ordinary differential equations of the first order. As a general rule, Hamilton's equations for specific problems are not easier to solve than Lagrange's equations. The great importance of the Hamiltonian formalism lies in providing a powerful, general and flexible method for the investigation of the deepest structural questions of classical mechanics and also in serving as the foundation of quantum mechanics and statistical mechanics. In this and the next chapters we shall study how to effect the transition from Lagrange's to Hamilton's formalism as well as the innumerable formal advantages inherent to the Hamiltonian description of classical mechanics.

7.1 Hamilton's Canonical Equations

Lagrange's equations for an n-degree-of-freedom system comprise a set of n second-order ordinary differential equations for the generalised coordinates $q_1(t), \ldots, q_n(t)$. The motion of the system is uniquely determined as long as $2n$ initial conditions are specified, namely the values of all qs and $\dot{q}$s at a particular instant t_0. The motion can be geometrically represented by a path in the n-dimensional configuration space spanned by the generalised coordinates $q_1, \ldots, q_n$.

In the formulation introduced by Hamilton, the picture is different: the equations of motion are $2n$ first-order ordinary differential equations for $2n$ independent variables. The motion can be represented by a curve described in **phase space**, a $2n$-dimensional space the coordinates of which are the independent variables just mentioned. Differently from configuration space in which a point defines only the positions of the particles of the system at a given instant, a point in phase space determines the **state** of the system – that is, its configuration (positions of the particles) and the time rate of change of that configuration (velocities of the particles) at the given instant.

Of course, Lagrange's equations can be trivially replaced by an equivalent first-order system with the number of equations doubled by merely introducing the variables $s_i = \dot{q}_i$, $i = 1, \ldots, n$ and treating $q_1, \ldots, q_n, s_1, \ldots, s_n$ as $2n$ independent variables. The equations of motion would be

$$\dot{q}_i = s_i, \qquad \frac{d}{dt}\left(\frac{\partial L}{\partial s_i}\right) - \frac{\partial L}{\partial q_i} = 0, \qquad i = 1, \ldots, n, \tag{7.1}$$

where $L(q, s, t)$ is the Lagrangian for the system. These equations, however, involve q_i and s_i very asymmetrically and are not particularly useful.

Canonical Momenta and Hamilton's Equations

William Rowan Hamilton showed in 1835 that a symmetric duplication of the number of independent variables is achieved by describing the dynamics by means of the $2n$ quantities $q_1, \ldots, q_n, p_1, \ldots, p_n$ where p_i is the **canonical momentum conjugate** to q_i, defined by

$$p_i = \frac{\partial L}{\partial \dot{q}_i}, \qquad i = 1, \ldots, n. \tag{7.2}$$

We assume that the **Hessian matrix** $\mathbf{W} = (W_{ij})$ with elements

$$W_{ij} = \frac{\partial^2 L}{\partial \dot{q}_i \partial \dot{q}_j} \tag{7.3}$$

is non-singular:

$$\det \mathbf{W} \neq 0. \tag{7.4}$$

In this case the implicit function theorem (Spivak, 1965; Protter & Morrey Jr., 1985) guarantees that Eqs. (7.2) can be solved for the generalised velocities. Thus, the Hamiltonian description involves the replacement of the variables $(q, \dot{q})$ by (q, p) in all mechanical quantities and substituting a function $H(q, p, t)$ for the Lagrangian $L(q, \dot{q}, t)$ as generator of the dynamics. Such a change of description is carried out by a **Legendre transformation**,[1] which in the present context consists in replacing the generalised velocities by the canonical momenta as basic variables and introducing **Hamilton's function** or the **Hamiltonian** $H(q, p, t)$ defined by

$$H(q, p, t) = \sum_{i=1}^{n} \dot{q}_i p_i - L(q, \dot{q}, t). \tag{7.5}$$

It is understood that on the right-hand side of this equation the velocities are expressed in the form $\dot{q}_i = f_i(q, p, t)$ as a result of having solved the n equations (7.2) for the n generalised velocities.

The most immediate consequences of introducing the Hamiltonian H can be found by taking the differential of Eq. (7.5):

---

[1] Legendre transformations are widely used in thermodynamics and also have an interesting geometric meaning (Callen, 1960; Arnold, 1989; Zia, Redish & McKay, 2009) which would take us too far afield to discuss here.

$$dH = \sum_{i=1}^{n} (\dot{q}_i dp_i + p_i d\dot{q}_i) - \left\{ \sum_{i=1}^{n} \left(\frac{\partial L}{\partial q_i} dq_i + \frac{\partial L}{\partial \dot{q}_i} d\dot{q}_i \right) + \frac{\partial L}{\partial t} dt \right\} . \qquad (7.6)$$

In virtue of definition (7.2) and of Lagrange's equations, this last equation reduces to

$$dH = \sum_{i=1}^{n} (\dot{q}_i dp_i - \dot{p}_i dq_i) - \frac{\partial L}{\partial t} dt , \qquad (7.7)$$

which makes it clear that, indeed, H depends only on the qs and ps. On the other hand,

$$dH = \sum_{i=1}^{n} \left(\frac{\partial H}{\partial q_i} dq_i + \frac{\partial H}{\partial p_i} dp_i \right) + \frac{\partial H}{\partial t} dt . \qquad (7.8)$$

Comparing these two last equations we find

$$\dot{q}_i = \frac{\partial H}{\partial p_i} , \qquad \dot{p}_i = -\frac{\partial H}{\partial q_i} , \qquad i = 1, \dots, n , \qquad (7.9)$$

and, as a byproduct,

$$\frac{\partial H}{\partial t} = -\frac{\partial L}{\partial t} . \qquad (7.10)$$

Equations (7.9) are known as **Hamilton's equations** or **Hamilton's canonical equations**, and constitute a set of $2n$ first-order ordinary differential equations equivalent to the system of n second-order Lagrange's equations (Hamilton, 1835). The quantities (q, p) are called **canonical variables**[2] and the $2n$-dimensional Cartesian space whose points are represented by the $2n$-tuples $(q, p) \equiv (q_1, \dots, q_n, p_1, \dots, p_n)$ is called the **phase space**. In general, the configuration space is a differentiable manifold and the phase space is its **cotangent bundle** or **cotangent manifold** (José & Saletan, 1998). A point in phase space defines the state of the system – positions and velocities of the particles – at a given instant. An existence and uniqueness theorem ensures that the specification of the state of the system at a time t_0 determines only one solution $(q(t), p(t))$ to Hamilton's equations. Thus, only one dynamical path $(q(t), p(t))$ goes through each phase space point and two distinct dynamical paths never cross or touch each other. As to (7.10), it is not an equation of motion but an important relationship between the explicit time dependence of the Lagrangian and the Hamiltonian. The first half of Hamilton's equations expresses the generalised velocities in terms of the canonical variables, which means that they are the inverses of the equations (7.2) that define the canonical momenta p_i. It must be stressed, however, that contrarily to the Lagrangian formulation in which there holds the *a priori* connexion $\dot{q}_i = dq_i/dt$, in Hamilton's dynamics there is no *a priori* connexion among the canonical variables: the qs and ps are entirely independent among themselves. This is why the two halves of Hamilton's equations (7.9) must be treated on equal footing, comprising the complete set of equations of motion for the system.

Except for certain cases in which the Hamiltonian can be written down directly, the construction of Hamilton's equations involves the following steps:

---

[2] The word "canonical" is used in the sense of "standard".

(a) Choose generalised coordinates and set up the Lagrangian $L(q, \dot{q}, t)$.
(b) Solve Eqs. (7.2) for the velocities $\dot{q}_i$ as functions of q, p, t.
(c) Construct $H(q, p, t)$ by substituting into (7.5) the $\dot{q}$s obtained in (b).
(d) Once in possession of $H(q, p, t)$, write down the equations of motion (7.9).

Hamiltonian and the Total Energy

The usual Lagrangian is $L = T - V$. If (1) T is a purely quadratic function of the velocities and (2) V does not depend on the velocities then Euler's theorem on homogeneous functions (Appendix C) yields $\sum_i \dot{q}_i \partial L / \partial \dot{q}_i = \sum_i \dot{q}_i \partial T / \partial \dot{q}_i = 2T$. As a consequence

$$H = T + V = E, \tag{7.11}$$

that is, *the Hamiltonian is the total energy expressed as a function of the coordinates and momenta*. Conditions (1) and (2) prevail in the vast majority of physically interesting cases, so the Hamiltonian has the extremely important physical meaning of being the total energy in most situations of physical relevance. Furthermore, these conditions are only sufficient, and even if they are not satisfied it is possible that H coincide with the total energy (see Example 7.2 below).

Example 7.1   Obtain Hamilton's equations for a particle in a central potential.

Solution

In spherical coordinates the Lagrangian is

$$L = T - V = \frac{m}{2}\left(\dot{r}^2 + r^2\dot{\theta}^2 + r^2 \sin^2\theta\,\dot{\phi}^2\right) - V(r), \tag{7.12}$$

whence

$$p_r = \frac{\partial L}{\partial \dot{r}} = m\dot{r}, \quad p_\theta = \frac{\partial L}{\partial \dot{\theta}} = mr^2\dot{\theta}, \quad p_\phi = \frac{\partial L}{\partial \dot{\phi}} = mr^2 \sin^2\theta\,\dot{\phi}. \tag{7.13}$$

Solving these equations for the velocities we find

$$\dot{r} = \frac{p_r}{m}, \quad \dot{\theta} = \frac{p_\theta}{mr^2}, \quad \dot{\phi} = \frac{p_\phi}{mr^2 \sin^2\theta}, \tag{7.14}$$

so that

$$H = \dot{r}\,p_r + \dot{\theta}\,p_\theta + \dot{\phi}\,p_\phi - L = \frac{1}{2m}\left(p_r^2 + \frac{p_\theta^2}{r^2} + \frac{p_\phi^2}{r^2 \sin^2\theta}\right) + V(r). \tag{7.15}$$

This Hamiltonian equals the total energy and Hamilton's equations are

$$\dot{r} = \frac{\partial H}{\partial p_r} = \frac{p_r}{m}, \quad \dot{\theta} = \frac{\partial H}{\partial p_\theta} = \frac{p_\theta}{mr^2}, \quad \dot{\phi} = \frac{\partial H}{\partial p_\phi} = \frac{p_\phi}{mr^2 \sin^2\theta}, \tag{7.16}$$

$$\dot{p}_r = -\frac{\partial H}{\partial r} = \frac{p_\theta^2}{mr^3} + \frac{p_\phi^2}{mr^3 \sin^2 \theta} - \frac{dV}{dr},$$

$$\dot{p}_\theta = -\frac{\partial H}{\partial \theta} = \frac{p_\phi^2 \cot \theta}{mr^2 \sin^2 \theta}, \quad \dot{p}_\phi = -\frac{\partial H}{\partial \phi} = 0. \tag{7.17}$$

According to a general remark previously made, Eqs. (7.16) are the inverses of (7.13). With their use Eqs. (7.17) become identical to the Lagrange equations generated by the Lagrangian (7.12).

Example 7.2 Construct the Hamiltonian and Hamilton's equations for a charged particle in an external electromagnetic field.

Solution

In Cartesian coordinates we have

$$L = \frac{m}{2}(\dot{x}^2 + \dot{y}^2 + \dot{z}^2) - e\phi(\mathbf{r}, t) + \frac{e}{c} \mathbf{v} \cdot \mathbf{A}(\mathbf{r}, t), \tag{7.18}$$

whence

$$p_x = \frac{\partial L}{\partial \dot{x}} = m\dot{x} + \frac{e}{c} A_x, \quad p_y = \frac{\partial L}{\partial \dot{y}} = m\dot{y} + \frac{e}{c} A_y, \quad p_z = \frac{\partial L}{\partial \dot{z}} = m\dot{z} + \frac{e}{c} A_z. \tag{7.19}$$

Thus,

$$\mathbf{v} = \frac{1}{m}\left(\mathbf{p} - \frac{e}{c}\mathbf{A}\right) \tag{7.20}$$

and, consequently,

$$H = \mathbf{v} \cdot \mathbf{p} - L = \frac{1}{2m}\left(\mathbf{p} - \frac{e}{c}\mathbf{A}\right)^2 + e\phi. \tag{7.21}$$

This Hamiltonian is the total energy if ϕ and $\mathbf{A}$ do not explicitly depend on time ($\mathbf{E}$ and $\mathbf{B}$ static fields). Hamilton's equations take the form

$$\dot{\mathbf{r}} = \frac{\partial H}{\partial \mathbf{p}} = \frac{1}{m}\left(\mathbf{p} - \frac{e}{c}\mathbf{A}\right), \tag{7.22a}$$

$$\dot{\mathbf{p}} = -\frac{\partial H}{\partial \mathbf{r}} \equiv -\nabla H = \frac{e}{mc}\left[\left(\mathbf{p} - \frac{e}{c}\mathbf{A}\right) \cdot \nabla \mathbf{A} + \left(\mathbf{p} - \frac{e}{c}\mathbf{A}\right) \times (\nabla \times \mathbf{A})\right] - e\nabla\phi, \tag{7.22b}$$

where we have used $\nabla(\mathbf{G} \cdot \mathbf{G}) = 2(\mathbf{G} \cdot \nabla)\mathbf{G} + 2\mathbf{G} \times (\nabla \times \mathbf{G})$ with $\mathbf{G} = \mathbf{p} - (e/c)\mathbf{A}$.

Exercise 7.1.1 Substituting (7.22a) into (7.22b), show that Hamilton's equations (7.22) are equivalent to

$$m\ddot{\mathbf{r}} = e\left(\mathbf{E} + \frac{\mathbf{v}}{c} \times \mathbf{B}\right), \tag{7.23}$$

which is the Newtonian equation of motion.

7.2 Symmetries and Conservation Laws

As defined in Section 2.5, a coordinate q_j is said to be cyclic or ignorable if it does not appear in the Lagrangian. As a consequence of the definition of the canonical momenta and of the Hamiltonian, q_j does not appear in H either. Therefore, it follows from Hamilton's equations that

$$\dot{p}_j = -\frac{\partial H}{\partial q_j} = 0 \qquad (7.24)$$

and the momentum conjugate to q_j is a constant of the motion. In a slightly modified form, the connexion between symmetries and conservation laws, already studied in the Lagrangian formalism, can be reformulated in Hamiltonian language.

Exercise 7.2.1 Consider a system of particles described in Cartesian coordinates by the Hamiltonian $H(\mathbf{r}_1, \ldots, \mathbf{r}_N, \mathbf{p}_1, \ldots, \mathbf{p}_N)$. (1) If H is invariant under the infinitesimal translation $\delta \mathbf{r}_k = \epsilon \hat{\mathbf{n}}$, $k = 1, \ldots N$, prove that the component along direction $\hat{\mathbf{n}}$ of the total canonical linear momentum $\mathbf{P} = \sum_k \mathbf{p}_k$ is a constant of the motion. (2) If H is invariant under the infinitesimal rotation $\delta \mathbf{r}_k = \delta\theta \hat{\mathbf{n}} \times \mathbf{r}_k$, $\delta\mathbf{p}_k = \delta\theta \hat{\mathbf{n}} \times \mathbf{p}_k$, prove that the component along the rotation axis of the total canonical angular momentum $\mathbf{L} = \sum_k \mathbf{r}_k \times \mathbf{p}_k$ is a constant of the motion.

In Lagrangian mechanics it is known that if $\partial L/\partial t = 0$ then the energy function h defined by (2.147) is a constant of the motion. On the other hand, the defining Eq. (7.5) shows that, although they depend on different variables, H and h have the same value. This being the case, there must be a corresponding conservation theorem for H. In fact, by Hamilton's equations,

$$\frac{dH}{dt} = \sum_i \left(\frac{\partial H}{\partial q_i} \dot{q}_i + \frac{\partial H}{\partial p_i} \dot{p}_i \right) + \frac{\partial H}{\partial t}$$

$$= \sum_i \left(\frac{\partial H}{\partial q_i} \frac{\partial H}{\partial p_i} - \frac{\partial H}{\partial p_i} \frac{\partial H}{\partial q_i} \right) + \frac{\partial H}{\partial t} = \frac{\partial H}{\partial t} . \qquad (7.25)$$

From this equation one infers a simple but important result.

Theorem 7.2.1 *If the Hamiltonian does not explicitly depend on time – that is, $\partial H/\partial t = 0$ – then H is a constant of the motion.*

It is worth stressing that according to (7.10) H explicitly depends on time if and only if the same happens to L. On the other hand, as we have already seen, in virtually all situations of physical interest H does not explicitly depend on time and coincides with the total energy of the system. Thus, Theorem 7.2.1 is effectively the law of conservation of energy in the vast majority of circumstances. In exceptional cases, however, H may happen to be conserved without being the total energy or, conversely, H may happen to be the total energy without being conserved.

Example 7.3 Reconsidering Example 1.17, obtain the Hamiltonian and study its conservation, as well as that of the energy.

Solution

As seen in Example 1.17, the Lagrangian for the system is

$$L = T = \frac{m}{2}(\dot{r}^2 + \omega^2 r^2),\tag{7.26}$$

which equals the bead's kinetic energy. In the present case, $p_r = m\dot{r}$ and

$$H = \frac{p_r^2}{2m} - \frac{m\omega^2}{2}r^2,\tag{7.27}$$

which is *not* the total (purely kinetic) energy of the bead. However, H is a constant of the motion because $\partial H/\partial t = 0$. On the other hand, the total energy

$$E = T = \frac{p_r^2}{2m} + \frac{m\omega^2}{2}r^2\tag{7.28}$$

is not conserved because the constraint force does work during a *real* displacement of the particle (verify this qualitatively). It is left for the reader to show that the Hamiltonian (7.27) is the "total energy" in the rotating reference frame.

Conversely, H may coincide with the total energy but not be conserved owing to an explicit time dependence (Goldstein, 1980).

There is an important distinction of behaviour between the Lagrangian and the Hamiltonian concerning the choice of coordinates. Different choices of generalised coordinates change the *functional form* of a Lagrangian, but its value remains the same. As to the Hamiltonian, however, both the *functional form* and the *value* of H depend on the set of generalised coordinates one has adopted. Thus, it may well happen that the quantity represented by H is conserved in one coordinate system but varies with time in another. This mutability of H, which at first glance may appear inconvenient, is highly advantageous and lies at the basis of the Hamilton-Jacobi theory, a powerful alternative method for integrating the equations of motion which will be studied in Chapter 9.

7.3 The Virial Theorem

Let f be a real-valued function of the real variable t. The mean value of f is defined by

$$\langle f \rangle = \lim_{\tau \to \infty} \frac{1}{\tau} \int_0^\tau f(t)\,dt\tag{7.29}$$

whenever this limit exists. In particular, if f is an integrable periodic function with period $\mathcal{T}$ the limit exists:

$$\lim_{\tau \to \infty} \frac{1}{\tau} \int_0^\tau f(t)\,dt = \frac{1}{\mathcal{T}} \int_0^{\mathcal{T}} f(t)\,dt.\tag{7.30}$$

Consider now an n-degree-of-freedom mechanical system described by the canonical variables (q, p) and by the Hamiltonian $H(q, p, t)$. The quantity

$$\mathcal{V} = -\sum_i q_i \frac{\partial H}{\partial q_i} \tag{7.31}$$

is a generalised version of the **virial of Clausius**, introduced in the kinetic theory of gases in the form $\mathcal{V} = \sum_k \mathbf{r}_k \cdot \mathbf{F}_k$ by Clausius.[3] Under certain circumstances the mean value of $\mathcal{V}$ can be related to the mean value of other dynamical quantities of the system.

Theorem 7.3.1 (Virial Theorem)   *If $q_i(t)$ and $p_i(t)$ are bounded functions and if the mean values of $\sum_i q_i \, \partial H/\partial q_i$ and $\sum_i p_i \, \partial H/\partial p_i$ exist separately, then they are equal:*

$$\left\langle \sum_i q_i \frac{\partial H}{\partial q_i} \right\rangle = \left\langle \sum_i p_i \frac{\partial H}{\partial p_i} \right\rangle. \tag{7.32}$$

Proof   Define the function

$$G(t) = \sum_i p_i(t) \, q_i(t). \tag{7.33}$$

Since q_i and p_i are bounded[4] functions, so is G. Differentiating (7.33) with respect to time we obtain

$$\frac{dG}{dt} = \sum_i p_i \dot{q}_i + \sum_i \dot{p}_i q_i = \sum_i p_i \frac{\partial H}{\partial p_i} - \sum_i q_i \frac{\partial H}{\partial q_i}, \tag{7.34}$$

where Hamilton's equations have been used. Taking the mean value of both sides of this equation there results

$$\left\langle \frac{dG}{dt} \right\rangle = \left\langle \sum_i p_i \frac{\partial H}{\partial p_i} - \sum_i q_i \frac{\partial H}{\partial q_i} \right\rangle. \tag{7.35}$$

However

$$\left\langle \frac{dG}{dt} \right\rangle = \lim_{\tau \to \infty} \frac{1}{\tau} \int_0^\tau \frac{dG}{dt} \, dt = \lim_{\tau \to \infty} \frac{G(\tau) - G(0)}{\tau} = 0 \tag{7.36}$$

because the numerator is bounded but the denominator grows indefinitely. With the second hypothesis in the statement of the theorem[5] we are led to Eq. (7.32), as we wished to show. $\quad\square$

Particle in a Central Potential

As a first application, consider a particle in a bounded orbit under a central force, such as a planet moving about the Sun or an electron orbiting a proton in Rutherford's model of the hydrogen atom. The Hamiltonian in Cartesian coordinates is

[3] Some authors refer to the quantity $-(1/2)\langle \sum_k \mathbf{r}_k \cdot \mathbf{F}_k \rangle$ as the virial of Clausius.
[4] A function f is said to be bounded if there exists a positive number M such that $|f(t)| \leq M$ for all t in the domain of f.
[5] It is worth noting that $\langle f + g \rangle = \langle f \rangle + \langle g \rangle$ holds true only when the mean values of f and g exist separately. Examine the following example: $f(t) = 1 + t$ and $g(t) = 1 - t$.

$$H = T + V = \frac{\mathbf{p}^2}{2m} + V(r) \tag{7.37}$$

and the virial theorem implies

$$\left\langle \mathbf{p} \cdot \frac{\mathbf{p}}{m} \right\rangle = \langle \mathbf{r} \cdot \nabla V(r) \rangle = \left\langle \mathbf{r} \cdot \left(\frac{dV}{dr} \frac{\mathbf{r}}{r} \right) \right\rangle = \left\langle r \frac{dV}{dr} \right\rangle. \tag{7.38}$$

For the potential

$$V(r) = \frac{A}{r^n} \tag{7.39}$$

Eq. (7.38) becomes

$$\langle T \rangle = -\frac{n}{2} \langle V \rangle. \tag{7.40}$$

In the case of the gravitational or electrostatic potential ($n = 1$), the virial theorem establishes that

$$2\langle T \rangle = -\langle V \rangle, \tag{7.41}$$

that is, the mean value of the potential energy is minus twice the mean value of the kinetic energy. In general, the energy is given by

$$E = \langle E \rangle = \langle T \rangle + \langle V \rangle = \frac{n-2}{n} \langle T \rangle. \tag{7.42}$$

For $n = 1$ (gravitational or electrostatic case), this last equation gives $E = -\langle T \rangle$, showing that the bound states (elliptic orbits) have negative total energy. In case $n < 0$, however, the bound states in the potential $V(r)$ have positive total energy. The virial theorem has other interesting applications, such as finding approximate expressions for the frequency of anharmonic oscillators (Sivardière, 1986).

Equation of State of an Ideal Gas

Allied to the energy equipartition theorem, the virial theorem affords a strikingly concise derivation of the equation of state of and ideal gas. Let us consider the gas as made up of a vast number of non-interacting particles confined in a container with volume V. The gas molecules experience forces only when they collide with the walls of the container, and the mean force on those that hit the area element dS is $d\mathbf{F} = -P\hat{\mathbf{n}}dS$, where P is the gas pressure and $\hat{\mathbf{n}}$ is unit normal vector *exterior* to the closed surface S bounding volume V. In Cartesian coordinates the Hamiltonian is $H = \sum_i \mathbf{p}_i^2/2m_i + U(\mathbf{r_1}, \ldots, \mathbf{r_N})$ where $U(\mathbf{r_1}, \ldots, \mathbf{r_N})$ plays the role of potential of the forces exerted by the walls of the container. The virial theorem (7.32) applied to the gas takes the form

$$\left\langle \sum_i \mathbf{p}_i \cdot \frac{\mathbf{p}_i}{m_i} \right\rangle = -\left\langle \sum_i \mathbf{r}_i \cdot \mathbf{F}_i \right\rangle = -\oint_S \mathbf{r} \cdot d\mathbf{F} = \oint_S P\mathbf{r} \cdot \hat{\mathbf{n}}\, dS, \tag{7.43}$$

where we have used $\partial H/\partial \mathbf{r}_i = \partial V/\partial \mathbf{r}_i = -\mathbf{F}_i$ and taken into account that the forces on the gas molecules are due to their collisions with the walls of the container. Since P is constant, the divergence theorem yields

$$2\langle T\rangle = P\int_V \nabla\cdot\mathbf{r}\,dV = P\int_V 3\,dV = 3PV\,. \tag{7.44}$$

By the energy equipartition theorem, the mean kinetic energy of each molecule is $3k\Theta/2$, where k is the Boltzmann constant and Θ is the absolute temperature. Thus, if N is the number of gas molecules, Eq. (7.44) takes the form

$$PV = Nk\Theta, \tag{7.45}$$

which is the ideal gas equation of state. If interactions between the gas molecules are taken into account, there appears an additional term on the right-hand side of Eq. (7.45) which depends on the interaction potential and on a function $g(r)$ that measures the probability of finding two molecules separated by the distance r (Pathria, 1972).

7.4 Relativistic Hamiltonian Formulation

Let us start with the non-covariant formalism, in which one works in a specific reference frame and the time, thought of as a parameter used to describe the evolution of the system, is treated distinctly from the spatial coordinates. For a particle in a velocity-independent potential V, we have

$$L = -mc^2\sqrt{1 - v^2/c^2} - V(\mathbf{r})\,, \tag{7.46}$$

so that

$$\mathbf{p} = \frac{\partial L}{\partial \mathbf{v}} = \frac{m\,\mathbf{v}}{\sqrt{1 - v^2/c^2}}\,. \tag{7.47}$$

Note that the canonical momentum $\mathbf{p}$ coincides with the relativistic momentum. Therefore,

$$H = \mathbf{v}\cdot\mathbf{p} - L = \frac{m\,v^2}{\sqrt{1 - v^2/c^2}} + mc^2\sqrt{1 - v^2/c^2} + V(\mathbf{r}) = \frac{m\,c^2}{\sqrt{1 - v^2/c^2}} + V\,, \tag{7.48}$$

and H is the total energy although the Lagrangian (7.46) does not have the form of the difference of the kinetic and potential energies. We still need to express H as a function of $(\mathbf{r},\mathbf{p})$. This is easily accomplished by resorting to the relativistic relation (6.120) between energy and momentum for a free particle, from which we get

$$H(\mathbf{r},\mathbf{p}) = \sqrt{p^2c^2 + m^2c^4} + V(\mathbf{r})\,. \tag{7.49}$$

In the case of a particle in an external electromagnetic field,

$$L = -mc^2\sqrt{1 - v^2/c^2} - e\,\phi(\mathbf{r},t) + \frac{e}{c}\mathbf{v}\cdot\mathbf{A}(\mathbf{r},t)\,, \tag{7.50}$$

whence

$$\mathbf{p} = \frac{\partial L}{\partial \mathbf{v}} = \frac{m\,\mathbf{v}}{\sqrt{1 - v^2/c^2}} + \frac{e}{c}\mathbf{A}\,. \tag{7.51}$$

Note that now the canonical momentum is no longer identical to the relativistic momentum. We have

$$H = \mathbf{v} \cdot \mathbf{p} - L = \frac{m\,c^2}{\sqrt{1 - v^2/c^2}} + e\,\phi\,, \tag{7.52}$$

and again H is the total energy if $\mathbf{E}$ and $\mathbf{B}$ are static fields. Making use once more of (6.120) with $\mathbf{p} - (e/c)\mathbf{A}$ substituted for the relativistic momentum, we obtain the Hamiltonian in its final form:

$$H(\mathbf{r}, \mathbf{p}, t) = \sqrt{\left(\mathbf{p} - \frac{e}{c}\mathbf{A}(\mathbf{r}, t)\right)^2 c^2 + m^2 c^4} + e\,\phi(\mathbf{r}, t)\,. \tag{7.53}$$

The free particle relativistic Hamiltonian is given by Eq. (7.49) with $V = 0$, and clearly the transiton to the Hamiltonian (7.53) for a particle in an external electromagnetic field is achieved by the prescription

$$H \longrightarrow H - e\,\phi\,, \qquad \mathbf{p} \longrightarrow \mathbf{p} - \frac{e}{c}\mathbf{A}\,, \tag{7.54}$$

known as **minimal coupling**.

As to the covariant formulation, taking as starting point the Lagrangian

$$L = \frac{m}{2}U^\alpha U_\alpha + \frac{e}{c}A^\alpha U_\alpha \tag{7.55}$$

for a charged particle in an external electromagnetic field, we obtain the canonical momenta

$$P^\mu = \frac{\partial L}{\partial U_\mu} = mU^\mu + \frac{e}{c}A^\mu. \tag{7.56}$$

The canonical four-momentum differs from the particle's four-momentum. The covariant Hamiltonian is given by

$$H = P^\alpha U_\alpha - L = \frac{1}{2m}\left(P^\alpha - \frac{e}{c}A^\alpha\right)\left(P_\alpha - \frac{e}{c}A_\alpha\right)\,. \tag{7.57}$$

The transition from the free-particle Hamiltonian to that for a particle in an eletromagnetic field is effected by the minimal coupling

$$P^\alpha \longrightarrow P^\alpha - \frac{e}{c}A^\alpha\,, \tag{7.58}$$

which is the covariant version of (7.54). The covariant Hamilton's equations are

$$\frac{dx^\mu}{d\tau} = \frac{\partial H}{\partial P_\mu}\,, \qquad \frac{dP^\mu}{d\tau} = -\frac{\partial H}{\partial x_\mu}\,. \tag{7.59}$$

Exercise 7.4.1 (a) Prove that the space part of Eqs. (7.59) is equivalent to the relativistic equation of motion $d(\gamma m\mathbf{v})/dt = e(\mathbf{E} + \mathbf{v} \times \mathbf{B}/c)$. (b) The Hamiltonian (7.57) is a scalar, so it cannot be equal to the particle's total energy. Show that, this notwithstanding, H is a constant of the motion and its value is $mc^2/2$.

Regarding the time components of (7.59), we have

$$U^0 = \frac{\partial H}{\partial P_0} = \frac{1}{m}\left(P^0 - \frac{e}{c}A^0\right)\,, \tag{7.60}$$

whence

$$P^0 = mU^0 + \frac{e}{c} A^0 = \frac{E + e\phi}{c} . \tag{7.61}$$

According to this equation, the total energy $E_{tot} = E + e\phi$ acts as the momentum conjugate to time.[6] We further have

$$\frac{dP_0}{d\tau} = -\frac{\partial H}{\partial x^0} = -\frac{1}{c} \frac{\partial H}{\partial t} = \frac{e}{mc^2} \left(p^\alpha - \frac{e}{c} A^\alpha \right) \frac{\partial A_\alpha}{\partial t} \tag{7.62}$$

or, equivalently,

$$\gamma \frac{d}{dt}(E + e\,\phi) = \frac{e}{mc} mU^\alpha \frac{\partial A_\alpha}{\partial t} = \frac{e}{c} \left[\gamma c \frac{\partial \phi}{\partial t} - \gamma \mathbf{v} \cdot \frac{\partial \mathbf{A}}{\partial t} \right] . \tag{7.63}$$

Cancelling the common γ factor this last equation can be recast in the form

$$\frac{dE}{dt} + e \left(\mathbf{v} \cdot \nabla \phi + \frac{\partial \phi}{\partial t} \right) = e \frac{\partial \phi}{\partial t} - \frac{ev}{c} \cdot \frac{\partial \mathbf{A}}{\partial t} \tag{7.64}$$

or, finally,

$$\frac{dE}{dt} = e \left(-\nabla \phi - \frac{1}{c} \frac{\partial \mathbf{A}}{\partial t} \right) \cdot \mathbf{v} = e\mathbf{E} \cdot \mathbf{v} = \mathbf{F} \cdot \mathbf{v} , \tag{7.65}$$

where $\mathbf{F} = e(\mathbf{E} + \mathbf{v} \times \mathbf{B}/c)$. Equation (7.65) correctly expresses the energy conservation law, completing the verification that the covariant Hamilton's equations (7.59) have the same content as the covariant equation of motion (6.106).

7.5 Hamilton's Equations in Variational Form

Hamilton's principle formulated in terms of variations of the path in configuration space is written

$$\delta S \equiv \delta \int_{t_1}^{t_2} L(q, \dot{q}, t)\, dt = 0 , \tag{7.66}$$

with $\delta q_i(t_1) = \delta q_i(t_2) = 0$. This variational principle can be reformulated in terms of variations of the path in phase space as follows:

$$\delta S \equiv \delta \int_{t_1}^{t_2} \left\{ \sum_{i=1}^{n} p_i \dot{q}_i - H(q, p, t) \right\} dt = 0 , \tag{7.67}$$

where $\delta q_i(t_1) = \delta q_i(t_2) = 0$ and the variations of the qs and ps are considered independent. Indeed, we have

$$\delta S = \int_{t_1}^{t_2} dt \sum_{i=1}^{n} \left(p_i\, \delta \dot{q}_i + \dot{q}_i\, \delta p_i - \frac{\partial H}{\partial q_i} \delta q_i - \frac{\partial H}{\partial p_i} \delta p_i \right) = 0 . \tag{7.68}$$

---

[6] In a certain sense this result is not peculiar to relativistic mechanics, as will be discussed in Section 7.6.

An integration by parts gives[7]

$$\int_{t_1}^{t_2} p_i\,\delta\dot{q}_i\,dt = p_i\,\delta q_i\Big|_{t_1}^{t_2} - \int_{t_1}^{t_2} \dot{p}_i\,\delta q_i\,dt = -\int_{t_1}^{t_2} \dot{p}_i\,\delta q_i\,dt\,, \tag{7.69}$$

so Eq. (7.68) becomes

$$\delta S = \int_{t_1}^{t_2} dt \sum_{i=1}^{n} \left\{ \left(\dot{q}_i - \frac{\partial H}{\partial p_i}\right)\delta p_i - \left(\dot{p}_i + \frac{\partial H}{\partial q_i}\right)\delta q_i \right\} = 0\,. \tag{7.70}$$

Since the variations of the qs and ps are arbitrary and independent of each other, it follows from (7.70) that

$$\dot{q}_i - \frac{\partial H}{\partial p_i} = 0\,, \quad \dot{p}_i + \frac{\partial H}{\partial q_i} = 0\,, \quad i = 1,\ldots,n\,, \tag{7.71}$$

which are Hamilton's equations.

A comparison of Hamilton's principle in phase space – Eq. (7.67) – with Hamilton's principle in configuration space – Eq. (7.66) – signals that the integrand in Eq. (7.67) works just like a Lagrangian in phase space. Let $\mathscr{L} = \mathscr{L}(q,p,\dot{q},\dot{p},t)$ be the **phase space Lagrangian** defined by

$$\mathscr{L}(q,p,\dot{q},\dot{p},t) = \sum_{k=1}^{n} p_k\,\dot{q}_k - H(q,p,t)\,. \tag{7.72}$$

In terms of $\mathscr{L}$ Hamilton's equations take the form of Lagrange's equations:

$$\frac{d}{dt}\left(\frac{\partial\mathscr{L}}{\partial\dot{q}_i}\right) - \frac{\partial\mathscr{L}}{\partial q_i} = 0 \implies \dot{p}_i + \frac{\partial H}{\partial q_i} = 0\,, \tag{7.73}$$

$$\frac{d}{dt}\left(\frac{\partial\mathscr{L}}{\partial\dot{p}_i}\right) - \frac{\partial\mathscr{L}}{\partial p_i} = 0 \implies -\dot{q}_i + \frac{\partial H}{\partial p_i} = 0\,. \tag{7.74}$$

According to Theorem 2.3.1, Lagrangians that differ only by a total time derivative yield the same equations of motion. A corresponding result holds for phase space Lagrangians.

Proposition 7.5.1 *Let $\mathscr{L}$ be the phase space Lagrangian given by (7.72). If $\Phi(q,p,t)$ is any infinitely differentiable function, the new phase space Lagrangian $\bar{\mathscr{L}}$ defined by*

$$\bar{\mathscr{L}}(q,p,\dot{q},\dot{p},t) = \mathscr{L}(q,p,\dot{q},\dot{p},t) + \frac{d}{dt}\Phi(q,p,t) \tag{7.75}$$

yields exactly the same Hamilton's equations as $\mathscr{L}$.

Proof   If $\bar{\mathscr{L}}$ and $\mathscr{L}$ are connected by (7.75) we explicitly have

$$\bar{\mathscr{L}}(q,p,\dot{q},\dot{p},t) = \mathscr{L}(q,p,\dot{q},\dot{p},t) + \sum_{k} \left(\frac{\partial\Phi}{\partial q_k}\dot{q}_k + \frac{\partial\Phi}{\partial p_k}\dot{p}_k\right) + \frac{\partial\Phi}{\partial t}\,, \tag{7.76}$$

---

[7] Since the integrand in (7.67) does not contain $\dot{p}$s, the variations δp_i are not required to vanish at the endpoints; they are completely arbitrary.

whence

$$
\frac{d}{dt}\left(\frac{\partial \bar{\mathscr{L}}}{\partial \dot{q}_i}\right) = \frac{d}{dt}\left(\frac{\partial \mathscr{L}}{\partial \dot{q}_i}\right) + \frac{d}{dt}\left(\frac{\partial \Phi}{\partial q_i}\right)
$$

$$
= \frac{d}{dt}\left(\frac{\partial \mathscr{L}}{\partial \dot{q}_i}\right) + \sum_k \left(\frac{\partial^2 \Phi}{\partial q_k \partial q_i}\dot{q}_k + \frac{\partial^2 \Phi}{\partial p_k \partial q_i}\dot{p}_k\right) + \frac{\partial^2 \Phi}{\partial t \partial q_i} \qquad (7.77)
$$

and

$$
\frac{\partial \bar{\mathscr{L}}}{\partial q_i} = \frac{\partial \mathscr{L}}{\partial q_i} + \sum_k \left(\frac{\partial^2 \Phi}{\partial q_i \partial q_k}\dot{q}_k + \frac{\partial^2 \Phi}{\partial q_i \partial p_k}\dot{p}_k\right) + \frac{\partial^2 \Phi}{\partial q_i \partial t}. \qquad (7.78)
$$

The equality of the mixed partial derivatives of the function Φ implies

$$
\frac{d}{dt}\left(\frac{\partial \bar{\mathscr{L}}}{\partial \dot{q}_i}\right) - \frac{\partial \bar{\mathscr{L}}}{\partial q_i} = \frac{d}{dt}\left(\frac{\partial \mathscr{L}}{\partial \dot{q}_i}\right) - \frac{\partial \mathscr{L}}{\partial q_i}. \qquad (7.79)
$$

A similar argument shows that

$$
\frac{d}{dt}\left(\frac{\partial \bar{\mathscr{L}}}{\partial \dot{p}_i}\right) - \frac{\partial \bar{\mathscr{L}}}{\partial p_i} = \frac{d}{dt}\left(\frac{\partial \mathscr{L}}{\partial \dot{p}_i}\right) - \frac{\partial \mathscr{L}}{\partial p_i} \qquad (7.80)
$$

and the proof is complete.   □

This proposition turns out to be particularly important in the study of canonical transformations, to be undertaken in the next chapter.

7.6 Time as a Canonical Variable

An interesting phenomenon takes place when we try to include the time in the set of generalised coordinates by using an auxiliary parameter θ to follow the evolution of the system. In the case of a single particle, θ could be the arc length along the particle's trajectory in space. The only restriction on θ is $dt/d\theta > 0$ – that is, t must be an increasing function of θ. Let us, therefore, set $q_{n+1} \equiv t$ and write the action in the form

$$
S = \int_{\theta_1}^{\theta_2}\left\{\sum_{i=1}^n p_i\frac{dq_i}{dt} - H(q,p,t)\right\}\frac{dt}{d\theta}\,d\theta = \int_{\theta_1}^{\theta_2}\left\{\sum_{i=1}^n p_i\frac{dq_i}{d\theta} - H\frac{dq_{n+1}}{d\theta}\right\}d\theta
$$
$$(7.81)$$

or, equivalently,

$$
S = \int_{\theta_1}^{\theta_2} d\theta \sum_{i=1}^{n+1} p_i q_i', \qquad (7.82)
$$

where

$$
q_i' \equiv \frac{dq_i}{d\theta}, \quad i = 1,\ldots,n+1, \quad p_{n+1} = -H. \qquad (7.83)
$$

Three features of Eqs. (7.82) and (7.83) stand out: (1) the momentum conjugate to time is $-H$; (2) in the **extended phase space** spanned by the $2(n+1)$ canonical variables

$(q,p) \equiv (q_1, \ldots, q_{n+1}, p_1, \ldots, p_{n+1})$ the Hamiltonian is zero; (3) the equation $p_{n+1} = -H$ is a constraint in the extended phase space.

Introducing the super-Hamiltonian[8] function

$$\mathcal{H}(q,p) = p_{n+1} + H(q_1, \ldots, q_{n+1}, p_1, \ldots, p_n), \qquad (7.84)$$

defined in the extended phase space, the constraint is expressed as

$$\mathcal{H}(q,p) = 0. \qquad (7.85)$$

Because of this constraint, in order to correctly formulate the equations of motion by a variational principle in the extended phase space, we must make use of a Lagrange multiplier λ and rewrite the action in the form (see the last paragraph of Section 2.4)

$$S[q,p,\lambda] = \int_{\theta_1}^{\theta_2} d\theta \left\{ \sum_{i=1}^{n+1} p_i \, q_i' - \lambda \mathcal{H}(q,p) \right\} \equiv \int_{\theta_1}^{\theta_2} d\theta \, F(q,p,\lambda,q',p',\lambda'). \qquad (7.86)$$

From $\delta S = 0$ there arise the Euler's equations

$$\frac{d}{d\theta}\left(\frac{\partial F}{\partial q_k'}\right) - \frac{\partial F}{\partial q_k} = 0, \qquad \frac{d}{d\theta}\left(\frac{\partial F}{\partial p_k'}\right) - \frac{\partial F}{\partial p_k} = 0, \qquad k = 1, \ldots, n+1, \qquad (7.87)$$

and also

$$\frac{d}{d\theta}\left(\frac{\partial F}{\partial \lambda'}\right) - \frac{\partial F}{\partial \lambda} = 0. \qquad (7.88)$$

Taking into account the explicit form of F defined by (7.86), these equations reduce to

$$p_k' + \lambda \frac{\partial \mathcal{H}}{\partial q_k} = 0, \qquad -q_k' + \lambda \frac{\partial \mathcal{H}}{\partial p_k} = 0, \qquad k = 1, \ldots, n+1, \qquad (7.89)$$

$$\mathcal{H}(q,p) = 0. \qquad (7.90)$$

Therefore, the constraint is automatically satisfied in consequence of the variational principle. Considering the $(n+1)$th component of Eqs. (7.89), we find

$$\lambda = \frac{dt}{d\theta} \qquad (7.91)$$

and also

$$\frac{dp_{n+1}}{d\theta} + \lambda \frac{\partial \mathcal{H}}{\partial q_{n+1}} = 0 \quad \Longrightarrow \quad -\frac{dH}{d\theta} = -\frac{dt}{d\theta}\frac{\partial H}{\partial t} \quad \Longrightarrow \quad \frac{dH}{dt} = \frac{\partial H}{\partial t}, \qquad (7.92)$$

which is nothing but Eq. (7.25). Finally, with the result (7.91), the first n components of (7.89) become

$$\frac{dq_k}{d\theta} = \lambda \frac{\partial H}{\partial p_k} \quad \Longrightarrow \quad \frac{dq_k}{dt}\frac{dt}{d\theta} = \frac{dt}{d\theta}\frac{\partial H}{\partial p_k} \quad \Longrightarrow \quad \dot{q}_k = \frac{\partial H}{\partial p_k}, \qquad (7.93)$$

$$\frac{dp_k}{d\theta} = -\lambda \frac{\partial H}{\partial q_k} \quad \Longrightarrow \quad \frac{dp_k}{dt}\frac{dt}{d\theta} = -\frac{dt}{d\theta}\frac{\partial H}{\partial q_k} \quad \Longrightarrow \quad \dot{p}_k = -\frac{\partial H}{\partial q_k}, \qquad (7.94)$$

and we recover Hamilton's equations.

---

[8] This name is borrowed from quantum cosmology.

In short, one can treat time in equal terms with the generalised coordinates as long as one is willing to deal with an extended phase space in which there holds a constraint – that is, the canonical variables are not mutually independent. A formal byproduct of this approach is that time and the Hamiltonian come to the fore as conjugate canonical variables. This result reverberates beyond classical mechanics, for it suggests the existence of an uncertainty relation between energy and time in quantum mechanics – which turns out to be the case.

Exercise 7.6.1   We can interpret Eq. (7.91) not as determining λ but as serving to determine θ. In other words, a choice of λ defines which parameter θ is being used to describe the unfolding of the path in phase space. For example, the choice $\lambda = 1$ is equivalent to taking $\theta = t$ and the usual time as the evolution parameter. Show that in this case Eqs. (7.89) and (7.90) coincide exactly with Hamilton's Eqs. (7.9) supplemented by (7.25).

Systems with Parameterised Time

The necessary and sufficient condition for the Hamiltonian to be zero is that the Lagrangian be a first-degree homogeneous function of the velocities. This is an immediate consequence of (7.2), (7.5) and of Euler's theorem on homogeneous functions (Appendix C). A system with parameterised time is defined as one whose equations of motion do not depend on the choice of evolution parameter or, equivalently, whose associated action is invariant under arbitrary transformations of the evolution parameter. A system with parameterised time is necessarily described by a Lagrangian which is a first-degree homogeneous function of the generalised velocities.

Exercise 7.6.2   A parameterised-time system is characterised by an action which is invariant under time reparameterisations, meaning arbitrary changes of the evolution parameter. Consider, therefore, a system with a Lagrangian $L(q, \dot{q})$ without explicit time dependence and suppose the associated action is invariant under the arbitrary infinitesimal transformation $t' = t + \epsilon X(t)$ of the evolution parameter. Use the Noether condition (2.167) to prove that L is a first-degree homogeneous function of the velocities.

In problems which are invariant under transformations of the evolution parameter, the procedure of going over from the extended phase space, where there holds the super-Hamiltonian constraint $\mathcal{H}(q, p) = 0$, to the **reduced phase space**, in which the canonical variables are mutually independent, consists in traversing the path that led to (7.82) in the reversed direction. Given an action of the form (7.86) one *picks* a canonical variable to play the role of time and solves the constraint equation $\mathcal{H}(q, p) = 0$ for the canonical variable conjugate to the one chosen as time. This variable conjugate to time, now expressed as a function of the remaining canonical variables, is to be introduced into the action. The canonical variable conjugate to time no longer appears in the resulting **reduced action**, which has the standard form

$$S = \int_{t_1}^{t_2} \left\{ \sum_{i=1}^{n} p_i \dot{q}_i - H(q, p, t) \right\} dt \qquad (7.95)$$

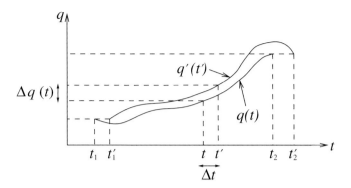

Fig. 7.1   Variation involved in the principle of Maupertuis.

$$I \equiv \int_{t_1}^{t_2} dt \sum_i p_i \dot{q}_i \qquad (7.121)$$

is a minimum or, more generally, is stationary.

Proof   For the actual path and any other comparison path the abbreviated action is given by

$$I \equiv \int_{t_1}^{t_2} dt \sum_i p_i \dot{q}_i = \int_{t_1}^{t_2} (L + H)\, dt = \int_{t_1}^{t_2} L\, dt + H\,(t_2 - t_1) \qquad (7.122)$$

since, according to hypotheses (1) and (2), H remains constant along the actual path and has the same constant value along the varied paths. Therefore,

$$\Delta I = \Delta \int_{t_1}^{t_2} L\, dt + H\,(\Delta t_2 - \Delta t_1), \qquad (7.123)$$

where $\Delta t_1 = \epsilon\, X(t_1)$ with similar definition for Δt_2. The Δ variation is a particular case of the infinitesimal transformations (2.160) admitted in Noether' theorem. Taking advantage of the calculations in Section 2.7 that led to equations (2.166)–(2.168), we find

$$\Delta \int_{t_1}^{t_2} L\, dt = \int_{t_1}^{t_2} \frac{d}{dt} \left\{ \sum_{i=1}^n \frac{\partial L}{\partial \dot{q}_i} \Delta q_i - h\Delta t \right\} dt = \left(\sum_{i=1}^n p_i \Delta q_i - h\Delta t \right) \Bigg|_{t_1}^{t_2}. \qquad (7.124)$$

Consequently, since $h = H$ and $\Delta q_i(t_1) = \Delta q_i(t_2) = 0$, Eq. (7.124) reduces to

$$\Delta \int_{t_1}^{t_2} L\, dt = -H\,(\Delta t_2 - \Delta t_1), \qquad (7.125)$$

where we once again have used the constancy of H. Inserted into (7.123), this last result leads immediately to (7.120), completing the proof. □

The Principle of Maupertuis and Geometrical Optics

It is illuminating to examine the content of the principle of Maupertuis in some particular cases. When the functions (1.71) which define the generalised coordinates do not explicitly depend on time, the kinetic energy is a quadratic function of the generalised velocities:

$$T = \frac{1}{2} \sum_{k,l} M_{kl}(q)\, \dot{q}_k\, \dot{q}_l \,. \tag{7.126}$$

If, besides, $L = T - V$ with V independent of the velocities,

$$\sum_i \dot{q}_i p_i = \sum_i \dot{q}_i \frac{\partial T}{\partial \dot{q}_i} = 2T \tag{7.127}$$

in virtue of Euler's theorem on homogeneous functions. In such circumstances the principle of Maupertuis asserts that

$$\Delta \int_{t_1}^{t_2} T\, dt = 0 \,. \tag{7.128}$$

For example, consider a particle or rigid body in the absence of external forces. The kinetic energy is constant and the principle of Maupertuis takes the special form

$$\Delta(t_2 - t_1) = 0 \,. \tag{7.129}$$

In words: out of all possible paths between two fixed configurations, the actual path travelled by the system is the one for which the transit time is least. The similarity with Fermat's principle of geometrical optics is apparent.

Jacobi's Principle

The form (7.126) taken by the kinetic energy suggests that the configuration space be thought of as a space endowed with a metric, the distance $d\rho$ between two infinitesimally close points being given by

$$d\rho^2 = \sum_{k,l} M_{kl}(q)\, dq_k\, dq_l \,. \tag{7.130}$$

$M_{kl}(q)$ acts as metric tensor and

$$T = \frac{1}{2} \left(\frac{d\rho}{dt} \right)^2 \tag{7.131}$$

or, equivalently,

$$dt = \frac{d\rho}{\sqrt{2T}} \,. \tag{7.132}$$

Using this, the principle of Maupertuis takes the form

$$\Delta \int_{\rho_1}^{\rho_2} \sqrt{T}\, d\rho = 0 \tag{7.133}$$

or, finally,

$$\Delta \int_{\rho_1}^{\rho_2} \sqrt{E - V(q)}\, d\rho = 0 \,. \tag{7.134}$$

This equation is called **Jacobi's principle** or Jacobi's form of Maupertuis' principle. Jacobi's principle is a statement not on the motion in time but on the shape of the curve in configuration space traversed by the system. In the single particle case $d\rho$ is proportional to the infinitesimal three-dimensional distance in arbitrary curvilinear coordinates, and Jacobi's principle becomes identical to Fermat's principle for the propagation of light in an optically heterogeneous medium with a refractive index proportional to $\sqrt{E - V}$. Other even more remarkable aspects of the analogy between classical mechanics and geometrical optics will be explored in Chapter 9.

When T is constant Jacobi's principle states that the system follows the path of least length in configuration space: the physical trajectory is a **geodesic** of configuration space. For a free particle in three dimensions the configuration space is the usual three-dimensional space: the geodesics are straight lines, corresponding to the fact that a free particle describes uniform rectilinear motion.

Exercise 7.7.1 An otherwise free particle is confined to a fixed smooth surface in three-dimensional space. Show that the particle's path is a geodesic of the surface.

Problems

7.1 A particle of mass m oscillates in a vertical plane suspended by a massless string that goes through a hole in a horizontal table. The string is pulled through the hole at a constant rate, so that the hanging length at time t is $r = \ell - \alpha t$, where ℓ and α are positive constants. Taking the angle the string makes with the vertical as a generalised coordinate, find the Hamiltonian for the system. Is the Hamiltonian the total energy? Is the Hamiltonian a constant of the motion?

7.2 Given the Hamiltonian $H = p^2/2m + \mathbf{a} \cdot \mathbf{p} + V$, with $\mathbf{a} = \mathbf{a}(\mathbf{r})$ and $V = V(\mathbf{r})$, find the corresponding Lagrangian. If $V = -\mathbf{F} \cdot \mathbf{r}$ with $\mathbf{a}$ and $\mathbf{F}$ constant vectors, solve Hamilton's equations with the initial conditions $\mathbf{r}(0) = 0$, $\mathbf{p}(0) = 0$.

7.3 Given the Hamiltonian $H(q_1, q_2, p_1, p_2) = q_1 p_1 - q_2 p_2 - a q_1^2 + b q_2^2$ with a and b constants, prove that

$$F_1 = \frac{p_2 - bq_2}{q_1}, \qquad F_2 = q_1 q_2, \qquad F_3 = q_1 e^{-t}$$

are constants of the motion. Identify a fourth constant of the motion independent of these three constants and, using them, obtain the general solution to the equations of motion – that is, $q_1(t), q_2(t), p_1(t), p_2(t)$ involving four arbitrary constants.

7.4 Consider a mechanical system whose kinetic energy T is a quadratic function of the generalised velocities and assume that the potential energy vanishes, so that the Lagrangian and the Hamiltonian are equal: $L = H = T$. Lagrange's equations give

$$\dot{p}_i = \frac{\partial L}{\partial q_i} = \frac{\partial T}{\partial q_i},$$

whereas Hamilton's equations give

$$\dot{p}_i = -\frac{\partial H}{\partial q_i} = -\frac{\partial T}{\partial q_i}.$$

Seemingly, from these equations it follows that $\dot{p}_i = 0$, so all canonical momenta are constants of the motion. However, a simple example such as that of a free particle in spherical coordinates shows that this conclusion is false (check it). Resolve the apparent paradox (Gangopadhyaya & Ramsey, 2013; Lemos, 2014b).

7.5 A one-degree-of-freedom mechanical system is described by the following Lagrangian:

$$L(Q, \dot{Q}, t) = \frac{\dot{Q}^2}{2} \cos^2 \omega t - \frac{\omega}{2} Q\dot{Q} \sin 2\omega t - \frac{\omega^2 Q^2}{2} \cos 2\omega t .$$

(a) Find the corresponding Hamiltonian. (b) Is this Hamiltonian a constant of the motion? (c) Show that the Hamiltonian expressed in terms of the new variable $q = Q \cos \omega t$ and its conjugate momentum does not explicitly depend on time. What physical system does it describe?

7.6 Consider the n-body problem in the the centre-of-mass frame. The Hamiltonian is given by $H = T + V$ where $T = \sum_i |\mathbf{p}_i|^2 / 2m_i$ and

$$V(\mathbf{r}_1, \ldots, \mathbf{r}_n) = -\frac{1}{2} \sum_{\substack{i,j \\ i \neq j}} \frac{Gm_i m_j}{|\mathbf{r}_i - \mathbf{r}_j|} .$$

(a) Introduce the quantity $I = \frac{1}{2} \sum_i m_i |\mathbf{r}_i|^2$ and prove that $\ddot{I} = E + T$. Hint: show that V is a homogeneous function and apply Euler's theorem from Appendix C. (b) Taking into account that T is always positive, perform two successive integrations to prove that $I(t) \geq I(0) + \dot{I}(0)t + Et^2/2$. Conclude, finally, that if the total energy is positive at least one of the bodies will escape to infinity in the limit $t \to \infty$.

7.7 A planet describes an elliptic orbit about the Sun. Let the spherical coordinates be so chosen that the plane of the motion is given by $\theta = \pi/2$. By Eq. (7.15), in polar coordinates r, ϕ in the orbital plane the Hamiltonian is written

$$H = T + V = \frac{1}{2m} \left(p_r^2 + \frac{p_\phi^2}{r^2} \right) - \frac{k}{r} ,$$

with $k > 0$. On the face of it, the virial theorem is valid no matter what canonical variables are employed. Show that in terms of the polar coordinates r, ϕ and their conjugate momenta, Eq. (7.32) takes the form

$$2\langle T \rangle = -\langle V \rangle - \left\langle \frac{p_\phi^2}{mr^2} \right\rangle .$$

Comparing this result with (7.41), show that $\langle p_\phi^2 / mr^2 \rangle = 0$. Using the fact that for elliptic orbits p_ϕ is a non-zero constant of the motion, conclude that $\langle 1/r^2 \rangle = 0$. Argue that this is impossible. Examine carefully the conditions for the validity of the virial theorem in the present case and explain why its use in polar coordinates leads to an absurdity (Chagas, das & Lemos, 1981).

7.8 According to Kotkin and Serbo (1971), the motion of certain turbulences in liquid helium is approximately described by the Hamiltonian $H(x, p) = A\sqrt{p} - Fx$, with A and F constants. (a) Find the general solution to Hamilton's equations. (b) Determine the Lagrangian associated with this Hamiltonian.

7.9 Construct the Hamiltonian associated with the Lagrangian of Problem 1.3. Is H the total energy? Is H a constant of the motion?

7.10 A particle of charge e moves in a central potential $V(r)$ superimposed onto a uniform magnetic field $\mathbf{B}$ whose vector potential is $\mathbf{A} = \mathbf{B} \times \mathbf{r}/2$. (a) Show that if $\mathbf{B}$ is a weak field, so that effects proportional to B^2 can be neglected, then

$$H(\mathbf{r}, \mathbf{p}) = \frac{p^2}{2m} + V(\mathbf{r}) - \boldsymbol{\mu} \cdot \mathbf{B},$$

where

$$\boldsymbol{\mu} = \frac{e}{2mc} \mathbf{r} \times \mathbf{p}.$$

(b) Write down Hamilton's equations in this approximation.

7.11 A physical system is described by the Lagrangian

$$L = \frac{m}{2}\left(\dot{\rho}^2 + \rho^2\dot{\phi}^2 + \dot{z}^2\right) + a\rho^2\dot{\phi},$$

where ρ, ϕ, z are cylindrical coordinates and a is a constant. (a) Construct H. (b) Identify three constants of the motion. (c) Show that the solution to the radial equation can be reduced to a quadrature in the form

$$t = \int \frac{d\rho}{\sqrt{\alpha - (\beta - a\rho^2)^2/m^2\rho^2}},$$

with α and β constants.

7.12 Starting from the Hamiltonian formulation, obtain another formulation of mechanics using as variables $p, \dot{p}$ and a function $Y(p, \dot{p}, t)$. By imitating the example of Eqs. (7.2) and (7.5), the Legendre transformation that replaces q by $\dot{p}$ must presuppose that the equations $\dot{p}_i = -\partial H/\partial q_i$ can be solved for the qs in terms of $p, \dot{p}, t$. Construct the Legendre transformation that takes from $H(q, p, t)$ to $Y(p, \dot{p}, t)$ and derive the equations of motion in terms of Y. Apply this approach to the particular case in which $H = p^2/2 + \omega^2 q^2/2$ and comment on the results obtained.

7.13 Find the Hamiltonian and Hamilton's equations for the mechanical system of Problem 1.17.

7.14 In Weber's electrodynamics the force between two electric charges in motion is given by the generalised potential

$$U(r, \dot{r}) = \frac{ee'}{r}\left(1 + \frac{\dot{r}^2}{2c^2}\right),$$

where r is the distance between the charges. Consider a charged particle moving in a plane subject to this potential. Using polar coordinates in the plane of the motion, find the Hamiltonian $H(r, \phi, p_r, p_\phi)$ and Hamilton's equations. Show that p_ϕ is constant and reduce the radial problem to a first-order differential equation for r.

7.15 The motion of a particle in a central potential $V(r)$ is described in a reference frame that rotates with constant angular velocity $\boldsymbol{\omega}$ with respect to an inertial frame. The rotation axis contains the centre of force. (a) Show that the momentum conjugate to the position vector $\mathbf{R}$ ($= \mathbf{r}$, but seen from the rotating frame) is $\mathbf{P} = m[\mathbf{V} + \boldsymbol{\omega} \times \mathbf{R}]$, where $\mathbf{V}$ is the velocity relative to the rotating frame. (b) Construct the Hamiltonian $H(\mathbf{R}, \mathbf{P}) = \mathbf{P} \cdot \mathbf{V} - L$, where $L = mv^2/2 - V(r)$. (c) Show that $H(\mathbf{R}, \mathbf{P})$ is conserved but is not the total energy in the rotating frame – that is, $H \neq mV^2/2 + V(R)$.

7.16 Consider the relativistic motion of a particle in a central potential $V(r)$. (a) Show that $l = \mathbf{r} \times \mathbf{p}$ is a constant of the motion, where $\mathbf{p}$ is the relativistic linear momentum, and that the Hamiltonian (7.49) can be cast in the form

$$H = c\sqrt{m^2 c^2 + p_r^2 + \frac{p_\phi^2}{r^2}} + V(r),$$

where r, ϕ are polar coordinates in the plane of the orbit. (b) Making use of Hamilton's equations, prove that

$$p_r = \frac{p_\phi}{r^2} \frac{dr}{d\phi}.$$

(c) Show that, in terms of the variable $u = 1/r$, the equation for the orbit is

$$\phi = \int \frac{l \, du}{\sqrt{\frac{1}{c^2}[E - V(1/u)]^2 - m^2 c^2 - l^2 u^2}},$$

where E and l denote the constant values of H and p_ϕ, respectively. (d) In the case of the relativistic Kepler problem, with $V(r) = -\kappa/r$, show that the equation for the orbit takes the form

$$r = \frac{\alpha}{1 + \epsilon \cos \gamma \phi}$$

with

$$\gamma = \sqrt{1 - \frac{\kappa^2}{c^2 l^2}}, \quad \alpha = \frac{c^2 l^2 \gamma^2}{\kappa E}, \quad \epsilon = \sqrt{1 + \frac{\gamma^2 (1 - m^2 c^4/E^2)}{1 - \gamma^2}}.$$

(e) In the case of bound motion, the orbit is not an ellipse with eccentricity ϵ because $\gamma \neq 1$, but behaves as a precessing ellipse. If γ is nearly equal to 1, show that the angular displacement $\Delta \phi$ between two successive minima of r is $\Delta \phi = \pi \kappa^2/c^2 l^2$.

7.17 Using the principle of Maupertuis, show that a projectile fired from near the surface of the earth follows a parabolic trajectory.

7.18 The principle of Maupertuis can be extended to non-conservative systems, mechanical systems whose Hamiltonian explicitly depends on time. With h defined by equation (2.147), show that the functional

$$I = \int_{t_1}^{t_2} \left\{ \sum_{i=1}^{n} p_i \dot{q}_i + t \frac{dh}{dt} \right\} dt$$

is stationary for the actual path as compared to neighbouring paths with the same terminal points for which h has the same terminal values (Whittaker, 1944).

7.19 Prove that the one-dimensional equation of motion

$$\ddot{q} + F(q)\dot{q}^2 + G(q) = 0$$

can be put in Hamiltonian form by taking

$$H(q,p) = \frac{p^2}{2f(q)} + g(q)$$

with

$$f(q) = \exp\left[2\int^q F(q')dq'\right]$$

and the function g suitably chosen.

7.20 From the Lagrangian of Problem 3.10, show that the Hamiltonian which describes the motion of a particle with respect to an arbitrary non-inertial frame is

$$H = \frac{\mathbf{p}^2}{2m} - \boldsymbol{\omega} \cdot (\mathbf{r} \times \mathbf{p}) + V + m\mathbf{a}_{o'} \cdot \mathbf{r}.$$

7.21 A one-degree-of-freedom mechanical system obeys the following equations of motion: $\dot{q} = pf(q)$, $\dot{p} = g(q,p)$. What restrictions must be imposed on the functions f and g in order that this system be Hamiltonian? If the system is Hamiltonian, what is the most general form of Hamilton's function H?

8   Canonical Transformations

> You boil it in sawdust: you salt it in glue:
> You condense it with locusts and tape:
> Still keeping one principal object in view –
> To preserve its symmetrical shape.
>
> Lewis Carroll, *The Hunting of the Snark*

Lagrange's equations are invariant under a general *coordinate* transformation in configuration space: the form of the equations of motion stays the same whatever the choice of generalised coordinates. In the Hamiltonian formulation the *coordinates and momenta* are independent variables, and we are naturally led to study changes of variables in *phase space* that preserve the form of Hamilton's equations, thus enormously enlarging the range of admissible transformations. This enlargement, in its turn, often makes possible the judicious choice of canonical variables that simplify the Hamiltonian, thereby making the equations of motion easy to solve.

8.1 Canonical Transformations and Generating Functions

A change of variables in phase space is of our interest if it preserves the canonical form of the equations of motion. More precisely, given the canonical variables (q, p), the Hamiltonian $H(q, p, t)$ and Hamilton's equations

$$\dot{q}_i = \frac{\partial H}{\partial p_i}, \qquad \dot{p}_i = -\frac{\partial H}{\partial q_i}, \qquad i = 1, \ldots, n, \tag{8.1}$$

we are interested in the invertible transformation

$$Q_i = Q_i(q, p, t), \qquad P_i = P_i(q, p, t), \qquad i = 1, \ldots, n, \tag{8.2}$$

as long as it is possible to find a function $K(Q, P, t)$ such that the equations of motion for the new variables take the Hamiltonian form:

$$\dot{Q}_i = \frac{\partial K}{\partial P_i}, \qquad \dot{P}_i = -\frac{\partial K}{\partial Q_i}, \qquad i = 1, \ldots, n. \tag{8.3}$$

It is worth stressing that the equations of motion for the new variables must have the Hamiltonian form *no matter what* the original Hamiltonian function $H(q, p, t)$ may be.[1]

---

[1] See the Remark at the end of Section 8.2.

The simultaneous validity of equations (8.1) and (8.3) implies the simultaneous validity of the variational principles

$$\delta \int_{t_1}^{t_2} \left\{ \sum_{i=1}^{n} p_i \dot{q}_i - H(q,p,t) \right\} dt = 0 \tag{8.4}$$

and

$$\delta \int_{t_1}^{t_2} \left\{ \sum_{i=1}^{n} P_i \dot{Q}_i - K(Q,P,t) \right\} dt = 0. \tag{8.5}$$

Of course, these variational principles will simultaneously hold if the respective integrands are equal, but this is not necessary.

The variational principles (8.4) and (8.5) give rise to the same Hamilton's equations expressed in terms of different sets of variables. Proposition 7.5.1 ensures that the same Hamilton's equations are obtained if the integrands involved in the respective variational principles differ by the total time derivative of a function $\Phi(q,p,t)$. Thus, the transformation (8.2) preserves the Hamiltonian form of the equations of motion if

$$\sum_{i=1}^{n} p_i \dot{q}_i - H = \sum_{i=1}^{n} P_i \dot{Q}_i - K + \frac{d\Phi}{dt}, \tag{8.6}$$

or, in terms of differential forms,

$$\sum_{i=1}^{n} (p_i \, dq_i - P_i \, dQ_i) + (K - H)dt = d\Phi. \tag{8.7}$$

This equation serves to characterise a canonical transformation.

Definition 8.1.1   *The invertible transformation (8.2) is said to be a* **canonical transformation** *if and only if there exist functions $K(Q,P,t)$ and $\Phi(q,p,t)$ such that Eq. (8.7) is satisfied.*

The form of (8.7) suggests considering Φ as a function of the old and the new generalised coordinates. Suppose the first n Eqs. (8.2) can be solved for the n ps in terms of (q,Q,t). In this case, the last n Eqs. (8.2) allow us to write the transformed momenta in terms of (q,Q,t). In other words, we can take (q,Q) as a set of $2n$ independent variables. Defining the **generating function** $F_1(q,Q,t)$ by

$$F_1(q,Q,t) = \Phi(q,p(q,Q,t),t), \tag{8.8}$$

from Eq. (8.7) one infers

$$p_i = \frac{\partial F_1}{\partial q_i}, \qquad P_i = -\frac{\partial F_1}{\partial Q_i}, \qquad i = 1,\ldots,n \tag{8.9}$$

and also

$$K(Q,P,t) = H(q,p,t) + \frac{\partial F_1}{\partial t}. \tag{8.10}$$

Note that, given a function $F_1(q, Q, t)$ such that

$$\det\left(\frac{\partial^2 F_1}{\partial q_i \partial Q_j}\right) \neq 0, \tag{8.11}$$

a canonical transformation is automatically defined by equations (8.9), with the trans-formed Hamiltonian given by (8.10). Indeed, solving the first n equations (8.9) for the Qs, which is possible owing to (8.11), we are led to $Q_i = Q_i(q, p, t)$. Inserting this result into the right-hand side of each of the last n equations (8.9), we get $P_i = P_i(q, p, t)$ and the phase space change of variables $(q, p) \longrightarrow (Q, P)$ is canonical by construction.

It may be inconvenient or even impossible to take (q, Q) as independent variables, as in the case of the identity transformation $Q_i = q_i, P_i = p_i$. But, if we can take (q, P) as $2n$ independent variables, there exists a generating function of the type $F_2(q, P, t)$. In fact, noting that

$$-\sum_i P_i \, dQ_i = -d\left(\sum_i P_i \, Q_i\right) + \sum_i Q_i \, dP_i, \tag{8.12}$$

introduce this identity into (8.7) and define

$$F_2 = \sum_i P_i \, Q_i + F_1 \tag{8.13}$$

or, more precisely,

$$F_2(q, P, t) = \sum_i P_i \, Q_i(q, P, t) + \Phi(q, p(q, P, t), t). \tag{8.14}$$

Then Eq. (8.7) takes the form

$$\sum_{i=1}^{n} (p_i \, dq_i + Q_i \, dP_i) + (K - H)dt = dF_2. \tag{8.15}$$

As a consequence,

$$p_i = \frac{\partial F_2}{\partial q_i}, \qquad Q_i = \frac{\partial F_2}{\partial P_i}, \qquad i = 1, \ldots, n, \tag{8.16}$$

$$K(Q, P, t) = H(q, p, t) + \frac{\partial F_2}{\partial t}. \tag{8.17}$$

As in the previous case, given a function $F_2(q, P, t)$ one solves the first n equations (8.16) for P_i to obtain $P_i = P_i(q, p, t)$ and inserts this result into the right-hand side of the last n equations (8.16) to get $Q_i = Q_i(q, p, t)$, the transformation so obtained being automatically canonical with the transformed Hamiltonian K and the original one H related by (8.17).

Exercise 8.1.1   (1) Suppose one can take (p, Q) as independent variables. Defining $F_3(p, Q, t) = -\sum_i q_i p_i + \Phi(q, p, t)$, derive

$$q_i = -\frac{\partial F_3}{\partial p_i}, \qquad P_i = -\frac{\partial F_3}{\partial Q_i}, \qquad i = 1, \ldots, n, \tag{8.18}$$

$$K(Q, P, t) = H(q, p, t) + \frac{\partial F_3}{\partial t}. \tag{8.19}$$

(2) Defining $F_4(p, P, t) = -\sum_i (q_i p_i - Q_i P_i) + \Phi(q, p, t)$, show that

$$q_i = -\frac{\partial F_4}{\partial p_i}, \qquad Q_i = \frac{\partial F_4}{\partial P_i}, \qquad i = 1, \ldots, n, \tag{8.20}$$

$$K(Q, P, t) = H(q, p, t) + \frac{\partial F_4}{\partial t}, \tag{8.21}$$

as long as it is possible to take (p, P) as independent variables.

Remark Conditions analogous to (8.11) must be required of the generating functions F_2, F_3 and F_4. It should be stressed that in most applications these conditions do not need to be satisfied everywhere; it is enough that they hold in a sufficiently large portion of the space of variables.

Of course, generating functions of hybrid types are possible: for a two-degree-of-freedom system a generating function such as $G(q_1, Q_1, q_2, P_2, t)$ mixes the two first types. In any event, functions of the old and new variables work as "bridges" between the old and new variables setting up a canonical transformation. The choice of a generating function defines a canonical transformation, but it is not possible to know beforehand whether such a transformation will be useful to simplify the Hamiltonian associated with some interesting physical system: it is a solution in quest of a problem. On the other hand, the exhibition of one generating function proves that a change of variables in phase space is canonical.

Example 8.1 Determine the canonical transformation generated by

$$F_2(q, P, t) = \sum_{k=1}^{n} q_k P_k. \tag{8.22}$$

Solution

From Eqs. (8.16) we have

$$p_i = \frac{\partial F_2}{\partial q_i} = P_i, \qquad Q_i = \frac{\partial F_2}{\partial P_i} = q_i, \qquad i = 1, \ldots, n \tag{8.23}$$

with $K(Q, P, t) = H(q, p, t)$, which is just the identity transformation.

The existence of a generating function characterises the identity transformation as canonical, an obvious result. Equations (8.23) do not allow us to choose (q, Q) as a set of $2n$ independent variables, so the identity transformation does not admit a generating function of the type F_1.

Exercise 8.1.2 Show that the canonical transformation generated by

$$F_1(q, Q, t) = \sum_{k=1}^{n} q_k Q_k \tag{8.24}$$

is

$$p_i = Q_i, \qquad P_i = -q_i, \qquad i = 1, \ldots, n, \tag{8.25}$$

which, up to a sign, amounts to an exchange of coordinates and momenta. Thus, in the context of the Hamiltonian formalism, there is no essential difference between coordinates and momenta. Show that the transformed Hamiltonian is $K(Q,P,t) = H(-P,Q,t)$. Does the transformation (8.25) admit a generating function of the type F_2?

Example 8.2   Find the canonical transformation generated by

$$F_1(q,Q,t) = \frac{m(q-Q)^2}{2t} \tag{8.26}$$

and apply it to solving the problem of the motion of a one-dimensional free particle, whose Hamiltonian is $H = p^2/2m$.

Solution

From Eqs. (8.9) we have

$$p = \frac{\partial F_1}{\partial q} = \frac{m(q-Q)}{t}, \qquad P = -\frac{\partial F_1}{\partial Q} = \frac{m(q-Q)}{t}, \tag{8.27}$$

or, in direct form,

$$Q = q - \frac{pt}{m}, \qquad P = p. \tag{8.28}$$

On the other hand,

$$K(Q,P,t) = H(q,p,t) + \frac{\partial F_1}{\partial t} = \frac{p^2}{2m} - \frac{m(q-Q)^2}{2t^2} = \frac{P^2}{2m} - \frac{P^2}{2m} = 0. \tag{8.29}$$

The new Hamiltonian is identically zero and the transformed Hamilton's equations are trivially solved:

$$\dot{Q} = \frac{\partial K}{\partial P} = 0 \implies Q = a, \qquad \dot{P} = -\frac{\partial K}{\partial Q} = 0 \implies P = b, \tag{8.30}$$

where a and b are arbitrary constants. Returning to the original variables we finally obtain

$$q = a + \frac{b}{m}t, \tag{8.31}$$

which is the general solution to the equation of motion for the free particle.

Example 8.3   Complete the phase space change of variables

$$P = \frac{1}{b^2}(p^2 + a^2q^2) \tag{8.32}$$

so as to bring forth a canonical transformation (a and b are constants). Next apply it to the problem of solving Hamilton's equations for the one-dimensional harmonic oscillator, whose Hamiltonian is $H = p^2/2m + m\omega^2 q^2/2$.

Solution

We need to find $Q = Q(q,p)$ so as to define a canonical transformation together with (8.32). Writing

$$p = (b^2 P - a^2 q^2)^{1/2} = b(P - c^2 q^2)^{1/2}, \qquad c = \frac{a}{b}, \tag{8.33}$$

and comparing this with (8.16) we find that it is appropriate to search for a generating function of the type $F_2(q, P)$. Therefore, we must have

$$p = \frac{\partial F_2}{\partial q} = b(P - c^2 q^2)^{1/2} \quad \Longrightarrow \quad F_2 = b \int \sqrt{P - c^2 q^2} \, dq. \tag{8.34}$$

This integral is elementary, but it is more convenient not to perform the integration at this stage because we are actually interested in the other half of the canonical transformation given by

$$Q = \frac{\partial F_2}{\partial P} = \frac{b}{2} \int \frac{dq}{(P - c^2 q^2)^{1/2}} = \frac{b}{2c} \sin^{-1}\left(\frac{cq}{\sqrt{P}}\right) = \frac{b^2}{2a} \sin^{-1}\left(\frac{aq}{b\sqrt{P}}\right), \tag{8.35}$$

where we have discarded a "constant" of integration $C(P)$ in order to deal with the simplest canonical transformation possible. Thus,

$$Q = \frac{b^2}{2a} \sin^{-1}\left\{\frac{aq}{(p^2 + a^2 q^2)^{1/2}}\right\}. \tag{8.36}$$

The inverse transformation has the form

$$q = \frac{b}{a}\sqrt{P}\sin\left(\frac{2aQ}{b^2}\right), \qquad p = b\sqrt{P}\cos\left(\frac{2aQ}{b^2}\right). \tag{8.37}$$

The Hamiltonian for the oscillator can be written in the form

$$H(q,p) = \frac{1}{2m}(p^2 + m^2\omega^2 q^2), \tag{8.38}$$

which suggests that we choose $a = m\omega$ in (8.32) and, in order to simplify equations (8.37), $b = \sqrt{2a} = \sqrt{2m\omega}$. Proceeding like this and taking into account that F_2 does not explicitly depend on time, we get

$$K(Q, P) = H(q,p) = \frac{p^2}{2m} + \frac{m\omega^2}{2}q^2 = \frac{\omega}{2m\omega}(p^2 + m^2\omega^2 q^2) = \omega P, \tag{8.39}$$

from which

$$\dot{Q} = \frac{\partial K}{\partial P} = \omega \quad \Longrightarrow \quad Q = \omega t + \delta, \tag{8.40a}$$

$$\dot{P} = -\frac{\partial K}{\partial Q} = 0 \quad \Longrightarrow \quad P = \text{constant} = \frac{E}{\omega}, \tag{8.40b}$$

where E denotes the total energy of the oscillator, the constant value of H. Returning to the original variable q with the help of (8.37), we finally obtain

$$q = \sqrt{\frac{2E}{m\omega^2}} \, \sin(\omega t + \delta), \tag{8.41}$$

which is the usual solution to the oscillator equation of motion.

Only a lunatic would resort to canonical transformations to solve the problems of the free particle or the harmonic oscillator, systems whose equations of motion are directly solved in a trivial fashion. The purpose of the two last examples was to illustrate, in elementary cases, how a carefully chosen canonical transformation may be able to drastically simplify the form of a given Hamiltonian and even reduce it to zero. The elaboration of a systematic scheme to construct a canonical transformation capable of making the transformed Hamiltonian vanish is the essence of the Hamilton-Jacobi theory, to be discussed in the next chapter.

Exercise 8.1.3 Displaying a generating function, show that a **point transformation** $Q_i = f_i(q_1, \ldots, q_n, t)$ is canonical and find the expression of the transformed momenta in terms of the original momenta.

8.2 Canonicity and Lagrange Brackets

The validity of (8.7) for variable t implies its validity for fixed t – that is,

$$\sum_{i=1}^{n} \left[p_i \, dq_i - P_i(q, p, t) \, dQ_i(q, p, t) \right] = d\Phi(q, p, t) \qquad \text{(fixed } t\text{)}, \tag{8.42}$$

where the differentials take into account only the variation of the qs and ps, with the time kept constant. Conversely, the validity of this equation for fixed t implies the validity of (8.7) for variable t (Problem 8.1). Putting to the test the canonical character of a transformation by the generating function method is not straightforward because it requires the inversion of one-half of equations (8.2) to obtain a set of $2n$ independent variables made up of n old variables and n new variables, followed by the explicit production of a generating function. Equation (8.42) allows us to formulate a general and direct canonicity criterion in terms of the *Lagrange brackets*, which we proceed to define.

Definition 8.2.1 *Let (η, ξ) be a set of $2n$ canonical variables which depend on the variables u, v. The **Lagrange bracket** of u and v with respect to (η, ξ), denoted by $[u, v]_{(\eta, \xi)}$, is defined by*

$$[u, v]_{(\eta, \xi)} = \sum_{k=1}^{n} \left(\frac{\partial \eta_k}{\partial u} \frac{\partial \xi_k}{\partial v} - \frac{\partial \eta_k}{\partial v} \frac{\partial \xi_k}{\partial u} \right). \tag{8.43}$$

The next result establishes the usefulness of the Lagrange bracket to characterise a transformation as canonical.

Theorem 8.2.1   *The transformation (8.2) is canonical if and only if*

$$[q_i, q_j]_{(Q,P)} = 0, \quad [p_i, p_j]_{(Q,P)} = 0, \quad [q_i, p_j]_{(Q,P)} = \delta_{ij}. \tag{8.44}$$

Proof   As we have already argued, the transformation (8.2) is canonical if and only if

$$\sum_{j=1}^{n} \left[p_j dq_j - P_j(q, p, t) \sum_{l=1}^{n} \left(\frac{\partial Q_j}{\partial q_l} dq_l + \frac{\partial Q_j}{\partial p_l} dp_l \right) \right] = d\Phi(q, p, t) \quad \text{(fixed } t\text{),} \tag{8.45}$$

that is,

$$\sum_{j=1}^{n} \left(A_j \, dq_j + B_j \, dp_j \right) = d\Phi(q, p, t) \quad \text{(fixed } t\text{),} \tag{8.46}$$

where

$$A_j = p_j - \sum_{l=1}^{n} P_l \frac{\partial Q_l}{\partial q_j}, \qquad B_j = -\sum_{l=1}^{n} P_l \frac{\partial Q_l}{\partial p_j}. \tag{8.47}$$

As proved in Appendix F, a necessary and sufficient condition for the left-hand side of Eq. (8.46) to be an exact differential is the equality of all cross-partial derivatives of the coefficients of the differentials dq_j and dp_j:

$$\frac{\partial A_j}{\partial q_i} = \frac{\partial A_i}{\partial q_j}, \qquad \frac{\partial B_j}{\partial p_i} = \frac{\partial B_i}{\partial p_j}, \qquad \frac{\partial A_i}{\partial p_j} = \frac{\partial B_j}{\partial q_i}. \tag{8.48}$$

The first of these equations yields

$$-\sum_{l=1}^{n} \left(\frac{\partial P_l}{\partial q_i} \frac{\partial Q_l}{\partial q_j} + P_l \frac{\partial^2 Q_l}{\partial q_i \partial q_j} \right) = -\sum_{l=1}^{n} \left(\frac{\partial P_l}{\partial q_j} \frac{\partial Q_l}{\partial q_i} + P_l \frac{\partial^2 Q_l}{\partial q_j \partial q_i} \right), \tag{8.49}$$

whence

$$[q_i, q_j]_{(Q,P)} = \sum_{l=1}^{n} \left(\frac{\partial Q_l}{\partial q_i} \frac{\partial P_l}{\partial q_j} - \frac{\partial Q_l}{\partial q_j} \frac{\partial P_l}{\partial q_i} \right) = 0. \tag{8.50}$$

In analogous fashion, the second of Eqs. (8.48) implies

$$[p_i, p_j]_{(Q,P)} = \sum_{l=1}^{n} \left(\frac{\partial Q_l}{\partial p_i} \frac{\partial P_l}{\partial p_j} - \frac{\partial Q_l}{\partial p_j} \frac{\partial P_l}{\partial p_i} \right) = 0. \tag{8.51}$$

Finally, the third of Eqs. (8.48) gives

$$\delta_{ij} - \sum_{l=1}^{n} \left(\frac{\partial P_l}{\partial p_j} \frac{\partial Q_l}{\partial q_i} + P_l \frac{\partial^2 Q_l}{\partial p_j \partial q_i} \right) = -\sum_{l=1}^{n} \left(\frac{\partial P_l}{\partial q_i} \frac{\partial Q_l}{\partial p_j} + P_l \frac{\partial^2 Q_l}{\partial q_i \partial p_j} \right), \tag{8.52}$$

which amounts to

$$[q_i, p_j]_{(Q,P)} = \sum_{l=1}^{n} \left(\frac{\partial Q_l}{\partial q_i} \frac{\partial P_l}{\partial p_j} - \frac{\partial Q_l}{\partial p_j} \frac{\partial P_l}{\partial q_i} \right) = \delta_{ij}. \tag{8.53}$$

This completes the proof of the theorem. $\square$

Example 8.4   Prove that the transformation

$$Q_1 = p_1^2, \quad Q_2 = p_2^2 + q_2, \quad P_1 = -\frac{q_1}{2p_1}, \quad P_2 = p_2 \tag{8.54}$$

is canonical.

Solution

We have

$$[q_1, q_2]_{(Q,P)} = \left(\frac{\partial Q_1}{\partial q_1}\frac{\partial P_1}{\partial q_2} - \frac{\partial Q_1}{\partial q_2}\frac{\partial P_1}{\partial q_1}\right) + \left(\frac{\partial Q_2}{\partial q_1}\frac{\partial P_2}{\partial q_2} - \frac{\partial Q_2}{\partial q_2}\frac{\partial P_2}{\partial q_1}\right) = 0 \tag{8.55}$$

and also

$$[p_1, p_2]_{(Q,P)} = \left(\frac{\partial Q_1}{\partial p_1}\frac{\partial P_1}{\partial p_2} - \frac{\partial Q_1}{\partial p_2}\frac{\partial P_1}{\partial p_1}\right) + \left(\frac{\partial Q_2}{\partial p_1}\frac{\partial P_2}{\partial p_2} - \frac{\partial Q_2}{\partial p_2}\frac{\partial P_2}{\partial p_1}\right) = 0, \tag{8.56}$$

with similar computations showing that $[q_1, p_2]_{(Q,P)} = [q_2, p_1]_{(Q,P)} = 0$. Finally,

$$[q_1, p_1]_{(Q,P)} = \left(\frac{\partial Q_1}{\partial q_1}\frac{\partial P_1}{\partial p_1} - \frac{\partial Q_1}{\partial p_1}\frac{\partial P_1}{\partial q_1}\right)$$
$$+ \left(\frac{\partial Q_2}{\partial q_1}\frac{\partial P_2}{\partial p_1} - \frac{\partial Q_2}{\partial p_1}\frac{\partial P_2}{\partial q_1}\right) = -2p_1\left(-\frac{1}{2p_1}\right) = 1 \tag{8.57}$$

and

$$[q_2, p_2]_{(Q,P)} = \left(\frac{\partial Q_1}{\partial q_2}\frac{\partial P_1}{\partial p_2} - \frac{\partial Q_1}{\partial p_2}\frac{\partial P_1}{\partial q_2}\right) + \left(\frac{\partial Q_2}{\partial q_2}\frac{\partial P_2}{\partial p_2} - \frac{\partial Q_2}{\partial p_2}\frac{\partial P_2}{\partial q_2}\right) = 1. \tag{8.58}$$

There is no need to compute the remaining Lagrange brackets because of their antisymmetry: $[u, v]_{(\eta,\xi)} = -[v, u]_{(\eta,\xi)}$. Since all conditions (8.44) are satisfied, the transformation is canonical.

Remark   A transformation may preserve the Hamiltonian form of the equations of motion for a particular Hamiltonian, but not for all of them (Problem 8.2). Such transformations are not canonical and seem to play no role in classical dynamics.

8.3 Symplectic Notation

There is a very convenient compact notation that condenses all of Hamilton's equations into a single matrix equation. This notation is particularly useful to simplify certain formulas and proofs that in the traditional notation are very cumbersome. Let $\mathbf{z}$ be a column matrix with $2n$ elements $z_1, \ldots, z_{2n}$ defined by

$$z_i = q_i, \quad z_{n+i} = p_i, \quad i = 1, \ldots, n. \tag{8.59}$$

In words: the first n zs are the the n qs, and the last n zs are the n ps. Analogously, $\partial H/\partial \mathbf{z}$ is a column matrix with the following elements:

$$\left(\frac{\partial H}{\partial \mathbf{z}}\right)_i = \frac{\partial H}{\partial q_i}, \qquad \left(\frac{\partial H}{\partial \mathbf{z}}\right)_{n+i} = \frac{\partial H}{\partial p_i}, \qquad i = 1, \ldots, n. \tag{8.60}$$

Finally, let $\mathbf{J}$ be the $(2n) \times (2n)$ matrix defined by

$$\mathbf{J} = \begin{pmatrix} \mathbf{0} & \mathbf{I} \\ -\mathbf{I} & \mathbf{0} \end{pmatrix} \tag{8.61}$$

where $\mathbf{0}$ is the $n \times n$ zero matrix and $\mathbf{I}$ is the $n \times n$ identity matrix. Hamilton's equations (7.9) can be tersely expressed as

$$\dot{\mathbf{z}} = \mathbf{J}\frac{\partial H}{\partial \mathbf{z}}, \tag{8.62}$$

known as the **symplectic form**[2] of Hamilton's equations.

Exercise 8.3.1 Show that (8.62) is equivalent to Hamilton's equations in their usual form.

The matrix $\mathbf{J}$ possesses the following proprerties, whose verification is left to the reader:
(1) $\mathbf{J}^2 = -\mathbf{I}$, where $\mathbf{I}$ is the $(2n) \times (2n)$ identity matrix;
(2) $\mathbf{J}^T\mathbf{J} = \mathbf{J}\mathbf{J}^T = \mathbf{I} \Longrightarrow \mathbf{J}^{-1} = \mathbf{J}^T = -\mathbf{J}$ ($\mathbf{J}$ is an orthogonal matrix);
(3) $\det \mathbf{J} = 1$.

With a similar grouping of the transformed variables (Q, P) to form the column matrix $\boldsymbol{\zeta}$ with $2n$ rows, a phase space change of variables is written

$$\boldsymbol{\zeta} = \boldsymbol{\zeta}(\mathbf{z}, t) \tag{8.63}$$

or, in components,

$$\zeta_k = \zeta_k(z_1, \ldots, z_{2n}, t), \qquad k = 1, \ldots, 2n. \tag{8.64}$$

With this notation we can write

$$[u, v]_{\mathbf{z}} = \left(\frac{\partial \mathbf{z}}{\partial u}\right)^T \mathbf{J}\frac{\partial \mathbf{z}}{\partial v} \equiv \sum_{r,s=1}^{2n} \frac{\partial z_r}{\partial u} J_{rs} \frac{\partial z_s}{\partial v}. \tag{8.65}$$

Exercise 8.3.2 Verify that (8.65) is identical to the Lagrange bracket $[u, v]_{(q,p)}$ given by Definition 8.2.1.

Lemma 8.3.1 *In symplectic notation the canonicity conditions (8.44) can be cast in the concise form $[z_r, z_s]_{\boldsymbol{\zeta}} = J_{rs}$.*

Proof If $r, s \in \{1, \ldots, n\}$ or $r, s \in \{n + 1, \ldots, 2n\}$ the left-hand side of $[z_r, z_s]_{\boldsymbol{\zeta}} = J_{rs}$ reduces to the Lagrange brackets of the qs among themselves or of the ps among themselves and the right-hand side vanishes, whereas if $r = i \in \{1, \ldots, n\}$ and $s = (n+j) \in \{n + 1, \ldots, 2n\}$ equation $[z_r, z_s]_{\boldsymbol{\zeta}} = J_{rs}$ reduces to $[q_i, p_j]_{(Q,P)} = J_{i,n+j} = \delta_{ij}$. The reversal of indices r, s repeats this last result owing to the antisymmetry of J_{rs}. $\square$

---

[2] The word "symplectic" was coined by the mathematician Hermann Weyl in 1939 and derives from a Greek root meaning "intertwined".

Let now $\mathbf{M} = (M_{rs})$ be the Jacobian matrix of transformation (8.63):

$$M_{rs} = \frac{\partial \zeta_r}{\partial z_s}. \tag{8.66}$$

We can write

$$[z_r, z_s]_\zeta = \sum_{k,l=1}^{2n} \frac{\partial \zeta_k}{\partial z_r} J_{kl} \frac{\partial \zeta_l}{\partial z_s} = \sum_{k,l=1}^{2n} M_{kr} J_{kl} M_{ls} \tag{8.67}$$

or, with the help of the definition of matrix product,

$$[z_r, z_s]_\zeta = \left(\mathbf{M}^T \mathbf{J} \mathbf{M}\right)_{rs}. \tag{8.68}$$

Theorem 8.3.1   *The transformation (8.63) is canonical if and only if its Jacobian matrix satisfies*

$$\mathbf{M}^T \mathbf{J} \mathbf{M} = \mathbf{J}. \tag{8.69}$$

Proof   Combine (8.68) with Lemma 8.3.1 and Theorem 8.2.1.   □

Any matrix that obeys the **symplectic condition** (8.69) is said to be a **symplectic matrix**. By the way, canonical transformations are called **symplectic transformations** by mathematicians. It is fitting to call attention to the resemblance between Eq. (8.69) and the condition (6.20) that characterises a Lorentz transformation. The symplectic condition can be used to test directly whether a given transformation is canonical.

Example 8.5   For a one-degree-of-freedom system, prove that the phase-space transformation $Q = (q - p)/\sqrt{2}$, $P = (q + p)/\sqrt{2}$ is canonical.

Solution

The straightforward computation

$$\mathbf{M}^T \mathbf{J} \mathbf{M} = \frac{1}{2} \begin{pmatrix} 1 & 1 \\ -1 & 1 \end{pmatrix} \begin{pmatrix} 0 & 1 \\ -1 & 0 \end{pmatrix} \begin{pmatrix} 1 & -1 \\ 1 & 1 \end{pmatrix}$$

$$= \frac{1}{2} \begin{pmatrix} 0 & 2 \\ -2 & 0 \end{pmatrix} = \begin{pmatrix} 0 & 1 \\ -1 & 0 \end{pmatrix} = \mathbf{J}$$

establishes the canonical nature of the transformation.

Theorem 8.3.2   *The determinant of the Jacobian matrix (the Jacobian) of a canonical transformation equals ± 1.*

Exercise 8.3.3   Taking the determinant of (8.69), prove Theorem 8.3.2.

It is possible to go further and prove that the Jacobian of any canonical transformation is $+1$ (Landau & Lifshitz, 1976), but we will not need this finer result.

Invariance of the Lagrange Bracket

The symplectic notation considerably simplifies the proof of an important invariance property of the Lagrange bracket.

Theorem 8.3.3   *The Lagrange bracket is invariant under canonical transformations – that is, if the transformation* $\mathbf{z} \rightarrow \boldsymbol{\zeta}$ *is canonical then*

$$[u, v]_{\boldsymbol{\zeta}} = [u, v]_{\mathbf{z}}. \tag{8.70}$$

Proof   Note first that by the chain rule

$$\frac{\partial \zeta_r}{\partial u} = \sum_{s=1}^{2n} \frac{\partial \zeta_r}{\partial z_s} \frac{\partial z_s}{\partial u} = \sum_{s=1}^{2n} M_{rs} \frac{\partial z_s}{\partial u} \tag{8.71}$$

or, in matrix notation,

$$\frac{\partial \boldsymbol{\zeta}}{\partial u} = \mathbf{M} \frac{\partial \mathbf{z}}{\partial u}. \tag{8.72}$$

Substituting this result into (8.65) we are led to

$$[u, v]_{\boldsymbol{\zeta}} = \left(\frac{\partial \boldsymbol{\zeta}}{\partial u}\right)^T \mathbf{J} \frac{\partial \boldsymbol{\zeta}}{\partial v} = \left(\frac{\partial \mathbf{z}}{\partial u}\right)^T \mathbf{M}^T \mathbf{J} \mathbf{M} \frac{\partial \mathbf{z}}{\partial v} = \left(\frac{\partial \mathbf{z}}{\partial u}\right)^T \mathbf{J} \frac{\partial \mathbf{z}}{\partial v} = [u, v]_{\mathbf{z}}, \tag{8.73}$$

where we have used (8.69). $\square$

Exercise 8.3.4   Using $\mathbf{J}^2 = -\mathbf{I}$ and the fact that the left inverse of a matrix coincides with the right inverse, show that

$$\mathbf{M} \mathbf{J} \mathbf{M}^T = \mathbf{J} \tag{8.74}$$

is a consequence of (8.69).

From equations (8.69) and (8.74) one easily deduces that the canonical transformations form a group: (1) the identity transformation is canonical; (2) the product (composition) of canonical transformations is associative; (3) two sucessive canonical transformations define a canonical transformation; (4) the inverse of a canonical transformation is a canonical transformation.

Exercise 8.3.5   Prove that the canonical transformations form a group.

The set of all $(2n) \times (2n)$ matrices which satisfy (8.69) form the **real symplectic group** $Sp_{2n}(\mathbb{R})$ over $\mathbb{R}^{2n}$.

Symplectic Form and Symplectic Manifolds

Taking the exterior derivative (Appendix B) of both sides of Eq. (8.42), we get

$$\sum_{i=1}^{n} dq_i \wedge dp_i = \sum_{i=1}^{n} dQ_i \wedge dP_i. \tag{8.75}$$

Therefore, the **symplectic form** ω defined by

$$\omega = \sum_{i=1}^{n} dq_i \wedge dp_i = \frac{1}{2} \sum_{r,s=1}^{2n} J_{rs} \, dz_r \wedge dz_s \tag{8.76}$$

is invariant under canonical transformations. The action of the 2-form ω on two vectors $\mathbf{X} = (X_1, \ldots, X_{2n})$, $\mathbf{Y} = (Y_1, \ldots, Y_{2n})$ in phase space is defined by

$$\omega(\mathbf{X}, \mathbf{Y}) = \frac{1}{2} \sum_{r,s=1}^{2n} J_{rs} X_r Y_s . \tag{8.77}$$

It is immediate that ω is a closed 2-form: $d\omega = 0$. Furthermore, ω is a non-degenerate 2-form: if $\omega(\mathbf{X}, \mathbf{Y}) = 0$ for all $\mathbf{Y}$ then $\mathbf{X} = 0$. In the case of a 2-form defined on a differentiable manifold, $\mathbf{X}$ and $\mathbf{Y}$ are elements of the tangent space to the manifold. Most generally, a symplectic form defined on a differentiable manifold $\mathcal{M}$ is a closed 2-form which is non-degenerate at each point of $\mathcal{M}$.

The phase space endowed with the 2-form ω defined by (8.76) constitutes a **symplectic manifold**. In general, the configuration space is a differentiable manifold $\mathcal{Q}$ and the phase space is the associated differentiable manifold $T^*\mathcal{Q}$, called the cotangent bundle or cotangent manifold. The existence of a closed non-degenerate 2-form defined at each point of $T^*\mathcal{Q}$ makes $T^*\mathcal{Q}$ a symplectic manifold. The abstract definition of symplectic manifold involves any differentiable manifold of even dimension.

Definition 8.3.1 *A **symplectic manifold** is an ordered pair $(\mathcal{M}, \omega)$ where $\mathcal{M}$ is an even-dimensional differentiable manifold and ω is a **symplectic form** on $\mathcal{M}$ – that is, a closed 2-form which is non-degenerate at each point $p \in \mathcal{M}$.*

Darboux's theorem – there is a proof in Arnold (1989) – ensures that for any $2n$-dimensional symplectic manifold $(\mathcal{M}, \omega)$ there exist local coordinates (q, p) such that ω takes the standard form $\omega = \sum_{i=1}^{n} dq_i \wedge dp_i$.

8.4 Poisson Brackets

Let $F(q, p, t)$ be an arbitrary **dynamical variable** – that is, an infinitely differentiable function of the canonical variables and time. Then

$$\frac{dF}{dt} = \sum_{k=1}^{n} \left(\frac{\partial F}{\partial q_k} \dot{q}_k + \frac{\partial F}{\partial p_k} \dot{p}_k \right) + \frac{\partial F}{\partial t} = \sum_{k=1}^{n} \left(\frac{\partial F}{\partial q_k} \frac{\partial H}{\partial p_k} - \frac{\partial F}{\partial p_k} \frac{\partial H}{\partial q_k} \right) + \frac{\partial F}{\partial t}, \tag{8.78}$$

where Hamilton's equations have been used.

Definition 8.4.1 *The **Poisson bracket** $\{F, G\}$ of two dynamical variables F and G is defined by*

$$\{F, G\} = \sum_{k=1}^{n} \left(\frac{\partial F}{\partial q_k} \frac{\partial G}{\partial p_k} - \frac{\partial F}{\partial p_k} \frac{\partial G}{\partial q_k} \right). \tag{8.79}$$

With this definition Eq. (8.78) takes the form

$$\frac{dF}{dt} = \{F, H\} + \frac{\partial F}{\partial t} .$$ (8.80)

In particular, taking first $F = q_i$ and next $F = p_i$ there result Hamilton's equations in terms of Poisson brackets

$$\dot{q}_i = \{q_i, H\}, \qquad \dot{p}_i = \{p_i, H\},$$ (8.81)

because, for both choices, F has no *explicit* time dependence.

Invariance of the Poisson Bracket

One of the main thoretical advantages of writing the equation of motion for an arbitrary dynamical variable in the form (8.80) lies in the property of the Poisson bracket of being invariant under canonical transformations. In other words, Eq. (8.80) does not depend on the set of canonical variables chosen to describe the dynamics. In order to shorten the proof of the invariance of the Poisson bracket, it is convenient to rewrite it in symplectic notation as

$$\{F, G\}_{\mathbf{z}} = \left(\frac{\partial F}{\partial \mathbf{z}}\right)^T \mathbf{J} \frac{\partial G}{\partial \mathbf{z}} \equiv \sum_{r,s=1}^{2n} \frac{\partial F}{\partial z_r} J_{rs} \frac{\partial G}{\partial z_s},$$ (8.82)

the subscript indicating computation with respect to the canonical variables $\mathbf{z} = (q, p)$.

Theorem 8.4.1   *If the transformation $\mathbf{z} \rightarrow \zeta$ is canonical then*

$$\{F, G\}_{\mathbf{z}} = \{F, G\}_{\zeta} .$$ (8.83)

Proof   According to the chain rule,

$$\frac{\partial F}{\partial z_r} = \sum_{s=1}^{2n} \frac{\partial \zeta_s}{\partial z_r} \frac{\partial F}{\partial \zeta_s} \qquad \Longrightarrow \qquad \frac{\partial F}{\partial \mathbf{z}} = \mathbf{M}^T \frac{\partial F}{\partial \zeta} .$$ (8.84)

Consequently,

$$\{F, G\}_{\mathbf{z}} = \left(\frac{\partial F}{\partial \mathbf{z}}\right)^T \mathbf{J} \frac{\partial G}{\partial \mathbf{z}} = \left(\frac{\partial F}{\partial \zeta}\right)^T \mathbf{M} \mathbf{J} \mathbf{M}^T \frac{\partial G}{\partial \zeta} = \left(\frac{\partial F}{\partial \zeta}\right)^T \mathbf{J} \frac{\partial G}{\partial \zeta} = \{F, G\}_{\zeta},$$ (8.85)

where we have used (8.74). $\square$

Remark   There is an abuse of notation in Eq. (8.83): the functions F and G that appear on the left-hand side are not the same functions that appear on the right-hand side, for the latter are the former expressed in terms of ζ. With the canonical transformation written in the form $\mathbf{z} = \mathbf{f}(\zeta, t)$, the functions on the right-hand side of (8.83) should be $F \circ \mathbf{f}$ and $G \circ \mathbf{f}$. This is the same abuse committed in the traditional way of expressing the chain rule of differentiation.

Canonicity and Poisson Brackets

As the reader probably suspects by now, the Lagrange and Poisson brackets satisfy a reciprocity relation.

Theorem 8.4.2 *Let $u_r = u_r(z_1, \ldots, z_{2n})$, $r = 1, \ldots, 2n$, be a set of $2n$ mutually independent functions, so that, conversely, $z_r = z_r(u_1, \ldots, u_{2n})$. Then*

$$\sum_{k=1}^{2n} \{u_r, u_k\}_\mathbf{z}\, [u_k, u_s]_\mathbf{z} = -\delta_{rs} . \tag{8.86}$$

Proof According to (8.65) and (8.82),

$$\Gamma_{rs} \equiv \sum_{k=1}^{2n} \{u_r, u_k\}_\mathbf{z}\, [u_k, u_s]_\mathbf{z} = \sum_{k=1}^{2n} \sum_{l,m=1}^{2n} \sum_{i,j=1}^{2n} \frac{\partial u_r}{\partial z_l} J_{lm} \frac{\partial u_k}{\partial z_m} \frac{\partial z_i}{\partial u_k} J_{ij} \frac{\partial z_j}{\partial u_s}$$

$$= \sum_{l,m=1}^{2n} \sum_{i,j=1}^{2n} \frac{\partial u_r}{\partial z_l} J_{lm} \frac{\partial z_i}{\partial z_m} J_{ij} \frac{\partial z_j}{\partial u_s} = \sum_{l,m=1}^{2n} \sum_{j=1}^{2n} \frac{\partial u_r}{\partial z_l} J_{lm} J_{mj} \frac{\partial z_j}{\partial u_s} , \tag{8.87}$$

where we have used the chain rule and $\partial z_i / \partial z_m = \delta_{im}$. On the other hand,

$$\sum_{m=1}^{2n} J_{lm} J_{mj} = \left(\mathbf{J}^2 \right)_{lj} = (-\mathbf{I})_{lj} = -\delta_{lj} , \tag{8.88}$$

whence

$$\Gamma_{rs} = -\sum_{l,j=1}^{2n} \frac{\partial u_r}{\partial z_l} \delta_{lj} \frac{\partial z_j}{\partial u_s} = -\sum_{l=1}^{2n} \frac{\partial u_r}{\partial z_l} \frac{\partial z_l}{\partial u_s} = -\frac{\partial u_r}{\partial u_s} = -\delta_{rs} \tag{8.89}$$

because, by hypothesis, the us are mutually independent. The proof of the theorem is complete. □

By virtue of this result, it is possible to reformulate the basic canonicity criterion for changes of variables in phase space in terms of Poisson brackets.

Theorem 8.4.3 *The transformation $\mathbf{z} \to \boldsymbol{\zeta}$ is canonical if and only if*

$$\{\zeta_r, \zeta_s\}_\mathbf{z} = J_{rs} , \tag{8.90}$$

or, in the traditional notation,

$$\{Q_i, Q_j\}_{(q,p)} = 0 , \quad \{P_i, P_j\}_{(q,p)} = 0 , \quad \{Q_i, P_j\}_{(q,p)} = \delta_{ij} . \tag{8.91}$$

Proof Let $\mathbf{X}$ and $\mathbf{Y}$ be matrices whose elements are $\{\zeta_r, \zeta_s\}_\mathbf{z}$ and $[\zeta_r, \zeta_s]_\mathbf{z}$, respectively. In matrix language Eq. (8.86) is equivalent to $\mathbf{X}\mathbf{Y} = -\mathbf{I}$. Since the inverse of a canonical transformation is canonical, Theorem 8.2.1 with the roles of the new and old variables exchanged asserts that the transformation $\mathbf{z} \to \boldsymbol{\zeta}$ is canonical if and only if $\mathbf{Y} = \mathbf{J}$. But since $\mathbf{J}^2 = -\mathbf{I}$, it immediately follows that $\mathbf{X} = \mathbf{J}$ if and only if $\mathbf{Y} = \mathbf{J}$, completing the proof. □

Exercise 8.4.1 Consider a one-degree-of-freedom mechanical system. Using the Poisson bracket criterion prove that the transformation

$$Q = \ln \left(\frac{\sin p}{q} \right), \qquad P = q \cot p \qquad (8.92)$$

is canonical.

Algebraic Properties of Poisson Brackets

The Poisson brackets have the following algebraic properties:

(**PB1**) Antisymmetry: $\{A, B\} = -\{B, A\}$ whence $\{A, A\} = 0$.
(**PB2**) Linearity: $\{A + \alpha B, C\} = \{A, C\} + \alpha \{B, C\}$, α independent of (q, p).
(**PB3**) $\{AB, C\} = A\{B, C\} + \{A, C\}B$; $\{A, BC\} = \{A, B\}C + B\{A, C\}$.
(**PB4**) Jacobi identity: $\{\{A, B\}, C\} + \{\{C, A\}, B\} + \{\{B, C\}, A\} = 0$.

Except for the Jacobi identity, the direct proof of which requires lengthy and tedious algebra, the easy proofs of the remaining properties are left as an exercise for the reader. We reserve for the next section a short proof of the Jacobi identity with virtually no calculations. An additional property that deserves to be mentioned is

$$\frac{\partial}{\partial \lambda} \{A, B\} = \left\{ \frac{\partial A}{\partial \lambda}, B \right\} + \left\{ A, \frac{\partial B}{\partial \lambda} \right\}, \qquad (8.93)$$

where λ is one of the qs, one of the ps, the time t or any other parameter.

The **fundamental Poisson brackets**

$$\{q_i, q_j\} = 0, \quad \{p_i, p_j\} = 0, \quad \{q_i, p_j\} = \delta_{ij}, \qquad (8.94)$$

together with properties PB1 to PB4, permit in several cases the calculation of Poisson brackets by purely algebraic means, without having to resort to formula (8.79). On the other hand, as the previous analysis has shown, the invariance of the fundamental Poisson brackets is a necessary and sufficient condition for a phase space transformation to be canonical.

A **Lie algebra** is a vector space of elements $X, Y, \dots$ on which a bilinear "product" $[X, Y]$ is defined which satisfies the property of antisymmetry and obeys the Jacobi identity (Arnold, 1989; Sternberg, 1994). Thus, the infinitely differentiable functions on phase space equipped with the Poisson bracket structure form a Lie algebra.

Poisson Manifolds

According to Darboux's theorem (Arnold, 1989), in a symplectic manifold (see Definition 8.3.1) one can always introduce canonical variables (q, p) such that the Poisson bracket takes the standard form (8.79). However, on an arbitrary differentiable manifold which does not need to be even dimensional it may be possible to introduce a Poisson bracket characterised by its algebraic properties. A differentiable manifold equipped with a Poisson bracket constitutes a **Poisson manifold**. In Appendix H we present some non-trivial examples of Poisson manifolds and describe how Hamiltonian equations of motion can be defined on them.

Poisson Brackets and Quantum Mechanics

Poisson brackets are extremely important because of the role they play in the transition from classical to quantum theory. The procedure known as **canonical quantisation** essentially consists in associating a self-adjoint operator $\hat{A}$ to each fundamental dynamical variable $A(q, p, t)$ in such a way that the commutator of any two such operators is the operator associated with the Poisson bracket of the corresponding dynamical variables multiplied by $i\hbar$. In the Heisenberg picture of quantum mechanics an operator $\hat{\mathcal{O}}_H$ satisfies

$$\frac{d\hat{\mathcal{O}}_H}{dt} = \frac{\partial\hat{\mathcal{O}}_H}{\partial t} + \frac{1}{i\hbar}[\hat{\mathcal{O}}_H, \hat{H}], \tag{8.95}$$

where $[\hat{A}, \hat{B}] = \hat{A}\hat{B} - \hat{B}\hat{A}$ is the commutator. The similarity between this equation and (8.80) is evident. Mechanics formulated in the language of Poisson brackets is the classical analogue of quantum theory in the Heisenberg picture, with the classical Poisson bracket corresponding to the quantum commutator divided by $i\hbar$. This correspondence is possible and consistent because the quantum commutator has exactly the same algebraic properties as the classical Poisson bracket. The quantisation rule that makes $i\hbar\{A, B\}$ correspond to $[\hat{A}, \hat{B}]$ was discovered by Dirac in 1926 (van der Waerden, 1967). It is worth noting that, under reasonable hypotheses, such a correspondence between classical dynamical variables and quantum operators cannot hold for *all* dynamical variables (Abraham and Marsden, 1978; Theorem 5.4.9).

8.5 Infinitesimal Canonical Transformations

An **infinitesimal canonical transformation** is of form

$$Q_i = q_i + \delta q_i \equiv q_i + \epsilon f_i(q, p, t), \qquad P_i = p_i + \delta p_i \equiv p_i + \epsilon g_i(q, p, t), \tag{8.96}$$

where ϵ is an infinitesimal parameter. Substituting this equation into (8.42) there results

$$\sum_i \{p_i\, dq_i - (p_i + \epsilon g_i)(dq_i + \epsilon df_i)\} = d\Phi, \tag{8.97}$$

or, neglecting second-order terms in ϵ and writing $\Phi = \epsilon F$,

$$\sum_i (g_i\, dq_i + p_i\, df_i) = -dF. \tag{8.98}$$

This equation can be rewritten in the form

$$\sum_i (g_i\, dq_i - f_i\, dp_i) = -dG, \tag{8.99}$$

More generally, the generator of rotations about the direction defined by the unit vector $\hat{\mathbf{n}}$ is $\mathbf{L} \cdot \hat{\mathbf{n}}$. (proof: the z-axis can always be chosen along the vector $\hat{\mathbf{n}}$). Therefore, for an arbitrary vector $\mathbf{F}$,

$$\delta \mathbf{F} = d\theta \, \{\mathbf{F}, \mathbf{L} \cdot \hat{\mathbf{n}}\} \,. \tag{8.125}$$

On the other hand we know that

$$\delta \mathbf{F} = d\theta \, \hat{\mathbf{n}} \times \mathbf{F} \,, \tag{8.126}$$

whence

$$\{\mathbf{F}, \mathbf{L} \cdot \hat{\mathbf{n}}\} = \hat{\mathbf{n}} \times \mathbf{F} \,, \tag{8.127}$$

which is the generic result we were looking for. It must be stressed that this last equation holds for any *system* vector $\mathbf{F}$, by which we mean a vector that can be expressed as a function of the system's canonical variables alone because $\delta \mathbf{F}$ is defined by Eq. (8.109). If $\mathbf{F}$ depends on some external vector, such as an external magnetic field, which is not affected by a rotation of the system's canonical variables, Eq. (8.127) does not apply.

From (8.127) several interesting results can be obtained. Picking, for example, $\mathbf{F} = \mathbf{p}$ and $\hat{\mathbf{n}} = \hat{\mathbf{z}}$ there results

$$\{\mathbf{p}, L_z\} = \hat{\mathbf{z}} \times \mathbf{p} = -p_y \hat{\mathbf{x}} + p_x \hat{\mathbf{z}} \,, \tag{8.128}$$

or, in components,

$$\{p_x, L_z\} = -p_y, \quad \{p_y, L_z\} = p_x, \quad \{p_z, L_z\} = 0 \,. \tag{8.129}$$

The Poisson brackets of $\mathbf{p}$ with L_x and L_y can be easily obtained in the same way.

Exercise 8.6.2   (1) Taking $\mathbf{F} = \mathbf{L}$ and $\hat{\mathbf{n}}$ successively equal to $\hat{\mathbf{x}}, \hat{\mathbf{y}}, \hat{\mathbf{z}}$, prove that

$$\{L_x, L_y\} = L_z \,, \tag{8.130}$$

the remaining Poisson brackets being obtained by cyclic permutation of xyz. (2) Prove the following theorem: if two components of the angular momentum are conserved, then the remaining component is necessarily conserved.[4]

As a last interesting result, note that a scalar is invariant under rotations, so

$$\{\mathbf{F} \cdot \mathbf{G}, \mathbf{L} \cdot \hat{\mathbf{n}}\} = 0 \tag{8.131}$$

for any two system vectors $\mathbf{F}$ and $\mathbf{G}$.

Exercise 8.6.3   Verify the identity $\{\mathbf{F} \cdot \mathbf{G}, A\} = \mathbf{F} \cdot \{\mathbf{G}, A\} + \{\mathbf{F}, A\} \cdot \mathbf{G}$ and use it to prove (8.131) directly.

---

[4] This theorem is true only in the absence of constraints. In the presence of constraints the theorem does not apply, as shown by counterexamples (Corben & Stehle, 1960, p. 229).

8.7 Lie Series and Finite Canonical Transformations

Given a dynamical variable $u(q, p)$ and an infinitesimal canonical transformation $(q, p) \rightarrow (Q, P)$ with generator $X(q, p)$, set $u = u(q, p)$ and $\bar{u} = u(Q, P)$. From Eq. (8.110) it follows that

$$\bar{u} = u + \epsilon \{u, X\}. \tag{8.132}$$

This equation can be written in the form

$$\bar{u} = u + \epsilon \{u, X\} \equiv u + \epsilon \mathcal{L}_X u \equiv (\mathbb{I} + \epsilon \mathcal{L}_X) u, \tag{8.133}$$

where $\mathbb{I}$ is the identity operator – characterised by $\mathbb{I} u = u$ for all u – and $\mathcal{L}_X$ is the **Lie operator** defined by

$$\mathcal{L}_X = \{\cdot, X\}. \tag{8.134}$$

The dot indicates where to insert the function on which $\mathcal{L}_X$ operates.

Exercise 8.7.1   Show that in terms of the Lie operator the Jacobi identity takes the form

$$[\mathcal{L}_X, \mathcal{L}_Y] = \mathcal{L}_{\{X,Y\}}, \tag{8.135}$$

where $[A, B] = AB - BA$ is the commutator of operators A and B.

Two consecutive infinitesimal canonical transformations produce the following effect:

$$\bar{u} = (\mathbb{I} + \epsilon \mathcal{L}_X)(\mathbb{I} + \epsilon \mathcal{L}_X) u = (\mathbb{I} + 2\epsilon \mathcal{L}_X) u + O(\epsilon^2), \tag{8.136}$$

where $O(\epsilon^2)$ denotes terms containing all powers of ϵ higher than the first. By induction, for any natural number N we have

$$(\mathbb{I} + \epsilon \mathcal{L}_X)^N = \mathbb{I} + N \epsilon \mathcal{L}_X + O(\epsilon^2). \tag{8.137}$$

Let κ be a *finite* parameter and let N be a very large natural number. Putting $\epsilon = \kappa/N$ and treating ϵ as infinitesimal, Eq. (8.137) suggests that the canonical transformation with finite parameter κ can be written in the form

$$\bar{u} = \lim_{N \to \infty} \left(\mathbb{I} + \frac{\kappa}{N} \mathcal{L}_X \right)^N u = e^{\kappa \mathcal{L}_X} u \tag{8.138}$$

or, equivalently, as the **Lie series**

$$\bar{u} = u + \kappa \mathcal{L}_X u + \frac{\kappa^2}{2!} \mathcal{L}_X^2 u + \frac{\kappa^3}{3!} \mathcal{L}_X^3 u + \cdots. \tag{8.139}$$

Note that repeated applications of the Lie operator involve sucessive Poisson brackets: explicitly, the Lie series (8.139) is written

$$\bar{u} = u + \kappa \{u, X\} + \frac{\kappa^2}{2!} \{\{u, X\}, X\} + \frac{\kappa^3}{3!} \{\{\{u, X\}, X\}, X\} + \cdots. \tag{8.140}$$

The meaning of the parameter κ depends on the physical or geometric meaning of the transformation generated by X.

Example 8.7 By means of the Lie series (8.139), express a finite rotation about the z-axis as a canonical transformation.

Solution

As shown in Section 8.6, the z-component of the angular momentum generates infinitesimal rotations about the z-axis. Therefore, for a clockwise finite rotation by angle θ, the Lie series (8.139) applies with $\kappa = \theta$ and $X = L_z = xp_y - yp_x$. Taking $u = x$ we have

$$\mathcal{L}_{L_z} x = \{x, xp_y - yp_x\} = -y\{x, p_x\} = -y; \tag{8.141}$$

$$\mathcal{L}_{L_z}^2 x = -\mathcal{L}_{L_z} y = -\{y, xp_y - yp_x\} = -x\{y, p_y\} = -x; \tag{8.142}$$

$$\mathcal{L}_{L_z}^3 x = -\mathcal{L}_{L_z} x = y; \qquad \mathcal{L}_{L_z}^4 x = \mathcal{L}_{L_z} y = x. \tag{8.143}$$

It is easy to conclude that, in general,

$$\mathcal{L}_{L_z}^{2n} x = (-1)^n x, \qquad \mathcal{L}_{L_z}^{2n+1} x = (-1)^{n+1} y. \tag{8.144}$$

Thus, the Lie series (8.139) takes the form

$$\bar{x} = \left(\sum_{n=0}^{\infty} (-1)^n \frac{\theta^{2n}}{(2n)!} \right) x - \left(\sum_{n=0}^{\infty} (-1)^n \frac{\theta^{2n+1}}{(2n+1)!} \right) y = x \cos\theta - y \sin\theta. \tag{8.145}$$

Similarly, one checks that $\bar{y} = x \sin\theta + y \cos\theta$. This establishes the validity of the Lie series (8.140) to express a finite rotation as a canonical transformation.

Time Evolution: Solution of the Equations of Motion

As seen in Section 8.5, the Hamiltonian is the generator of the infinitesimal canonical transformation that evolves the system from instant t to instant $t + dt$. Therefore, if τ is a finite time, taking $\kappa = \tau$ and $X = H$, the Lie series (8.140) yields

$$\bar{u} = u(t + \tau) = u(t) + \tau \{u, H\} + \frac{\tau^2}{2!} \{\{u, H\}, H\} + \frac{\tau^3}{3!} \{\{\{u, H\}, H\}, H\} + \cdots, \tag{8.146}$$

where all Poisson brackets are calculated at time t. Setting $t = 0$ and writing t for τ, we have

$$u(t) = u(0) + t\{u, H\}_0 + \frac{t^2}{2!} \{\{u, H\}, H\}_0 + \frac{t^3}{3!} \{\{\{u, H\}, H\}, H\}_0 + \cdots. \tag{8.147}$$

This represents the formal solution of Hamilton's equations for u as a power series in t in terms of the initial conditions at $t = 0$.

Notice that Eq. (8.147) can be cast in the form

$$u(t) = e^{t \mathcal{L}_H} u(0), \tag{8.148}$$

in which $e^{t \mathcal{L}_H}$ plays the role of the **time evolution operator** in classical mechanics. The reader familiar with quantum theory will not fail to recognise the structural similarity between $e^{t \mathcal{L}_H}$ and the time evolution operator of quantum mechanics.

Example 8.8 Given a particle in one dimension subject to a constant force F with Hamilton function $H = p^2/2m - Fx$, find $x(t)$ by using the Lie series (8.147).

Solution

We have

$$\{x, H\} = \frac{1}{2m}\{x, p^2\} = \frac{1}{2m}2p\{x, p\} = \frac{p}{m}, \tag{8.149}$$

$$\{\{x, H\}, H\} = \frac{1}{m}\{p, H\} = \frac{-F}{m}\{p, x\} = \frac{F}{m}. \tag{8.150}$$

Since $F = $ constant, all the next Poisson brackets vanish and the Lie series (8.147) with $u = x$ gives

$$x(t) = x_0 + t\,\{x, H\}_0 + \frac{t^2}{2!}\,\{\{x, H\}, H\}_0 = x_0 + \frac{p_0}{m}t + \frac{F}{2m}t^2, \tag{8.151}$$

which is the well-known elementary solution.

The previous example is exceptional in that the Lie series terminates and reduces to a polynomial. In other simple cases the Lie series is infinite but can be explicitly summed.

Example 8.9 (Farina de Souza & Gandelman, 1990) Obtain the solution to the equation of motion for a charged particle in a uniform magnetic field by means of the Lie series (8.147).

Solution

With $\mathbf{B} = B\hat{\mathbf{z}}$ and

$$\mathbf{A} = \frac{1}{2}\mathbf{B} \times \mathbf{r} = \frac{B}{2}(-y\hat{\mathbf{x}} + x\hat{\mathbf{y}}), \tag{8.152}$$

the Hamiltonian is

$$H = \frac{1}{2m}\left(\mathbf{p} - \frac{e}{c}\mathbf{A}\right)^2 = \frac{1}{2m}\left[\left(p_x + \frac{eB}{2c}y\right)^2 + \left(p_y - \frac{eB}{2c}x\right)^2 + p_z^2\right]. \tag{8.153}$$

For the purpose of an algebraic computation of the pertinent Poisson brackets, it is convenient to express the Hamiltonian in the following form:

$$H = \frac{1}{2m}\left(p_x^2 + p_y^2 + p_z^2\right) + \frac{\omega}{2}\left(yp_x - xp_y\right) + \frac{m\omega^2}{8}\left(x^2 + y^2\right), \qquad \omega = \frac{eB}{mc}. \tag{8.154}$$

For the z-coordinate we have:

$$\mathcal{L}_H z = \{z, H\} = \frac{p_z}{m} = v_z; \quad \mathcal{L}_H^2 z = \{\{z, H\}, H\} = \frac{1}{m}\{p_z, H\} = 0. \tag{8.155}$$

It follows that $\mathcal{L}_H^n z = 0$ for all $n \geq 2$ and, therefore,

$$z = z_0 + v_{0z}t. \tag{8.156}$$

The particle moves with constant velocity in the z-direction.

is clearly singular. The canonical momenta are given by

$$p_x = \frac{\partial L}{\partial \dot{x}} = \dot{x} - \dot{y}, \qquad p_y = \frac{\partial L}{\partial \dot{y}} = \dot{y} - \dot{x}. \tag{8.180}$$

These equations are not independent and from them one derives

$$\phi = p_x + p_y = 0, \tag{8.181}$$

which is the only primary constraint.

The Hamiltonian is defined in the usual way by $H = \sum_i p_i \dot{q}_i - L$ and Eq. (7.7) remains valid, showing that even in the presence of constraints the Hamiltonian is a function only of the qs and ps. As we already know, in the presence of primary constraints we must employ the modified Hamiltonian

$$H_T = H + \sum_{m=1}^{M} \lambda_m \phi_m, \tag{8.182}$$

which incorporates the constraints by means of the Lagrange multipliers λ_m. Hamilton's equations brought about by the variational principle

$$\delta \int_{t_1}^{t_2} \left(\sum_i p_i \dot{q}_i - H_T \right) dt = \delta \int_{t_1}^{t_2} \left(\sum_i p_i \dot{q}_i - H - \sum_m \lambda_m \phi_m \right) dt = 0 \tag{8.183}$$

with arbitrary and independent variations of the qs, ps and λs are

$$\dot{q}_i = \frac{\partial H}{\partial p_i} + \sum_m \lambda_m \frac{\partial \phi_m}{\partial p_i}, \tag{8.184a}$$

$$\dot{p}_i = -\frac{\partial H}{\partial q_i} - \sum_m \lambda_m \frac{\partial \phi_m}{\partial q_i}, \tag{8.184b}$$

with the subsidiary conditions $\phi_m = 0$.

In order to write Eqs. (8.184) in the compact notation of Poisson brackets, it is useful to introduce the notion of **weak equality**, denoted by the symbol $\approx$. We start by writing the constraint equations in the form

$$\phi_m(q, p) \approx 0 \tag{8.185}$$

to stress that each function ϕ_m is restricted to be zero but does not vanish identically as a function on the entire phase space. In particular, the ϕ_m have non-vanishing Poisson brackets with the canonical variables. More generally, two functions F, G that coincide on the surface in phase space defined by the constraints $\phi_m \approx 0$ are said to be **weakly equal**, and we write $F \approx G$. On the other hand, an equality that holds on the whole phase space, and not only on the surface $\phi_m \approx 0$, is said to be a **strong equality** and the standard equality symbol is used to denote it. In the case of the system described by the Lagrangian (8.178), for example, the functions $F(x, y, p_x, p_y) = x + y + (x^2 - y^2) \tan(p_x + p_y)$ and $G(x, y, p_x, p_y) = (x + y)e^{p_x + p_y}$ are distinct over the whole phase space, but coincide on the surface defined by the constraint (8.181), so we write $F \approx G$.

With these precautions, the equation of motion for any function $F(q,p)$ can be written in the form

$$\dot{F} = \{F,H\} + \sum_m \lambda_m \{F,\phi_m\} \approx \{F,H_T\}, \tag{8.186}$$

because each term $\{\lambda_m, F\}\phi_m$, being proportional to one of the constraints, is weakly zero. Equation (8.186) shows that the constraint equations (8.177) can only be used *after* the calculation of all relevant Poisson brackets.

Consistency Conditions and the Dirac-Bergmann Algorithm

If we take F as one of the ϕ_m we must have $\dot{\phi}_m \approx 0$ for consistency: the primary contraints must be preserved in time. This gives rise to the **consistency conditions**

$$\{\phi_m, H\} + \sum_{m'} \lambda_{m'} \{\phi_m, \phi_{m'}\} \approx 0. \tag{8.187}$$

Three distinct cases may occur.

   Case (1). The consistency conditions are identically satisfied. In this case the only constraints of the theory are the primary constraints ϕ_m and the Lagrange multipliers are completely arbitrary. The dynamics contain arbitrary functions of time.

Example 8.11 Discuss the consistency conditions and obtain the general solution to Hamilton's equations for the system described by the Lagrangian of Example 8.10.

Solution

The Hamiltonian associated with the Lagrangian (8.178) is

$$H = \dot{x}p_x + \dot{y}p_y - L = (\dot{x} - \dot{y})p_x - \frac{1}{2}p_x^2 = \frac{1}{2}p_x^2, \tag{8.188}$$

where we have used (8.180) and (8.181). The consistency condition

$$\{\phi, H\} + \lambda\{\phi, \phi\} \approx 0 \tag{8.189}$$

for the constraint (8.181) is identically satisfied because $\{\phi, H\} = 0$. There are no other constraints and Hamilton's equations (8.186) take the form

$$\dot{x} \approx \{x,H\} + \lambda\{x,\phi\} = p_x + \lambda, \qquad \dot{p}_x \approx \{p_x,H\} + \lambda\{p_x,\phi\} = 0, \tag{8.190a}$$

$$\dot{y} \approx \{y,H\} + \lambda\{y,\phi\} = \lambda, \qquad \dot{p}_y \approx \{p_y,H\} + \lambda\{p_y,\phi\} = 0. \tag{8.190b}$$

The general solution to these equations is $p_x = -p_y = a$, $x(t) = y(t) + at + b$ with a,b arbitrary constants and $y(t)$ an arbitrary function.

Exercise 8.9.1 Show that the general solution to the Lagrange equations derived from Lagrangian (8.178) coincides with the general solution to Hamilton's equations.

Case (2). The consistency conditions uniquely determine the Lagrange multipliers. This takes place if the constraint matrix $\| \{\phi_m, \phi_{m'}\} \|$ is non-singular, that is, if

$$\det \| \{\phi_m, \phi_{m'}\} \| \napprox 0. \tag{8.191}$$

In this case, if $\| C_{mm'} \|$ is the inverse of $\| \{\phi_m, \phi_{m'}\} \|$ we have

$$\sum_{m''} C_{mm''} \{\phi_{m''}, \phi_{m'}\} = \delta_{mm'}. \tag{8.192}$$

Then the consistency conditions (8.187) yield

$$\lambda_m \approx -\sum_{m'} C_{mm'} \{\phi_{m'}, H\}, \tag{8.193}$$

whence

$$\dot{F} \approx \{F, H\} - \sum_{m,m'} \{F, \phi_m\} C_{mm'} \{\phi_{m'}, H\}. \tag{8.194}$$

Defining the **Dirac bracket**

$$\{F, G\}^D = \{F, G\} - \sum_{m',m''} \{F, \phi_{m'}\} C_{m'm''} \{\phi_{m''}, G\}, \tag{8.195}$$

the equation of motion for an arbitrary function $F(q, p)$ reduces to

$$\dot{F} = \{F, H\}^D. \tag{8.196}$$

We are entitled to use the strong equality sign in this last equation because the Dirac brackets of the constraints with any function $F(q, p)$ are zero:

$$\{F, \phi_m\}^D = \{F, \phi_m\} - \sum_{m',m''} \{F, \phi_{m'}\} C_{m'm''} \{\phi_{m''}, \phi_m\}$$
$$= \{F, \phi_m\} - \sum_{m'} \{F, \phi_{m'}\} \delta_{m'm} = 0. \tag{8.197}$$

It follows that the constraints can be set equal to zero in the Hamiltonian H_T *before* computing the Dirac brackets, so $H_T = H$ and the weak equalities can be taken as strong equalities. The Dirac brackets have the same algebraic properties as the Poisson brackets and obey the Jacobi identity (Sudarshan & Mukunda, 1983).

Case (3). The consistency conditions generate **secondary constraints**

$$\chi_s(q, p) = 0, \quad s = 1, \ldots, S. \tag{8.198}$$

The difference between primary and secondary constraints is that the former arise only from the form of the Lagrangian and the definition of the canonical momenta, whereas the latter require the use of the equations of motion. For consistency we impose

$$\dot{\chi}_s \approx \{\chi_s, H\} + \sum_{m'} \lambda_{m'} \{\chi_s, \phi_{m'}\} \approx 0. \tag{8.199}$$

If we fall back to cases (1) or (2) the process ends at this stage. If this does not happen, further secondary constraints turn up and we repeat the procedure, known as the **Dirac-Bergmann algorithm**. After a finite number of stages the process ends and we are left with a set of secondary constraints denoted by

$$\phi_k(q,p) \approx 0, \qquad k = M+1, \ldots, M+K, \tag{8.200}$$

where K is the total number of secondary constraints. As you can see, from now on the primary and secondary constraints are encompassed in the following uniform notation:

$$\phi_j(q,p) \approx 0, \qquad j = 1, \ldots, M+K = J. \tag{8.201}$$

The final consistency conditions are

$$\{\phi_j, H\} + \sum_{m=1}^{M} \lambda_m \{\phi_j, \phi_m\} \approx 0, \qquad j = 1, \ldots, J \tag{8.202}$$

and they impose restrictions on the λs inasmuch as they do not give rise to new constraints. The system (8.202) of J inhomogeneous linear equations for the $M < J$ unknowns λ_m must be soluble, otherwise the dynamics described by the original Lagrangian would be inconsistent, a possibility we exclude. The general solution of (8.202) takes the form

$$\lambda_m = U_m + \sum_{a=1}^{A} v_a V_m^{(a)}, \tag{8.203}$$

where U_m is a particular solution of the inhomogeneous equations – that is,

$$\sum_m \{\phi_j, \phi_m\} U_m \approx -\{\phi_j, H\}, \tag{8.204}$$

and $V_m^{(a)}$, $a = 1, \ldots, A$ are linearly independent solutions of the homogeneous equations:

$$\sum_{m=1}^{M} \{\phi_j, \phi_m\} V_m^{(a)} \approx 0. \tag{8.205}$$

The coefficients v_a in (8.203) are completely arbitrary.

Substituting (8.203) into (8.182) we are led to the final Hamiltonian of the theory which, with Dirac, we call the **total Hamiltonian**:

$$H_T = H + \sum_m U_m \phi_m + \sum_{m,a} v_a V_m^{(a)} \phi_m \equiv H' + \sum_a v_a \Phi_a, \tag{8.206}$$

where

$$H' = H + \sum_{m=1}^{M} U_m \phi_m \tag{8.207}$$

and, by definition,

$$\Phi_a = \sum_{m=1}^{M} V_m^{(a)} \phi_m. \tag{8.208}$$

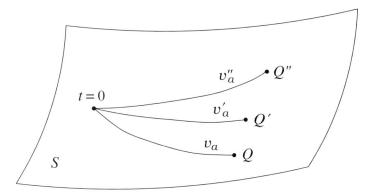

Fig. 8.3   Geometric representation of the dynamics of a constrained Hamiltonian system.

In terms of the total Hamiltonian the equations of motion are simply

$$\dot{F} \approx \{F, H_T\}\,. \tag{8.209}$$

The general picture can now be summarised. The coefficients v_a are arbitrary, so the general solution to the equations of motion contains arbitrary functions of time. The initial conditions do not uniquely determine the dynamical variables in future times, but the *physical* state of the system must needs be uniquely determined by the initial state. This is typical of gauge-invariant theories, which are generally described by a higher number of variables than the number of physical degrees of freedom.[8] Given the initial state $(q(0), p(0))$ compatible with the constraints, the physical state of the system at time t is represented by the equivalence class of the points $(q(t), p(t))$ that differ among themselves only by distinct choices of the coefficients v_a.

Geometrically, the situation can be described as follows (see Fig. 8.3). The dynamics unfold on the surface S of phase space defined by the constraints $\phi_j(q, p) \approx 0$, $j = 1, \ldots, J$. In general the initial values of (q, p) at $t = 0$ do not uniquely determine their values at time t because different choices for the coefficients v_a bring about different time evolutions for the canonical variables. This reflects the fact that the number of physical degrees of freedom is smaller than the number of variables used to describe the system's dynamics, so that the points Q, Q', Q'' represent the same physical state. A specific choice of the arbitrary coefficients v_a is called a **gauge fixing**, and makes the dynamics unique. The gauge fixing can be accomplished by imposing additional constraints whose consistency conditions completely determine the coefficients v_a.

First- and Second-Class Constraints

A classification of the constraints that is of fundamental importance, especially for the transition to the quantum theory, is the one that distinguishes **first-class constraints** from

---

[8] The archetypical gauge-invariant theory is Maxwell's electrodynamics described in terms of the scalar and vector potentials.

second-class constraints. A function $F(q, p)$ is said to be **first class** if its Poisson bracket with any constraint is weakly zero:

$$\{F, \phi_j\} \approx 0, \quad j = 1, \dots, J. \tag{8.210}$$

We say that F is **second class** if its Poisson bracket with at least one constraint is not weakly zero. The total Hamiltonian (8.206) is first class because it is the sum of H', which is first class, with a linear combination with arbitrary coefficients of the first-class constraints Φ_a defined by (8.208). The theory contains as many arbitrary functions of time as there are first-class constraints.

Exercise 8.9.2   Prove that H' and Φ_a, respectively defined by (8.207) and (8.208), are first-class functions.

The first-class primary constraints have an important physical meaning, which we proceed to discuss. Let initial conditions be given at $t = 0$ and consider

$$F_{\delta t} = F_0 + \dot{F} \delta t = F_0 + \{F, H_T\} \delta t = F_0 + \delta t \left[\{F, H'\} + \sum_a v_a \{F, \Phi_a\} \right]. \tag{8.211}$$

Since the vs are arbitrary, a choice of different values v'_a for these coefficients causes a change in $F_{\delta t}$:

$$\Delta F_{\delta t} = \delta t \sum_a (v'_a - v_a)\{F, \Phi_a\} = \sum_a \epsilon_a \{F, \Phi_a\}, \tag{8.212}$$

where

$$\epsilon_a = (v'_a - v_a)\delta t. \tag{8.213}$$

Since physical quantities must be insensitive to different choices of the coefficients v_a, both $F_{\delta t}$ and $F_{\delta t} + \Delta F_{\delta t}$ correspond to the same physical state. But (8.212) is a sum of infinitesimal canonical transformations, each generated by Φ_a with the associated infinitesimal parameter ϵ_a. Therefore, *the first class primary constraints generate gauge transformations, infinitesimal canonical transformations that change the qs and ps but do not change the system's physical state.*

Example 8.12   (Sundermeyer, 1982) Apply the Dirac-Bergmann algorithm to the Lagrangian

$$L(q_1, q_2, \dot{q}_1, \dot{q}_2) = \frac{1}{2}\dot{q}_1^2 + q_2\dot{q}_1 + (1 - \alpha)q_1\dot{q}_2 + \frac{\beta}{2}(q_1 - q_2)^2 \tag{8.214}$$

where α and β are constants.

Solution

The Hessian matrix

$$\mathbf{W} = \begin{pmatrix} 1 & 0 \\ 0 & 0 \end{pmatrix} \tag{8.215}$$

is singular. The canonical momenta are

$$p_1 = \frac{\partial L}{\partial \dot{q}_1} = \dot{q}_1 + q_2 \,, \qquad p_2 = \frac{\partial L}{\partial \dot{q}_2} = (1 - \alpha) q_1 \,, \tag{8.216}$$

and give rise to the primary constraint

$$\phi_1 = p_2 + (\alpha - 1) q_1 \approx 0 \,. \tag{8.217}$$

The Hamiltonian is given by

$$H = \frac{1}{2} (p_1 - q_2)^2 - \frac{\beta}{2} (q_1 - q_2)^2 \,. \tag{8.218}$$

The requirement that the primary constraint be preserved in time takes the form of the consistency condition

$$0 \approx \{\phi_1, H\} + \lambda \{\phi_1, \phi_1\} = \alpha (p_1 - q_2) - \beta (q_1 - q_2) \,. \tag{8.219}$$

Situation (a): $\alpha = \beta = 0$. The consistency condition is identically satisfied and there are no secondary constraints. The total Hamiltonian is $H_T = H + \lambda \phi_1$ and Hamilton's equations are

$$\dot{q}_1 = p_1 - q_2 \,, \qquad \dot{q}_2 = \lambda \,, \qquad \dot{p}_1 = \lambda \,, \qquad \dot{p}_2 = p_1 - q_2 \,. \tag{8.220}$$

Since λ is arbitrary, it follows that q_2 is an arbitrary function of time. On the other hand, from (8.220) one derives

$$\ddot{q}_1 = \lambda - \lambda = 0 \quad \Longrightarrow \quad q_1 = at + b \,, \tag{8.221}$$

with a and b arbitrary constants. It is easy to understand the why and wherefore this is the general solution to the equations of motion by noting that

$$L = \frac{1}{2} \dot{q}_1^2 + \frac{d}{dt} (q_1 q_2) \,. \tag{8.222}$$

Inasmuch as Lagrangians that differ by a total time derivative yield the same equations of motion, we are led to conclude that the system is equivalent to a free particle with only one degree of freedom. The variable q_2 does not represent a physical degree of freedom of the system: it is arbitrary and can be discarded. The appearence of an arbitrary function of time is in agreement with the general theory, for when there is only one constraint it is necessarily first class.

Situation (b): $\alpha = 0, \beta \neq 0$. In this case, from (8.219) we get the secondary constraint

$$\phi_2 = q_1 - q_2 \approx 0 \,. \tag{8.223}$$

The consistency condition applied to ϕ_2 gives

$$\{\phi_2, H\} + \lambda \{\phi_2, \phi_1\} = 0 \quad \Longrightarrow \quad p_1 - q_2 - \lambda = 0 \quad \Longrightarrow \quad \lambda = p_1 - q_2 \,. \tag{8.224}$$

Thus the Lagrange multiplier is determined and the total Hamiltonian becomes

$$H_T = \frac{1}{2} (p_1 - q_2)^2 + (p_1 - q_2)(p_2 - q_1) - \frac{\beta}{2} (q_1 - q_2)^2 \,. \tag{8.225}$$

The constraints ϕ_1 and ϕ_2 are second class because $\{\phi_1, \phi_2\} = 1$. An elementary calculation furnishes the constraint matrix and its inverse:

$$\| \{\phi_a, \phi_b\} \| = \begin{pmatrix} 0 & 1 \\ -1 & 0 \end{pmatrix} \implies \| C_{ab} \| = \| \{\phi_a, \phi_b\} \|^{-1} = \begin{pmatrix} 0 & -1 \\ 1 & 0 \end{pmatrix}. \qquad (8.226)$$

The Dirac bracket is given by

$$\{F, G\}^D = \{F, G\} - \sum_{a,b=1}^{2} \{F, \phi_a\} C_{ab} \{\phi_b, G\}$$

$$= \{F, G\} + \{F, \phi_1\}\{\phi_2, G\} - \{F, \phi_2\}\{\phi_1, G\}. \qquad (8.227)$$

With the use of the Dirac bracket the constraints can be set equal to zero in the Hamiltonian (8.225), which reduces to

$$H = H_T = \frac{1}{2}(p_1 - q_2)^2. \qquad (8.228)$$

The equations of motion (8.196) yield

$$\dot{q}_1 = p_1 - q_2, \quad \dot{q}_2 = p_1 - q_2, \quad \dot{p}_1 = p_1 - q_2, \quad \dot{p}_2 = p_1 - q_2. \qquad (8.229)$$

With the use of the Dirac bracket the constraints $p_2 - q_1 = 0$ and $q_2 - q_1 = 0$ become strong equations and can be substituted into (8.229) to give

$$\dot{q}_1 = p_1 - q_1, \quad \dot{p}_1 = p_1 - q_1. \qquad (8.230)$$

The general solution to these equations is

$$q_1(t) = q_2(t) = p_2(t) = at + b - a, \quad p_1(t) = at + b, \qquad (8.231)$$

with a and b arbitrary constants. The general analysis is confirmed that no arbitrary functions of time are left when all constraints are second class.

If $\alpha \neq 0$ the secondary constraint derived from the consistency condition (8.219) takes the form

$$\phi_2 = \alpha(p_1 - q_2) - \beta(q_1 - q_2) \approx 0. \qquad (8.232)$$

With the use of the Hamiltonian (8.218) the consistency condition associated with this constraint is

$$\{\phi_2, H\} + \lambda\{\phi_2, \phi_1\} = -\beta\left[(p_1 - q_2) - \alpha(q_1 - q_2)\right] + (\beta - \alpha^2)\lambda \approx 0. \qquad (8.233)$$

Situation (c): $\alpha \neq 0$, $\beta = \alpha^2$. The secondary constraint (8.232) assumes the simpler form

$$\phi_2 \equiv p_1 - q_2 - \alpha(q_1 - q_2) \approx 0. \qquad (8.234)$$

Using the Hamiltonian (8.218) with $\beta = \alpha^2$, the consistency condition applied to ϕ_2 does not generate new constraints because

$$\{\phi_2, H\} + \lambda\{\phi_2, \phi_1\} = -\alpha\phi_2 \approx 0 \qquad (8.235)$$

is identically satisfied owing to the constraint $\phi_2 \approx 0$. The equations of motion (8.209) give

$$\dot{q}_1 = p_1 - q_2 + (\alpha - 1)\lambda, \qquad \dot{q}_2 = \lambda, \qquad (8.236a)$$

$$\dot{p}_1 = \alpha^2(q_1 - q_2) - (\alpha - 1)\lambda, \qquad \dot{p}_2 = p_1 - q_2 - \alpha^2(q_1 - q_2). \qquad (8.236b)$$

It follows that λ (therefore q_2) is an arbitrary function of time. This is in agreement with the fact that the constraint ϕ_1 is first class.

Situation (d): $\alpha \neq 0$, $\beta \neq \alpha^2$. The Lagrange multiplier λ is uniquely determined by Eq. (8.233):

$$\lambda = \frac{\beta}{\beta - \alpha^2}\left[(p_1 - q_2) - \alpha(q_1 - q_2)\right] = \frac{\beta}{\alpha}(q_1 - q_2), \qquad (8.237)$$

where we have used (8.232). The constraints ϕ_1, ϕ_2 are second class and the equations of motion are

$$\dot{q}_1 = p_1 - q_2, \qquad \dot{q}_2 = \frac{\beta}{\alpha}(q_1 - q_2), \qquad (8.238a)$$

$$\dot{p}_1 = \frac{\beta}{\alpha}(q_1 - q_2), \qquad \dot{p}_2 = \frac{(1 - \alpha)\beta}{\alpha}(q_1 - q_2). \qquad (8.238b)$$

There are no arbitrary functions of time.

Exercise 8.9.3 In situation (a) of Example 8.12, what kind of infinitesimal canonical transformation is generated by the first-class primary constraint ϕ_1? Verify that this transformation affects only arbitrary quantities, so it is a gauge transformation which does not change the system's physical state. In situation (d) of Example 8.12, determine the explicit form of the Dirac bracket in terms of the Poisson bracket and obtain the equations of motion (8.238).

The examples discussed here are purely academic, but illustrate well the several situations encountered in the physically relevant models, which consist in gauge field theories that we are not yet prepared to take up.[9] Other important aspects of the theory of constrained Hamiltonian systems, such as the role of the first-class secondary constraints, the extended Hamiltonian, Dirac's conjecture, gauge fixing, the independent degrees of freedom and Dirac's quantisation rules are beyond the scope of this introductory treatment. For these and other issues the interested reader should consult specialised works (Dirac, 1964; Sundermeyer, 1982; Henneaux & Teitelboim, 1992).

Problems

8.1 (a) From Eq. (8.8) in the form $\Phi(q, p, t) = F_1(q, Q(q, p, t), t)$ prove that $\partial\Phi/\partial t = -\sum_i P_i \partial Q_i/\partial t + \partial F_1/\partial t$. (b) Using $d\Phi(q, p, t) = d\Phi(q, p, t)|_{\text{fixed } t} + (\partial\Phi/\partial t)\, dt$ and a similar equation for $dQ_i(q, p, t)$, show that if (8.42) holds then (8.7) is satisfied for variable t with $K(Q, P, t)$ *defined* by (8.10).

---

[9] In Chapter 11 we present a brief discussion, by means of simple examples, of constrained field theories.

8.2 Consider a unit-mass, one-dimensional free particle described by the Hamiltonian $H = p^2/2$. (a) Show that the transformation

$$Q = \frac{1}{2}(q + p^3), \qquad P = \frac{1}{2}(q - p^3)$$

preserves the Hamiltonian form of the equations of motion for the free particle by finding the transformed Hamiltonian $K(Q,P)$. (b) With the help of the Lagrange bracket criterion (8.44) show that the transformation in (a) is not canonical.

8.3 In the one-degree-of-freedom case, for what values of the constants α and β is the transformation

$$Q = q^\alpha \cos \beta p, \qquad P = q^\alpha \sin \beta p$$

canonical?

8.4 Find the canonical transformation generated by

$$F_1(q, Q) = \frac{1}{2\lambda} \sum_{k=1}^{n} (q_k - Q_k)^2,$$

where λ is a constant. To what does the transformation reduce if $\lambda = 0$? Is your result compatible with the statement in the paragraph just following Example 8.1 that the identity transformation does not admit a generating function of the type F_1?

8.5 A certain mechanical system with a single degree of freedom has Hamiltonian

$$H = \frac{1}{2}\left[(p - aq)^2 + \omega^2(q + bt)^2 \right],$$

with ω, a and b constants. (a) Prove that the transformation $Q = q + bt$, $P = p - aq + b$ is canonical and determine a generating function. (b) Show that the transformed Hamiltonian is $K = (P^2 + \omega^2 Q^2)/2$. (b) Using the well-known solution to the harmonic oscillator equation of motion, return to the original variables to obtain $q(t)$. (c) Prove that

$$R(q, p, t) = \frac{1}{2}(p - aq + b)^2 + \frac{\omega^2}{2}(q + bt)^2$$

is a constant of the motion.

8.6 A particle in one-dimensional motion under a constant force F is described by the Hamiltonian $H = p^2/2m - Fq$. (a) Find a canonical transformation such that the transformed Hamiltonian is $K(Q, P) = P$. (b) Solve Hamilton's equations for the new canonical variables and return to the original variables to obtain $q(t)$.

8.7 The Hamiltonian for a one-degree-of-freedom system is

$$H = \frac{1}{2}\left(q^4 p^2 + \frac{1}{q^2} \right).$$

(a) Write down Hamilton's equations and derive from them a second-order differential equation involving q alone. (b) Invent a canonical transformation that reduces H to a harmonic oscillator Hamiltonian: $K = (P^2 + Q^2)/2$. (c) From the knowledge that the general solution to the oscillator problem is $Q(t) = A\cos(\omega t + \delta)$, determine

$P(t)$. (d) Returning to the original variables, find $q(t)$. This function is, therefore, the general solution to the complicated differential equation obtained in part (a).

8.8   Solve the equation of motion for the one-dimensional harmonic oscillator by means of the Lie series (8.147), assuming generic initial conditions.

8.9   A one-degree-of-freedom system is described by the Hamiltonian

$$H = \frac{[p - g(q)]^2}{2f'(q)^2} - kf(q),$$

where k is a constant, f and g being given functions with f invertible. Find a canonical transformation that reduces this Hamiltonian to that for a unit-mass particle subject to a constant force. Solve Hamilton's equations for the transformed canonical variables and show that the motion in terms of the original variable q is $q(t) = f^{-1}(a + bt + kt^2/2)$, where a, b are arbitrary constants and f^{-1} is the inverse of function f.

8.10   According to Theorem 2.3.1, Lagrange's equations do not change if one replaces the Lagrangian $L(q, \dot{q}, t)$ by $\bar{L}(q, \dot{q}, t) = L(q, \dot{q}, t) + df(q, t)/dt$, where $f(q, t)$ is any sufficiently differentiable function. Prove that this transformation is canonical and find a generating function.

8.11   Consider the Hamiltonian

$$H = \frac{p_1^2}{2m} + \frac{(p_2 - kq_1)^2}{2m}.$$

(a) Determine the constants A and B such that the transformation

$$Q_1 = Ap_1, \quad P_1 = p_2 - kq_1, \quad Q_2 = B(p_1 - kq_2), \quad P_2 = p_2$$

is canonical. (b) Using this canonical transformation and solving the equations of motion for the transformed variables, find $q_1(t)$ and $q_2(t)$.

8.12   The Hamiltonian for a one-degree-of-freedom system is $H = \omega^2 p(q+t)^2$, where ω is a positive constant. (a) Show that the phase-space transformation $Q = q + t$, $P = p$ is canonical and find a generating function. (b) Using this canonical transformation solve Hamilton's equations for $q(t)$ and $p(t)$.

8.13   A particle of mass m moves on a straight line under the potential $V = mk/x^2$. Given the initial conditions $x(0) = x_0$, $\dot{x}(0) = 0$, obtain $x(t)$ by means of the Lie series (8.147) for $u = x^2$.

8.14   The so-called conformal mechanics (Cadoni, Carta & Mignemi, 2000) is a one-degree-of-freedom system defined by the Hamiltonian

$$H(q, p) = \frac{p^2}{2f(qp)^2},$$

where $f : \mathbb{R} \to \mathbb{R}$ is a differentiable function that never vanishes. (a) Prove that the transformation

$$Q = qf(qp), \quad P = \frac{p}{f(qp)}$$

is canonical and, with its help, prove that the general solution to Hamilton's equations for conformal mechanics is

$$q(t) = \frac{\alpha t + \beta}{f(\alpha^2 t + \alpha\beta)}, \qquad p(t) = \alpha f(\alpha^2 t + \alpha\beta),$$

with α and β arbitrary constants. (b) Show that with a suitable choice of the function f one gets the Hamiltonian of Problem 8.13 and compare the above solution for $q(t)$ with the results obtained in Problems 2.23 and 8.13.

8.15 In the one-degree-of-freedom case, show that the transformation

$$Q = \ln\left(1 + \sqrt{q}\cos p\right), \qquad P = 2\left(1 + \sqrt{q}\cos p\right)\sqrt{q}\sin p$$

is canonical and that a generating function is $F_3(p, Q) = -(e^Q - 1)^2 \tan p$.

8.16 (a) Find the Hamiltonian associated with the one-dimensional Lagrangian $L = e^{\lambda t}(m\dot{q}^2/2 - m\omega^2 q^2/2)$, which describes a damped harmonic oscillator (Problem 1.16). (b) Prove that the transformation $Q = qe^{\lambda t/2}$, $P = pe^{-\lambda t/2} + (m\lambda/2)qe^{\lambda t/2}$ is canonical and produce a generating function. (c) Prove that the transformation

$$Q = \frac{\bar{P}}{m\Omega}\sin\left(\frac{m\Omega\bar{Q}}{\bar{P}}\right), \qquad P = \bar{P}\cos\left(\frac{m\Omega\bar{Q}}{\bar{P}}\right)$$

is canonical and reduces the Hamiltonian for a harmonic oscillator of angular frequency Ω in the variables Q, P to the Hamiltonian for a free particle in the variables $\bar{Q}, \bar{P}$. (d) Taking $\Omega = (\omega^2 - \lambda^2/4)^{1/2}$, apply in succession the above canonical transformations to solve, in the case of weak damping, the equation of motion for the damped oscillator. (e) What modifications are necessary to deal with the cases of strong damping ($\lambda > 2\omega$) and critical damping ($\lambda = 2\omega$)?

8.17 The Hamiltonian for a certain one-degree-of-freedom system is $H = q + te^p$. (a) Show that the transformation $Q = q + te^p$, $P = p$ is canonical. (b) Find a generating function. (c) Find the transformed Hamiltonian $K(Q, P, t)$.

8.18 A two-dimensional isotropic harmonic oscillator has Hamiltonian

$$H = \frac{1}{2m}(p_x^2 + p_y^2) + \frac{m\omega^2}{2}(x^2 + y^2).$$

Given that $A = p_x^2 + m^2\omega^2 x^2$ and $L = xp_y - yp_x$ are constants of the motion, use Poisson's theorem to show that $B = p_xp_y + m^2\omega^2 xy$ and $C = p_x^2 - p_y^2 + m^2\omega^2(x^2 - y^2)$ are also constants of the motion.

8.19 A two-degree-of-freedom system is described by the Hamiltonian

$$H = p_1^2 + p_2^2 + \frac{1}{2}(q_1 - q_2)^2 + \frac{1}{8}(q_1 + q_2)^2.$$

Show that the transformation

$$q_1 = \sqrt{Q_1}\cos P_1 + \sqrt{2Q_2}\cos P_2, \qquad q_2 = -\sqrt{Q_1}\cos P_1 + \sqrt{2Q_2}\cos P_2,$$
$$p_1 = \sqrt{Q_1}\sin P_1 + \sqrt{Q_2/2}\sin P_2, \qquad p_2 = -\sqrt{Q_1}\sin P_1 + \sqrt{Q_2/2}\sin P_2$$

is canonical and obtain the transformed Hamiltonian in terms of the new canonical variables. Use your results to find q_1, q_2, p_1, p_2 as functions of time.

8.20 Consider the following change of variables in phase space:

$$Q_i = q_i \sum_{j=1}^{n} p_j^2 - 2p_i \sum_{j=1}^{n} q_j p_j, \qquad P_i = \frac{p_i}{\sum_{j=1}^{n} p_j^2}.$$

Prove that this transformation is canonical because it admits the generating function

$$F_2(q, P) = \frac{\sum_{i=1}^{n} q_i P_i}{\sum_{j=1}^{n} P_j^2}.$$

8.21 Contrary to what may seem, the specification of the function $\Phi(q, p, t)$ in Eq. (8.42) does not determine the canonical transformation uniquely. In the case of a one-degree-of-freedom system, show that all canonical transformations corresponding to $\Phi(q, p, t) \equiv 0$ are $Q = f(q)$, $P = p/f'(q)$. The canonical transformations with $\Phi = 0$ are known as Mathieu transformations (Whittaker, 1944).

8.22 Consider a charged particle in a crossed uniform electromagnetic field. Choosing the axes in such a way that $\mathbf{E} = E\hat{\mathbf{y}}$ and $\mathbf{B} = B\hat{\mathbf{z}}$, the scalar potential is $\phi = -Ey$ and the Hamiltonian for this problem is

$$H = H_B - eEy,$$

where H_B is the Hamiltonian (8.153) of Example 8.9. By means of the Lie series (8.147) show that the motion in the z-direction is uniform. Show also that (Farina de Souza & Gandelman, 1990)

$$x(t) = x_0 + \frac{v_{y0}}{\omega} + \frac{eEt}{m\omega} + \left(\frac{v_{x0}}{\omega} - \frac{eE}{m\omega^2} \right) \sin \omega t - \frac{v_{y0}}{\omega} \cos \omega t,$$

$$y(t) = y_0 - \frac{v_{x0}}{\omega} + \frac{eE}{m\omega^2} + \left(\frac{v_{x0}}{\omega} - \frac{eE}{m\omega^2} \right) \cos \omega t + \frac{v_{y0}}{\omega} \sin \omega t,$$

with $\omega = eB/mc$.

8.23 This problem consists in an alternative proof of Liouville's theorem. (a) Let $\mathbf{B}$ be a matrix such that $\mathbf{B} = \mathbf{I} + \epsilon \mathbf{A}$. Show that

$$\det \mathbf{B} = 1 + \epsilon \operatorname{tr} \mathbf{A} + O(\epsilon^2).$$

(b) Consider the system of ordinary differential equations

$$\frac{dx_k}{dt} = X_k(x(t), t), \qquad k = 1, \dots, N$$

and let $\mathcal{R}_t$ be a region in the phase space $(x_1, \dots, x_N)$ at instant t with volume

$$\Omega(t) = \int_{\mathcal{R}_t} dx_1(t) \cdots dx_N(t).$$

Prove that

$$\frac{d\Omega}{dt} = \int_{\mathcal{R}_t} \nabla \cdot \mathbf{X} \, dx_1(t) \cdots dx_N(t)$$

where

$$\nabla \cdot \mathbf{X} = \sum_{k=1}^{N} \frac{\partial X_k}{\partial x_k}.$$

(c) Prove, finally, that $\nabla \cdot \mathbf{X} = 0$ for any Hamiltonian system, which establishes Liouville's theorem.

Hint: Show that, up to the first order in ϵ,

$$x_k(t + \epsilon) = x_k(t) + \epsilon X_k(x(t), t)$$

and note that

$$\Omega(t + \epsilon) = \int_{\mathcal{R}_{t+\epsilon}} dx_1(t + \epsilon) \cdots dx_N(t + \epsilon)$$

$$= \int_{\mathcal{R}_t} \left| \frac{\partial(x_1(t + \epsilon), \dots, x_N(t + \epsilon))}{\partial(x_1(t), \dots, x_N(t))} \right| dx_1(t) \cdots dx_N(t).$$

Next, using the result of part (a), conclude that the Jacobian

$$\frac{\partial(x_1(t + \epsilon), \dots, x_N(t + \epsilon))}{\partial(x_1(t), \dots, x_N(t))} = 1 + \epsilon \sum_{k=1}^{N} \frac{\partial X_k(x(t), t)}{\partial x_k(t)} + O(\epsilon^2)$$

and derive the desired expression for $d\Omega/dt$ from the definition of derivative.

8.24 Consider the Lagrangian

$$L = (\dot{y} - z)(\dot{x} - y).$$

(a) If $\epsilon(t)$ is an infinitesimal function, what is the effect of the infinitesimal transformation

$$\delta x = \epsilon, \qquad \delta y = \dot{\epsilon}, \qquad \delta z = \ddot{\epsilon}$$

on the Lagrangian? On the equations of motion? (b) Find the primary constraints. (c) Construct the canonical Hamiltonian and the total Hamiltonian. (d) Apply the Dirac-Bergmann algorithm and determine all constraints of the theory. (e) Show that all constraints are first class. (f) Show that Hamilton's equations coincide with the original Lagrange equations.

8.25 A simple mechanical model which has gauge invariance properties analogous to those of quantum electrodynamics and quantum chromodynamics is described by the Lagrangian (Christ & Lee, 1980; Lee, 1981)

$$L = \frac{1}{2} \left[\dot{r}^2 + r^2(\dot{\theta} - \xi)^2 \right] - V(r),$$

where (r, θ) are polar coordinates in the plane and ξ is an additional degree of freedom. This Lagrangian formally describes a particle in a plane observed from a reference frame that rotates with angular velocity ξ, with ξ treated as a dynamical variable. (a) Show that the Lagrangian is invariant under the finite gauge transformation

$$\theta \rightarrow \theta + \alpha(t), \qquad \xi \rightarrow \xi + \dot{\alpha}(t),$$

where $\alpha(t)$ is an arbitrary function. (b) Find the primary constraints. (c) Obtain the canonical Hamiltonian and the total Hamiltonian. (d) Apply the Dirac-Bergmann algorithm and find all constraints of the model. (e) Show that all constraints are first class. (f) Show that Hamilton's equations coincide with the original Lagrange equations.

8.26 A measurable quantity associated with a mechanical system must depend only on the system's *physical state*, not on the particular coordinates chosen to describe it. In a theory with only first-class constraints, an *observable* is defined as any function of the canonical variables whose Poisson bracket with each of the constraints is weakly zero. Justify this definition in the light of Eq. (8.212).

8.27 Instead of the action (7.97), a quantum theory of the free relativistic particle inspired by string theory takes for starting point the new action

$$\bar{S}[x, e] = -\frac{1}{2} \int_{\theta_1}^{\theta_2} (e^{-1} \dot{x}^\alpha \dot{x}_\alpha + m^2 c^2 e) \, d\theta \,,$$

where $\dot{x}^\alpha = dx^\alpha/d\theta$. The action $\bar{S}$ contains an additional degree of freedom represented by e, which is usually called the *einbein* ("one leg", in German). (a) Show that Lagrange's equations for x^α and e are equivalent to $e = \sqrt{\dot{x}^\alpha \dot{x}_\alpha}/mc$ and to Eq. (7.107). (b) With the replacement $e = \sqrt{\dot{x}^\alpha \dot{x}_\alpha}/mc$ show that $\bar{S}$ becomes identical to the action (7.97). (c) Consider a time reparameterisation $\theta \to s = s(\theta)$. If e and $\dot{x}^\alpha$ change the same way, that is, if the *einbein* $e(\theta)$ transforms as $e'(s) = (d\theta/ds)e(\theta(s))$, prove that $\bar{S}' = \bar{S}$ and the action is invariant. (d) Find the primary constraint that arises in the passage to the Hamiltonian formulation, determine the only secondary constraint and show that both constraints are first class. (e) Check that, together with the constraints, Hamilton's equations are equivalent to Lagrange's equations. Careful: because of the negative sign in the definition (7.100) of P^α, the correct form of Hamilton's equations is $\dot{x}^\alpha = -\partial H/\partial P_\alpha$, $\dot{P}^\alpha = \partial H/\partial x_\alpha$.

9 The Hamilton-Jacobi Theory

It is impossible to explain honestly the beauties of the laws of nature without some deep understanding of mathematics.

Richard P. Feynman, *The Character of Physical Law*

In the previous chapter several examples were considered in which Hamilton's equations for simple mechanical systems could be solved with the help of canonical transformations. Nevertheless, those transformations were obtained by inspection, they did not emerge from any generic and systematic procedure. The Hamilton-Jacobi theory is a constructive method that, in many favourable cases, permits us to find a canonical transformation capable of drastically simplifying the equations of motion for a system described by a given Hamiltonian. As a consequence, the resolution of the original equations of motion becomes trivial because it is reduced to a mere algebraic process. Pinnacle of classical mechanics, the Hamilton-Jacobi theory provides the most powerful method for the integration of Hamilton's equations and for the elucidation of the deepest structural problems of mechanics.

9.1 The Hamilton-Jacobi Equation

Given a mechanical system described by the canonical variables (q, p) and by the Hamiltonian $H(q, p, t)$, let us perform a canonical transformation by means of a generating function $S(q, P, t)$ and assume that it can be so chosen that the transformed Hamiltonian vanishes: $K(Q, P, t) = 0$. Under these circumstances, the transformed Hamilton's equations are trivially solved:

$$\dot{Q}_i = \frac{\partial K}{\partial P_i} = 0 \quad \Longrightarrow \quad Q_i = \beta_i, \tag{9.1a}$$

$$\dot{P}_i = -\frac{\partial K}{\partial Q_i} = 0 \quad \Longrightarrow \quad P_i = \alpha_i, \tag{9.1b}$$

where the αs and βs are constants.

According to (8.17) we have

$$0 = K = H(q, p, t) + \frac{\partial S}{\partial t}, \tag{9.2}$$

and, from the transformation equations (8.16),

$$p_i = \frac{\partial S}{\partial q_i}\,. \tag{9.3}$$

Equations (9.2) and (9.3) conduce to the **Hamilton-Jacobi equation**:

$$H\left(q_1,\ldots,q_n,\frac{\partial S}{\partial q_1},\ldots,\frac{\partial S}{\partial q_n},t\right) + \frac{\partial S}{\partial t} = 0\,. \tag{9.4}$$

This is a first-order partial differential equation in the $n+1$ variables $q_1,\ldots,q_n,t$. Fortunately we do not need its general solution, which involves an arbitrary function and is hard to get owing to the equation's nonlinearity. As will be seen by means of numerous examples, it may be possible to find *particular* solutions of the Hamilton-Jacobi equation containing $n+1$ arbitrary constants $\alpha_1,\ldots,\alpha_{n+1}$. Since S is not directly involved in the equation but only its derivatives, one of the constants, say α_{n+1}, is merely additive. In other words, any solution containing $n+1$ parameters is of the form $S + \alpha_{n+1}$. The additive constant α_{n+1} can be dropped since it does not modify the transformation generated by S. The considerations in the previous paragraph suggest that, once we are in possession of a solution to the Hamilton-Jacobi equation of the form $S(q_1,\ldots,q_n,\alpha_1,\ldots,\alpha_n,t)$ where the αs are nonadditive constants of integration, we make the identification $\alpha_i \equiv P_i$ and the function $S(q,P,t)$ so constructed will generate a canonical transformation that reduces the new Hamiltonian to zero. It follows that the motion of the system in terms of the original canonical variables is determined by the equations

$$\beta_i = \frac{\partial S}{\partial \alpha_i}\,, \qquad i = 1,\ldots,n\,, \tag{9.5}$$

which result from the combination of (8.16) and (9.1a), with the identification $\alpha_i = P_i$. The n equations (9.5) can be solved for $q_1,\ldots,q_n$ leading to

$$q_i = q_i(\alpha,\beta,t)\,. \tag{9.6}$$

This result, together with (9.3), allows us to write

$$p_i = p_i(\alpha,\beta,t)\,. \tag{9.7}$$

These two last equations represent the general solution of the original Hamilton's equations involving $2n$ constants of integration whose values are determined by the initial conditions.

In view of the heuristic nature of the above discussion, it is worth making it rigorous by means of precise definitions and proofs of the results that have just been intimated.

Definition 9.1.1 *A* **complete solution** *or* **complete integral** *of the Hamilton-Jacobi equation is a solution of the form $S(q_1,\ldots,q_n,\alpha_1,\ldots,\alpha_n,t)$ containing n parameters $\alpha_1,\ldots,\alpha_n$ and such that*

$$\det\left(\frac{\partial^2 S}{\partial q_i \partial \alpha_j}\right) \neq 0\,. \tag{9.8}$$

Condition (9.8) characterises the parameters $\alpha_1,\ldots,\alpha_n$ as mutually independent and, at the same time, ensures that Eqs. (9.5) can be solved for the qs in the form (9.6). Now we can sum up the central result of the Hamilton-Jacobi theory in the theorem that follows (Jacobi, 1837).

Theorem 9.1.1 (Jacobi)   *Let $S(q, \alpha, t)$ be a complete integral of the Hamilton-Jacobi equation. Then the qs and ps determined by equations (9.5) and (9.3) satisfy Hamilton's equations*

$$\dot{q}_i = \frac{\partial H}{\partial p_i}, \qquad \dot{p}_i = -\frac{\partial H}{\partial q_i}, \qquad i = 1, \ldots, n. \tag{9.9}$$

Proof   Differentiation of (9.5) with respect to time yields

$$0 = \dot{\beta}_i = \sum_{j=1}^{n} \frac{\partial^2 S}{\partial q_j \partial \alpha_i} \dot{q}_j + \frac{\partial^2 S}{\partial t \partial \alpha_i}. \tag{9.10}$$

On the other hand, making use of (9.3) we can write the Hamilton-Jacobi equation (9.4) in the form

$$H(q, p(q, \alpha, t), t) + \frac{\partial S}{\partial t} = 0. \tag{9.11}$$

Differentiating this equation with respect to α_i we obtain

$$\sum_{j=1}^{n} \frac{\partial H}{\partial p_j} \frac{\partial p_j}{\partial \alpha_i} + \frac{\partial^2 S}{\partial \alpha_i \partial t} = \sum_{j=1}^{n} \frac{\partial H}{\partial p_j} \frac{\partial^2 S}{\partial \alpha_i \partial q_j} + \frac{\partial^2 S}{\partial \alpha_i \partial t} = 0, \tag{9.12}$$

where we have used (9.3) anew. Taking into account that the order of differentiation is immaterial and substituting (9.12) into (9.10) we are left with

$$\sum_{j=1}^{n} \frac{\partial^2 S}{\partial q_j \partial \alpha_i} \left(\dot{q}_j - \frac{\partial H}{\partial p_j} \right) = 0. \tag{9.13}$$

Since the matrix whose elements are $\partial^2 S / \partial q_j \partial \alpha_i$ is non-singular by (9.8), the homogeneous system of linear equations (9.13) has only the trivial solution

$$\dot{q}_j - \frac{\partial H}{\partial p_j} = 0, \qquad j = 1, \ldots, n. \tag{9.14}$$

Taking the total time derivative of (9.3) we find

$$\dot{p}_i = \sum_{j=1}^{n} \frac{\partial^2 S}{\partial q_j \partial q_i} \dot{q}_j + \frac{\partial^2 S}{\partial t \partial q_i}. \tag{9.15}$$

But from (9.11) one derives, differentiating with respect to q_i,

$$\frac{\partial H}{\partial q_i} + \sum_{j=1}^{n} \frac{\partial H}{\partial p_j} \frac{\partial^2 S}{\partial q_i \partial q_j} + \frac{\partial^2 S}{\partial q_i \partial t} = 0. \tag{9.16}$$

The substitution of this last result into (9.15) leads to

$$\dot{p}_i = -\frac{\partial H}{\partial q_i} + \sum_{j=1}^{n} \frac{\partial^2 S}{\partial q_i \partial q_j} \left(\dot{q}_j - \frac{\partial H}{\partial p_j} \right) = -\frac{\partial H}{\partial q_i} \tag{9.17}$$

by virtue of (9.14), completing the proof of the theorem.   □

9.2 One-Dimensional Examples

The discussion of some elementary examples will serve not only to illustrate the use of the Hamilton-Jacobi technique but also to motivate certain observations of a general character that will play an important role in the application of the formalism.

Example 9.1   Solve the equation of motion for the free particle in one dimension by the Hamilton-Jacobi method.

Solution

Since $H = p^2/2m$, the Hamilton-Jacobi equation takes the form

$$\frac{1}{2m}\left(\frac{\partial S}{\partial q}\right)^2 + \frac{\partial S}{\partial t} = 0. \tag{9.18}$$

A complete integral of this equation can be obtained by separation of variables in the form of a sum:

$$S = W(q) + T(t). \tag{9.19}$$

Inserting S of this form into (9.18) yields

$$\frac{1}{2m}\left(\frac{dW}{dq}\right)^2 = -\frac{dT}{dt}. \tag{9.20}$$

If we fix t and vary q the left-hand side of (9.20) should vary. However, it cannot vary because it is equal to the right-hand side, which remains fixed. Therefore, both sides of (9.20) are equal to the same positive constant we denote by α. Thus, we are left with the two ordinary differential equations

$$\frac{dW}{dq} = \sqrt{2m\alpha}, \qquad -\frac{dT}{dt} = \alpha, \tag{9.21}$$

whence

$$S(q, \alpha, t) = \sqrt{2m\alpha}\, q - \alpha t, \tag{9.22}$$

where we have abandoned merely additive constants of integration. The solution of the equation of motion for q is obtained from

$$\beta = \frac{\partial S}{\partial \alpha} = \sqrt{\frac{m}{2\alpha}}\, q - t, \tag{9.23}$$

and from

$$p = \frac{\partial S}{\partial q} = \sqrt{2m\alpha} = \text{constant} = p_0. \tag{9.24}$$

Therefore,

$$q = \sqrt{\frac{2\alpha}{m}}\, \beta + \sqrt{\frac{2\alpha}{m}}\, t \equiv q_0 + \frac{p_0}{m}\, t, \tag{9.25}$$

which is the elementary solution to the free particle problem.

As usual, the harmonic oscillator serves as a testing ground for any sophisticated method for integrating the equations of motion.

Example 9.2 Solve the equation of motion for the one-dimensional harmonic oscillator by the Hamilton-Jacobi method.

Solution

Now $H = p^2/2m + m\omega^2 q^2/2$ and the Hamilton-Jacobi equation is written

$$\frac{1}{2m}\left(\frac{\partial S}{\partial q}\right)^2 + \frac{m\omega^2}{2}q^2 + \frac{\partial S}{\partial t} = 0. \tag{9.26}$$

As in the free particle case, Eq. (9.26) is separable in the form

$$S = W(q) - \alpha t \tag{9.27}$$

where W obeys

$$\frac{1}{2m}\left(\frac{dW}{dq}\right)^2 + \frac{m\omega^2}{2}q^2 = \alpha. \tag{9.28}$$

The positive constant α coincides with the constant value of the Hamiltonian (equal to the total energy, in the present case) inasmuch as $dW/dq = \partial S/\partial q = p$. A solution to (9.28) is

$$W = \int \sqrt{2m\alpha - m^2\omega^2 q^2}\, dq. \tag{9.29}$$

As a consequence,

$$S(q,\alpha,t) = \int \sqrt{2m\alpha - m^2\omega^2 q^2}\, dq - \alpha t, \tag{9.30}$$

and the solution of the equation of motion for q is obtained from

$$\beta = \frac{\partial S}{\partial \alpha} = m\int \frac{dq}{\sqrt{2m\alpha - m^2\omega^2 q^2}} - t = \frac{1}{\omega}\sin^{-1}\left(\sqrt{\frac{m\omega^2}{2\alpha}}q\right) - t. \tag{9.31}$$

Solving this equation for q we find, with $\delta = \omega\beta$,

$$q(t) = \sqrt{\frac{2\alpha}{m\omega^2}}\sin(\omega t + \delta), \tag{9.32}$$

which is the well-known general solution to the oscillator equation of motion.

Our last one-dimensional example reveals that the Hamilton-Jacobi method may be applicable to certain non-conservative systems.

Example 9.3 (Lemos, 1979) Use the Hamilton-Jacobi theory to solve the equation of motion

$$\ddot{q} + \lambda\dot{q} + \omega^2 q = 0 \tag{9.33}$$

that governs the damped harmonic oscillator.

Solution

Equation (9.33) is generated by the Lagrangian (see Problem 1.16)

$$L = e^{\lambda t}\left(\frac{m\dot{q}^2}{2} - \frac{m\omega^2}{2}q^2\right),\tag{9.34}$$

whose corresponding Hamiltonian is

$$H = e^{-\lambda t}\frac{p^2}{2m} + \frac{m\omega^2}{2}q^2 e^{\lambda t}.\tag{9.35}$$

The Hamilton-Jacobi equation associated with this Hamiltonian is

$$\frac{e^{-\lambda t}}{2m}\left(\frac{\partial S}{\partial q}\right)^2 + \frac{m\omega^2}{2}q^2 e^{\lambda t} + \frac{\partial S}{\partial t} = 0.\tag{9.36}$$

Owing to the explicit time dependence of H, this last equation does not admit separation of the time variable in the form of the two previous examples (check it). However, the form of H suggests that the transformation

$$Q = qe^{\lambda t/2}, \qquad P = pe^{-\lambda t/2}\tag{9.37}$$

is likely to simplify the Hamiltonian. It turns out that this transformation is canonical with a generating function $F_2(q,P,t) = e^{\lambda t/2}qP$, and the transformed Hamiltonian

$$K(Q,P,t) = H(q,p,t) + \frac{\partial F_2}{\partial t} = \frac{P^2}{2m} + \frac{m\omega^2}{2}Q^2 + \frac{\lambda}{2}QP\tag{9.38}$$

does not explicitly depend on time. Thus, the same technique of the two previous examples is applicable to the system described by the new canonical variables (Q,P) and the transformed Hamiltonian K. The Hamilton-Jacobi equation associated with K is written

$$\frac{1}{2m}\left(\frac{\partial S}{\partial Q}\right)^2 + \frac{m\omega^2}{2}Q^2 + \frac{\lambda}{2}Q\frac{\partial S}{\partial Q} + \frac{\partial S}{\partial t} = 0,\tag{9.39}$$

which has a solution of the form

$$S = W(Q) - \alpha t,\tag{9.40}$$

where

$$\frac{1}{2m}\left(\frac{dW}{dQ}\right)^2 + \frac{m\omega^2}{2}Q^2 + \frac{\lambda}{2}Q\frac{dW}{dQ} = \alpha.\tag{9.41}$$

Note that α is equal to the constant value of the Hamiltonian K. Solving this second-degree algebraic equation for dW/dQ we obtain

$$\frac{dW}{dx} = -ax \pm \left[b^2 - (1-a^2)x^2\right]^{1/2}\tag{9.42}$$

with $x = (m\omega)^{1/2}Q$, $a = \lambda/2\omega$, $b = (2\alpha/\omega)^{1/2}$. As far as solving (9.33) is concerned, the choice of sign in (9.42) is irrelevant – any complete integral will do. So we take

$$W = -\frac{ax^2}{2} + \int \left[b^2 - (1-a^2)x^2\right]^{1/2}dx.\tag{9.43}$$

Let us consider the case $a < 1$ – that is, $\lambda < 2\omega$. Defining $\gamma = (1 - a^2)^{1/2}$ there results

$$S = -\alpha t - \frac{ax^2}{2} + \int \left(\frac{2\alpha}{\omega} - \gamma^2 x^2 \right)^{1/2} dx, \qquad (9.44)$$

whence

$$\beta = \frac{\partial S}{\partial \alpha} = -t + \frac{1}{\omega} \int \frac{dx}{(b^2 - \gamma^2 x^2)^{1/2}} = -t + \frac{1}{\omega\gamma} \sin^{-1} \left(\frac{\gamma x}{b} \right). \qquad (9.45)$$

Solving for x, returning to the variable Q and, finally, to the variable q, we find

$$q(t) = A e^{-\lambda t/2} \sin(\Omega t + \delta), \qquad \Omega = (\omega^2 - \lambda^2/4)^{1/2}, \qquad (9.46)$$

where A and δ are constants determined by the initial conditions. Equation (9.46) is the usual solution for the damped oscillator in the weak damping case.

Exercise 9.2.1   Complete Example 9.3 by treating the cases $\lambda = 2\omega$ (critical damping) and $\lambda > 2\omega$ (strong damping).

9.3 Separation of Variables

A simplifying aspect of the Hamilton-Jacobi method is the needlessness of discovering the general solution of the Hamilton-Jacobi equation in order to solve the dynamical problem: it suffices to find a complete integral, which is only a particular solution. Particular solutions to partial differential equations can usually be obtained by separation of variables. The nature of the Hamilton-Jacobi equation makes it possible, in many cases, to effect the separation of variables in the form of a sum. A few remarks of a general character, however simple, facilitate considerably the task of separating variables in most situations of relevance.

1. Cyclic coordinates. If a certain variable, say q_n, is a cyclic coordinate of H, it is at once possible to separate it from the others. In fact, suppose H does not depend on q_n. Then we can write

$$S = \alpha_n q_n + \bar{S}(q_1, \ldots, q_{n-1}, t), \qquad (9.47)$$

where, in virtue of (9.4), $\bar{S}$ satisfies

$$H\left(q_1, \ldots, q_{n-1}, \frac{\partial \bar{S}}{\partial q_1}, \ldots, \frac{\partial \bar{S}}{\partial q_{n-1}}, \alpha_n, t \right) + \frac{\partial \bar{S}}{\partial t} = 0. \qquad (9.48)$$

This equation involves only the variables $q_1, \ldots, q_{n-1}, t$. The motivation for searching a solution to the Hamilton-Jacobi equation in the form (9.47) comes from

$$\frac{\partial S}{\partial q_n} = p_n = \alpha_n, \qquad (9.49)$$

since p_n is a constant of the motion whenever q_n is a cyclic coordinate. The solution of this last differential equation for S is precisely (9.47). An additional justification is that we are searching for a canonical transformation that makes all of the Ps constants of the motion. But p_n is already a constant of the motion, so as far as the pair (q_n, p_n) is concerned it suffices to perform the identity transformation, whose generating function is $F_2 = q_n P_n \equiv q_n \alpha_n$ owing to the identification $\alpha_i \equiv P_i$. If there are more cyclic coordinates, the first term on the right-hand side of (9.47) becomes a sum over all cyclic coordinates, whereas $\bar{S}$ involves the non-cyclic coordinates alone. For example, if q_n and q_{n-1} are cyclic we have

$$S = \alpha_n q_n + \alpha_{n-1} q_{n-1} + \bar{S}(q_1, \ldots, q_{n-2}, t), \tag{9.50}$$

where

$$H\left(q_1, \ldots, q_{n-2}, \frac{\partial \bar{S}}{\partial q_1}, \ldots, \frac{\partial \bar{S}}{\partial q_{n-2}}, \alpha_{n-1}, \alpha_n, t\right) + \frac{\partial \bar{S}}{\partial t} = 0. \tag{9.51}$$

2. Time variable. As the examples in Section 9.2 indicate, if H does not explicitly depend on time the Hamilton-Jacobi equation

$$H\left(q_1, \ldots, q_n, \frac{\partial S}{\partial q_1}, \ldots, \frac{\partial S}{\partial q_n}\right) + \frac{\partial S}{\partial t} = 0 \tag{9.52}$$

admits separation of the variable t in the form

$$S = W(q_1, \ldots, q_n) - \alpha_1 t, \tag{9.53}$$

where W – called **Hamilton's characteristic function** (Hamilton, 1834) – satisfies

$$H\left(q_1, \ldots, q_n, \frac{\partial W}{\partial q_1}, \ldots, \frac{\partial W}{\partial q_n}\right) = \alpha_1. \tag{9.54}$$

This equation for W does not contain the time and α_1 is equal to the constant value of the Hamiltonian. A complete integral $S(q, \alpha, t)$ of (9.52) is determined as soon as one finds a complete integral $W(q, \alpha)$ of the **time-independent Hamilton-Jacobi equation** (9.54). The function $W(q, \alpha)$ by itself already gives rise to a canonical transformation that completely solves the equations of motion for the qs.

Exercise 9.3.1   Suppose H does not explicitly depend on time and let $W(q, \alpha)$ be a complete integral of (9.54). Taking W as the generating function of a canonical transformation with $\alpha_i \equiv P_i$, show that the transformed Hamilton's equations are

$$\dot{P}_i = 0 \quad \Longrightarrow \quad P_i = \alpha_i, \quad i = 1, \ldots, n, \tag{9.55a}$$

$$\dot{Q}_1 = 1 \quad \Longrightarrow \quad Q_1 = t + \beta_1, \tag{9.55b}$$

$$\dot{Q}_i = 0 \quad \Longrightarrow \quad Q_i = \beta_i, \quad i = 2, \ldots, n. \tag{9.55c}$$

In other words, conclude that by means of the equations

$$\frac{\partial W}{\partial \alpha_1} = t + \beta_1, \quad \frac{\partial W}{\partial \alpha_i} = \beta_i, \quad i = 2, \ldots, n, \tag{9.56}$$

the $q_i(\alpha, \beta, t)$ are determined which satisfy Hamilton's equations and contain $2n$ constants of integration whose values are fixed by the initial conditions.

The separability of the Hamilton-Jacobi equation depends on the physical problem and on the chosen set of generalised coordinates. The systems for which the Hamilton-Jacobi equation is separable in some coordinate system belong to the class of *integrable systems* (see Section 9.7). The problem of three bodies under mutual gravitational attraction is one of the most famous examples of a non-integrable system. No completely general criterion is known to decide whether there exists some coordinate system in which the Hamilton-Jacobi equation is soluble by separation of variables. For orthogonal generalised coordinates a theorem due to Stäckel establishes necessary and sufficient conditions for separability (Pars, 1965). Here, however, we will circumscribe ourselves to treating a few examples of physical interest that illustrate the techniques for separation of variables customarily employed.

Example 9.4   Find a complete integral of the Hamilton-Jacobi equation for a particle in the potential

$$V(r,\theta) = a(r) + \frac{b(\theta)}{r^2} \tag{9.57}$$

where r, θ, ϕ are spherical coordinates while $a(r)$ and $b(\theta)$ are known functions.

Solution

The Hamiltonian for this problem is given by (7.15) with $V(r)$ replaced by $V(r,\theta)$ defined by (9.57):

$$H = \frac{1}{2m}\left(p_r^2 + \frac{p_\theta^2}{r^2} + \frac{p_\phi^2}{r^2\sin^2\theta}\right) + a(r) + \frac{b(\theta)}{r^2}. \tag{9.58}$$

The Hamilton-Jacobi equation takes the form

$$\frac{1}{2m}\left[\left(\frac{\partial S}{\partial r}\right)^2 + \frac{1}{r^2}\left(\frac{\partial S}{\partial \theta}\right)^2 + \frac{1}{r^2\sin^2\theta}\left(\frac{\partial S}{\partial \phi}\right)^2\right] + a(r) + \frac{b(\theta)}{r^2} + \frac{\partial S}{\partial t} = 0. \tag{9.59}$$

Since H does not explicitly depend on time and ϕ is a cyclic coordinate, we can write

$$S = -\alpha_1 t + \alpha_\phi \phi + W(r,\theta), \tag{9.60}$$

where W satisfies

$$\frac{1}{2m}\left[\left(\frac{\partial W}{\partial r}\right)^2 + \frac{1}{r^2}\left(\frac{\partial W}{\partial \theta}\right)^2 + \frac{\alpha_\phi^2}{r^2\sin^2\theta}\right] + a(r) + \frac{b(\theta)}{r^2} = \alpha_1. \tag{9.61}$$

By trying a solution to this equation of the form

$$W(r,\theta) = W_1(r) + W_2(\theta), \tag{9.62}$$

we are led to

$$\frac{1}{2m}\left[\left(\frac{dW_1}{dr}\right)^2 + \frac{1}{r^2}\left(\frac{dW_2}{d\theta}\right)^2 + \frac{\alpha_\phi^2}{r^2\sin^2\theta}\right] + a(r) + \frac{b(\theta)}{r^2} = \alpha_1. \tag{9.63}$$

Multiplying the above equation by r^2 we arrive at

$$r^2 \left[\frac{1}{2m} \left(\frac{dW_1}{dr} \right)^2 + a(r) - \alpha_1 \right] = - \left[\frac{1}{2m} \left(\frac{dW_2}{d\theta} \right)^2 + \frac{\alpha_\phi^2}{2m \sin^2 \theta} + b(\theta) \right] = -\frac{\alpha_\theta^2}{2m},$$

(9.64)

with the variables r and θ separated and the separation constant conveniently denoted by $-\alpha_\theta^2/2m$. Equation (9.64) amounts to the two ordinary differential equations

$$\frac{dW_1}{dr} = \sqrt{2m \left[\alpha_1 - a(r) \right] - \frac{\alpha_\theta^2}{r^2}},$$

(9.65a)

$$\frac{dW_2}{d\theta} = \sqrt{\alpha_\theta^2 - 2mb(\theta) - \frac{\alpha_\phi^2}{\sin^2 \theta}}.$$

(9.65b)

Integrating these equations and inserting the results into (9.62) and (9.60) there results a complete integral of (9.59) in the form

$$S = -\alpha_1 t + \alpha_\phi \phi + \int \left[2m \left[\alpha_1 - a(r) \right] - \frac{\alpha_\theta^2}{r^2} \right]^{1/2} dr + \int \left[\alpha_\theta^2 - 2mb(\theta) - \frac{\alpha_\phi^2}{\sin^2 \theta} \right]^{1/2} d\theta.$$

(9.66)

The solution of the equations of motion is given by

$$\beta_1 = \frac{\partial S}{\partial \alpha_1} = -t + m \int \frac{dr}{\left[2m \left[\alpha_1 - a(r) \right] - \alpha_\theta^2/r^2 \right]^{1/2}},$$

(9.67)

$$\beta_\theta = \frac{\partial S}{\partial \alpha_\theta} = - \int \frac{\alpha_\theta dr}{r^2 \left[2m \left[\alpha_1 - a(r) \right] - \alpha_\theta^2/r^2 \right]^{1/2}}$$

$$+ \int \frac{\alpha_\theta d\theta}{\left[\alpha_\theta^2 - 2mb(\theta) - \alpha_\phi^2/\sin^2 \theta \right]^{1/2}},$$

(9.68)

$$\beta_\phi = \frac{\partial S}{\partial \alpha_\phi} = \phi - \int \frac{\alpha_\phi d\theta}{\sin^2 \theta \left[\alpha_\theta^2 - 2mb(\theta) - \alpha_\phi^2/\sin^2 \theta \right]^{1/2}}.$$

(9.69)

It is hardly possible not to wonder at the impressive swiftness and simplicity with which the solution of this problem is reduced to quadratures by the Hamilton-Jacobi method. Equation (9.67) furnishes $r(t)$ which, after substitution into (9.68), determines $\theta(t)$. Having found $\theta(t)$, by means of (9.69) we obtain $\phi(t)$, completing the resolution of the equations of motion. If one is only interested in the shape of the geometric trajectory described by the particle, it suffices to use the two last equations to express the equation of the curve in the form $r = r(\theta)$, $\phi = \phi(\theta)$ or $\theta = \theta(\phi)$, $r = r(\phi)$.

Although central potentials are the most important for physical applications, the potential $V(r, \theta) = A/r + B/r^2 \sin^2 \theta$, A and B constants, has been proposed (Hartmann, 1972) to describe the motion of an electron in ring-shaped molecules such as the benzene molecule. Hartmann's potential is exactly of the form assumed in the example we have just considered. The motion of a charged particle in the field of a fixed electric dipole, to be investigated in the next section, also involves a potential of the form (9.57).

Example 9.5   Solve the equations of motion for a projectile in space by the Hamilton-Jacobi method.

Solution

With the z-axis oriented vertically upwards,

$$H = \frac{1}{2m}\left(p_x^2 + p_y^2 + p_z^2\right) + mgz,$$ (9.70)

and the Hamilton-Jacobi equation is

$$\frac{1}{2m}\left[\left(\frac{\partial S}{\partial x}\right)^2 + \left(\frac{\partial S}{\partial y}\right)^2 + \left(\frac{\partial S}{\partial z}\right)^2\right] + mgz + \frac{\partial S}{\partial t} = 0.$$ (9.71)

Since x and y are cyclic coordinates and H does not explicitly depend on time,

$$S = -\alpha_1 t + \alpha_x x + \alpha_y y + W(z),$$ (9.72)

whence

$$\left(\frac{dW}{dz}\right)^2 = 2m(\alpha_1 - mgz) - \alpha_x^2 - \alpha_y^2.$$ (9.73)

Integrating this equation we find

$$S = -\alpha_1 t + \alpha_x x + \alpha_y y - \frac{1}{3m^2 g}\left[2m(\alpha_1 - mgz) - \alpha_x^2 - \alpha_y^2\right]^{3/2}.$$ (9.74)

The motion of the particle is determined by means of

$$\beta_1 = \frac{\partial S}{\partial \alpha_1} = -t - \frac{1}{mg}\left[2m(\alpha_1 - mgz) - \alpha_x^2 - \alpha_y^2\right]^{1/2},$$ (9.75a)

$$\beta_2 = \frac{\partial S}{\partial \alpha_x} = x + \frac{\alpha_x}{m^2 g}\left[2m(\alpha_1 - mgz) - \alpha_x^2 - \alpha_y^2\right]^{1/2},$$ (9.75b)

$$\beta_3 = \frac{\partial S}{\partial \alpha_y} = y + \frac{\alpha_y}{m^2 g}\left[2m(\alpha_1 - mgz) - \alpha_x^2 - \alpha_y^2\right]^{1/2}.$$ (9.75c)

The resolution of these three last equations for x, y, z yields

$$x = A + \frac{\alpha_x}{m}t,$$ (9.76a)

$$y = B + \frac{\alpha_y}{m}t,$$ (9.76b)

$$z = C + Dt - \frac{gt^2}{2},$$ (9.76c)

where the constants A, B, C, D are given in terms of the αs and βs. Equations (9.76) coincide with the usual solution of this problem by elementary means.

Separation of Variables: Possible Incompleteness of the Theory

The separation of variables technique provides solutions to the Hamilton-Jacobi equation of the form $S(q, \alpha, t)$. According to the implicit function theorem, condition (9.8) ensures that the n equations (9.5) can be solved for the coordinates as $q_i(t) = f_i(\alpha, \beta, t)$. Insertion of these functions into the right-hand side of (9.3) gives $p_i(t) = g_i(\alpha, \beta, t)$. Thus, the coordinates and momenta are found as functions of time and $2n$ arbitrary constants, namely the n alphas and n betas. Since the initial conditions are also $2n$ in number, to wit the values of the n coordinates and n momenta at some initial time $t = t_0$, the alphas and betas can always be so chosen as to accomodate all possible initial conditions, so that the Hamilton-Jacobi theory yields *all* solutions of Hamilton's equations. But, in order for this to be the case, it must be emphasised that condition (9.8) is crucial, as we proceed to explain further.

With $t = t_0$ Eq. (9.3) becomes

$$p_{0i} = \frac{\partial S}{\partial q_i}(q_0, \alpha, t_0),\tag{9.77}$$

where (q_0, p_0) are the initial values of (q, p). Condition (9.8) guarantees that these equations can be solved for $\alpha_1, \ldots, \alpha_n$ in terms of q_0, p_0, t_0. Insertion of the alphas so obtained into the right-hand side of (9.5) determines $\beta_1, \ldots, \beta_n$ in terms of q_0, p_0, t_0. Therefore, each set of initial conditions corresponds to a unique set of alphas and betas. Nonetheless, if condition (9.8) fails for certain values of $\alpha_1, \ldots, \alpha_n, q_{01} \ldots, q_{0n}$, some initial conditions may be inaccessible whatever the choice of alphas and betas, which means that some solutions to the equations of motion are missed. Under such circumstances the theory is incomplete (Lemos, 2014c). It is worth stressing that condition (9.8) may fail because either $\det(\partial^2 S/\partial \alpha_i \partial q_j)$ is zero or does not exist for some values of $\alpha_1, \ldots, \alpha_n, q_1, \ldots, q_n$.

There is no guarantee that the solution $S(q, \alpha, t)$ obtained by separation of variables will automatically satisfy (9.8) for all values of α, q. In most problems of physical interest in which (9.8) is not satisfied everywhere, the missed solutions are trivial and immediately identifiable by inspection – but there are exceptions. The problem of the motion of a charged particle in the presence of a fixed electric dipole is an anomalous case in which condition (9.8) is not satisfied everywhere and the solution $S(q, \alpha, t)$ obtained by separation of variables in spherical coordinates is unable to generate some non-trivial solutions of Hamilton's equations.

9.4 Incompleteness of the Theory: Point Charge in Dipole Field

Let us investigate the motion of a charged particle in the electric field created by a point electric dipole fixed at the origin. Since Lagrange's and Hamilton's equations are equivalent, we first study the equations of motion from the Lagrangian point of view. Next, we take up the problem of solving the equations of motion by the Hamilton-Jacobi method (Lemos, 2014c).

Lagrangian Treatment

The electrostatic potential energy of a particle with electric charge q in the field of a point electric dipole of magnitude p_0 fixed at the origin and oriented in the z-direction is $V(r,\theta) = k\cos\theta/r^2$, where r,θ,ϕ are spherical coordinates and $k = qp_0/4\pi\epsilon_0$ in SI units. Therefore, the particle's motion is described by the Lagrangian

$$L = T - V = \frac{m}{2}\left(\dot{r}^2 + r^2\dot{\theta}^2 + r^2\sin^2\theta\,\dot{\phi}^2\right) - \frac{k\cos\theta}{r^2}\,. \tag{9.78}$$

Lagrange's equations are

$$m\ddot{r} - mr\dot{\theta}^2 - mr\sin^2\theta\,\dot{\phi}^2 - \frac{2k\cos\theta}{r^3} = 0\,, \tag{9.79a}$$

$$mr^2\ddot{\theta} + 2mr\dot{r}\dot{\theta} - mr^2\sin\theta\cos\theta\,\dot{\phi}^2 - \frac{k\sin\theta}{r^2} = 0\,, \tag{9.79b}$$

$$\frac{d}{dt}\left(mr^2\sin^2\theta\,\dot{\phi}\right) = 0 \implies mr^2\sin^2\theta\,\dot{\phi} = \ell = \text{constant}\,, \tag{9.79c}$$

where ℓ denotes the constant value of L_z, the z-component of the particle's angular momentum with respect to the origin.

If $\ell = L_z = 0$ there is a remarkable solution (Jones, 1995) to these equations of motion, namely $\phi = 0$, $r = r_0$ with the angle θ satisfying

$$mr_0^2\ddot{\theta} - \frac{k\sin\theta}{r_0^2} = 0 \tag{9.80}$$

and

$$mr_0\dot{\theta}^2 = -\frac{2k\cos\theta}{r_0^3}\,. \tag{9.81}$$

For $\phi = 0$ the range of values of θ must be redefined to $[0, 2\pi)$ in order that the coordinates (r,θ) cover the entire xz-plane. Assuming $k > 0$, Eq. (9.81) requires $\frac{\pi}{2} \leq \theta \leq \frac{3\pi}{2}$. Putting $\psi = \theta - \pi$ we have $-\frac{\pi}{2} \leq \psi \leq \frac{\pi}{2}$ and Eq. (9.80) becomes

$$\ddot{\psi} + \frac{k}{mr_0^4}\sin\psi = 0\,, \tag{9.82}$$

which is the equation of a pendulum. From Eq. (9.81) it follows that the turning points are $\theta = \pi/2$ and $\theta = 3\pi/2$, which correspond to $\psi = -\pi/2$ and $\psi = \pi/2$, respectively. The particle moves in a semicircular path on the half-plane $z \leq 0$ exactly like a pendulum with length r_0 and amplitude $\pi/2$, oscillating periodically between the point with coordinate r_0 and the point with coordinate $-r_0$, both on the x-axis (Fig. 9.1). From (9.81) it also follows that this solution has zero energy:

$$E = T + V = \frac{mr_0^2}{2}\dot{\theta}^2 + \frac{k\cos\theta}{r_0^2} = -\frac{k\cos\theta}{r_0^2} + \frac{k\cos\theta}{r_0^2} = 0\,. \tag{9.83}$$

Other suprising solutions to the equations of motion (9.79) with $r = r_0$ but $L_z \neq 0$, such that the charged particle moves on the surface of a sphere, are exhibited in Gutiérrez-López et al. (2008).

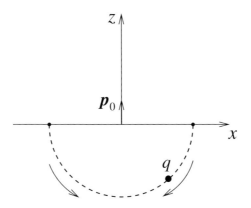

Fig. 9.1   Pendular motion of a charged particle in the presence of an electric dipole.

Hamilton-Jacobi Approach

The Hamiltonian associated with the Lagrangian (9.78) is

$$H = \frac{1}{2m}\left(p_r^2 + \frac{p_\theta^2}{r^2} + \frac{p_\phi^2}{r^2\sin^2\theta}\right) + \frac{k\cos\theta}{r^2}, \tag{9.84}$$

which has the form (9.58) with $a(r) = 0$ and $b(\theta) = k\cos\theta$. According to Example 9.4, the Hamilton-Jacobi equation can be solved by separation of variables in the form

$$S = -Et + \alpha_\phi\phi + W_1(r) + W_2(\theta), \tag{9.85}$$

where the separation constant E represents the total energy (for convenience, we write E instead of α_1) and the second separation constant α_ϕ is the z-component of the angular momentum. By Eq. (9.64) the functions W_1 and W_2 satisfy

$$r^2\left[\frac{1}{2m}\left(\frac{dW_1}{dr}\right)^2 - E\right] = -\left[\frac{1}{2m}\left(\frac{dW_2}{d\theta}\right)^2 + \frac{\alpha_\phi^2}{2m\sin^2\theta} + k\cos\theta\right] = -\frac{\alpha_\theta}{2m}, \tag{9.86}$$

where we prefer to write the third separation constant as $\alpha_\theta/2m$ (instead of $\alpha_\theta^2/2m$). Thus, we obtain a complete integral $S(r,\theta,\phi,E,\alpha_\theta,\alpha_\phi,t)$ given by

$$S = -Et + \alpha_\phi\phi + \int\left[2mE - \frac{\alpha_\theta}{r^2}\right]^{1/2}dr + \int\left[\alpha_\theta - 2mk\cos\theta - \frac{\alpha_\phi^2}{\sin^2\theta}\right]^{1/2}d\theta. \tag{9.87}$$

This complete integral solves the equations of motion by quadratures:

$$\beta_1 = \frac{\partial S}{\partial E} = -t + \int\frac{m\,dr}{\left[2mE - \alpha_\theta/r^2\right]^{1/2}}; \tag{9.88}$$

$$\beta_2 = \frac{\partial S}{\partial \alpha_\theta} = -\frac{1}{2} \int \frac{dr}{r^2 \left[2mE - \alpha_\theta/r^2\right]^{1/2}}$$
$$+ \frac{1}{2} \int \frac{d\theta}{\left[\alpha_\theta - 2mk\cos\theta - \alpha_\phi^2/\sin^2\theta\right]^{1/2}} ; \tag{9.89}$$

$$\beta_3 = \frac{\partial S}{\partial \alpha_\phi} = \phi - \int \frac{\alpha_\phi d\theta}{\sin^2\theta \left[\alpha_\theta - 2mk\cos\theta - \alpha_\phi^2/\sin^2\theta\right]^{1/2}} . \tag{9.90}$$

Apparently, Eq. (9.88) allows us to completely determine the radial motion $r(t)$. There are three cases to consider.

Case (1): $E > 0$. In this case we can write (9.88) in the form

$$\beta_1 = -t + \sqrt{\frac{m}{2E}} \int \frac{rdr}{\sqrt{r^2 - a}} = \sqrt{\frac{m}{2E}} \sqrt{r^2 - a}, \tag{9.91}$$

whence

$$r = \left[a + \frac{2E}{m}(t + \beta_1)^2\right]^{1/2}, \qquad a = \frac{\alpha_\theta}{2mE}. \tag{9.92}$$

Case (2): $E = 0$. This requires $\alpha_\theta < 0$ and Eq. (9.88) leads to

$$\beta_1 = -t + \frac{m}{|\alpha_\theta|^{1/2}} \int rdr = -t + \frac{m}{|\alpha_\theta|^{1/2}} \frac{r^2}{2}, \tag{9.93}$$

which implies

$$r = \sqrt{\frac{2|\alpha_\theta|^{1/2}}{m}}(t + \beta_1)^{1/2} . \tag{9.94}$$

Case (3): $E < 0$. This also requires $\alpha_\theta < 0$. As a consequence, Eq. (9.88) becomes

$$\beta_1 = -t + \sqrt{\frac{m}{2|E|}} \int \frac{rdr}{\sqrt{a - r^2}} = -t - \sqrt{\frac{m}{2|E|}} \sqrt{a - r^2}, \tag{9.95}$$

and it follows that

$$r = \left[a - \frac{2|E|}{m}(t + \beta_1)^2\right]^{1/2}, \qquad a = \frac{|\alpha_\theta|}{2m|E|}. \tag{9.96}$$

Note that in no case does one find a constant r. If attacked by the Hamilton-Jacobi method alone, the problem of the motion of a charged particle in an electric dipole field does not reveal the beautiful solution found by Jones (1995).

The trouble arises from the circumstance that the pendulum-like motion takes place with $E = 0$, $\alpha_\phi = 0$ and, because of Eq. (9.86), $\alpha_\theta = 0$, since $dW_1/dr = p_r = m\dot{r} = 0$ for $r = r_0$. For these values of the separation constants Eq. (9.90) gives the correct result $\phi = \beta_3 = $ constant, but Eqs. (9.88) and (9.89) are not defined. This, in turn, results from the failure of condition (9.8).

Indeed, setting $q_1 = r$, $q_2 = \theta$, $q_3 = \phi$, $\alpha_1 = E$, $\alpha_2 = \alpha_\theta$, $\alpha_3 = \alpha_\phi$, it follows immediately from Eqs. (9.88)–(9.90) that

$$\left(\frac{\partial^2 S}{\partial \alpha_i \partial q_j} \right) = \begin{pmatrix} \dfrac{m}{\left(2mE - \alpha_\theta/r^2\right)^{1/2}} & 0 & 0 \\[2ex] -\dfrac{1}{2r^2 \left(2mE - \alpha_\theta/r^2\right)^{1/2}} & \dfrac{1}{2\left(\alpha_\theta - 2mk\cos\theta - \alpha_\phi^2/\sin^2\theta\right)^{1/2}} & 0 \\[2ex] 0 & -\dfrac{\alpha_\phi}{\sin^2\theta\left(\alpha_\theta - 2mk\cos\theta - \alpha_\phi^2/\sin^2\theta\right)^{1/2}} & 1 \end{pmatrix}, \quad (9.97)$$

from which one finds

$$\det\left(\frac{\partial^2 S}{\partial \alpha_i \partial q_j} \right) = \frac{m}{2} \left[2mE - \alpha_\theta/r^2\right]^{-1/2} \left[\alpha_\theta - 2mk\cos\theta - \alpha_\phi^2/\sin^2\theta\right]^{-1/2}. \quad (9.98)$$

This determinant is not defined (it is formally infinite) if $E = 0$ and $\alpha_\theta = 0$, whatever the values of $\alpha_\phi, r, \theta, \phi$. The pendulum-like motion corresponds to a situation in which the particular solution to the Hamilton-Jacobi equation obtained by separation of variables fails to be a complete integral.

Although, strictly speaking, the pendulum-like motion with $r = r_0$ and $\phi = 0$ cannot be obtained directly from Eqs. (9.88)–(9.90), it is possible to recover it by a special limiting procedure (Lemos, 2014c).

From this singular example a lesson is learnt: once a solution of the Hamilton-Jacobi equation is found by separation of variables, one must pay close attention to the portion of the space of the variables $q_1, \ldots, q_n, \alpha_1, \ldots, \alpha_n$ in which condition (9.8) fails to be sure that no non-trivial solution of the equations of motion has been missed.

9.5 Action as a Function of the Coordinates

As noted in Section 8.5, the action acts as the generating function of the canonical transformation that brings the state of the system at time t to its state at a past fixed time t_0. This means that the transformed canonical variables are constants equal to the values of the qs and ps at time t_0. Thus, both the action and a complete integral of the Hamilton-Jacobi equation perform a canonical transformation with the same properties. It is to be expected, therefore, that the action and a complete integral S should differ at most by a constant – this explains the use of the same notation for both quantities.

Taking the total time derivative of $S(q, \alpha, t)$, we find

$$\frac{dS}{dt} = \sum_i \frac{\partial S}{\partial q_i}\dot{q}_i + \frac{\partial S}{\partial t} = \sum_i p_i \dot{q}_i - H = L, \quad (9.99)$$

where we have used (9.3) and (9.4). Therefore,

$$S = \int L\, dt, \quad (9.100)$$

showing that, seen as a function of time, S and the action differ only by a constant. It was Hamilton who discovered that the action, considered as a function of the coordinates, satisfies (9.4), which for this reason is also known as **Hamilton's partial differential equation**.

Hamilton's principal function, name customarily reserved for the action expressed in terms of the coordinates, is constructed by the procedure that follows. Let $q(\tau)$ be the unique segment of physical path connecting an initial configuration $q_0 = (q_{01}, \ldots, q_{0n})$ at $\tau = t_0$ to the configuration $q = (q_1, \ldots, q_n)$ at $\tau = t$. The corresponding action can be expressed in terms of q and t with q_0, t_0 as fixed parameters:

$$S(q_0, t_0; q, t) = \int_{t_0}^{t} L\left(q(\tau), \frac{dq(\tau)}{d\tau}, \tau\right) d\tau. \tag{9.101}$$

The requirement that there exists only one physical path segment linking (q_0, t_0) to (q, t) ensures that the function S is well defined by Eq. (9.101).

Theorem 9.5.1 *Hamilton's principal function $S(q_0, t_0; q, t)$, defined by (9.101), satisfies equation (9.4).*

Proof Let us first see how $S(q_0, t_0; q, t)$ varies with q for fixed t. If the final configuration at time t is varied from q to $q + \delta q$, the physical path will be varied, becoming $\bar{q}(\tau)$, as shown in Fig. 9.2. Consequently, according to Eq. (2.45) with $\dot{q}_i$ denoting $dq_i(\tau)/d\tau$,

$$\delta S = \int_{t_0}^{t} d\tau \sum_i \left[\frac{\partial L}{\partial q_i} - \frac{d}{d\tau}\left(\frac{\partial L}{\partial \dot{q}_i}\right)\right]\delta q_i(\tau) + \sum_i \frac{\partial L}{\partial \dot{q}_i}\delta q_i(\tau)\Big|_{\tau=t_0}^{\tau=t} = \sum_i p_i \delta q_i, \tag{9.102}$$

inasmuch as $\delta q_i(t_0) = 0$ and the physical path obeys Lagrange's equations. From (9.102) one derives

$$p_i = \frac{\partial S}{\partial q_i}. \tag{9.103}$$

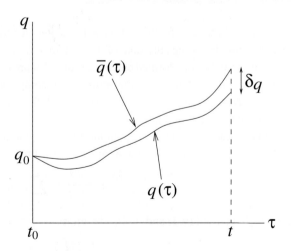

Fig. 9.2 Variation involved in the proof of Theorem 9.5.1.

Let us now study how S varies as a function of t. On the one hand, it is clear from (9.101) that

$$\frac{dS}{dt} = L.$$ (9.104)

On the other hand,

$$\frac{dS}{dt} = \sum_i \frac{\partial S}{\partial q_i} \dot{q}_i + \frac{\partial S}{\partial t} = \sum_i p_i \dot{q}_i + \frac{\partial S}{\partial t},$$ (9.105)

whence

$$\frac{\partial S}{\partial t} = L - \sum_i p_i \dot{q}_i = -H.$$ (9.106)

Combining this with (9.103), one concludes that $S(q_0, t_0; q, t)$ satisfies (9.4). □

Example 9.6 Construct Hamilton's principal function for a particle subject to a constant force F.

Solution

The physical motion with $x(0) = x_0$ is

$$x(\tau) = x_0 + v_0 \tau + \frac{F}{2m} \tau^2.$$ (9.107)

Imposing the condition $x(t) = x$ we obtain

$$v_0 = \frac{x - x_0}{t} - \frac{F}{2m} t,$$ (9.108)

whence

$$x(\tau) = x_0 + (x - x_0)\frac{\tau}{t} + \frac{F\tau}{2m}(\tau - t).$$ (9.109)

Therefore, with $t_0 = 0$,

$$S(x, x_0, t) = \int_0^t \left[\frac{m}{2} \left(\frac{dx}{d\tau} \right)^2 + Fx(\tau) \right] d\tau = \frac{m(x - x_0)^2}{2t} + \frac{x + x_0}{2} Ft - \frac{F^2 t^3}{24m}.$$ (9.110)

The reader is invited to check that the Hamilton-Jacobi equation

$$\frac{1}{2m} \left(\frac{\partial S}{\partial x} \right)^2 - Fx + \frac{\partial S}{\partial t} = 0$$ (9.111)

is satisfied by Hamilton's principal function (9.110).

With the identification $q_{0i} \equiv P_i$, Hamilton's principal function generates a canonical transformation that reduces the transformed canonical variables to constants of the motion. Thus, finding $S(q_0, t_0; q, t)$ is tantamount to completely solving Hamilton's equations for the original canonical variables. It is worth stressing that the construction of a unique function $S(q_0, t_0; q, t)$ is only possible if there exists just one dynamical path going through q_0 at time t_0 and q at time t. But this can only be guaranteed in general for $|t - t_0|$ sufficiently small because then the physical paths emanating from q_0 at $t = t_0$ do not cross. This is why the existence of a complete integral for the Hamilton-Jacobi equation is only ensured *locally*. A much more delicate question is that of constructing a complete integral valid *globally* (for all finite t), which is only possible for *integrable systems* (see Section 9.7).

Historically, it was Hamilton who realised that, if expressed in terms of the initial and final points of a physical path segment, the action satisfies Eq. (9.4). However, as far as solving the equations of motion is concerned, Hamilton's train of thought is a vicious circle, for the construction of Hamilton's principal function requires the previous knowledge of the solution of the equations of motion. The merit for breaking this vicious circle belongs to Jacobi, who was the first to understand that any complete integral of Eq. (9.4) provides the solution of the dynamical problem.[1] This does not mean that Hamilton's ideas were fruitless. The analogy between optics and mechanics developed by him turned out to be extremely important for Schrödinger's formulation of wave mechanics (see Section 9.13). In this analogy, the function $S(q_0, t_0; q, t)$ plays a crucial role, as well as in Feynman's formulation of quantum mechanics, in which physical quantities are expressed by means of path integrals (Feynman & Hibbs, 1965).

9.6 Action-Angle Variables

Periodic systems are of great importance in practically all branches of physics. There is a powerful method for finding the frequencies of such systems that is based on the Hamilton-Jacobi theory and does not require the detailed solution of the equations of motion. The method consists in the introduction of certain *action variables*, to be defined shortly. First, however, it is necessary to specify the class of problems treatable by the method.

Multiply Periodic Systems

Action-angle variables are specially adapted to describe separable multiply periodic systems, which we proceed to define.

Definition 9.6.1 *A system whose Hamiltonian does not explicitly depend on time is said to be* **separable** *if for some set of generalised coordinates $q_1, \ldots, q_n$ there exists a complete integral of the time-independent Hamilton-Jacobi equation (9.54) of the form*

$$W(q_1, \ldots, q_n, \alpha_1, \ldots, \alpha_n) = W_1(q_1, \alpha_1, \ldots, \alpha_n) + \cdots + W_n(q_n, \alpha_1, \ldots, \alpha_n). \quad (9.112)$$

For a separable system we have

$$p_i = \frac{\partial S}{\partial q_i} = \frac{\partial W_i}{\partial q_i} = f_i(q_i, \alpha_1, \ldots, \alpha_n). \quad (9.113)$$

[1] For a historical account of the origins of the Hamilton-Jacobi theory the reader is referred to Nakane and Fraser (2002).

This integral is easily calculated by the substitution $x = \sqrt{2\alpha_x/k_x}\,\sin\theta$, giving

$$J_x = \frac{1}{2\pi}2\int_{-\pi/2}^{\pi/2}\sqrt{mk_x}\,\frac{2\alpha_x}{k_x}\cos^2\theta\,d\theta = \frac{1}{\pi}\sqrt{mk_x}\,\frac{2\alpha_x}{k_x}\frac{\pi}{2} = \sqrt{\frac{m}{k_x}}\,\alpha_x. \tag{9.131}$$

The generic procedure outlined above is not necessary in the case of the harmonic oscillator. The previous result for J_x can be obtained virtually without calculations by noting that Eq. (9.115a) can be written in the form

$$\frac{x^2}{a^2} + \frac{p_x^2}{b^2} = 1, \quad a = \sqrt{2\alpha_x/k_x}, \quad b = \sqrt{2m\alpha_x}, \tag{9.132}$$

which is the standard equation of an ellipse in the (x, p_x) phase plane. The action variable J_x multiplied by 2π is the area of this ellipse:

$$J_x = \frac{1}{2\pi}\pi ab = \sqrt{\frac{m}{k_x}}\,\alpha_x. \tag{9.133}$$

Similarly, $J_y = \sqrt{m/k_y}\,\alpha_y$ and, as a consequence,

$$H = \alpha_1 = \alpha_x + \alpha_y = \sqrt{\frac{k_x}{m}}\,J_x + \sqrt{\frac{k_y}{m}}\,J_y, \tag{9.134}$$

from which

$$\omega_x = \frac{\partial H}{\partial J_x} = \sqrt{\frac{k_x}{m}}, \quad \omega_y = \frac{\partial H}{\partial J_y} = \sqrt{\frac{k_y}{m}}, \tag{9.135}$$

which is the correct result for the frequencies of the oscillations in the x, y-directions.

Example 9.9 In the case of the one-dimensional harmonic oscillator, obtain the canonical transformation from the usual canonical variables (x, p) to the action-angle variables (ϕ, J).

Solution

By (9.29), $W = \int (2m\alpha - m^2\omega^2 x^2)^{1/2}\,dx$. According to Example 9.8, $J = \sqrt{m/k}\,\alpha$ or $E = \alpha = (k/m)^{1/2}J = \omega J$. Thus, $W(x, J) = \int (2m\omega J - m^2\omega^2 x^2)^{1/2}\,dx$, whence

$$\phi = \frac{\partial W}{\partial J} = \int \frac{m\omega\,dx}{\sqrt{2m\omega J - m^2\omega^2 x^2}} = \sin^{-1}\left(\sqrt{\frac{m\omega}{2J}}\,x\right), \tag{9.136a}$$

$$p = \frac{\partial W}{\partial x} = \sqrt{2m\omega J - m^2\omega^2 x^2}. \tag{9.136b}$$

Therefore,

$$x = \sqrt{\frac{2J}{m\omega}}\,\sin\phi, \quad p = \sqrt{2m\omega J}\,\cos\phi \tag{9.137}$$

is the desired canonical transformation.

9.7 Integrable Systems: The Liouville-Arnold Theorem

The notion of **integrability** of a mechanical system refers to the possibility of explicitly solving its equations of motion. For instance, in the case of a single-particle one-dimensional conservative system, conservation of energy allows us to write the solution of the equation of motion in the form

$$t - t_0 = \sqrt{\frac{m}{2}} \int_{x_0}^{x} \frac{d\xi}{\sqrt{E - V(\xi)}} \, . \tag{9.138}$$

An additional inversion suffices to get the solution $x(t)$: the problem is solved by a quadrature. The symmetric top with one point fixed is a three-degree-of-freedom system that possesses three independent constants of the motion: p_ϕ, p_ψ and H. The Poisson bracket of any two of these constants of the motion vanishes. As shown in Section 4.9, the equations of motion for the symmetric top are solved by quadratures.

Definition 9.7.1   *The m dynamical variables $F_1(q,p), \ldots, F_m(q,p)$ are said to be in* **involution** *if the Poisson bracket of any two of them is zero:*

$$\{F_k, F_l\} = 0, \qquad k, l = 1, \ldots, m \, . \tag{9.139}$$

In Example 9.5 that deals with the projectile in three dimensions, p_x, p_y and H are constants of the motion in involution. A complete integral of the associated Hamilton-Jacobi equation was easily found by separation of variables and the equations of motion were solved by quadratures. In general, if by any means we can find a complete integral $S(q, \alpha, t) = W(q, \alpha) - \alpha_1 t$ of the Hamilton-Jacobi equation for an n-degree-of-freedom conservative system, the transformed canonical momenta $P_k = \alpha_k(q, p)$, obtained by inversion of $p_k = \partial W/\partial q_k$, form a set of n constants of the motion clearly in involution and the equations of motions are solved by quadratures. It appears, therefore, that the integrability by quadratures of an n-degree-of-freedom conservative Hamiltonian system is intimately connected to the existence of n constants of the motion in involution. Poisson, in 1837, and Liouville, in 1840, showed that if a two-degree-of-freedom conservative Hamiltonian system has two independent constants of the motion in involution, then its equations of motion can be solved by quadratures. The theorem was subsequently generalised by Liouville, in 1855, and, in the twentieth century, enriched by Arnold's contribution.

Lemma 9.7.1   *If a conservative Hamiltonian system with n degrees of freedom admits n constants of the motion in involution $F_1(q, p), \ldots, F_n(q, p)$ and the matrix* **W** *with elements $W_{kl} = \partial F_k/\partial p_l$ is non-singular, then there exist new canonical variables $(\phi_1, \ldots, \phi_n, J_1, \ldots, J_n)$ such that $H = H(J_1, \ldots, J_n)$.*

Proof   We follow Pars (1965) and Berry (1978). Let $F_1, \ldots, F_n$ be constants of the motion in involution with $\det(\partial F_k/\partial p_l) \neq 0$. Define the would-be new momenta $J_1, \ldots, J_n$ by

$$J_k = F_k(q, p), \qquad k = 1, \ldots, n \, . \tag{9.140}$$

According to the implicit function theorem, these equations can be solved for the ps in the form

$$p_k = \psi_k(q, J), \quad k = 1, \ldots, n. \tag{9.141}$$

Therefore,

$$F_k(q, \psi(q, J)) \equiv J_k, \quad k = 1, \ldots, n, \tag{9.142}$$

whence, differentiating with respect to q_l, we find

$$\frac{\partial F_k}{\partial q_l} + \sum_s \frac{\partial F_k}{\partial p_s} \frac{\partial \psi_s}{\partial q_l} = 0 \implies \frac{\partial F_k}{\partial q_l} = -\sum_s W_{ks} \frac{\partial \psi_s}{\partial q_l}. \tag{9.143}$$

On the other hand, writing $\{F_k, F_l\} = 0$ explicitly we have

$$\sum_j \left(\frac{\partial F_k}{\partial q_j} \frac{\partial F_l}{\partial p_j} - \frac{\partial F_k}{\partial p_j} \frac{\partial F_l}{\partial q_j} \right) = 0, \tag{9.144}$$

which, with the use of (9.143), becomes

$$-\sum_{j,s} W_{ks} \frac{\partial \psi_s}{\partial q_j} W_{lj} + \sum_{j,s} W_{kj} \frac{\partial \psi_s}{\partial q_j} W_{ls} = 0, \tag{9.145}$$

or, with the exchange $j \leftrightarrow s$ of the dummy indices in the second sum,

$$\sum_{j,s} W_{ks} \left(\frac{\partial \psi_s}{\partial q_j} - \frac{\partial \psi_j}{\partial q_s} \right) W_{lj} = 0. \tag{9.146}$$

Defininig the matrix **X** with elements

$$X_{rs} = \frac{\partial \psi_r}{\partial q_s} - \frac{\partial \psi_s}{\partial q_r}, \tag{9.147}$$

we can express (9.146) in matrix notation as

$$\mathbf{W X W}^T = 0. \tag{9.148}$$

Since **W** has an inverse, multiplication of this equation from the left by $\mathbf{W}^{-1}$ and from the right by $(\mathbf{W}^T)^{-1}$ gives $\mathbf{X} = 0$ – that is,

$$\frac{\partial \psi_r}{\partial q_s} = \frac{\partial \psi_s}{\partial q_r}. \tag{9.149}$$

As shown in Appendix F, this implies that there exists a function $\Phi(q, J)$ such that

$$\psi_r(q, J) = \frac{\partial \Phi(q, J)}{\partial q_r}. \tag{9.150}$$

Thus, by Eq. (8.16), the transformation $(q, p) \to (\phi, J)$ defined by

$$\phi_r = \frac{\partial \Phi}{\partial J_r}, \quad p_r = \frac{\partial \Phi}{\partial q_r} \tag{9.151}$$

is canonical with generating function $\Phi(q, J)$. Since the Js are constants of the motion, the transformed Hamilton's equations $\dot{J}_k = -\partial H/\partial \phi_k$ imply that $\partial H/\partial \phi_k = 0$. Therefore H does not depend on the ϕs and the proof is complete. $\qquad \square$

Definition 9.7.2   *A conservative Hamiltonian system with n degrees of freedom and Hamiltonian $H(q,p)$ is said to be* **completely integrable** *or, simply,* **integrable** *if there exist n independent constants of the motion in involution, that is:*

(a) $\{F_i, H\} = 0, \quad i = 1, \ldots, n$;

(b) $\{F_i, F_j\} = 0, \quad i, j = 1, \ldots, n$;

(c) the vectors ∇F_i, where $\nabla = (\partial/\partial q_1, \ldots, \partial/\partial q_n, \partial/\partial p_1, \ldots, \partial/\partial p_n)$ is the nabla operator in phase space, are linearly independent at each point of phase space.

The adjective "integrable" will also be used for the Hamiltonian of an integrable system. We are now ready to state the central result concerning integrable Hamiltonian systems.

Theorem 9.7.1 (Liouville-Arnold)   *Consider an integrable Hamiltonian system with n degrees of freedom and Hamiltonian $H(q,p)$. Then:*

(a) There exist new canonical variables $(\phi_1, \ldots, \phi_n, J_1, \ldots, J_n)$ in terms of which $H = H(J_1, \ldots, J_n)$, with the consequence that the solution of the equations of motion for the new variables is

$$J_k = C_k, \quad \phi_k = \phi_k(0) + \omega_k t, \qquad k = 1, \ldots n, \qquad (9.152)$$

where each C_k and each $\omega_k(J) = \partial H/\partial J_k$ is a constant.

(b) Hamilton's equations for the original variables (q,p) can be solved by quadratures.[4]

(c) If the set of level surfaces of the constants of the motion in involution defined by

$$\mathcal{M}_C = \left\{ (q,p) \mid F_k(q,p) = C_k, \ k = 1, \ldots, n \right\}, \qquad (9.153)$$

is compact and connected,[5] the canonical variables (J, ϕ) can be chosen as action-angle variables and the motion is multiply periodic with frequencies $\omega_k = \partial H/\partial J_k$.

Proof   A result in linear algebra (Dettman, 1986; Theorem 2.6.5) and some moments of reflection show that condition (c) in Definition 9.7.2 is equivalent to saying that one of the square matrices with elements $\partial F_i/\partial z_s$, where s takes n values in the set $\{1, 2, \ldots, 2n\}$, is non-singular. If some of the zs coincide with some of the qs we can make a canonical transformation that exchanges each of such qs by its conjugate momentum (see Exercise 8.1.2). Therefore, without loss of generality, we can assume that $s = n + 1, \ldots, 2n$, so that $\det(\partial F_i/\partial p_j) \neq 0$. By Lemma 9.7.1, there exist canonical variables (ϕ, J), with $J_k = F_k(q,p)$, such that the transformed Hamiltonian depends only on the Js and Hamilton's equations for the new variables are solvable in the form (9.152). This proves (a). In order to obtain the generating function $\Phi(q, J)$ of the canonical transformation $(q,p) \rightarrow (\phi, J)$ all one has to do is solve the system of equations (9.140) for the ps and integrate along a straight line a function constructed with the known functions $\psi_k(q, J)$ – see Eq. (F.7) in Appendix F. Thus, the determination of the original canonical variables as functions of time only requires the inversion and integration of known functions, so the original Hamilton's equations can be solved by quadratures. This proves (b). Part (c) refers

---

[4] In the case in which H explicitly depends on time, if there exist n constants of the motion $F_1(q,p,t), \ldots, F_n(q,p,t)$, numbers c_{ij}^k such that $\{F_i, F_j\} = \sum_k c_{ij}^k F_k$ and certain technical conditions are fulfilled, Hamilton's equations are solvable by quadratures (Kozlov, 1983).

[5] In the present context, compact means the same as closed and bounded. A set is connected if it is *not* the union of disjoint open sets.

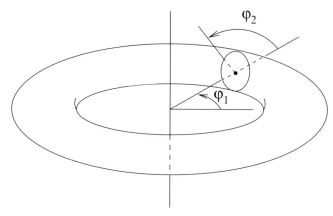

Fig. 9.5   Invariant torus in the case $n = 2$.

to the global structure of the phase space, and its proof will be omitted because it is not only very long but also demands a topological, group theoretic, differential geometric machinery with which the reader is not assumed to be familiar. The reader interested in a detailed proof of part (c) should consult the pertinent literature (Arnold, 1989; Thirring, 1997; Fasano & Marmi, 2006).   □

If the motion of an integrable system is limited, the accessible region of phase space is bounded and $\mathcal{M}_C$ is a compact manifold. By part (c) of the theorem, $\mathcal{M}_C$ is described by n angles and is topologically identical to an n-dimensional torus (see Fig. 9.5). Using the canonical variables defined above, the equations $J_k = C_k$ specify the torus on which the system moves while the ϕs are coordinates on the torus. An orbit that starts on one of those n-tori remains on it forever. This is why they are called **invariant tori**.

Since the Hamiltonian of a conservative system is a constant of the motion, in order that such an n-degree-of-freedom system be integrable, it is enough that there exist only $n - 1$ constants of the motion in involution, provided they are mutually independent and independent of H. In the case of a system with solely two degrees of freedom, the existence of an additional constant of the motion independent of H is sufficient for the system to be integrable.

Example 9.10   Consider the two-degree-of-freedom system with Hamiltonian

$$H = \frac{1}{2}\left(p_1^2 + p_2^2\right) + \frac{1}{4}\left(q_1^4 + q_2^4\right) + \frac{a}{2}q_1^2 q_2^2. \tag{9.154}$$

For the values of a listed below, accompanied by the corresponding additional constant of the motion, the system is integrable:

$$a = 0, \qquad F = \frac{1}{2}\left(p_1^2 - p_2^2\right) + \frac{1}{4}\left(q_1^4 - q_2^4\right), \tag{9.155a}$$

$$a = 1, \qquad F = p_1 q_2 - p_2 q_1, \tag{9.155b}$$

$$a = 3, \qquad F = p_1 p_2 + q_1 q_2\left(q_1^2 + q_2^2\right). \tag{9.155c}$$

It can be proved that this system is integrable only when the parameter a takes one of the values mentioned above (Goriely, 2001).

9.8 Non-integrability Criteria

A conservative Hamiltonian system with n degrees of freedom is integrable if it has n independent constants of the motion in involution, one of them being the Hamiltonian itself. But how to know whether the system admits other constants of the motion besides the Hamiltonian? No systematic procedure is known to produce such constants of the motion, even when there is strong numerical evidence that they exist. One might, for example, search for constants of the motion in the form of polynomial or rational functions of the qs and ps, but this does not always work because the constants of the motion may be transcendental functions of the canonical variables (Problem 9.19). Furthermore, if there are no constants of the motion other than H, any attempt to find them will be a wasted effort. Hence the importance of seeking practical criteria that allow one to decide whether or not there exist constants of the motion in addition to the Hamiltonian.

The typical non-integrability theorems establish that, under certain conditions, there exists no constant of the motion (integral) independent of the Hamiltonian that is an analytic function of the canonical variables. For systems with n degrees of freedom, the non-integrability criteria in these theorems are intricate and difficult to apply (Kozlov, 1983; Ziglin, 1983; Yoshida, 1989; Goriely, 2001; Morales-Ruiz & Ramis, 2001; Nakagawa & Yoshida, 2001). Even for systems with only two degrees of freedom, there are relatively simple results only when the Hamiltonian is of the form $H = T + V$ and the potential energy V is a homogeneous function of integer degree, that is,

$$H = \frac{1}{2}\left(p_x^2 + p_y^2\right) + V(x, y), \quad V(\lambda x, \lambda y) = \lambda^k V(x, y) \ \forall \lambda > 0, \quad k \in \mathbb{Z}. \quad (9.156)$$

For the sake of simplicity, in the statement of the next theorem we use the shorthand notation V_x, V_y, V_{xx} and V_{yy} for the first and second partial derivatives of V.

Theorem 9.8.1 (Yoshida) *Given a Hamiltonian system of the form (9.156) with $k \neq 0, \pm 2$, let $\mathbf{c} = (c_1, c_2)$ be a solution of the system of equations*

$$c_1 = V_x(c_1, c_2), \qquad c_2 = V_y(c_1, c_2). \quad (9.157)$$

*Define the **integrability coefficient** λ by*

$$\lambda = V_{xx}(c_1, c_2) + V_{yy}(c_1, c_2) - k + 1. \quad (9.158)$$

If

$$\cos\left(\frac{2\pi}{k}\right) + 2\cos^2\left[\frac{\pi}{2k}\sqrt{(k-2)^2 + 8k\lambda}\right] > 1 \quad (9.159)$$

then there is no second analytic integral which is independent of H. In this sense, the system is not integrable.

The proof of this theorem is far from being elementary and will be omitted. A proof as well as a detailed specification of the values of λ for which (9.159) holds can be found in Yoshida (1987). It is worth emphasising that (9.159) is only a *sufficient* condition for non-integrability: if it is violated one cannot conclude that the system is integrable.

Example 9.11 Let us apply the above theorem to the potential

$$V(x, y) = \frac{a}{2} x^2 y^2. \tag{9.160}$$

The homogeneity degree of this potential is $k = 4$. Equations (9.157) are

$$c_1 = a c_1 c_2^2, \qquad c_2 = a c_1^2 c_2, \tag{9.161}$$

which admit the solution $c_1^2 = c_2^2 = 1/a$. For this solution the integrability coefficient is

$$\lambda = a c_2^2 + a c_1^2 - 4 + 1 = -1. \tag{9.162}$$

It follows that, with $k = 4$ and $\lambda = -1$,

$$\cos\left(\frac{2\pi}{k}\right) + 2\cos^2\left[\frac{\pi}{2k}\sqrt{(k-2)^2 + 8k\lambda}\right] = 2\cos^2\left(\frac{i\sqrt{7}\,\pi}{4}\right) = 2\cosh^2\left(\frac{\sqrt{7}\,\pi}{4}\right) > 1, \tag{9.163}$$

since $\cos(ix) = \cosh x$ and $\cosh x \geq 1$ for all $x \in \mathbb{R}$. Therefore, the Hamiltonian system (9.156) with the potential (9.160) is not integrable (Ichtiaroglou, 1997).

Although it is directly applicable only to two-degree-of-freedom systems, in certain favourable cases the above theorem allows us to prove the non-integrability of systems with three or more degrees of freedom.

Example 9.12 (Störmer's problem) Energetic charged particles impinging on the Earth from outer space can be entrapped by our planet's magnetic field. To the first approximation, the terrestrial magnetic field can be described as the field of a point magnetic dipole. The motion of a charged particle in the field of a point magnetic dipole was extensively studied by Störmer (1907) and plays a central role in the theoretical understanding of the origin of the Van Allen belts (Braun, 1981; Dilão & Alves-Pires, 2007). The vector potential of a point magnetic dipole fixed at the origin is

$$\mathbf{A}(\mathbf{r}) = \frac{\mathbf{M} \times \mathbf{r}}{r^3}, \tag{9.164}$$

where $\mathbf{M}$ is a constant vector. Since the scalar potential ϕ is zero, the Lagrangian (1.143) for a unit-mass particle with electric charge e in the field created by the magnetic dipole is

$$L = \frac{v^2}{2} + \frac{e}{c} \frac{\mathbf{v} \cdot (\mathbf{M} \times \mathbf{r})}{r^3}. \tag{9.165}$$

In cylidrical coordinates, with $\mathbf{M} = M\hat{\mathbf{z}}$, $\mathbf{r} = \rho\hat{\boldsymbol{\rho}} + z\hat{\mathbf{z}}$ and $\mathbf{v} = \dot{\rho}\hat{\boldsymbol{\rho}} + \rho\dot{\phi}\hat{\boldsymbol{\phi}} + \dot{z}\hat{\mathbf{z}}$, the Lagrangian (9.165) takes the form

$$L = \frac{1}{2}(\dot{\rho}^2 + \rho^2\dot{\phi}^2 + \dot{z}^2) + a\frac{\rho^2\dot{\phi}}{(\rho^2 + z^2)^{3/2}}, \qquad a = \frac{eM}{c}. \tag{9.166}$$

The Hamiltonian is

$$H = \frac{1}{2}(p_\rho^2 + p_z^2) + \frac{1}{2\rho^2}\left[p_\phi - \frac{a\rho^2}{(\rho^2 + z^2)^{3/2}}\right]^2. \tag{9.167}$$

Two immediate constants of the motion are H and p_ϕ. Is there a third constant of the motion independent of these two? If the Hamiltonian system (9.167) admits a third constant of the motion independent of H and p_ϕ, the two-degree-of-freedom system obtained by setting $p_\phi = 0$ will also have one constant of the motion independent of H, that is, the reduced Hamiltonian system

$$H = \frac{1}{2}(p_\rho^2 + p_z^2) + \frac{a^2}{2}\frac{\rho^2}{(\rho^2 + z^2)^3}. \tag{9.168}$$

will be integrable. This Hamiltonian is of the form (9.156) with $k = -4$. Equations (9.157) become

$$c_1 = \frac{a^2 c_1(c_2^2 - 2c_1^2)}{(c_1^2 + c_2^2)^4}, \qquad c_2 = -\frac{3a^2 c_1^2 c_2}{(c_1^2 + c_2^2)^4}, \tag{9.169}$$

which admit the solution $c_1 = (-2a^2)^{1/6}$, $c_2 = 0$. For this solution the integrability coefficient is

$$\lambda = V_{\rho\rho}(c_1, c_2) + V_{zz}(c_1, c_2) - k + 1 = \frac{10a^2}{c_1^6} - \frac{3a^2}{c_1^6} + 4 + 1 = \frac{3}{2}. \tag{9.170}$$

For this value of λ and $k = -4$ it is immediately checked that condition (9.159) is fulfilled. Therefore, the Hamiltonian system (9.168) does not possess a second analytic constant of the motion independent of H. It follows that Störmer's problem, characterised by the Hamiltonian (9.167), is not integrable (Almeida, Moreira & Yoshida, 1992).

Yoshida's theorem does not apply if $k = -2$, but it can be proved directly that this is an integrable case (Problem 9.20).

9.9 Integrability, Chaos, Determinism and Ergodicity

According to equations (9.152), small deviations from the initial conditions of an integrable system grow linearly with time, and the motion of integrable systems is said to be **regular**. Differently, non-integrable systems can, in certain regions of phase space, present a **sensitive dependence on initial conditions**: small deviations from the initial conditions may grow exponentially with time, making it impossible in practice to predict even the short-term behaviour of the system. This kind of motion is called **irregular** or **chaotic**. Thus,

chaos is an exclusive property of non-integrable systems. By way of illustration: Störmer's problem, which was shown to be non-integrable in Example 9.12, displays chaotic behaviour (Braun, 1981; Dilão & Alves-Pires, 2007). It is not known, however, if every non-integrable Hamiltonian system is necessarily chaotic in some region of phase space.

It is often stated that the discovery of chaos has debunked determinism in classical mechanics. This is a misconception that stems from confusing a matter of principle with a matter of practice, from the failure to distinguish between determinism and predictability. Given exactly known initial conditions, Hamilton's equations have a unique solution that exactly predicts the future states of any mechanical system. Therefore, Laplace's famous remark, that nothing would be uncertain for a sufficiently vast intelligence that could perfectly apprehend all forces that animate nature as well as the initial conditions of its constituents and solve the equations of motion, is unimpeachable in the context of classical mechanics. Laplace himself recognised that we shall always remain infinitely removed from this imaginary intellect, and our imperfect knowledge of the present state of the world prevents us from an exact prediction of its evolution (Laplace, 1840). Doubtlessly chaos has a strong impact on the practical issue of predictability, but does not change the fact that classical mechanics is deterministic (Bricmont, 1996).

In statistical mechanics, in order to avoid the need to calculate time averages which require the knowledge of the exact solution of the equations of motion, a fundamental assumption is made – the famous **ergodic hypothesis**. According to this hypothesis, in the course of its time evolution the system explores the whole phase space region that is energetically available (called the **energy surface**) and eventually covers this region uniformly. This allows the replacement of time averages by averages over the energy surface in phase space, which are much easier to calculate. In the case of an integrable system, the existence of n independent constants of the motion implies that each path can explore at most an n-dimensional manifold in the $2n$-dimensional phase space. Except for the trivial case $n = 1$, such a manifold is smaller than the energy surface, whose dimensionality is $2n - 1$. Therefore, the ergodic hypothesis is generally false.

A natural question is the following: is integrability the rule or the exception? The exactly solved problems in mechanics textbooks are all integrable, which may give rise to the false impression that the integrable systems are generic. It is known that this is not true: generic Hamiltonian systems are neither integrable nor ergodic (Markus & Meyer, 1974). Integrable systems are the ones usually discussed in mechanics textbooks just because they are analytically solvable, but they constitute the exception rather than the rule. Perhaps the most emblematic example of non-integrability is the celebrated three-body problem of celestial mechanics: Poincaré proved that the restricted three-body problem does not admit any analytic integral independent of the Hamiltonian, and then extended his non-integrability result to the general problem (Whittaker, 1944; Chenciner, 2007).[6] What

---

[6] Non-integrability is not to be confused with unsolvability. Between 1906 and 1912 the Finnish mathematician Karl Sundman proved that the general three-body problem admits exact solutions for the coordinates of the three bodies as power series that converge for all $t \in \mathbb{R}$ (Barrow-Green, 2010). Sundman's results are fully described by Siegel and Moser (1971); see Saari (1990) for an elementary account. In 1991, the Chinese mathematician Qiudong Wang finally solved the general n-body problem (Diacu, 1996). Alas, the power series obtained by Sundman and Wang have no practical utility because they converge excruciatingly slowly.

is rather surprising is that non-integrability can manifest itself in mechanical systems as simple as the double pendulum (Shinbrot *et al.*, 1992; Hand & Finch, 1998).

9.10 Prelude to the KAM Theorem

An issue related to the genericity of integrable systems but somewhat different is whether integrability resists small perturbations. George Birkhoff and Enrico Fermi, as well as most other mathematicians and theoretical physicists from the first half of the twentieth century, believed that the slightest perturbation of an integrable system would destroy integrability, which would tend to validate the ergodic hypothesis and, consequently, statistical mechanics. What really happens, however, is that "most" tori survive a perturbation, albeit in distorted form. If the perturbation is sufficiently small, the tori which are only slightly deformed fill phase space almost completely. The destroyed tori distribute themselves among the preserved ones in a very irregular fashion, and the volume of phase space that they occupy tends to zero together with the strength of the perturbation. This was proved rigorously in 1962–1963 by Vladimir I. Arnold and Jürgen Moser, independently, after a proof sketched by A. N. Kolmogorov in 1954, and has become known as the Kolmogorov-Arnold-Moser theorem, or just the "KAM theorem".

The manifold on which the motion of an integrable system with Hamiltonian $H_0(J)$ unfolds is an n-torus T^n. Assume the frequencies $\omega_1,\ldots,\omega_n$ are incommensurate or, more precisely, are **rationally independent** or **non-resonant**, which means that given n integer numbers $m_1,\ldots,m_n$ it so happens that

$$\sum_{k=1}^{n} m_k \omega_k = 0 \iff m_1 = \cdots = m_n = 0. \tag{9.171}$$

After a sufficiently long time the orbit returns to an arbitrarily small neighbourhood of its starting point on T^n but does not close: the orbits are *dense* on T^n. The motion is said to be **quasiperiodic** and the torus travelled by the system is called a **non-resonant torus**. When the frequencies are rationally dependent, the tori are said to be **resonant**.

Let $H_0(J)$ be the Hamiltonian of an integrable system expressed in terms of action variables. Consider now the perturbed Hamiltonian

$$H(J,\phi;\epsilon) = H_0(J) + \epsilon F(J,\phi;\epsilon), \tag{9.172}$$

where ϵ is small, and suppose the perturbation breaks the system's integrability. The fundamental problem consists in determining if, and in what sense, this system is stable: does the perturbation only slightly modify the motions described by $H_0(J)$ or destroy them completely? General arguments and specific examples (Arnold, 1989; Ott, 1993) indicate that, typically, the resonant tori are destroyed no matter how small the perturbation is. As to the non-resonant tori, the question is to a large extent answered by the theorem of Kolmogorov, Arnold and Moser.

Theorem 9.10.1 (KAM) *If the frequencies of an integrable Hamiltonian H_0 are rationally independent and sufficiently irrational, and if H_0 obeys a non-degeneracy condition,[7] then, for ϵ small enough, the solutions of the perturbed system (9.172) are predominantly quasiperiodic and differ only slightly from those of the unperturbed system. The majority of the non-resonant tori of H_0 are only slightly deformed and the perturbed system also possesses non-resonant tori with the same frequencies as those of the unperturbed system.*

At this point, the theorem has been stated in a somewhat qualitative and vague form in order not to drown the reader in technicalities, some of which will be addressed in Section 10.5. As pointed out by Berry (1978), the rigorous proof of the KAM theorem is "long, intricate and subtle". In Dumas (2014) the reader will find an excellent exposition of the subject without technical details, with emphasis on its historical context. Simplified proofs – but still very difficult – of particular cases of the theorem can be found in Thirring (1997) and Chandre and Jauslin (1998).

There are actually several varieties of theorems of the KAM type, so it is less appropriate to speak of the KAM theorem than of the KAM theory (Arnold, Kozlov & Neishtadt, 2006). In the next chapter, in Section 10.5, the Kolmogorov-Arnold-Moser theory is re-examined in more depth. The version of the KAM theorem discussed here is formulated in precise terms and some aspects of the rigorous proof are discussed without entirely shunning the mathematical complexity inherent in the problem.

The KAM theorem is a great mathematical feat but its physical relevance is considered somewhat doubtful. Unfortunately, the stability of the solar system is not unequivocally established because the theorem seems to require extremely small perturbations, much smaller than those suggested by numerical experiments (Tabor, 1989), besides being silent on the fate of the motions with rationally dependent frequencies. The orbits to which the theorem does not apply are densely distributed throughout the set of orbits encompassed by the theorem, so that Hamiltonian systems generally display an intricate admixture of regular and irregular (chaotic) motions. Typically, in the regions in which tori are destroyed, there develops the **homoclinic tangle** discovered by Poincaré in 1889 (Barrow-Green, 1994, 1997; Andersson, 1994; Diacu & Holmes 1996), a staggeringly convoluted structure (Tabor, 1989; José & Saletan, 1998; Lam, 2014) which will not be discussed here. The homoclinic tangle is the hallmark of chaos in Hamiltonian systems.

The so-called *Kirkwood gaps*, discovered in 1866, are empty spaces in the asteroid belt between Mars and Jupiter at distances from the Sun such that the unperturbed periods of revolution are simple fractions of Jupiter's period, such as 1/2, 1/3, 2/5 or 3/7. They were at first interpreted on the basis of the KAM theorem as due to the destruction of invariant tori in the neighbourhood of resonant frequencies (Berry, 1978). This explanation, however, is considered neither conclusive nor the most plausible. There is still no satisfactory explanation of all Kirkwood gaps, but strong numerical evidence indicates that at least one of them is attributable to the chaotic behaviour of the solar system (Wisdom, 1987; Peterson, 1993; Lissauer, 1999).

---

[7] The precise characterisation of what is meant by sufficiently irrational frequencies and non-degenerate Hamiltonian H_0 can be found in Section 10.5.

In fact, computational simulations carried out during the last thirty years or so have shown that the solar system is chaotic: it is impossible to make precise predictions of the trajectories of the planets beyond a few tens of million years because the errors in the initial positions are multiplied by roughly 10 each 10 million years. In a scale of five billion years, before the Sun becomes a red giant and engulfs the Earth, there is a small but non-negligible probability (about 1%) that Mars escapes from the solar system and that Mercury's orbit gets so eccentric as to make possible a collision with Venus (Laskar, 2013).

Historical note   By initiative of the Swedish mathematician Gösta Mittag-Leffler, in 1884 King Oscar II of Sweden and Norway agreed to sponsor a mathematical competition with the prize to be awarded on 21 January 1889, in celebration of the King's sixtieth birthday, and publication in the prestigious journal *Acta Mathematica*, of which Mittag-Leffler was editor-in-chief (Barrow-Green, 1994, 1997; Andersson, 1994; Diacu & Holmes, 1996). The prize was first announced in July 1885 and declared the papers entering the contest should be sent before 1 June 1888. Four questions were proposed, and the prize of 2500 crowns and a gold medal would go for the best work submitted, according to the judgement of a committee consisting of Karl Weierstrass, Charles Hermite and Mittag-Leffler himself. The first question, the *n*-body problem formulated by Weierstrass, is the one that concerns us:

> For a system of arbitrarily many mass points that attract each other according to Newton's law, assuming that no two points ever collide, find a series expansion of the coordinates of each point in known functions of time converging uniformly for any period of time.[8]

Jules Henri Poincaré (1854–1912) won the prize with an investigation of the three-body-problem. Although he did not solve the problem as proposed, Poincaré obtained remarkable results, among them a proof of non-integrability of the three-body-problem, the discovery of integral invariants and of the recurrence theorem. As if that were not enough, he introduced new qualitative and geometric methods in celestial mechanics. In a report sent to Mittag-Leffler, Weierstrass asserted: "I have no difficulty in declaring that the memoir in question deserves the prize. You may tell your Sovereign that this work cannot, in truth, be considered as supplying a complete solution to the question we originally proposed, but that it is nevertheless of such importance that its publication will open a new era in the history of celestial mechanics. His Majesty's goal in opening the contest can be considered attained."[9]

During the preparation of the paper for publication, the editor Edvard Phragmén asked Poincaré about certain obscure passages. Poincaré clarified them, corrected some minor errors and the paper went to press. But the doubts raised by Phragmén led Poincaré to reflect more deeply on his results and he eventually found an error in a stability theorem for the restricted three-body-problem caused by his incomplete understanding of the intricate geometric properties of the dynamical trajectories. Poincaré sent a telegram to Mittag-Leffler asking him to stop the printing of the paper because of the *importante erreur* that he had detected. Upon receiving it, Mittag-Leffler immediately wrote to several

---

[8] Translation taken from http://www.mittag-leffler.se/library/prize-competition.
[9] Translation taken from Diacu and Holmes (1996).

distinguished mathematicians and astronomers requesting that they return the preliminary copy of volume 13 of *Acta Mathematica* that had been mailed to them. The entire original edition was destroyed and Poincaré had to pay 3500 crowns for a new corrected edition, 1000 crowns more than the prize he had won. In the paper that was finally published in 1890, with 270 pages,[10] Poincaré discovered chaos in the Newtonian model of the solar system: a slight perturbation of the initial conditions may drastically change the final state. Chaos was largely ignored by other scientists, and Poincaré himself moved on to other subjects. The subsequent revolutions of relativity and quantum mechanics diverted physicists and mathematicians from classical mechanics and smothered whatever impact Poincaré's discovery might have made. It was only around the middle of the twentieth century that the seminal work of Poincaré was rediscovered, resumed and developed.

9.11 Action Variables in the Kepler Problem

For the role it played in the advancement of quantum mechanics, the treatment of the bound motion of a charged particle in the Coulomb potential by action variables is of considerable historical interest. The potential

$$V(r) = -\frac{k}{r}, \qquad k > 0, \tag{9.173}$$

is of the form considered in Example 9.4, with $a(r) = -k/r$ and $b(\theta) = 0$. Therefore the system is separable and multiply periodic, and the action variables are written

$$J_\phi = \frac{1}{2\pi} \oint p_\phi d\phi = \frac{\alpha_\phi}{2\pi} \int_0^{2\pi} d\phi = \alpha_\phi, \tag{9.174}$$

$$J_\theta = \frac{1}{2\pi} \oint p_\theta d\theta = \frac{1}{2\pi} \oint \frac{\partial W}{\partial \theta} d\theta = \frac{1}{2\pi} \oint \sqrt{\alpha_\theta^2 - \alpha_\phi^2 / \sin^2 \theta}\, d\theta, \tag{9.175}$$

$$J_r = \frac{1}{2\pi} \oint p_r dr = \frac{1}{2\pi} \oint \frac{\partial W}{\partial r} dr = \frac{1}{2\pi} \oint \sqrt{2m\alpha_1 + \frac{2mk}{r} - \frac{\alpha_\theta^2}{r^2}}\, dr, \tag{9.176}$$

where Eqs. (9.65) have been used. The integrals (9.175) and (9.176) can be elegantly calculated by techniques of contour integration in the complex plane (Goldstein, 1980; Saletan & Cromer, 1971), but we prefer to do it by elementary methods.

Necessarily $\alpha_\theta^2 - \alpha_\phi^2 / \sin^2 \theta \geq 0$, so $\theta_0 \leq \theta \leq \pi - \theta_0$ where $\sin \theta_0 = \alpha_\phi / |\alpha_\theta|$. For the calculation of the integral (9.175) we must take the positive square root ($p_\theta > 0$) as θ increases from θ_0 to $\pi - \theta_0$ and the negative square root ($p_\theta < 0$) as θ decreases from $\pi - \theta_0$ to θ_0. Thus,

$$J_\theta = \frac{1}{2\pi} 2 \int_{\theta_0}^{\pi - \theta_0} \sqrt{\alpha_\theta^2 - \alpha_\phi^2 / \sin^2 \theta}\, d\theta = \frac{|\alpha_\theta|}{\pi} \int_{\theta_0}^{\pi - \theta_0} \left(1 - \frac{\sin^2 \theta_0}{\sin^2 \theta}\right)^{1/2} d\theta. \tag{9.177}$$

---

[10] The suppressed original publication had 158 pages.

In order to make the integration interval symmetric, it is convenient to introduce the new variable ψ defined by $\theta = \psi + \pi/2$ and the angle $\psi_0 = \pi/2 - \theta_0$ so that

$$J_\theta = \frac{|\alpha_\theta|}{\pi} \int_{-\psi_0}^{\psi_0} \left(1 - \frac{\cos^2 \psi_0}{\cos^2 \psi}\right)^{1/2} d\psi = \frac{2|\alpha_\theta|}{\pi} \int_0^{\psi_0} \frac{\sqrt{\sin^2 \psi_0 - \sin^2 \psi}}{\cos \psi} d\psi, \quad (9.178)$$

where

$$\cos \psi_0 = \sin \theta_0 = \frac{\alpha_\phi}{|\alpha_\theta|}. \quad (9.179)$$

The form of the integrand in (9.178) suggests the change of variable

$$\sin \psi = \sin \psi_0 \sin u, \quad (9.180)$$

that leads to

$$J_\theta = \frac{2|\alpha_\theta|}{\pi} \int_0^{\pi/2} \frac{\sin^2 \psi_0 \cos^2 u}{1 - \sin^2 \psi_0 \sin^2 u} du = \frac{2|\alpha_\theta|}{\pi} \int_0^{\pi/2} \left(1 - \frac{\cos^2 \psi_0}{1 - \sin^2 \psi_0 \sin^2 u}\right) du. \quad (9.181)$$

With the use of the formula (Gradshteyn & Ryzhik, 1980)

$$\int \frac{dx}{1 - k^2 \sin^2 x} = (1 - k^2)^{-1/2} \tan^{-1} \left(\sqrt{1 - k^2} \tan x\right), \quad |k| < 1,$$

we finally get

$$J_\theta = |\alpha_\theta|(1 - \cos \psi_0) = |\alpha_\theta| - \alpha_\phi. \quad (9.182)$$

As to the radial action variable, we have

$$J_r = \frac{1}{2\pi} 2 \int_{r_-}^{r_+} \sqrt{2m\alpha_1 + \frac{2mk}{r} - \frac{\alpha_\theta^2}{r^2}} \, dr = \frac{1}{\pi} \int_{r_-}^{r_+} \frac{\sqrt{2m\alpha_1 r^2 + 2mkr - \alpha_\theta^2}}{r} \, dr, \quad (9.183)$$

where r_+ and r_- are the roots of

$$2m\alpha_1 + \frac{2mk}{r} - \frac{\alpha_\theta^2}{r^2} = 0. \quad (9.184)$$

Of course, we are considering $E < 0$ since only in this case is the motion periodic and the radial variable bounded. Writing $\alpha_1 = E = -\mathcal{E}$ with $\mathcal{E} > 0$, the solutions of (9.184) are

$$r_\pm = \frac{k \pm \sqrt{k^2 - 2\mathcal{E}\alpha_\theta^2/m}}{2\mathcal{E}}. \quad (9.185)$$

Using the formula (Gradshteyn & Ryzhik, 1980)

$$\int \frac{\sqrt{a + bx + cx^2}}{x} \, dx = \sqrt{a + bx + cx^2} - \sqrt{-a} \sin^{-1} \left(\frac{2a + bx}{x\sqrt{\Delta}}\right)$$

$$- \frac{b}{2\sqrt{-c}} \sin^{-1} \left(\frac{2cx + b}{\sqrt{\Delta}}\right),$$

valid for $a < 0$, $c < 0$ and $\Delta = b^2 - 4ac > 0$, after some algebraic work we arrive at

$$J_r = -|\alpha_\theta| + \left(-\frac{2m}{E}\right)^{1/2} \frac{k}{2}. \tag{9.186}$$

Combining this result with Eqs. (9.174) and (9.182) we obtain

$$H = E = -\frac{mk^2/2}{(J_r + J_\theta + J_\phi)^2}, \tag{9.187}$$

whence

$$\omega_r = \frac{\partial H}{\partial J_r} = \frac{mk^2}{(J_r + J_\theta + J_\phi)^3} = \left(-\frac{8E^3}{mk^2}\right)^{1/2}. \tag{9.188}$$

There is a complete degeneracy since the other frequencies ω_θ and ω_ϕ are equal to ω_r. Being all equal, the three frequencies are obviously commensurate. This would inevitably have to occur owing to the fact that the motion is periodic: for $E < 0$ the orbits are closed (circles or ellipses).

9.12 Adiabatic Invariants

Consider a mechanical system with a characteristic parameter λ and suppose that, under the influence of an external agency, λ varies very slowly ("adiabatically"). For example, a mass point m suspended by a string of length l may be executing small oscillations with period τ. If l is gradually reduced by pulling the string through a hole in the ceiling in such a way that sensible changes of l occur only in a time scale $T \gg \tau$, we say that the pendulum is undergoing an **adiabatic transformation**. A physical quantity that remains essentially unchanged during an adiabatic transformation is called an **adiabatic invariant**.

Let us show that for a harmonic oscillator with slowly varying frequency the quantity E/ω is an adiabatic invariant. The total energy of a harmonic oscillator is

$$E = \frac{m}{2}\dot{x}^2 + \frac{m\omega^2}{2}x^2. \tag{9.189}$$

The angular frequency ω might be varying slowly because $\omega^2 = g/l$, as in the case of the pendulum, with l varying adiabatically. Whatever the cause, we can write

$$\frac{dE}{dt} = m\dot{x}\ddot{x} + m\omega^2 x\dot{x} + m\omega x^2 \frac{d\omega}{dt}. \tag{9.190}$$

This equation contains fast and slow terms. During the time it takes for the slow terms E and ω to vary significantly, the fast terms perform a huge number of oscillations, so we can replace them by their average values in an oscillation period. Proceeding this way, we are left with

$$\frac{dE}{dt} = m\langle \dot{x}\ddot{x}\rangle + m\omega^2\langle x\dot{x}\rangle + m\omega\langle x^2\rangle\frac{d\omega}{dt}. \tag{9.191}$$

The first two averages are zero and $\langle x^2 \rangle = E/m\omega^2$, so

$$\frac{dE}{E} = \frac{d\omega}{\omega} \implies \frac{E}{\omega} = \text{constant} = \text{adiabatic invariant}, \qquad (9.192)$$

which is the intended result.

If the frequency ω is a function of the product ϵt, explicit examples show, and it can be proved in general, that as $\epsilon \to 0$ the time rate of change of E/ω tends to zero at least as fast as ϵ^2.

Example 9.13 For a harmonic oscillator with time-dependent frequency, using the equation of motion

$$\ddot{x} + \omega^2(t)x = 0 \qquad (9.193)$$

it is easy to show that

$$\frac{d}{dt}\left(\frac{E}{\omega}\right) = \frac{m\dot{\omega}}{2\omega^2}(\omega^2 x^2 - \dot{x}^2). \qquad (9.194)$$

Consider the time-dependent frequency

$$\omega(t) = \omega_0 e^{\epsilon t}. \qquad (9.195)$$

By taking ϵ sufficiently small one can make $\omega(t)$ vary with time as slowly as one pleases. A solution to the equation of motion (9.193) is

$$x(t) = CJ_0\left(\frac{\omega_0}{\epsilon}e^{\epsilon t}\right), \qquad (9.196)$$

where C is an arbitrary constant and J_ν is the Bessel function of the first kind and order ν. Making use of $J_0' = -J_1$ it follows that

$$\frac{d}{dt}\left(\frac{E}{\omega}\right) = \epsilon\frac{m\omega C^2}{2}\left[J_0\left(\frac{\omega_0}{\epsilon}e^{\epsilon t}\right)^2 - J_1\left(\frac{\omega_0}{\epsilon}e^{\epsilon t}\right)^2\right]. \qquad (9.197)$$

For $z \to \infty$ the Bessel function $J_\nu(z)$ takes the asymptotic form (Whittaker & Watson, 1927)

$$J_\nu(z) = \sqrt{\frac{2}{\pi z}}\left[\cos\left(z - \frac{\nu\pi}{2} - \frac{\pi}{4}\right) + O\left(\frac{1}{z}\right)\right], \qquad (9.198)$$

and simple algebra allows us to conclude that, as $\epsilon \to 0$,

$$\frac{d}{dt}\left(\frac{E}{\omega}\right) = \epsilon^2\left[\frac{mC^2}{\pi}\sin\left(\frac{2\omega_0}{\epsilon}e^{\epsilon t}\right) + O(\epsilon)\right]. \qquad (9.199)$$

Therefore, the departure of the time rate of change of E/ω from zero as $\epsilon \to 0$ is a second-order effect.

effect – that is, the influence of a uniform electric field on the hydrogen atom. Neglecting the motion of the proton and choosing the z-axis parallel to $\mathbf{E}$, show that $V = -k/\sqrt{\rho^2 + z^2} - eEz$, $k > 0$. Verify that, expressed in parabolic coordinates, this potential is of the form above and identify the functions $a(\xi)$ and $b(\eta)$. Finally, making use of the complete integral already found, reduce the solution of the equations of motion for the electron to quadratures.

9.3 Consider the Kepler problem in the plane – that is, a particle of mass m in the potential $V(r) = k/r$, where $r = \sqrt{x^2 + y^2}$. Write down the Hamilton-Jacobi equation in terms of the variables $u = r + x$, $v = r - x$ and find a complete integral. Determine the equation of the orbit in terms of the coordinates u, v.

9.4 A particle with positive energy moves in a straight line under the influence of the potential $V(x) = F|x|$, where F is a positive constant. Use action-angle variables to find the period of the motion as a function of the energy. What is the energy spectrum predicted by the Wilson-Sommerfeld quantisation rules?

9.5 Let $H(q, p, t)$ be the Hamiltonian of an n-degree-of-freedom system and suppose S satisfies the partial differential equation

$$H\left(-\frac{\partial S}{\partial p_1}, \ldots, -\frac{\partial S}{\partial p_n}, p_1, \ldots, p_n, t\right) + \frac{\partial S}{\partial t} = 0,$$

known as the Hamilton-Jacobi equation in the momentum representation. What kind of generating function S must be considered in order to arrive at this equation? Write down the equations and describe the procedure that allows us to obtain the solution of the equations of motion as soon as a complete integral $S(p, \alpha, t)$ of the above equation is found. Apply this formulation to the one-dimensional problem of a particle subject to a constant force F, whose Hamiltonian is $H = p^2/2m - Fq$, and determine $q(t)$. Why is the Hamilton-Jacobi equation in the momentum representation almost always more difficult to solve than in the usual representation?

9.6 A particle of mass m and electric charge e moves in the xy-plane in the presence of a uniform magnetic field perpendicular to the plane. (a) Show that a Hamiltonian for this system is

$$H = \frac{1}{2m}p_x^2 + \frac{1}{2m}\left(p_y - \frac{eB}{c}x\right)^2$$

and obtain the solution $x(t), y(t), p_x(t), p_y(t)$ of the equations of motion by the Hamilton-Jacobi method. (b) In Example 8.9 the following alternative Hamiltonian was used for this system:

$$H = \frac{1}{2m}\left(p_x + \frac{eB}{2c}y\right)^2 + \frac{1}{2m}\left(p_y - \frac{eB}{2c}x\right)^2.$$

Verify that its associated Hamilton-Jacobi equation is not separable in the usual way. This shows that the separability of the Hamilton-Jacobi equation depends not only on the choice of coordinates, but also on the choice of gauge. Separate the variables by putting $W = Cxy + \alpha_y y + X(x)$, where α_y is an arbitrary constant and C is a suitably chosen constant. Find the general solution of the equations of motion and compare with the results obtained in part (a).

9.7　A particle in one dimension subject to a time-dependent force has Hamiltonian

$$H = \frac{p^2}{2m} - F(t)\,x.$$

Solve the equations of motion by the Hamilton-Jacobi method.

9.8　Show that the Hamilton-Jacobi equation (9.18) for a free particle in one dimension admits separation of variables in the form of a product – that is, it has solutions of the form $S(q,t) = W(q)T(t)$. Using this separation of variables technique, find a complete integral and use it to solve the equation of motion for $q(t)$.

9.9　A particle of mass m moves in a straight line in the potential $V(x) = V_0/\cos^2(x/\ell)$, $V_0 > 0$, $\ell > 0$. (a) Sketch a graph of V versus x for $x \in (-\ell\pi/2, \ell\pi/2)$. Show that for any classically allowed value of the energy, the motion is periodic. (b) Use the method of action-angle variables to find the oscillation period as a function of the energy. (c) What are the energy levels according to the Wilson-Sommerfeld quantisation rules?

9.10　The motion of a two-degree-of-freedom system is described by the Hamiltonian

$$H = p_x p_y \cos \omega t + \frac{1}{2}(p_x^2 - p_y^2) \sin \omega t.$$

(a) Find a complete integral of the associated Hamilton-Jacobi equation. (b) Obtain the general solution of the equation of motion for $x(t)$ and $y(t)$.

9.11　The motion of a two-degree-of-freedom system is described by the Hamiltonian

$$H = \frac{1}{2}\frac{p_x^2 + p_y^2}{x^2 + y^2} + \frac{k}{x^2 + y^2},$$

where x, y are Cartesian coordinates and $k > 0$. (a) Use the Hamilton-Jacobi equation to get an equation for the trajectory. (b) Show that the orbit is a conic section (ellipse, parabola or hyperbola).

9.12　Reconsider the swinging Atwood's machine of Problem 1.5 in the special case $M = 3m$. Show that the change of variables

$$\xi = \sqrt{r\left(1 + \sin\frac{\theta}{2}\right)}, \qquad \eta = \sqrt{r\left(1 - \sin\frac{\theta}{2}\right)}$$

with inverse

$$r = \frac{\xi^2 + \eta^2}{2}, \qquad \theta = 2\tan^{-1}\frac{\xi^2 - \eta^2}{2\xi\eta}$$

transforms the Lagrangian given in Problem 1.5 into

$$L = 2m(\xi^2 + \eta^2)(\dot{\xi}^2 + \dot{\eta}^2) - 2mg\frac{\xi^4 + \eta^4}{\xi^2 + \eta^2},$$

which gives rise to the Hamiltonian

$$H = \frac{1}{8m}\frac{p_\xi^2 + p_\eta^2}{\xi^2 + \eta^2} + 2mg\frac{\xi^4 + \eta^4}{\xi^2 + \eta^2}.$$

Show that the corresponding Hamilton-Jacobi equation is separable and find a complete integral (Tufillaro, 1986). It can be shown that if $M < m$ or $M > 3m$ the swinging Atwood's machine is a non-integrable system (Casasayas, Nunes & Tufillaro, 1990).

9.13 Give an elementary proof of the result in Exercise 9.12.1 by calculating the work done by the external force to vary the distance between the walls by dL and taking into account that the variation of the particle's energy must equal this work. Hint: for an adiabatic process the external force must counterbalance the average force the particle exerts on the walls due to the repeated collisions.

9.14 A particle of mass m moves in a plane in the gravitational field of two much larger masses m_1 and m_2 that remain fixed on the x-axis at the positions $(\pm a, 0)$. Show that in *elliptic coordinates* u, v defined by

$$x = a \cosh u \cos v, \qquad y = a \sinh u \sin v,$$

the Hamiltonian for this problem is

$$H = \frac{1}{2ma^2} \frac{p_u^2 + p_v^2}{\cosh^2 u - \cos^2 v} + \frac{k}{\cosh u - \cos v} + \frac{k'}{\cosh u + \cos v}.$$

Show that the corresponding Hamilton-Jacobi equation is separable and reduce the solution of the equations of motion to quadratures.

9.15 A particle of mass m with negative energy moves in a straight line in the potential $V = -V_0 / \cosh^2(\alpha x)$, where V_0 and α are positive constants. If V_0 varies slowly with time, prove that $\sqrt{V_0} - \sqrt{-E}$ is an adiabatic invariant.

9.16 Consider the system described in Problem 2.12 and assume the string is pulled so as to reduce very slowly the radius r of the circle traced out with speed v by the particle of mass m_2. Show that the product vr is an adiabatic invariant. Present also an elementary proof of the result without resorting to action variables.

9.17 In the development of the optico-mechanical analogy, the attentive reader has certainly noticed that the phase velocity (9.218) is not equal to the particle's velocity – it is, in truth, inversely proportional to it. This mismatch appears to make unviable the interpretation of S as the phase of a wave process associated with the particle. Nevertheless, it can be shown (Jackson, 1999) that the velocity of a wave packet formed by superposing waves within a narrow band of frequencies is the group velocity given by $v_g = d\omega/dk$, where $\mathbf{k}$ is the wave vector. Starting from the expression for $\omega(k)$ determined by combining (9.214), (9.216), the Planck relation $E = \hbar \omega$ and the de Broglie relation $\mathbf{p} = \hbar \mathbf{k}$, show that the group velocity is identical to the particle's velocity.

9.18 In quantum mechanics the notion of *complete set of commuting observables* plays an important role. Its classical analogue for an n-degree-of-freedom system is a set $\{R_1(q, p), \ldots, R_n(q, p)\}$ of n independent functions in involution. Prove the following theorem (Sharan, 1982): if $F(q, p)$ has zero Poisson bracket with all the $R_k(q, p)$ then F is a function of the R_k, that is, there exists a function $\Phi : \mathbb{R}^n \to \mathbb{R}$ such that $F(q, p) = \Phi(R_1(q, p), \ldots, R_n(q, p))$. Hint: Lemma 9.7.1.

9.19 Constants of the motion associated with algebraically simple Hamiltonians may involve transcendental functions. Consider a two-degree-of-freedom system with Hamiltonian (Hietarinta, 1984)

$$H = \frac{1}{2}(p_x^2 + p_y^2) + 2yp_xp_y - x.$$

(a) Write down Hamilton's equations. (b) Let F_1 and F_2 be defined by

$$F_1 = p_y e^{p_x^2}, \qquad F_2 = -ye^{-p_x^2} + \frac{\sqrt{2\pi}}{4}p_y e^{p_x^2}\,\mathrm{erf}(\sqrt{2}\,p_x),$$

where

$$\mathrm{erf}(x) = \frac{2}{\sqrt{\pi}}\int_0^x e^{-t^2}\,dt$$

is the *error function*. Prove that F_1 and F_2 are constants of the motion. (c) Are F_1 and F_2 in involution?

9.20 Let a two-degree-of-freedom system be described by the Hamiltonian

$$H = \frac{1}{2}(p_x^2 + p_y^2) + V(x, y)$$

and suppose the potential energy V is a homogenous function of degree -2:

$$V(\lambda x, \lambda y) = \lambda^{-2}V(x, y) \qquad \forall \lambda > 0.$$

Show that

$$\Phi = (xp_y - yp_x)^2 + 2(x^2 + y^2)V(x, y)$$

is a second constant of the motion independent of the Hamiltonian (Yoshida, 1987). Therefore, this system is integrable.

9.21 Consider a two-degree-of-freedom system described by the Hamiltonian

$$H = \frac{1}{2}(p_1^2 + p_2^2) + a(q_1^2 - q_2^2)^2$$

where $a \neq 0$ is a constant. By means of Yoshida's non-integrability theorem (Theorem 9.8.1) prove that this system is non-integrable (Almeida, Moreira & Santos, 1998).

9.22 A particle of mass m moves on the x-axis confined to the interval $(-a, a)$ by the potential

$$V(x) = V_0 \tan^2\left(\frac{\pi x}{2a}\right), \qquad V_0 > 0.$$

(a) In the semiclassical regime, in which the quantisation rule $J = n\hbar$ holds where J is the action variable, show that the energy levels are

$$E_n = V_0\left[\left(1 + \frac{nh}{4a\sqrt{2mV_0}}\right)^2 - 1\right].$$

(b) If V_0 varies adiabatically, show that $\sqrt{E + V_0} - \sqrt{V_0}$ is an adiabatic invariant.

9.23 In spherical coordinates an isotropic harmonic oscillator has potential energy $V(r) = kr^2/2$. (a) Find the action variables for the isotropic harmonic oscillator and show that

$$E = \sqrt{\frac{k}{m}}\,(2J_r + J_\theta + J_\phi)\,.$$

For the computation of J_r use the definite integral

$$\int_a^b \frac{\sqrt{(x-a)(b-x)}}{x}\,dx = \frac{\pi}{2}(a+b) - \pi\sqrt{ab}\,, \qquad b > a \geq 0\,.$$

(b) Determine the frequencies and conclude that they are commensurate. Was this expected? Can you explain why the radial frequency is twice as large as the frequencies associated with the angles?

10

Hamiltonian Perturbation Theory

C'est par la logique qu'on démontre, c'est par l'intuition qu'on invente.

Henri Poincaré, *Science et Méthode*

Given that non-integrability is the rule, rare are the mechanical systems whose equations of motion can be completely solved in closed form. Nevertheless, many mechanical systems of great importance, especially in celestial mechanics, differ little from systems we know how to solve exactly. Among these, the quasi-integrable systems stand out, which are characterised by a Hamiltonian that consists in an integrable Hamiltonian plus a "small" perturbation. In such cases, from the known solution of the integrable problem, one seeks the solution of the quasi-integrable problem by a scheme of successive approximations. This chapter is a brief introduction to what Poincaré considered to be the "*problème général de la dynamique*": the perturbation theory of quasi-integrable Hamiltonian systems.

10.1 Statement of the Problem

Perturbative methods have been largely employed in celestial mechanics ever since the investigations of pioneers such as Lagrange and Laplace. A spectacular triumph of perturbation theory was the discovery of the planet Neptune, in 1846, from deviations of the expected orbit of Uranus calculated independently by Urbain Le Verrier and John Couch Adams.

Let us start with two simple examples of mechanical systems that differ little from a corresponding integrable system.

Example 10.1 Suppose one wishes to study approximately the influence exerted by Jupiter on the Earth's motion around the Sun. In a restricted version of the three-body problem, let us make use of an XYZ reference frame with origin at the centre of the Sun. We assume that Jupiter describes a circular orbit about the Sun and that the Earth (of mass m) moves in the same XY-plane of Jupiter's orbit (see Fig. 10.1). Since the mass of the Earth is about three hundred thousand times smaller than that of the Sun, it is reasonable to neglect the deviations of Jupiter's orbit caused by the Earth. Let us introduce a new xyz frame that rotates about the Z-axis with the same angular velocity Ω as that of Jupiter. According to Problem 7.20 the Hamiltonian is

$$h(\mathbf{r}, \mathbf{p}) = \frac{\mathbf{p}^2}{2m} - \boldsymbol{\omega} \cdot (\mathbf{r} \times \mathbf{p}) - \frac{GmM_S}{r} - \frac{GmM_J}{|\mathbf{r} - \mathbf{r}_J|} + \frac{GmM_J}{r_J^3}\mathbf{r}_J \cdot \mathbf{r}, \qquad (10.1)$$

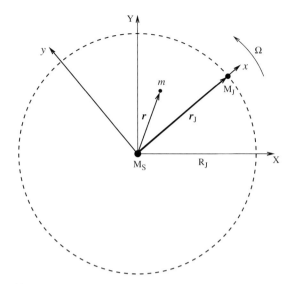

Fig. 10.1 Restricted three-body problem.

where $\boldsymbol{\omega} = \Omega\hat{\mathbf{z}}$. The right-most term in Eq. (10.1) takes into account that the origin of the chosen frame of reference (the centre of the Sun) has an acceleration $\mathbf{a}_{o'} = GM_J\mathbf{r}_J/r_J^3$ with respect to an external inertial reference frame due to the gravitational attraction of Jupiter. In terms of polar coordinates in the xy-plane we have

$$h(q,p) = \frac{1}{2m}\left(p_r^2 + \frac{p_\phi^2}{r^2}\right) - \Omega p_\phi - \frac{GmM_S}{r} - \frac{GmM_J}{|\mathbf{r}-\mathbf{r}_J|} + \frac{GmM_J}{R_J^2}r\cos\phi, \quad (10.2)$$

where $(q,p) = (r,\phi,p_r,p_\phi)$ and $\mathbf{r}_J = (R_J,0)$ is a constant vector in the rotating frame. Since Jupiter's mass is about 1,000 times smaller than the Sun's, on the right-hand side of (10.2) the ratio of the fifth and fourth terms to the third term is small. We can write

$$h(q,p) = h_0(q,p) + \epsilon f(q,p), \quad (10.3)$$

where

$$h_0(q,p) = \frac{1}{2m}\left(p_r^2 + \frac{p_\phi^2}{r^2}\right) - \Omega p_\phi - \frac{GmM_S}{r} \quad (10.4)$$

and

$$f(q,p) = -\frac{GmM_S}{\sqrt{r^2 + R_J^2 - 2rR_J\cos\phi}} + \frac{GmM_S}{R_J^2}r\cos\phi, \quad \epsilon = \frac{M_J}{M_S} \approx 10^{-3}. \quad (10.5)$$

In this case, the small parameter is the ratio of the mass of Jupiter to that of the Sun. Since p_ϕ is a constant of the motion of the unperturbed system, the two-degree-of-freedom Hamiltonian system described by h_0 is integrable.

Example 10.2 The Hamiltonian of a simple pendulum – with zero potential gravitational energy at its lowest point – is

$$h(q, p) = \frac{p^2}{2ml^2} + mgl(1 - \cos q). \tag{10.6}$$

Expanding the cosine in a power series there results

$$h(q, p) = \frac{p^2}{2ml^2} + \frac{mgl}{2}q^2 - \frac{mgl}{4!}q^4 + \frac{mgl}{6!}q^6 - \dots . \tag{10.7}$$

The small oscillations approximation consists in keeping only the quadratic term in q, so that the unperturbed Hamiltonian represents a harmonic oscillator:

$$h_0(q, p) = \frac{p^2}{2ml^2} + \frac{mgl}{2}q^2. \tag{10.8}$$

If the pendulum's energy is such that $E \ll mgl$, the terms containing q^4 and higher powers are successively smaller perturbations but no small parameter appears naturally to characterise the "size" of the perturbation. In order to keep track of the order of the corrections, it is convenient to artificially introduce a parameter ϵ and write

$$\epsilon f(q, p; \epsilon) = -\epsilon \frac{mgl}{4!}q^4 + \epsilon^2 \frac{mgl}{6!}q^6 - \dots . \tag{10.9}$$

A second-order perturbative calculation, for instance, must be conducted up to the first order in the term containing ϵ^2 and up to the second order in the term linear in ϵ. At the end of the computations, one sets $\epsilon = 1$. The complete Hamiltonian is of the form $h(q, p; \epsilon) = h_0(q, p) + \epsilon f(q, p; \epsilon)$. Differently from what happens in most cases of interest, the exact simple pendulum described by the complete Hamiltonian (10.6) is an integrable system as much as the unperturbed system with Hamiltonian (10.8). The simple pendulum is a valuable testing ground for perturbative methods because the exact solution is known.

The Hamiltonian perturbation theory – or canonical perturbation theory – aims primordially at the quasi-integrable systems.

Definition 10.1.1 *A mechanical system is said to be* **quasi-integrable** *if its Hamiltonian is of the form*

$$h(q, p; \epsilon) = h_0(q, p) + \epsilon f(q, p; \epsilon) \tag{10.10}$$

where h_0 is the Hamiltonian of an integrable system, ϵ is a small parameter and

$$f(q, p; \epsilon) = f_1(q, p) + \epsilon f_2(q, p) + \epsilon^2 f_3(q, p) + \dots . \tag{10.11}$$

Introducing action-angle variables (J, ϕ) for the unperturbed integrable system, the Hamiltonian (10.10) takes the form

$$H(J, \phi; \epsilon) = H_0(J) + \epsilon F(J, \phi; \epsilon), \tag{10.12}$$

where

$$F(J, \phi; \epsilon) = F_1(J, \phi) + \epsilon F_2(J, \phi) + \epsilon^2 F_3(J, \phi) + \dots . \tag{10.13}$$

Of course, $H_0(J)$ and $F(J, \phi; \epsilon)$ are, respectively, the functions $h_0(q, p)$ and $f(q, p; \epsilon)$ expressed in terms of the action-angle variables for the unperturbed system. By hypothesis, the perturbation F is a periodic function of each angle variable with period 2π. The angle variables $\phi_1, \ldots, \phi_n$ are interpreted as coordinates on the n-dimensional torus T^n.

The fundamental problem of Hamiltonian perturbation theory can be characterised as follows: find a canonical transformation $(J, \phi) \rightarrow (J', \phi')$ that eliminates the dependence of the Hamiltonian on the new angles up to some desired order in ϵ; if possible, to all orders.

10.2 Generating Function Method

Consider a canonical transformation $(J, \phi) \rightarrow (J', \phi')$ with generating function

$$W(J', \phi; \epsilon) = J' \cdot \phi + \epsilon W^{(1)}(J', \phi) + \epsilon^2 W^{(2)}(J', \phi) + O(\epsilon^3), \tag{10.14}$$

where $J' \cdot \phi = \sum_{k=1}^{n} J'_k \phi_k$ and each $W^{(k)}$ is a 2π-periodic function of each of the angle variables $\phi_1, \ldots, \phi_n$. The goal of this canonical transformation is to transform $H(J, \phi; \epsilon)$ into a new Hamiltonian $H'(J', \phi'; \epsilon)$ of the form

$$H'(J', \phi'; \epsilon) = H'_0(J') + \epsilon H'_1(J') + \epsilon^2 H'_2(J') + \epsilon^3 F'(J', \phi'; \epsilon). \tag{10.15}$$

In words: the canonical transformation generated by (10.14) must eliminate the angular dependence of the Hamiltonian up to the second order in the small parameter.

According to (8.16), the canonical transformation induced by the generating function (10.14) is

$$J_k = \frac{\partial W}{\partial \phi_k} = J'_k + \epsilon \frac{\partial W^{(1)}}{\partial \phi_k}(J', \phi) + \epsilon^2 \frac{\partial W^{(2)}}{\partial \phi_k}(J', \phi) + O(\epsilon^3), \tag{10.16a}$$

$$\phi'_k = \frac{\partial W}{\partial J'_k} = \phi_k + \epsilon \frac{\partial W^{(1)}}{\partial J'_k}(J', \phi) + \epsilon^2 \frac{\partial W^{(2)}}{\partial J'_k}(J', \phi) + O(\epsilon^3). \tag{10.16b}$$

The generating function (10.14) does not explicitly depend on time, so

$$H'(J', \phi'; \epsilon) = H(J, \phi; \epsilon). \tag{10.17}$$

Combining this with (10.12), (10.13), (10.15) and (10.16a) we have

$$H_0\big(J' + \epsilon \nabla_\phi W^{(1)} + \epsilon^2 \nabla_\phi W^{(2)}\big) + \epsilon F_1(J' + \epsilon \nabla_\phi W^{(1)}, \phi)$$
$$+ \epsilon^2 F_2(J', \phi) + O(\epsilon^3) = H'_0(J') + \epsilon H'_1(J') + \epsilon^2 H'_2(J') + O(\epsilon^3). \tag{10.18}$$

Expanding H_0 and F_1 in Taylor series about J', this last equation becomes

$$H_0(J') + \epsilon \nabla_{J'} H_0(J') \cdot \nabla_\phi W^{(1)} + \frac{\epsilon^2}{2} \frac{\partial^2 H_0}{\partial J'_k \partial J'_l} \frac{\partial W^{(1)}}{\partial \phi_k} \frac{\partial W^{(1)}}{\partial \phi_l}$$
$$+ \epsilon^2 \nabla_{J'} H_0(J') \cdot \nabla_\phi W^{(2)} + \epsilon F_1(J', \phi) + \epsilon^2 \nabla_{J'} F_1(J', \phi) \cdot \nabla_\phi W^{(1)}$$
$$+ \epsilon^2 F_2(J', \phi) + O(\epsilon^3) = H'_0(J') + \epsilon H'_1(J') + \epsilon^2 H'_2(J') + O(\epsilon^3), \tag{10.19}$$

where the convention of sum over repeated indices is used. Equating the coefficients of the same powers of ϵ on both sides of this last equation, to zeroth order we obtain

$$H_0'(J') = H_0(J') . \tag{10.20}$$

This is an obvious result: to zeroth order the new Hamiltonian coincides with the old one. The equality of the first-order terms yields

$$\omega(J') \cdot \nabla_\phi W^{(1)}(J', \phi) + F_1(J', \phi) = H_1'(J') \tag{10.21}$$

where $\omega = \nabla_J H_0$ is the unperturbed frequencies vector:

$$\omega_k(J') = \frac{\partial H_0}{\partial J_k'}(J') = \frac{\partial H_0'}{\partial J_k'}(J') . \tag{10.22}$$

Equating the second-order terms we get

$$\omega(J') \cdot \nabla_\phi W^{(2)}(J', \phi) + \frac{1}{2} \frac{\partial^2 H_0}{\partial J_k' \partial J_l'} \frac{\partial W^{(1)}}{\partial \phi_k} \frac{\partial W^{(1)}}{\partial \phi_l}$$
$$+ \nabla_{J'} F_1(J', \phi) \cdot \nabla_\phi W^{(1)}(J', \phi) + F_2(J', \phi) = H_2'(J') . \tag{10.23}$$

The differential Eq. (10.23) for $W^{(2)}$ has the same form as the differential Eq. (10.21) for $W^{(1)}$. From the third order on the equations become prohibitively complicated and will not be written down. But to all orders the perturbative scheme always demands the solution of an equation of the form (10.21), which, therefore, is called the **fundamental equation of perturbation theory**. Before discussing how to solve it in general, it is worthwhile to examine first the case of systems with only one degree of freedom, which do not suffer from certain pathologies that can afflict systems with two or more degrees of freedom.

10.3 One Degree of Freedom

If $n = 1$, Eq. (10.21) reduces to

$$\omega(J') \frac{\partial W^{(1)}}{\partial \phi}(J', \phi) + F_1(J', \phi) = H_1'(J') . \tag{10.24}$$

Integrating this equation over an angular period and taking into account that $W^{(1)}$ is a periodic function of ϕ with period 2π, we get

$$H_1'(J') = \frac{1}{2\pi} \int_0^{2\pi} F_1(J', \phi) d\phi \overset{\text{def}}{=} \langle F_1 \rangle . \tag{10.25}$$

This determines the first-order correction to the unperturbed Hamiltonian as the average of the perturbation in a period of the unperturbed motion, just like the perturbation theory of quantum mechanics.

Having found H_1', if $\omega(J') \neq 0$ an integration immediately gives $W^{(1)}$ in the form

$$W^{(1)}(J', \phi) = \frac{1}{\omega(J')} \int_0^\phi \left[H_1'(J') - F_1(J', \psi) \right] d\psi , \tag{10.26}$$

where we have discarded an irrelevant "constant" of integration $C(J')$. With H_1' and $W^{(1)}$ already determined, the second-order correction to the unperturbed Hamiltonian is obtained by taking the average of the one-dimensional version of Eq. (10.23):

$$H_2'(J') = \langle F_2 \rangle + \left\langle \frac{\partial F_1}{\partial J'} \frac{\partial W^{(1)}}{\partial \phi} \right\rangle + \frac{1}{2} \frac{\partial \omega}{\partial J'} \left\langle \left(\frac{\partial W^{(1)}}{\partial \phi} \right)^2 \right\rangle. \qquad (10.27)$$

Substituting (10.26) and (10.27) into the one-dimensional version of (10.23), one determines $W^{(2)}$ by a direct integration, just like we did to find $W^{(1)}$.

Example 10.3 Determine the corrections to the frequency of a simple pendulum up to the second order.

Solution

From Example 10.2, up to the second order beyond the small oscillations approximation, the Hamiltonian for the pendulum is

$$h(q, p; \epsilon) = \frac{p^2}{2ml^2} + \frac{ml^2\omega_0^2}{2}q^2 - \epsilon \frac{ml^2\omega_0^2}{4!}q^4 + \epsilon^2 \frac{ml^2\omega_0^2}{6!}q^6, \qquad (10.28)$$

where $\omega_0 = \sqrt{g/l}$. By Example 9.9, the action-angle variables for the unperturbed system are given by

$$q = \sqrt{\frac{2J}{ml^2\omega_0}} \sin\phi, \qquad p = \sqrt{2m\omega_0 l^2 J} \cos\phi. \qquad (10.29)$$

Introducing these expressions into Eq. (10.28) we are led to

$$H(J, \phi; \epsilon) = \omega_0 J - \epsilon \frac{J^2}{6ml^2} \sin^4\phi + \epsilon^2 \frac{J^3}{90m^2l^4\omega_0} \sin^6\phi. \qquad (10.30)$$

Comparing this with (10.12) and (10.13) we identify

$$H_0(J) = \omega_0 J, \qquad (10.31)$$

as well as

$$F_1(J, \phi) = -\frac{J^2}{6ml^2} \sin^4\phi, \qquad F_2(J, \phi) = \frac{J^3}{90m^2l^4\omega_0} \sin^6\phi. \qquad (10.32)$$

Therefore, in accordance with (10.25) and using

$$\langle \cos^{2n} \rangle = \langle \sin^{2n} \rangle = \frac{1}{2\pi} \int_0^{2\pi} \sin^{2n}\phi \, d\phi = \frac{(2n)!}{(2^n n!)^2}, \qquad (10.33)$$

we obtain

$$H_1'(J') = -\frac{J'^2}{6ml^2} \langle \sin^4 \rangle = -\frac{J'^2}{16ml^2}. \qquad (10.34)$$

Since $\omega = \omega_0$, we have $\partial\omega/\partial J' = 0$ and Eq. (10.27) reduces to

$$H_2'(J') = \langle F_2 \rangle + \left\langle \frac{\partial F_1}{\partial J'} \frac{\partial W^{(1)}}{\partial \phi} \right\rangle. \qquad (10.35)$$

We have

$$\langle F_2 \rangle = \frac{J'^3}{90m^2l^4\omega_0} \langle \sin^6 \rangle = \frac{J'^3}{288m^2l^4\omega_0} \tag{10.36}$$

and, with the help of (10.26),

$$\left\langle \frac{\partial F_1}{\partial J'} \frac{\partial W^{(1)}}{\partial \phi} \right\rangle = -\frac{J'}{3ml^2\omega_0} \left\langle \sin^4 \phi \left[H_1'(J') - F_1(J',\phi) \right] \right\rangle$$

$$= -\frac{J'^3}{3m^2l^4\omega_0} \left[-\frac{\langle \sin^4 \rangle}{16} + \frac{\langle \sin^8 \rangle}{6} \right] = -\frac{17J'^3}{2304m^2l^4\omega_0}. \tag{10.37}$$

Introducing these two last results into Eq. (10.35) we arrive at

$$H_2'(J') = -\frac{J'^3}{256m^2l^4\omega_0}. \tag{10.38}$$

It follows that the second-order transformed Hamiltonian is

$$H'(J') = H_0'(J') + \epsilon H_1'(J') + \epsilon^2 H_2'(J') = \omega_0 J' - \epsilon \frac{J'^2}{16ml^2} - \epsilon^2 \frac{J'^3}{256m^2l^4\omega_0}. \tag{10.39}$$

As a consequence, for the second-order frequency we get

$$\omega' = \frac{\partial H'}{\partial J'} = \omega_0 - \epsilon \frac{J'}{8ml^2} - \epsilon^2 \frac{3J'^2}{256m^2l^4\omega_0}. \tag{10.40}$$

Physically, it makes much more sense to express ω' as a function of the energy, which up to the second order is given by

$$E = H' = \omega_0 J' - \epsilon \frac{J'^2}{16ml^2} - \epsilon^2 \frac{J'^3}{256m^2l^4\omega_0}. \tag{10.41}$$

The first-order solution of this equation for J' is

$$J' = \frac{E}{\omega_0} + \epsilon \frac{E^2}{16ml^2\omega_0^3}. \tag{10.42}$$

Inserting this into (10.40) one obtains, up to the second order in ϵ,

$$\omega' = \omega_0 - \epsilon \frac{E}{8ml^2\omega_0} - \epsilon^2 \frac{5E^2}{256m^2l^4\omega_0^3}. \tag{10.43}$$

Finally, setting $\epsilon = 1$,

$$\omega' = \omega_0 \left[1 - \frac{1}{8} \frac{E}{ml^2\omega_0^2} - \frac{5}{256} \left(\frac{E}{ml^2\omega_0^2} \right)^2 \right]. \tag{10.44}$$

This result is consistent with the exact value of the frequency (Problem 10.1).

10.4 Several Degrees of Freedom

Since the term $\omega \cdot \nabla_\phi W^{(1)}$ in Eq. (10.21) has zero average over the torus T^n, we have

$$\frac{1}{(2\pi)^n} \int_0^{2\pi} \cdots \int_0^{2\pi} \left[H_1'(J') - F_1(J', \phi) \right] d\phi_1 \ldots d\phi_n = 0 \, . \qquad (10.45)$$

This gives the first-order correction to the Hamiltonian (in self-explanatory notation):

$$H_1'(J') = \frac{1}{(2\pi)^n} \oint_{T^n} F_1(J', \phi) \, d^n\phi \stackrel{\text{def}}{=} \langle F_1 \rangle \, . \qquad (10.46)$$

Availing ourselves of the periodicity of the functions $W^{(1)}$ and F_1 on T^n we can develop each of them in a Fourier series:

$$W^{(1)}(J', \phi) = \sum_{m \in \mathbb{Z}^n} \hat{W}_m(J') e^{im \cdot \phi} \, , \qquad (10.47)$$

$$F_1(J', \phi) = \sum_{m \in \mathbb{Z}^n} \hat{F}_m(J') e^{im \cdot \phi} \, , \qquad (10.48)$$

where $m \cdot \phi = m_1 \phi_1 + \ldots + m_n \phi_n$. Substituting (10.47) and (10.48) into Eq. (10.21) we find

$$im \cdot \omega(J') \hat{W}_m(J') + \hat{F}_m(J') = 0 \, , \qquad m \neq (0, \ldots, 0) \equiv \mathbf{0} \, , \qquad (10.49)$$

whence

$$\hat{W}_m(J') = \frac{i \hat{F}_m(J')}{m \cdot \omega(J')} \, , \qquad m \neq \mathbf{0} \, . \qquad (10.50)$$

The Fourier coefficient $\hat{W}_0(J')$ is not determined by Eq. (10.21) and we can simply take $\hat{W}_0(J') = 0$. The equality of the $m = \mathbf{0}$ Fourier coefficient on both sides of (10.21) gives

$$\hat{F}_0(J') = H_1'(J') \, , \qquad (10.51)$$

which is exactly the same thing as (10.46).

Once $W^{(1)}$ has been found, the second-order correction to the transformed Hamiltonain is obtained by taking the average of Eq. (10.23) over the torus T^n:

$$H_2'(J') = \langle F_2 \rangle + \left\langle \nabla_{J'} F_1 \cdot \nabla_\phi W^{(1)} \right\rangle + \frac{1}{2} \frac{\partial^2 H_0}{\partial J_k' \partial J_l'} \left\langle \frac{\partial W^{(1)}}{\partial \phi_k} \frac{\partial W^{(1)}}{\partial \phi_l} \right\rangle \, . \qquad (10.52)$$

Next $W^{(2)}(J', \phi)$ and $F_2(J', \phi)$ are to be expanded in Fourier series like (10.47) and (10.48), the insertion of which in Eq. (10.23) allows the Fourier coefficients of $W^{(2)}$ with $m \neq \mathbf{0}$ to be determined.

In most cases, Eq. (10.49) can only be solved for the Fourier coefficients of $W^{(1)}$ if the factors $m \cdot \omega(J')$ do not vanish for any $m \neq \mathbf{0}$. If this is not the case, perturbation theory may fail because $W^{(1)}$ may not exist. Thus, as a general rule, the perturbative series does not exist if the frequencies of the unperturbed system are such that $m \cdot \omega(J') = 0$ for some $m \neq \mathbf{0}$.

Definition 10.4.1 *Let $\mathbb{Z}^n \setminus \{\mathbf{0}\}$ denote $\mathbb{Z}^n$ without the point $\mathbf{0} = (0,\dots,0)$. The frequencies $\omega \in \mathbb{R}^n$ are said to be* **non-resonant** *if, for all $m \in \mathbb{Z}^n \setminus \{\mathbf{0}\}$,*

$$m \cdot \omega = m_1\omega_1 + \dots + m_n\omega_n \neq 0. \tag{10.53}$$

Otherwise, that is, if there exists $m \in \mathbb{Z}^n \setminus \{\mathbf{0}\}$ such that $m \cdot \omega = 0$, the frequencies are said to be **resonant**.

In order for the frequencies $\omega = (\omega_1,\dots,\omega_n) \in \mathbb{R}^n$ to be non-resonant, at least one of them has to be irrational.

Example 10.4 The frequencies $(1, \pi, \pi^2)$ are non-resonant because π is a transcendental number – that is, it is not the root of any polynomial equation with integer coefficients. The frequencies $(1, \sqrt{2}, 2\sqrt{2})$, on the other hand, are clearly resonant: taking $m = (0, -2, 1)$ one has $m \cdot \omega = 0$.

Of course, the requirement that the unperturbed frequencies be non-resonant guarantees the existence of the Fourier coefficients of $W^{(1)}$ but does not ensure by itself the convergence of the series (10.47).

In particular cases the presence of resonant frequencies may not cause any trouble. It is quite possible that the Fourier coefficient $\hat{F}_m(J')$ be zero whenever $m \cdot \omega(J') = 0$. Then Eq. (10.49) is identically satisfied and one can simply choose $\hat{W}_m(J') = 0$. It may also happen that only a finite number of Fourier coefficients are non-zero, so that the expansion (10.47) reduces to a trigonometric polynomial and the convergence issue does not arise.

Example 10.5 Given two coupled harmonic oscillators with Hamiltonian

$$H = \frac{1}{2}\left(p_1^2 + \omega_1^2 q_1^2\right) + \frac{1}{2}\left(p_2^2 + \omega_2^2 q_2^2\right) + \epsilon q_1^2 q_2^2, \tag{10.54}$$

determine: (a) the Hamiltonian and the frequencies corrected up to the first order; (b) the function $W^{(1)}$.

Solution

(a) In terms of the action-angle variables defined by

$$q_k = \sqrt{\frac{2J_k}{\omega_k}} \sin\phi_k, \qquad p_k = \sqrt{2\omega_k J_k}\cos\phi_k, \qquad k = 1,2, \tag{10.55}$$

the Hamiltonian (10.54) becomes

$$H = \omega_1 J_1 + \omega_2 J_2 + \epsilon \frac{4J_1 J_2}{\omega_1 \omega_2} \sin^2\phi_1 \sin^2\phi_2. \tag{10.56}$$

Therefore,

$$F_1(J,\phi) = \frac{4J_1 J_2}{\omega_1 \omega_2}\sin^2\phi_1 \sin^2\phi_2, \quad F_2 = F_3 = \dots = 0 \tag{10.57}$$

Proposition 10.5.1   *If F_1 is infinitely differentiable and J is such that $\omega(J)$ satisfies (10.77), then the Fourier series*

$$A^{(0)}(\psi) = - \sum_{\substack{m \in \mathbb{Z}^n \\ m \neq 0}} \frac{m \hat{F}_{1m}(J)}{\omega(J) \cdot m} e^{im \cdot \psi} \tag{10.79}$$

converges uniformly to an infinitely differentiable function.

Proof   With $||G|| = \sqrt{|G_1|^2 + \ldots + |G_n|^2}$ we have

$$\left|\left| \sum_{\substack{m \in \mathbb{Z}^n \\ m \neq 0}} \frac{m \hat{F}_{1m}(J)}{\omega(J) \cdot m} e^{im \cdot \psi} \right|\right| \leq \sum_{\substack{m \in \mathbb{Z}^n \\ m \neq 0}} \frac{||m|| \, |\hat{F}_{1m}(J)|}{|\omega(J) \cdot m|} . \tag{10.80}$$

On the one hand, we have $||m|| \leq |m|$. On the other hand, since F_1 is infinitely differentiable, for any natural number s there exists $C > 0$ such that $|\hat{F}_{1m}(J)| \leq C|m|^{-s}$ (see Appendix I). These observations combined with (10.77) lead to

$$\left|\left| \sum_{\substack{m \in \mathbb{Z}^n \\ m \neq 0}} \frac{m \hat{F}_{1m}(J)}{\omega(J) \cdot m} e^{im \cdot \psi} \right|\right| \leq M \sum_{\substack{m \in \mathbb{Z}^n \\ m \neq 0}} \frac{1}{|m|^{s-\mu-1}} . \tag{10.81}$$

where $M = C/\gamma$. The summation can be performed in the following way:

$$\sum_{\substack{m \in \mathbb{Z}^n \\ m \neq 0}} \frac{1}{|m|^{s-\mu-1}} = \sum_{r=1}^{\infty} \sum_{\substack{m \in \mathbb{Z}^n \\ |m|=r}} \frac{1}{r^{s-\mu-1}} . \tag{10.82}$$

The number of possible values of each $m_k \in \mathbb{Z}$ such that $|m_1| + \ldots + |m_n| = r$ surely does not exceed $4r$. So, taking $s > \mu + n + 3$,

$$\left|\left| \sum_{\substack{m \in \mathbb{Z}^n \\ m \neq 0}} \frac{m \hat{F}_{1m}(J)}{\omega(J) \cdot m} e^{im \cdot \psi} \right|\right| \leq M \sum_{r=1}^{\infty} \frac{(4r)^n}{r^{s-\mu-1}} \leq \tilde{M} \sum_{r=1}^{\infty} \frac{1}{r^2} < \infty, \tag{10.83}$$

which proves the uniform convergence of the series (10.79). By the same argument one proves the uniform convergence of the series obtained by differentiating the series (10.79) term by term any number of times. Therefore, $A^{(0)}$ is infinitely differentiable.   □

Thus, if both H_0 and the perturbation F given by (10.13) are infinitely differentiable, all functions $A^{(k)}$ and $B^{(k)}$ obey equations of the form (10.70) with the right-hand side infinitely differentiable because it consists of polynomials in H_0, $F_1 \ldots, F_{k-1}$, $A^{(0)} \ldots, A^{(k-1)}$, $B^{(0)} \ldots, B^{(k-1)}$ and their derivatives. Therefore, under the hypothesis (10.77), $A^{(k)}$ and $B^{(k)}$ exist and are infinitely differentiable functions to all orders of perturbation theory. If H_0 and the perturbation F are analytic functions, the Fourier coefficients of F_k decay exponentially – see Eq. (I.12) in Appendix I – and one proves that the functions $A^{(k)}$ and $B^{(k)}$ are also analytic.

Convergence of the Lindstedt Series

An incomparably more difficult problem is to prove the convergence of the Lindstedt series. Weierstrass suspected that certain formal power series in the n-body problem, analogous to the Lindstedt series, were convergent, but could not find a rigorous proof. In contrast, Poincaré, in his extensive investigation of the three-body problem carried out in the course of nearly thirty years, strongly argued that those series are divergent for many initial conditions and conjectured that they are divergent for generic initial conditions (Barrow-Green, 1997; Dumas, 2014).

That is essentially how matters stood until 1954, when Andrei Kolmogorov announced, without presenting a detailed proof, a theorem on the persistence of invariant tori under small perturbations.[1] He conceived of an iterative process, similar to Newton's method for finding zeros of a function, which converges much faster than the series of conventional perturbation theory and indirectly establishes the convergence of the Lindstedt series.[2] The ideas of Kolmogorov (1954, 1957) were converted into a full-fledged rigorous proof by his pupil Vladimir Arnold (1963) and, independently, by Jürgen Moser (1962).

Theorem 10.5.1 (The KAM Theorem)   *Let there be given an n-degree-of-freedom quasi-integrable system such that its Hamiltonian H is an analytic function and the unperturbed system is non-degenerate in the open set $U \in \mathbb{R}^n$.*

(a) If ϵ is sufficiently small, for each $J_0 \in U$ such that $\omega(J_0)$ satisfies (10.77), there exists a deformation of the torus $\{J_0\} \times T^n$ in invariant tori for the perturbed quasi-integrable system with Hamiltonian (10.12).

(b) Suppose the open set $U \subset \mathbb{R}^n$ is bounded and denote by $\Sigma_\epsilon \subset U \times T^n$ the corresponding set of surviving tori, those which undergo only a deformation in invariant tori. Then,

$$\lim_{\epsilon \to 0} \frac{volume\,(\Sigma_\epsilon)}{volume\,(U \times T^n)} = 1 \,, \tag{10.84}$$

where the volume of a subset of $\mathbb{R}^{2n}$ is defined by its Lebesgue measure.

The proof of the KAM theorem does not consist in directly proving that the Lindstedt series are convergent for each J_0 specified in (a). In the original formulation of Kolmogorov (1954) what is proved is that for ϵ sufficiently small there exists a near-identity canonical transformation

$$J = J' + \epsilon\, U(J', \phi', \epsilon), \qquad \phi = \phi' + \epsilon\, V(J', \phi', \epsilon) \tag{10.85}$$

such that

$$H'(J', \phi', \epsilon) = C + \omega_0 \cdot (J' - J_0) + R(J', \phi', \epsilon) \tag{10.86}$$

---

[1] Although Kolmogorov never published a detailed proof, the mathematician Ya. G. Sinai reportedly said that in 1957 he attended lectures in which Kolmogorov gave a complete proof of his stability theorem (Yandell, 2002).

[2] It took over three decades after Kolmogorov's announcement for a direct proof of convergence of the Lindstedt series to be discovered (Eliasson, 1996). The interested reader is referred to Bartucelli and Gentile (2002) for a review of this approach, which consists in facing head-on the King Oscar II prize problem posed by Weierstrass.

where C is a constant, $\omega_0 = \omega(J_0)$ and the function R is of order $||J' - J_0||^2$ – that is,

$$R(J', \phi', \epsilon) = \sum_{k,l} C_{kl}(J', \phi', \epsilon)(J' - J_0)_k(J' - J_0)_l . \qquad (10.87)$$

The transformed Hamilton's equations are

$$\dot{\phi}' = \omega_0 + \frac{\partial R}{\partial J'}, \qquad \dot{J}' = -\frac{\partial R}{\partial \phi'} . \qquad (10.88)$$

These equations admit the solution $J'(t) = J_0$, $\phi'(t) = \phi'_0 + \omega_0 t$ because both $\partial R / \partial \phi'$ and $\partial R / \partial J'$ vanish for $J' = J_0$ inasmuch as R depends quadratically on the difference $J' - J_0$. In terms of the original canonical variables expressed by (10.85), this exact solution of the perturbed Hamilton's equations takes precisely the form of Eqs. (10.64) and (10.65). It is, therefore, a small deformation of the invariant torus $\{J_0\} \times T^n$ associated with H_0.

Part (b) of the KAM theorem is the mathematically precise way of saying that "almost all" invariant tori of the unperturbed system are only slightly deformed by a small perturbation: if ϵ is sufficiently small the phase space volume occupied by the tori which are destroyed is comparatively very much smaller than the volume occupied by the preserved tori.

The reader interested in the details of the proof of the KAM theorem, which is too long and involved to be included here, is referred to the original sources (Kolmogorov, 1954, 1957; Arnold, 1963; Moser, 1962, 1967, 1973). In Benettin *et al.* (1984) there is a proof of part (a) of the KAM theorem with technical simplifications achieved by the use of Lie series to implement the iterative scheme of perturbation theory, an approach that is described in outline in Section 10.7.

The action variables J_0 such that the frequencies $\omega(J_0)$ obey condition (10.77) constitute an extremely complex set with fractal structure, a Cantor-like set[3] of positive measure. The tori that are destroyed by the perturbation lie densely interwoven with the tori that are only slightly deformed, in such a way that small changes in the initial conditions can lead to widely different outcomes: generic Hamiltonian systems display an intricate admixture of regular and irregular (chaotic) motions.

10.6 Stability: Eternal or Long Term

The Kolmogorov-Arnold-Moser theory establishes *eternal stability* (in an infinite time scale) of *a fraction* of the trajectories of the unperturbed system, those whose frequencies are sufficiently irrational. The fate of the remaining trajectories can be highly complex: chaotic behaviour may take place even when the perturbation is small, if the time scale considered is very long. Hamiltonian perturbation theory is powerful enough to prove stability of *all* trajectories of generic quasi-integrable systems as long as one accepts stability in a *finite time scale*. If the time scale involved is sufficiently long (of the order of the age of the Universe for astronomical systems) the result is physically more relevant than

---

[3] For the definition and main properties of the Cantor set, consult, for instance, Boas, Jr. (1996).

the KAM theorem. In the decade of the 1950s, soon after the announcement of the KAM theorem, Moser and Littlewood conjectured that the stability time scale grows more rapidly than any power of $1/\epsilon$; in 1968 Arnold put forward a similar conjecture (Dumas, 2014).

The most important and far-reaching results on long-term stability have been obtained by N. N. Nekhoroshev, a student of Arnold, and confirm the conjectures of Moser, Littlewood and Arnold: the stability time scale depends exponentially on an inverse power of the small parameter that measures the strength of the perturbation.

Theorem 10.6.1 (Nekhoroshev)   *Consider a quasi-integrable system such that the Hamiltonian H is an analytic function and H_0 satisfies a steepness condition.[4] Then, there exist positive constants a, b, A, B and ϵ_0 such that, for $0 < \epsilon < \epsilon_0$, any solution $J(t), \phi(t)$ of the equations of motion associated with the perturbed Hamiltonian (10.12) satisfies*

$$||J(t) - J(0)|| < B\epsilon^b \qquad whenever \qquad 0 \le t \le \frac{A}{\epsilon} \exp\left(\epsilon^{-a}\right). \qquad (10.89)$$

The proof of this theorem (Nekhoroshev, 1977), which is technically highly complex, will be omitted: it is no less a titanic tour de force than the proof of the KAM theorem.

The conclusion of Nekhoroshev's theorem that $||J(t) - J(0)|| < B\epsilon^b$ in a finite time scale is weaker than the eternal confinement of trajectories to invariant tori established by the KAM theorem. As a compensation, the conclusion of Nekhoroshev's theorem holds for *all* initial conditions. Furthermore, the time scale $T = (A/\epsilon) \exp\left(\epsilon^{-a}\right)$ grows extremely fast as the strength of the perturbation decreases and can, for all practical purposes, be as long as eternity. Indeed, a recent computer-aided study of the Sun-Jupiter-Saturn system indicates perturbative stability in the time scale of the age of the universe (Giorgilli, Locatelli & Sansottera, 2013).

The theories of Kolmogorov-Arnold-Moser and Nekhoroshev are stupendous accomplishments but do not solve the stability problem of the solar system. It is possible that the solar system is in a dynamical region external to the stable tori predicted by the KAM theory. At present, computer simulations only allow probabilistic predictions on the fate of the solar system: there is a small but non-negligible probability (about 1%) of disaggregation of the solar system in a time scale of a few billion years (Laskar, 2013).

10.7 Lie Series Method

Traditional canonical perturbation theory makes use of a generating function W, which depends on old and new canonical variables. This inconvenient mixing of variables makes the equations of perturbation theory forbiddingly unwieldy from the third order on. If the relevant canonical transformation is expressed as a Lie series, the new canonical variables are given directly in terms of the old ones and vice-versa, which makes it easier to extend perturbative calculations to higher orders.

---

[4] For the definition of *steepness*, which is a generalisation of convexity, consult Nekhoroshev (1977) or Dumas (2014).

Consider a mechanical system with Hamiltonian of the form

$$h(q,p;\epsilon) = h_0(q,p) + \epsilon f_1(q,p) + \epsilon^2 f_2(q,p) + \epsilon^3 f_3(q,p) + \ldots, \tag{10.90}$$

where h_0 is the Hamiltonian of an integrable system. For the time being it will not be assumed that the Hamiltonian is expressed in terms of the action-angle variables for the unperturbed system.

Let us search for a canonical transformation $q = u(q',p')$, $p = v(q',p')$ such that the Hamiltonian expressed in terms of the new variables is free of the first order term in ϵ. The transformed Hamiltonian h' is just the original Hamiltonian written as a function of the new variables: $h'(q',p') = h(q,p) = h(u(q',p'), v(q',p'))$. If X is the generator of the canonical transformation with parameter ϵ, Eq. (8.138) allows us to write

$$h'(q',p') = \left(e^{\epsilon \mathcal{L}_X} h\right)(q',p'), \tag{10.91}$$

$$(q,p) = e^{\epsilon \mathcal{L}_X}(q',p'). \tag{10.92}$$

Therefore, we have

$$h' = \left(\mathbb{I} + \epsilon \mathcal{L}_X + \frac{\epsilon^2}{2}\mathcal{L}_X^2 + \ldots\right)(h_0 + \epsilon f_1 + \epsilon^2 f_2 + \ldots)$$

$$= h_0 + \epsilon\left(\mathcal{L}_X h_0 + f_1\right) + \epsilon^2\left(\frac{1}{2}\mathcal{L}_X^2 h_0 + \mathcal{L}_X f_1 + f_2\right) + O(\epsilon^3). \tag{10.93}$$

Recalling that $\mathcal{L}_X f = \{f, X\}$, the removal of the first-order term requires that X satisfy the following first-order partial differential equation:

$$\{h_0, X\} = -f_1. \tag{10.94}$$

With X so determined, the transformed Hamiltonian is

$$h' = h_0 + \epsilon^2 h_2' + O(\epsilon^3) \tag{10.95}$$

where

$$h_2' = \frac{1}{2}\mathcal{L}_X^2 h_0 + \mathcal{L}_X f_1 + f_2. \tag{10.96}$$

Using (10.94) this reduces to

$$h_2' = \frac{1}{2}\mathcal{L}_X f_1 + f_2. \tag{10.97}$$

In order to remove the second order term, we perform a new canonical transformation $(q',p') \to (q'',p'')$ with generator X' such that

$$h'' = \left(e^{\epsilon^2 \mathcal{L}_{X'}} h'\right)(q'',p''), \qquad (q',p') = e^{\epsilon^2 \mathcal{L}_{X'}}(q'',p''). \tag{10.98}$$

It follows that

$$h'' = \left(\mathbb{I} + \epsilon^2 \mathcal{L}_{X'} + \frac{\epsilon^4}{2}\mathcal{L}_{X'}^2 + \ldots\right)(h_0 + \epsilon^2 h_2' + \ldots)$$

$$= h_0 + \epsilon^2\left(\mathcal{L}_{X'} h_0 + h_2'\right) + O(\epsilon^3). \tag{10.99}$$

The removal of the second-order term is achieved by requiring that X' obey the equation

$$\{h_0, X'\} = -h_2'. \tag{10.100}$$

Iterating this procedure we can rid the Hamiltonian of all perturbation terms up to any desired order. It is not hard to see that at every order of the iterative perturbation scheme one must solve an equation of the form (10.94).

An important difference between the Lie series method and the generating function method deserves mention: each function $W^{(k)}$ in the perturbative series (10.14) generates an exact canonical transformation, whereas the Lie series truncated at any order yields a transformation that is only approximately canonical.

Equations of the form (10.94) are easy to solve only if h_0 depends on the ps alone. This is automatically the case when the Hamiltonian is expressed in terms of the action-angle variables for the unperturbed system. Let us illustrate the procedure in an elementary problem, whose exact solution is trivially found.

Example 10.6   A particle in one-dimensional motion subject to a constant force of magnitude $|\epsilon|$, treated as a perturbation, is described by the Hamiltonian

$$h(q,p;\epsilon) = \frac{p^2}{2m} - \epsilon q \equiv h_0 + \epsilon f. \qquad (10.101)$$

Eliminate the perturbation from h up to the first order and find the corresponding solution of the equations of motion for the canonical variables (q,p).

Solution

We have $h_0 = p^2/2m$ and $f_1 = -q$, with $f_2 = f_3 = \ldots = 0$. Equation (10.94) takes the form

$$-\frac{p}{m}\frac{\partial X}{\partial q} = q, \qquad (10.102)$$

which admits the solution

$$X(q,p) = -\frac{mq^2}{2p}. \qquad (10.103)$$

The transformation of the canonical variables is given by

$$q = q' + \epsilon\{q', X(q',p')\} = q' + \epsilon\frac{mq'^2}{2p'^2}, \qquad (10.104a)$$

$$p = p' + \epsilon\{p', X(q',p')\} = p' + \epsilon\frac{mq'}{p'}. \qquad (10.104b)$$

Since $h' = h_0$ up to the first order in ϵ, it follows that

$$h'(q',p') = \frac{p'^2}{2m}. \qquad (10.105)$$

The solution of the transformed equations of motion is trivial:

$$p' = p'_0, \qquad q' = q'_0 + \frac{p'_0}{m}t. \qquad (10.106)$$

Substituting this into (10.104) we get

$$q = q'_0 + \frac{p'_0}{m}t + \epsilon\frac{m}{2(p'_0)^2}\left(q'_0 + \frac{p'_0}{m}t\right)^2, \qquad p = p'_0 + \epsilon\frac{m}{p'_0}\left(q'_0 + \frac{p'_0}{m}t\right). \qquad (10.107)$$

Up to the first order in ϵ the inverse of (10.104) is

$$q' = q - \epsilon \frac{mq^2}{2p^2}, \qquad p' = p - \epsilon \frac{mq}{p}, \tag{10.108}$$

and the initial conditions are expressed in terms of the original canonical variables as follows:

$$q'_0 = q_0 - \epsilon \frac{mq_0^2}{2p_0^2}, \qquad p'_0 = p_0 - \epsilon \frac{mq_0}{p_0}. \tag{10.109}$$

The substitution of (10.109) into (10.107), with neglect of the quadratic terms in ϵ, yields

$$q = q_0 + \frac{p_0}{m}t + \frac{\epsilon}{2m}t^2, \qquad p = p_0 + \epsilon t. \tag{10.110}$$

The first-order perturbation theory gives the exact result.

This example raises a question: granted that the first-order result is exact, is it the case that all higher-order corrections vanish and perturbation theory is exact, or do the higher-order terms introduce spurious corrections?

Consider the second-order corrections to the system with Hamiltonian (10.101). Start by determining h'_2 by means of Eq. (10.97):

$$h'_2(q',q') = \frac{1}{2}\{f_1, X\}_{(q',p')} = -\frac{1}{2}\{q, X\}_{(q',p')} = -\frac{1}{2}\frac{\partial X}{\partial p}(q',p') = -\frac{mq'^2}{4p'^2}. \tag{10.111}$$

The generator X' is determined with the use of (10.100):

$$\{h_0, X'\}_{(q',p')} = -h'_2 = \frac{mq'^2}{4p'^2} \implies X' = -\frac{m^2q'^3}{12p'^3}. \tag{10.112}$$

Transformation (10.98) becomes, up to the second order,

$$q' = q'' + \epsilon^2\{q', X'\}_{(q'',p'')} = q'' + \epsilon^2 \frac{\partial X'}{\partial p'}\bigg|_{(q'',p'')} = q'' + \epsilon^2 \frac{m^2q''^3}{4p''^4}, \tag{10.113}$$

$$p' = p'' + \epsilon^2\{p', X'\}_{(q'',p'')} = p'' - \epsilon^2 \frac{\partial X'}{\partial q'}\bigg|_{(q'',p'')} = p'' + \epsilon^2 \frac{m^2q''^2}{4p''^3}. \tag{10.114}$$

Now $h'' = h_0$ up to the second order and the solution of the equations of motion for (q'', p'') is

$$p'' = p''_0, \qquad q'' = q''_0 + \frac{p''_0}{m}t. \tag{10.115}$$

Combining Eqs. (10.114) and (10.104) we obtain

$$p = p''_0 + \epsilon \frac{mq'}{p'} + \epsilon^2 \frac{m^2q''^2}{4p''^3}. \tag{10.116}$$

Since the difference between (q',p') and (q'',p'') is of the second order, this last equation can be rewritten in the form

$$p = p''_0 + \epsilon \frac{mq''}{p''_0} + \epsilon^2 \frac{m^2q''^2}{4p''^3_0}, \tag{10.117}$$

which includes all terms up to the second order in ϵ. Owing to (10.115), the second-order term gives p a quadratic time dependence, which is incorrect. Perturbation theory only provides a good approximation to the true motion if the quadratic term is small compared to the linear term – that is, if

$$\left| \epsilon^2 \frac{m^2 q''^2}{4 p_0''^3} \right| \ll \left| \epsilon \frac{m q''}{p_0''} \right| \implies m|\epsilon q''| \ll p_0''^2, \tag{10.118}$$

In order to clear up the meaning of this condition, let us take, without loss of generality, the initial condition $q_0'' = 0$. In this case, $q'' = (p_0''/m)t$ and condition (10.118) reduces to

$$|\epsilon t| \ll |p_0''|. \tag{10.119}$$

Second-order perturbation theory is a good approximation only in a time scale such that the impulse delivered by the force is small compared with the initial momentum. As Nekhoroshev's theorem indicates, perturbation theory gives good results in a time scale which is the longer the smaller the strength of the perturbation.

Things Not Treated

We have limited our discussion to the most elementary aspects of perturbation theory via Lie series. When the motions of the unperturbed as well as of the perturbed system are bounded, the Lie series perturbative technique can give rise to **secular terms**, spurious corrections that grow indefinitely with time. For a more extensive and sophisticated treatment of the subject, which includes time dependent perturbations, elimination of the secular terms by averaging and numerical estimates of the degree of accuracy of the method, the reader is referred to Cary (1981), José and Saletan (1998) and Sussman and Wisdom (2001). As we have noted before, in Benettin *et al.* (1984) there is a proof of part (a) of the KAM theorem with some technical aspects made simpler by the use of Lie series to implement the iterative procedure of perturbation theory.

Problems

10.1 According to Problem 5.2, the angular frequency $\omega' = 2\pi/T$ of the pendulum is given by

$$\omega' = \frac{\pi \omega_0}{2} K(\kappa)^{-1}$$

where $\omega_0 = \sqrt{g/l}$ and

$$\kappa = \sin\left(\frac{\theta_0}{2}\right) = \sqrt{\frac{E}{2mgl}}.$$

Taking into acccount that the complete elliptic integral of the first kind admits the power series

$$K(\kappa) = \frac{\pi}{2}\left[1 + \frac{1}{4}k^2 + \frac{9}{64}k^4 + \dots\right],$$

show that the result (10.44) is correct up to second order in the energy.

10.2 Consider the one-dimensional anharmonic oscillator with Hamiltonian

$$H = \frac{p^2}{2m} + \frac{m\omega^2}{2}q^2 + \epsilon q^3.$$

What is the first-order correction to the frequency? Find the frequency corrected up to the second order as a function of the energy.

10.3 A one-dimensional harmonic oscillator has Hamiltonian

$$H = \frac{p^2}{2m} + \frac{m\omega_0^2(1+\epsilon)}{2}q^2.$$

Treating the term proportional to ϵ as a small perturbation, determine the oscillator's frequency corrected up to the second order and compare with the exact result.

10.4 Show that, including only the lowest-order correction term and discarding a constant, the Hamiltonian for a one-dimensional relativistic oscillator is

$$H = \frac{p^2}{2m} + \frac{m\omega^2}{2}q^2 - \frac{p^4}{8m^3c^2}.$$

Find the first-order relativistic correction to the oscillator's frequency as a function of the energy.

10.5 The Hamiltonian of a pendulum is

$$H = \frac{p_\theta^2}{2ml^2} - mgl\cos\theta,$$

where the horizontal plane of zero gravitational potential energy contains the suspension point. If the pendulum energy is such that $E \gg mgl$, the kinetic energy is much larger than the potential energy, which can be treated as a perturbation. Applying the first-order Lie series pertubation theory, show that the generator is

$$X = -\frac{m^2gl^3}{p_\theta}\sin\theta.$$

Show that the approximate solution of the equations of motion for p_θ is

$$p_\theta = p_0 + \frac{m^2gl^3}{p_0}\left[\cos\left(\theta_0 + \frac{p_0}{ml^2}t\right) - \cos\theta_0\right],$$

where (θ_0, p_0) are the initial values of (θ, p_θ). Obtain the corresponding solution for θ. Show that the condition that ensures the validity of pertubation theory is $p_0^2 \gg m^2gl^3$ and that this is equivalent to $E \gg mgl$.

10.6 Reexamine the previous problem of obtaining an approximate description of the pendulum's motion with energy $E \gg mgl$, this time by making use of the traditional pertubation theory expounded in Sections 10.2 and 10.3. Show that, up to the first order in the perturbation, the result is identical to the one found in Problem 10.5 by the Lie series technique.

11

Classical Field Theory

That one body may act upon another at a distance through a vacuum, without the mediation of any thing else, by and through which their action and force may be conveyed from one to the other, is to me so great an absurdity, that I believe no man who has in philosophical matters a competent faculty of thinking, can ever fall into it.

Isaac Newton, *Letter to Bentley*

Continuous systems possess infinitely many degrees of freedom and are described by fields. It is a remarkable fact that virtually all field theories of physical interest can be described by the formalisms of Lagrange and Hamilton. The interactions of elementary particles, the basic constituents of matter, are expressed by quantum field theories. More often than not, the construction of the quantum field theories of nature's fundamental interactions crucially depends on first formulating them as classical field theories in the Lagrangian and Hamiltonian languages, which is the object of the present chapter.

11.1 Lagrangian Field Theory

A mecanical system with a finite number of degrees of freedom is described by generalised coordinates $q_k(t)$. The simplest continuous system is described by a single coordinate $\varphi_{\mathbf{x}}(t)$ associated with each point $\mathbf{x}$ of space – that is, the discrete index k is replaced by the continuous index $\mathbf{x}$. We first consider fields in one spatial dimension and, instead of using the space coordinate as a subscript, we use the traditional notation $\varphi(x, t)$. For instance, $\varphi(x, t)$ could represent the transversal displacement at time t of the point x of a vibrating string.

The Lagrangian for a discrete system involves a sum over all degrees of freedom, so for a continuous system the Lagrangian must be expressed in terms of the space integral of a function $\mathcal{L}$, called the **Lagrangian density**. The Lagrangian density has to contain a kinetic term, therefore it necessarily depends on $\dot{\varphi} \equiv \partial\varphi/\partial t$. In contrast with the idea of action at a distance, we assume that the field φ at a point x interacts with itself only in an infinitesimal neighbourhood of this point, with the consequence that $\mathcal{L}$ must depend on $\varphi(x, t)$ and $\varphi(x + dx, t)$. Alternatively, instead of the latter quantity, it is more fruitful to use $\varphi' \equiv \partial\varphi/\partial x$. Admitting a possible explicit dependence on x and t, the most general[1] action for a one-dimensional field theory has the form

---

[1] Assuming $\mathcal{L}$ does not depend on higher-order derivatives of the fields.

$$S = \int_{t_1}^{t_2} L dt = \int_{t_1}^{t_2} dt \int_{x_1}^{x_2} dx \, \mathcal{L} \left(\varphi, \frac{\partial \varphi}{\partial x}, \frac{\partial \varphi}{\partial t}, x, t \right) . \tag{11.1}$$

Lagrange's equation for φ arises fom Hamilton's principle

$$\delta S = \delta \int_{t_1}^{t_2} dt \int_{x_1}^{x_2} dx \, \mathcal{L} = 0 . \tag{11.2}$$

As the example of the vibrating string with clamped ends suggests, the variation of the field must vanish at the time and space endpoints:

$$\delta \varphi(x, t_1) = \delta \varphi(x, t_2) = 0 , \qquad \delta \varphi(x_1, t) = \delta \varphi(x_2, t) = 0 . \tag{11.3}$$

Carrying out the variation of the action (11.1) we get

$$\delta S = \int_{t_1}^{t_2} dt \int_{x_1}^{x_2} dx \left\{ \frac{\partial \mathcal{L}}{\partial \varphi} \delta \varphi + \frac{\partial \mathcal{L}}{\partial \dot{\varphi}} \delta \dot{\varphi} + \frac{\partial \mathcal{L}}{\partial \varphi'} \delta \varphi' \right\} . \tag{11.4}$$

Using

$$\delta \dot{\varphi} = \frac{\partial (\delta \varphi)}{\partial t} , \qquad \delta \varphi' = \frac{\partial (\delta \varphi)}{\partial x} , \tag{11.5}$$

performing integrations by parts and taking account of (11.3), we obtain

$$\int_{t_1}^{t_2} dt \int_{x_1}^{x_2} dx \frac{\partial \mathcal{L}}{\partial \dot{\varphi}} \delta \dot{\varphi} = \int_{x_1}^{x_2} dx \int_{t_1}^{t_2} dt \frac{\partial \mathcal{L}}{\partial \dot{\varphi}} \frac{\partial (\delta \varphi)}{\partial t}$$

$$= \int_{x_1}^{x_2} dx \frac{\partial \mathcal{L}}{\partial \dot{\varphi}} \delta \varphi \Big|_{t_1}^{t_2} - \int_{x_1}^{x_2} dx \int_{t_1}^{t_2} dt \frac{\partial}{\partial t} \left(\frac{\partial \mathcal{L}}{\partial \dot{\varphi}} \right) \delta \varphi$$

$$= - \int_{t_1}^{t_2} dt \int_{x_1}^{x_2} dx \frac{\partial}{\partial t} \left(\frac{\partial \mathcal{L}}{\partial \dot{\varphi}} \right) \delta \varphi \tag{11.6}$$

and, similarly,

$$\int_{t_1}^{t_2} dt \int_{x_1}^{x_2} dx \frac{\partial \mathcal{L}}{\partial \varphi'} \delta \varphi' = \int_{t_1}^{t_2} dt \frac{\partial \mathcal{L}}{\partial \varphi'} \delta \varphi \Big|_{x_1}^{x_2} - \int_{t_1}^{t_2} dt \int_{x_1}^{x_2} dx \frac{\partial}{\partial x} \left(\frac{\partial \mathcal{L}}{\partial \varphi'} \right) \delta \varphi$$

$$- - \int_{t_1}^{t_2} dt \int_{x_1}^{x_2} dx \frac{\partial}{\partial x} \left(\frac{\partial \mathcal{L}}{\partial \varphi'} \right) \delta \varphi . \tag{11.7}$$

Substituting these results into (11.4), Hamilton's principle becomes

$$\int_{t_1}^{t_2} dt \int_{x_1}^{x_2} dx \left\{ \frac{\partial \mathcal{L}}{\partial \varphi} - \frac{\partial}{\partial t} \left(\frac{\partial \mathcal{L}}{\partial (\partial \varphi / \partial t)} \right) - \frac{\partial}{\partial x} \left(\frac{\partial \mathcal{L}}{\partial (\partial \varphi / \partial x)} \right) \right\} \delta \varphi = 0 . \tag{11.8}$$

By the same argument used in Section 2.1, the arbitrariness of $\delta \varphi$ implies the Lagrange equation

$$\frac{\partial}{\partial t} \left(\frac{\partial \mathcal{L}}{\partial (\partial \varphi / \partial t)} \right) + \frac{\partial}{\partial x} \left(\frac{\partial \mathcal{L}}{\partial (\partial \varphi / \partial x)} \right) - \frac{\partial \mathcal{L}}{\partial \varphi} = 0 . \tag{11.9}$$

Example 11.1 A vibrating string with clamped ends has linear mass density σ. If τ is the string tension, show that the Lagrangian density

$$\mathcal{L} = \frac{1}{2}\sigma\left(\frac{\partial\varphi}{\partial t}\right)^2 - \frac{1}{2}\tau\left(\frac{\partial\varphi}{\partial x}\right)^2 \qquad (11.10)$$

generates the correct equation of motion for small transversal vibrations of the string.

Solution

Writing $\mathcal{L} = \sigma\dot\varphi^2/2 - \tau\varphi'^2/2$ one immediately gets

$$\frac{\partial\mathcal{L}}{\partial(\partial\varphi/\partial t)} \equiv \frac{\partial\mathcal{L}}{\partial\dot\varphi} = \sigma\dot\varphi, \qquad \frac{\partial\mathcal{L}}{\partial(\partial\varphi/\partial x)} \equiv \frac{\partial\mathcal{L}}{\partial\varphi'} = -\tau\varphi', \qquad \frac{\partial\mathcal{L}}{\partial\varphi} = 0. \qquad (11.11)$$

Putting these results into the Lagrange equation (11.9) we get

$$\sigma\frac{\partial^2\varphi}{\partial t^2} - \tau\frac{\partial^2\varphi}{\partial x^2} = 0, \qquad (11.12)$$

which is the well-known one-dimensional wave equation for the vibrating string.

Transition from the Discrete to the Continuous

Many field theories (but not all) can be interpreted as the continuous limit of a discrete mechanical system with a finite number of degrees of freedom. The vibrating string is one of these cases, and it is instructive and enlightening to regard the string as the continuous limit of a discrete system according to the description that follows. Suppose N equal masses are connected by massless stretchable strings, as depicted in Fig. 11.1. In the equilibrium situation the tension τ in the strings is constant and the masses are equally spaced by a. One expects to obtain the continuous string by taking the limit as $a \to 0$ and $m \to 0$ in such a way that the linear mass density remains finite.

The generalised coordinates are the transversal displacements $\eta_1, \ldots, \eta_N$ with the restrictions $\eta_0 = \eta_{N+1} = 0$, corresponding to clamped ends. Figure 11.2 and Newton's second law allow us to write the equation of motion of the ith particle in the form

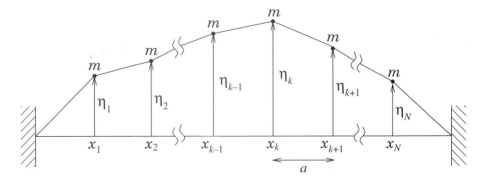

Fig. 11.1   Discrete model for transversal vibrations of a string.

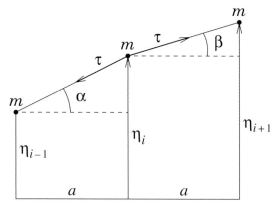

Forces on the ith particle of the discrete model for the vibrating string.

$$m\ddot{\eta}_i = \tau \sin \beta - \tau \sin \alpha \approx \tau(\tan \beta - \tan \alpha), \qquad (11.13)$$

assuming small transversal displacements.[2] Using

$$\tan \alpha = \frac{\eta_i - \eta_{i-1}}{a}, \qquad \tan \beta = \frac{\eta_{i+1} - \eta_i}{a}, \qquad (11.14)$$

Eq. (11.13) becomes

$$m\ddot{\eta}_i = \frac{\tau}{a} \left[(\eta_{i+1} - \eta_i) - (\eta_i - \eta_{i-1}) \right]. \qquad (11.15)$$

These equations of motion arise from the Lagrangian

$$L = \sum_{k=1}^{N} \frac{m}{2} \dot{\eta}_k^2 - \sum_{k=0}^{N} \frac{\tau}{2a} (\eta_{k+1} - \eta_k)^2. \qquad (11.16)$$

Exercise 11.1.1   Show that the Lagrange equations generated by the Lagrangian (11.16) are exactly Eqs. (11.15).

Setting $\Delta x = a$, denoting by x the abscissa of the kth particle and resorting to the change of notation

$$\eta_k(t) \to \varphi(x, t), \qquad \eta_{k+1}(t) - \eta_k(t) \to \varphi(x + \Delta x, t) - \varphi(x, t), \qquad (11.17)$$

we can rewrite the Lagrangian (11.16) in the convenient form

$$L = \sum \Delta x \left[\frac{1}{2} \frac{m}{\Delta x} \left(\frac{\partial \varphi(x, t)}{\partial t} \right)^2 \right] - \sum \Delta x \frac{\tau}{2} \left(\frac{\varphi(x + \Delta x, t) - \varphi(x, t)}{\Delta x} \right)^2. \qquad (11.18)$$

The passage to the continuous limit is effected by letting $\Delta x \to 0$ and $m \to 0$ in such a way that $m/\Delta x = \sigma$ and $Na = l$, which brings about the Lagrangian

---

[2] We are justified in using the equilibrium tension τ in (11.13) because it already appears multiplied by terms of the first order in the small quantities. The deformation of each massless string is a second-order effect and is also disregarded.

$$L = \int_0^l dx \left\{ \frac{\sigma}{2} \left(\frac{\partial \varphi}{\partial t} \right)^2 - \frac{\tau}{2} \left(\frac{\partial \varphi}{\partial x} \right)^2 \right\} , \tag{11.19}$$

from which one immediately identifies the Lagrangian density (11.10).

Several Fields in Three Spatial Dimensions

In the case of N fields in three spatial dimensions, collectively represented by $\varphi = (\varphi_1, \ldots, \varphi_N)$, the Lagrange equations arising from Hamilton's principle[3]

$$\delta S = \delta \int_\Omega d^4x \, \mathcal{L} \left(\varphi, \frac{\partial \varphi}{\partial t}, \nabla \varphi, \mathbf{x}, t \right) = 0 \tag{11.20}$$

are written

$$\frac{\partial}{\partial t} \left(\frac{\partial \mathcal{L}}{\partial(\partial \varphi_\alpha/\partial t)} \right) + \nabla \cdot \left(\frac{\partial \mathcal{L}}{\partial(\nabla \varphi_\alpha)} \right) - \frac{\partial \mathcal{L}}{\partial \varphi_\alpha} = 0, \quad \alpha = 1, \ldots, N. \tag{11.21}$$

In order to derive this result, let us consider

$$\delta S = \int_{t_1}^{t_2} dt \int_V d^3x \sum_{\alpha=1}^N \left\{ \frac{\partial \mathcal{L}}{\partial \varphi_\alpha} \delta \varphi_\alpha + \frac{\partial \mathcal{L}}{\partial \dot\varphi_\alpha} \delta \dot\varphi_\alpha + \frac{\partial \mathcal{L}}{\partial(\nabla \varphi_\alpha)} \cdot \delta(\nabla \varphi_\alpha) \right\} , \tag{11.22}$$

where the variations of the φ_α are mutually independent and vanish at the endpoints of the time integration and on the surface Σ that bounds the three-dimensional region V. Using $\delta \dot\varphi_\alpha = \partial(\delta \varphi_\alpha)/\partial t$, an integration by parts as in the one-dimensional case yields

$$\int_{t_1}^{t_2} dt \int_V d^3x \frac{\partial \mathcal{L}}{\partial \dot\varphi_\alpha} \delta \dot\varphi_\alpha = - \int_{t_1}^{t_2} dt \int_V d^3x \frac{\partial}{\partial t} \left(\frac{\partial \mathcal{L}}{\partial \dot\varphi_\alpha} \right) \delta \varphi_\alpha . \tag{11.23}$$

Using now $\delta(\nabla \varphi_\alpha) = \nabla(\delta \varphi_\alpha)$ and relying on the identity

$$\mathbf{A} \cdot \nabla f = \nabla \cdot (f\mathbf{A}) - f \nabla \cdot \mathbf{A}, \tag{11.24}$$

we can perform an integration by parts with the help of the divergence theorem to get

$$\int_{t_1}^{t_2} dt \int_V d^3x \frac{\partial \mathcal{L}}{\partial(\nabla \varphi_\alpha)} \cdot \delta(\nabla \varphi_\alpha) = \int_{t_1}^{t_2} dt \int_V d^3x \frac{\partial \mathcal{L}}{\partial(\nabla \varphi_\alpha)} \cdot \nabla(\delta \varphi_\alpha)$$

$$= \int_{t_1}^{t_2} dt \oint_\Sigma \mathbf{da} \cdot \frac{\partial \mathcal{L}}{\partial(\nabla \varphi_\alpha)} \delta \varphi_\alpha - \int_{t_1}^{t_2} dt \int_V d^3x \nabla \cdot \left(\frac{\partial \mathcal{L}}{\partial(\nabla \varphi_\alpha)} \right) \delta \varphi_\alpha$$

$$= - \int_{t_1}^{t_2} dt \int_V d^3x \nabla \cdot \left(\frac{\partial \mathcal{L}}{\partial(\nabla \varphi_\alpha)} \right) \delta \varphi_\alpha , \tag{11.25}$$

inasmuch as the variations $\delta \varphi_\alpha$ vanish on Σ. Introducing (11.23) and (11.25) into (11.22), Hamilton's principle takes the form

$$\delta S = \int_{t_1}^{t_2} dt \int_V d^3x \sum_{\alpha=1}^N \left\{ \frac{\partial \mathcal{L}}{\partial \varphi_\alpha} - \frac{\partial}{\partial t} \left(\frac{\partial \mathcal{L}}{\partial \dot\varphi_\alpha} \right) - \nabla \cdot \left(\frac{\partial \mathcal{L}}{\partial(\nabla \varphi_\alpha)} \right) \right\} \delta \varphi_\alpha = 0, \tag{11.26}$$

[3] We use the notation $d^3x \equiv dxdydz$ and $d^4x \equiv dtdxdydz$.

whence there immediately follow Lagrange's equations (11.21) because the variations $\delta\varphi_\alpha$ are independent and arbitrary.

Complex Fields

Certain theories naturally require complex fields. A complex field is tantamount to two independent real fields. Equivalently, instead of the real and imaginary parts of a complex field ψ, we can consider the field itself and its complex conjugate ψ^* as independent fields. In this case, for each complex field there will be a pair of Lagrange equations of the form (11.21) by first taking $\varphi_1 = \psi$ and next $\varphi_2 = \psi^*$.

Exercise 11.1.2 Let ψ be a complex field and consider the Lagrangian density

$$\mathcal{L} = i\hbar\psi^*\dot\psi - \frac{\hbar^2}{2m}\nabla\psi \cdot \nabla\psi^* - V(\mathbf{x},t)\psi^*\psi \,. \tag{11.27}$$

Taking ψ and ψ^* as independent fields, show that the corresponding Lagrange equations are the Schrödinger equation

$$i\hbar\frac{\partial\psi}{\partial t} = -\frac{\hbar^2}{2m}\nabla^2\psi + V(\mathbf{x},t)\psi \tag{11.28}$$

and its complex conjugate.

11.2 Relativistic Field Theories

Lagrange's equations (11.21) are invariant under a scale transformation of the coordinates $\mathbf{x}, t$. In particular, setting $x^0 = ct$ we have

$$\frac{\partial}{\partial t}\left(\frac{\partial\mathcal{L}}{\partial(\partial\varphi_\alpha/\partial t)}\right) = \frac{\partial}{\partial x^0}\left(\frac{\partial\mathcal{L}}{\partial(\partial\varphi_\alpha/\partial x^0)}\right). \tag{11.29}$$

From now on we assume that the time derivative is always with respect to $x^0 = ct$, so that, in terms of the covariant notation of Chapter 6, Lagrange's equations (11.21) are written

$$\partial_\mu\left(\frac{\partial\mathcal{L}}{\partial(\partial_\mu\varphi_\alpha)}\right) - \frac{\partial\mathcal{L}}{\partial\varphi_\alpha} = 0, \quad \alpha = 1,\dots,N, \tag{11.30}$$

where

$$\partial_\mu \equiv \frac{\partial}{\partial x^\mu} = \left(\frac{1}{c}\frac{\partial}{\partial t}, \frac{\partial}{\partial x}, \frac{\partial}{\partial y}, \frac{\partial}{\partial z}\right) = \left(\frac{\partial}{\partial x^0}, \nabla\right). \tag{11.31}$$

If φ_α is a scalar field for a given value of α, then $\partial_\mu\varphi_\alpha$ is a covariant four-vector. In this case, if the Lagrangian[4] $\mathcal{L}$ is a scalar quantity it follows that $\partial\mathcal{L}/\partial(\partial_\mu\varphi_\alpha)$ is a contravariant four-vector and the first term on the left-hand side of Eq. (11.30) is a scalar. In general, in order for the Lagrange equations (11.30) to be manifestly covariant, it suffices

---

[4] Being aware of the abuse of language, we henceforward refer to the Lagrangian density simply as the Lagrangian, as is costumary in field theory.

to require that the Lagrangian be a scalar. By the way, since the four-dimensional volume element d^4x is invariant under Lorentz transformations, the action $S = \int d^4x \mathcal{L}$ is also a scalar whenever $\mathcal{L}$ is a scalar.

Klein-Gordon Field

As the first example of a relativistic field theory, consider the theory of a scalar meson characterised by the Lagrangian

$$\mathcal{L} = \frac{1}{2}\partial_\mu\phi\partial^\mu\phi - \frac{m^2}{2}\phi^2, \tag{11.32}$$

where m is the particle's mass (in units such that $\hbar = c = 1$). This Lagrangian is a scalar under Lorentz transformations, since ϕ is a scalar field. In order to make the computations more transparent, let us introduce the notation $\phi_\mu \equiv \partial_\mu\phi$, in terms of which we can write

$$\frac{\partial\mathcal{L}}{\partial(\partial_\mu\phi)} = \frac{\partial}{\partial\phi_\mu}\left(\frac{1}{2}g^{\nu\lambda}\phi_\nu\phi_\lambda\right) = \frac{1}{2}g^{\nu\lambda}\delta_\nu^\mu\phi_\lambda + \frac{1}{2}g^{\nu\lambda}\phi_\nu\delta_\lambda^\mu$$

$$= \frac{1}{2}\left[g^{\mu\lambda}\phi_\lambda + g^{\nu\mu}\phi_\nu\right] = \phi^\mu. \tag{11.33}$$

Inserting this result into (11.30) we obtain the Klein-Gordon equation

$$(\Box + m^2)\phi = 0, \tag{11.34}$$

where we use the d'Alembertian operator defined by Eq. (6.67). In quantum field theory this equation describes uncharged scalar mesons, which are spinless particles.

Electromagnetic Field

An even more important example of relativistic field theory is the electromagnetic field in the presence of charges and currents described by the four-current J^μ. As independent fields we take the components of the four-potential A_μ. Let us show that the Lagrangian[5]

$$\mathcal{L} = -\frac{1}{4}F_{\mu\nu}F^{\mu\nu} - \frac{1}{c}J^\mu A_\mu, \tag{11.35}$$

with

$$F_{\mu\nu} = \partial_\mu A_\nu - \partial_\nu A_\mu, \tag{11.36}$$

yields the correct field equations. Now $\varphi_\alpha \equiv A_\alpha$ and α is a four-vector index. In order to make the calculations easier, let us introduce the auxiliary tensor field

$$A_{\mu\alpha} \equiv \partial_\mu A_\alpha. \tag{11.37}$$

Since $\partial\mathcal{L}/\partial A_\alpha = -(1/c)J^\alpha$, Lagrange's equations reduce to

$$\partial_\mu\left(\frac{\partial\mathcal{L}}{\partial A_{\mu\alpha}}\right) + \frac{1}{c}J^\alpha = 0. \tag{11.38}$$

---

[5] The Lagrangian (11.35) is appropriate in the Heaviside-Lorentz system of units (Jackson, 1999), widely used in quantum field theory. In the Gaussian system, the factor $1/4$ must be replaced by $1/16\pi$.

From (11.35) one finds

$$\frac{\partial \mathcal{L}}{\partial A_{\mu\alpha}} = -\frac{1}{4}\left(\frac{\partial F_{\beta\gamma}}{\partial A_{\mu\alpha}}F^{\beta\gamma} + F_{\beta\gamma}\frac{\partial F^{\beta\gamma}}{\partial A_{\mu\alpha}}\right) = -\frac{1}{2}\frac{\partial F_{\beta\gamma}}{\partial A_{\mu\alpha}}F^{\beta\gamma}. \tag{11.39}$$

On the other hand, from $F_{\beta\gamma} = A_{\beta\gamma} - A_{\gamma\beta}$ it immediately follows that

$$\frac{\partial F_{\beta\gamma}}{\partial A_{\mu\alpha}} = \delta^\mu_\beta \delta^\alpha_\gamma - \delta^\mu_\gamma \delta^\alpha_\beta. \tag{11.40}$$

As a consequence,

$$\frac{\partial \mathcal{L}}{\partial A_{\mu\alpha}} = -\frac{1}{2}(\delta^\mu_\beta \delta^\alpha_\gamma - \delta^\mu_\gamma \delta^\alpha_\beta)F^{\beta\gamma} = -\frac{1}{2}(F^{\mu\alpha} - F^{\alpha\mu}) = -F^{\mu\alpha} \tag{11.41}$$

and the field equations become

$$\partial_\mu F^{\mu\alpha} = \frac{1}{c}J^\alpha, \tag{11.42}$$

which are identical to (6.80), except for the absence of the factor 4π on the right-hand side because we are using Heaviside-Lorentz units.

Maxwell's homogeneous equations are automatically satisfied when the electromagnetic field is described by the potentials ϕ and $\mathbf{A}$. In the manifestly covariant formalism the homogeneous equations do not arise from the Lagrangian, but from the *identities*

$$\partial^\alpha F^{\beta\gamma} + \partial^\beta F^{\gamma\alpha} + \partial^\gamma F^{\alpha\beta} = 0. \tag{11.43}$$

As seen in Section 6.5, these equations coincide with Maxwell's homogeneous equations.

Exercise 11.2.1   Prove that Eqs. (11.43) are identically satisfied by virtue of the definition $F_{\mu\nu} = \partial_\mu A_\nu - \partial_\nu A_\mu$.

11.3 Functional Derivatives

In classical field theory we are naturally led to deal with functionals, and the notion of functional derivative plays a salient role. Let $x = (x_1, \ldots, x_n)$ be a point of $\mathbb{R}^n$ and $f : \mathbb{R}^n \to \mathbb{R}$ be a sufficiently differentiable function. From Section 2.1 we know that a functional F is a rule that assigns a real number $F[f]$ to each function f of a certain class. The **functional derivative** of F with respect to f at the point $x \in \mathbb{R}^n$, denoted by $\delta F/\delta f(x)$, is defined by the equality

$$\left(\frac{d}{d\epsilon}F[f + \epsilon\sigma]\right)_{\epsilon=0} = \int_{\mathbb{R}^n} d^n x\, \sigma(x)\frac{\delta F}{\delta f(x)}, \tag{11.44}$$

here $\sigma : \mathbb{R}^n \to \mathbb{R}$ is an arbitrary infinitely differentiable function that vanishes on the boundary of the integration region.[6] The functional derivative $\delta F/\delta f(x)$ measures the rate

---

[6] In rigorous mathematical language, σ must be any infinitely differentiable function of compact support and the functional derivative is a distribution (Schwartz, 2008).

of change of the functional F with respect to changes of the function f in a neighbourhood of the point x. From this implicit definition one easily proves that functional differentiation satisfies

$$\frac{\delta}{\delta f(x)}(c_1 F_1 + c_2 F_2) = c_1 \frac{\delta F_1}{\delta f(x)} + c_2 \frac{\delta F_2}{\delta f(x)}, \qquad (11.45)$$

where c_1 an c_2 are real constants, and obeys the Leibniz rule for derivative of a product:

$$\frac{\delta}{\delta f(x)}(F_1 F_2) = \frac{\delta F_1}{\delta f(x)} F_2 + F_1 \frac{\delta F_2}{\delta f(x)}. \qquad (11.46)$$

One also proves that if $\Phi = \Phi(F)$ is a differentiable function of the functional F then the chain rule of functional differentiation holds in the form

$$\frac{\delta \Phi}{\delta f(x)} = \frac{d\Phi}{dF} \frac{\delta F}{\delta f(x)}. \qquad (11.47)$$

Let us see some simple examples of computation of functional derivatives. If the functional F is defined by

$$F[f] = \int_{\mathbb{R}^n} d^n x f(x), \qquad (11.48)$$

then

$$F[f + \epsilon\sigma] = F[f] + \epsilon \int_{\mathbb{R}^n} d^n x \sigma(x) \implies \left(\frac{d}{d\epsilon} F[f + \epsilon\sigma]\right)_{\epsilon=0} = \int_{\mathbb{R}^n} d^n x \sigma(x). \qquad (11.49)$$

From definition (11.44) it follows that

$$\frac{\delta F}{\delta f(x)} = 1. \qquad (11.50)$$

A more interesting example is

$$F[f] = \int d^n y \, d^n z \, K(y,z) f(y) f(z), \qquad (11.51)$$

where $K(y,z)$ is a known function. Now we have, leaving out the term quadratic in ϵ, which does not contribute to the functional derivative,

$$F[f + \epsilon\sigma] = F[f] + \epsilon \int d^n y d^n z K(y,z)\sigma(y)f(z) + \epsilon \int d^n y d^n z K(y,z)f(y)\sigma(z). \qquad (11.52)$$

Therefore,

$$\left(\frac{d}{d\epsilon} F[f + \epsilon\sigma]\right)_{\epsilon=0} = \int d^n y d^n z K(y,z)\sigma(y)f(z) + \int d^n y d^n z K(y,z)f(y)\sigma(z). \qquad (11.53)$$

Renaming the integration variables $y \to x$ in the first integral, $z \to x$ in the second and comparing the resulting equation with (11.44), we get

$$\frac{\delta F}{\delta f(x)} = \int d^n z K(x,z)f(z) + \int d^n y K(y,x)f(y), \qquad (11.54)$$

which is the same thing as

$$\frac{\delta F}{\delta f(x)} = \int d^n y \left[K(x, y) + K(y, x) \right] f(y). \tag{11.55}$$

Functionals depending on parameters are a frequent occurrence. An example of interest is the functional F_x defined by

$$F_x[f] = \int_{\mathbb{R}^n} d^n x' \, K(x, x') f(x'). \tag{11.56}$$

A calculation as simple as those above shows that the functional derivative of F_x is

$$\frac{\delta F_x}{\delta f(y)} = K(x, y). \tag{11.57}$$

With the use of the Dirac delta "function" one can think of $f(x)$ as a functional of the form (11.56):

$$f(x) = \int_{\mathbb{R}^n} d^n x' \, \delta(x - x') f(x'). \tag{11.58}$$

By (11.57) we have

$$\frac{\delta f(x)}{\delta f(y)} = \delta(x - y). \tag{11.59}$$

Note that by using this last result Eq. (11.54) can be quickly derived as follows:

$$\frac{\delta}{\delta f(x)} \int d^n y \, d^n z \, K(y, z) f(y) f(z) = \int d^n y \, d^n z \, K(y, z) \frac{\delta}{\delta f(x)} \left[f(y) f(z) \right]$$

$$= \int d^n y \, d^n z \, K(y, z) \left[\frac{\delta f(y)}{\delta f(x)} f(z) + f(y) \frac{\delta f(z)}{\delta f(x)} \right]$$

$$= \int d^n y \, d^n z \, K(y, z) \delta(y - x) f(z)$$

$$+ \int d^n y \, d^n z \, K(y, z) \delta(z - x) f(y)$$

$$= \int d^n z K(x, z) f(z) + \int d^n y K(y, x) f(y). \tag{11.60}$$

As a rule, one can take the functional derivative under the integral sign:

$$\frac{\delta}{\delta f(x)} \int d^n y \, \cdots = \int d^n y \, \frac{\delta}{\delta f(x)} \, \cdots . \tag{11.61}$$

Exercise 11.3.1 The derivative of the Dirac delta "function" is formally defined by the following property:

$$\int_{-\infty}^{\infty} \delta'(x) f(x) dx = -f'(0). \tag{11.62}$$

Using this definition, show that

$$\frac{\delta f'(x)}{\delta f(y)} = \delta'(x - y). \tag{11.63}$$

Consider the functional involved in the simplest problem of the calculus of variations, namely

$$J[f] = \int_a^b F(f(y), f'(y), y)\, dy\,, \tag{11.64}$$

where $F : \mathbb{R}^3 \to \mathbb{R}$ is at least twice continuously differentiable. Equation (11.61) and the chain rule give

$$\frac{\delta J}{\delta f(x)} = \int_a^b \left[\frac{\partial F}{\partial f}\frac{\delta f(y)}{\delta f(x)} + \frac{\partial F}{\partial f'}\frac{\delta f'(y)}{\delta f(x)}\right] dy. \tag{11.65}$$

Conclude that

$$\frac{\delta J}{\delta f(x)} = (D_1 F)\,(f(x), f'(x), x) - \frac{d}{dx}\,(D_2 F)\,(f(x), f'(x), x)\,, \tag{11.66}$$

where D_1 and D_2 respectively denote partial derivatives with respect to the first and second arguments of F. Thus, Euler's equation (2.14) can be written in the following concise form:

$$\frac{\delta J}{\delta y(x)} = 0\,. \tag{11.67}$$

Check that (11.66) is exactly the same as the result that ensues from the direct application of definition (11.44).

If $F[f_1, \ldots, f_N]$ is a functional of N variables the functional derivative $\delta F/\delta f_k(x)$ is defined by a natural extension of (11.44), to wit

$$\left(\frac{d}{d\epsilon} F[f_1 + \epsilon\sigma_1, \ldots, f_N + \epsilon\sigma_N]\right)_{\epsilon=0} = \int d^n x \sum_{k=1}^N \sigma_k(x)\,\frac{\delta F}{\delta f_k(x)}\,. \tag{11.68}$$

Comparing this definition of functional derivative with the notion of variation of a functional introduced in Section 2.2, we can write

$$\delta F = \int d^n x \sum_{k=1}^N \frac{\delta F}{\delta f_k(x)}\,\delta f_k(x)\,. \tag{11.69}$$

In other words, the functional derivative of F is given by the first-order term of the expansion of $F[f + \delta f]$ in powers of δf. Equation (11.59) acquires the generalised form

$$\frac{\delta f_i(x)}{\delta f_j(y)} = \delta_{ij}\,\delta(x - y)\,. \tag{11.70}$$

Inspecting (11.26) and comparing it with (11.69), we are led to

$$\frac{\delta S}{\delta\varphi_\alpha(x)} = \frac{\partial\mathcal{L}}{\partial\varphi_\alpha(x)} - \frac{\partial}{\partial t}\left(\frac{\partial\mathcal{L}}{\partial\dot\varphi_\alpha(x)}\right) - \nabla\cdot\left(\frac{\partial\mathcal{L}}{\partial(\nabla\varphi_\alpha(x))}\right)\,, \tag{11.71}$$

where $x = (\mathbf{x}, t)$. Therefore, in terms of functional derivatives of the action, Lagrange's equations can be written in the compact form

$$\frac{\delta S}{\delta\varphi_\alpha(x)} = 0\,. \tag{11.72}$$

Functional derivatives are particularly useful in the Hamiltonian formulation of field theories, which we proceed to discuss.

where we have used (11.97) and (11.99), while d/dx^μ denotes what is legitimate to call the "total partial derivative" with respect to the coordinate x^μ,

$$\frac{d\mathcal{L}}{dx^\mu} = \frac{\partial\mathcal{L}}{\partial\varphi_\alpha}\frac{\partial\varphi_\alpha}{\partial x^\mu} + \frac{\partial\mathcal{L}}{\partial\varphi_{\alpha;\beta}}\frac{\partial\varphi_{\alpha;\beta}}{\partial x^\mu} + \frac{\partial\mathcal{L}}{\partial x^\mu}, \tag{11.102}$$

which takes into account the explicit dependence and the implicit dependence through the fields.

On the other hand, by (11.94a) and part (a) of Problem 8.23, we have

$$d^4x' = \frac{\partial(x'^0,x'^1,x'^2,x'^3)}{\partial(x^0,x^1,x^2,x^3)}d^4x = \left(1 + \frac{\partial\Delta x^\mu}{\partial x^\mu}\right)d^4x. \tag{11.103}$$

Substituting (11.101) and (11.103) into (11.100), and neglectiong terms of order higher than the first in the variation, there results

$$\Delta S = \int_\Omega d^4x\left[\delta\mathcal{L} + \frac{d\mathcal{L}}{dx^\mu}\Delta x^\mu + \mathcal{L}\frac{\partial\Delta x^\mu}{\partial x^\mu}\right] = \int_\Omega d^4x\left[\delta\mathcal{L} + \frac{d}{dx^\mu}\left(\mathcal{L}\,\Delta x^\mu\right)\right]. \tag{11.104}$$

Using the field equations of motion

$$\frac{\partial\mathcal{L}}{\partial\varphi_\alpha} = \frac{d}{dx^\mu}\left(\frac{\partial\mathcal{L}}{\partial\varphi_{\alpha;\mu}}\right) \tag{11.105}$$

we get

$$\begin{aligned}
\delta\mathcal{L} &= \frac{\partial\mathcal{L}}{\partial\varphi_\alpha}\delta\varphi_\alpha + \frac{\partial\mathcal{L}}{\partial\varphi_{\alpha;\beta}}\delta\varphi_{\alpha;\beta} \\
&= \frac{d}{dx^\mu}\left(\frac{\partial\mathcal{L}}{\partial\varphi_{\alpha;\mu}}\right)\delta\varphi_\alpha + \frac{\partial\mathcal{L}}{\partial\varphi_{\alpha;\beta}}\frac{\partial}{\partial x^\beta}(\delta\varphi_\alpha) = \frac{d}{dx^\mu}\left(\frac{\partial\mathcal{L}}{\partial\varphi_{\alpha;\mu}}\delta\varphi_\alpha\right),
\end{aligned} \tag{11.106}$$

where, in order to arrive at the last equality, the dummy index β has been replaced by μ. Inserting (11.106) into (11.104) the invariance condition for the action becomes

$$\Delta S = \int_\Omega d^4x\,\frac{d}{dx^\mu}\left\{\frac{\partial\mathcal{L}}{\partial\varphi_{\alpha;\mu}}\delta\varphi_\alpha + \mathcal{L}\,\Delta x^\mu\right\} = 0, \tag{11.107}$$

which implies a conservation law in the covariant form (6.73). Indeed, the arbitrariness of the integration region requires that the integrand in Eq. (11.107) be zero. This is equivalent to the continity equation for the four-current defined by the expression within braces.

Conserved Quantities

It turns out to be more convenient to express the four-current conservation in terms of the infinitesimal parameters that characterise the transformation. Suppose the transformation (11.94) is specified by R independent infinitesimal parameters $\epsilon_1,\ldots,\epsilon_R$ in the form

$$\Delta x^\mu = \sum_{r=1}^R X^{\mu\,(r)}\epsilon_r \equiv X^{\mu\,(r)}\epsilon_r, \qquad \Delta\varphi_\alpha = \sum_{r=1}^R \Psi_\alpha^{(r)}\epsilon_r \equiv \Psi_\alpha^{(r)}\epsilon_r. \tag{11.108}$$

The indices α and r of the fields and of the transformation parameters may or may not have tensor character, and we adhere to the convention that repeated indices of any kind are to be summed over. Substituting (11.108) into (11.97) we can write

$$\Delta x^\mu = X^{\mu\,(r)}\epsilon_r\,, \qquad \delta\varphi_\alpha = \left(\Psi_\alpha^{(r)} - \varphi_{\alpha;\nu}X^{\nu\,(r)}\right)\epsilon_r\,, \tag{11.109}$$

which, once inserted into (11.107), lead to

$$\Delta S = -\int_\Omega d^4x\,\epsilon_r\,\frac{d}{dx^\mu}\,\Theta^{\mu\,(r)} = 0\,, \tag{11.110}$$

where

$$\Theta^{\mu\,(r)} = -\frac{\partial\mathcal{L}}{\partial\varphi_{\alpha;\mu}}\left(\Psi_\alpha^{(r)} - \varphi_{\alpha;\nu}X^{\nu\,(r)}\right) - \mathcal{L}X^{\mu\,(r)}\,. \tag{11.111}$$

Since the integration region and the parameters ϵ_r are arbitrary, from (11.110) one infers the R local conservation laws[8]

$$\partial_\mu\,\Theta^{\mu\,(r)} = 0\,, \qquad r = 1,\dots,R\,. \tag{11.112}$$

Putting $\Theta^{\mu\,(r)} = (\Theta^{0\,(r)}, \mathbf{\Theta}^{(r)})$ we can write (11.112) in the usual form

$$\partial_0\,\Theta^{0\,(r)} + \nabla\cdot\mathbf{\Theta}^{(r)} = 0\,, \tag{11.113}$$

whence

$$\frac{d}{dx^0}\int_V d^3x\,\Theta^{0\,(r)} = \int_V d^3x\,\partial_0\,\Theta^{0\,(r)} = -\int_V d^3x\,\nabla\cdot\mathbf{\Theta}^{(r)} = -\oint_\Sigma \mathbf{\Theta}^{(r)}\cdot\mathbf{da}\,, \tag{11.114}$$

where Σ is the boundary surface of the three-dimensional region V and the divergence theorem has been used. If V is the whole three-dimensional space and the fields tend to zero sufficiently fast at spatial infinity, the surface integral vanishes and the volume integrals

$$C^{(r)} = \int d^3x\,\Theta^{0\,(r)} \tag{11.115}$$

are conserved quantities, since they are time independent. It is thus proved that to each R-parametric infinitesimal transformation of coordinates and fields that leaves the action invariant there correspond R conserved quantites $C^{(r)}$ given by (11.115), known as **Noether charges**.

Conservation of Four-Momentum

As the first important example, consider the space-time translation

$$x'^\mu = x^\mu + \epsilon^\mu\,, \tag{11.116}$$

which does not modify the fields – that is, $\Delta\varphi_\alpha = 0$. Since this transformation has unit Jacobian, equations (11.100) and (11.101) show that the action is invariant provided the Lagrangian does not explicitly depend on the space-time coordinates. The index r of the transformation parameters now has four-vector character and, according to (11.108), we have

$$X^{\mu(\beta)} = g^{\mu\beta}\,, \qquad \Psi_\alpha^{(\beta)} = 0\,. \tag{11.117}$$

---

[8] There is no ambiguity any more, so we revert to the usual symbol ∂_μ for the total partial derivative with respect to x^μ.

Substituting these expressions into (11.111) we obtain the **canonical energy-momentum tensor** or just **energy-momentum tensor**

$$T^{\mu\nu} = \frac{\partial \mathcal{L}}{\partial \varphi_{\alpha;\mu}} \varphi_{\alpha;\beta} g^{\beta\nu} - \mathcal{L} g^{\mu\nu} = \frac{\partial \mathcal{L}}{\partial \varphi_{\alpha;\mu}} \frac{\partial \varphi_\alpha}{\partial x_\nu} - \mathcal{L} g^{\mu\nu}. \tag{11.118}$$

The conserved Noether charges constitute the four-vector

$$P^\nu = \int d^3x \, T^{0\nu}. \tag{11.119}$$

The zeroth component of this four-vector is the space integral of the Hamiltonian density – that is, it is the field energy. Covariance considerations, and the fact that the conservation of linear momentum is associated with invariance under space translations, establish that P^ν is the field four-momentum, hence calling $T^{\mu\nu}$ the energy-momentum tensor.

Exercise 11.5.1   Show that the energy-momentum tensor of the scalar field described by the Klein-Gordon Lagrangian (11.32) is

$$T^{\mu\nu} = \partial^\mu \phi \, \partial^\nu \phi - \frac{1}{2} \left(\partial_\alpha \phi \, \partial^\alpha \phi - m^2 \phi^2 \right) g^{\mu\nu} \tag{11.120}$$

and find the expression for the (conserved) field momentum.

Gauge Invariance and Conservation of Electric Charge

Consider, now, a complex scalar field described by the Lagrangian

$$\mathcal{L} = \partial_\mu \phi^* \, \partial^\mu \phi - m^2 \phi^* \phi, \tag{11.121}$$

which is obviously invariant under the one-parameter transformation

$$\phi' = e^{i\lambda} \phi, \qquad \phi^{*'} = e^{-i\lambda} \phi^*, \tag{11.122}$$

where λ is an arbitrary real number. This transformation is known as a **global gauge transformation** or **gauge transformation of the first kind**. Its infinitesimal version with $\lambda = \epsilon$ is

$$\Delta\phi = i\epsilon \, \phi, \qquad \Delta\phi^* = -i\epsilon \, \phi^*, \tag{11.123}$$

whence

$$X^{\mu\,(1)} = 0, \qquad \Psi_1^{(1)} = i, \qquad \Psi_2^{(1)} = -i, \tag{11.124}$$

with $\varphi_1 = \phi$, $\varphi_2 = \phi^*$ and the index r taking the single value $r = 1$. Since the coordinates remain unchanged, invariance of the Lagrangian implies invariance of the action. Noether's current (11.111) takes the form

$$J^\mu = -\frac{\partial \mathcal{L}}{\partial \varphi_{1;\mu}} \Psi_1^{(1)} - \frac{\partial \mathcal{L}}{\partial \varphi_{2;\mu}} \Psi_2^{(1)} = -i \left(\phi^* \, \partial^\mu \phi - \phi \, \partial^\mu \phi^* \right). \tag{11.125}$$

The conserved Noether charge, interpreted as electric charge, is

$$Q = i \int d^3x \left(\phi \, \dot{\phi}^* - \dot{\phi} \, \phi^* \right). \tag{11.126}$$

A real scalar field cannot interact with the electromagnetic field because the current density J^μ is zero (see Problem 11.10 for more details). Therefore, complex scalar fields are needed to describe spinless charged particles.

Exercise 11.5.2 Show that the Lagrangian (11.27) for the Schrödinger field is invariant under a gauge transformation of the first kind. Find the Noether current and the conserved Noether charge – the probability of finding the particle somewhere in space, in quantum mechanics.

11.6 Solitary Waves and Solitons

Field configurations with finite and localised energy that travel without change of shape or diminution of speed are called **solitary waves**. Such waves typically occur in nonlinear field theories (Lee, 1981) and were originally discussed by J. Scott Russell (1844), who thus reported his first encounter with the phenomenon in 1834:

> I was observing the motion of a boat which was rapidly drawn along a narrow channel by a pair of horses, when the boat suddenly stopped – not so the mass of water in the channel which it had put in motion; it accumulated round the prow of the vessel in a state of violent agitation, then suddenly leaving it behind, rolled forward with great velocity, assuming the form of a large solitary elevation, a rounded, smooth and well-defined heap of water, which continued its course along the channel apparently without change of form or diminution of speed. I followed it on horseback, and overtook it still rolling on at a rate of some eight or nine miles an hour, preserving its original figure some thirty feet long and a foot to a foot and a half in height. Its height gradually diminished, and after a chase of one or two miles I lost it in the windings of the channel. Such, in the month of August 1834, was my first chance interview with that singular and beautiful phenomenon . . .

As an example of a model that admits solutions of the type described above, consider a real scalar field φ in one spatial dimension with Lagrangian

$$\mathcal{L} = \frac{1}{2}\partial^\mu\varphi\partial_\mu\varphi - \frac{\lambda^2}{8}(\varphi^2 - a^2)^2 = \frac{1}{2}\left[\left(\frac{\partial\varphi}{\partial t}\right)^2 - \left(\frac{\partial\varphi}{\partial x}\right)^2\right] - \frac{\lambda^2}{8}(\varphi^2 - a^2)^2, \quad (11.127)$$

where λ and a are positive constants, with units chosen so that $c = 1$. Lagrange's equation for φ is

$$\frac{\partial^2\varphi}{\partial t^2} - \frac{\partial^2\varphi}{\partial x^2} + \frac{\lambda^2}{2}\varphi(\varphi^2 - a^2) = 0. \quad (11.128)$$

Using $\pi = \partial\mathcal{L}/\partial\dot{\varphi} = \dot{\varphi}$, the Hamiltonian density is found to be

$$\mathcal{H} = \pi\dot{\varphi} - \mathcal{L} = \frac{1}{2}\left[\left(\frac{\partial\varphi}{\partial t}\right)^2 + \left(\frac{\partial\varphi}{\partial x}\right)^2\right] + \frac{\lambda^2}{8}(\varphi^2 - a^2)^2. \quad (11.129)$$

Note that $\varphi = a$ or $\varphi = -a$ are static solutions of the field equation (11.128) with zero energy (because $\mathcal{H} = 0$ for these solutions), which is the lowest energy possible. Thus, this

The matrix (11.152) has both discrete and continuous indices, and by analogy with (8.192) its inverse, if it exists, has elements $C_{rs}(\mathbf{x}, \mathbf{y})$ defined by

$$\int d^3z \sum_{j=1}^{2} C_{rj}(\mathbf{x}, \mathbf{z}) \{\phi_j(\mathbf{z}, t), \phi_s(\mathbf{y}, t)\} = \delta_{rs} \delta(\mathbf{x} - \mathbf{y}). \tag{11.154}$$

This definition is justified because, as concerns the continuous indices, the elements of the identity matrix are $\delta(\mathbf{x} - \mathbf{y})$. Noting that $\epsilon^2 = -\mathbf{I}$, one easily checks that (11.154) is satisfied by

$$C_{rs}(\mathbf{x}, \mathbf{y}) = \frac{1}{i\hbar} \epsilon_{rs} \delta(\mathbf{x} - \mathbf{y}). \tag{11.155}$$

The existence of the inverse of the matrix whose elements are (11.152) shows that the constraints are second class. Introducing the Dirac bracket

$$\{X, Y\}^D = \{X, Y\} - \sum_{rs} \int d^3x\, d^3y \{X, \phi_r(\mathbf{x}, t)\} C_{rs}(\mathbf{x}, \mathbf{y}) \{\phi_s(\mathbf{y}, t), Y\} \tag{11.156}$$

or, explicitly,

$$\begin{aligned} \{X, Y\}^D = \{X, Y\} &- \frac{1}{i\hbar} \int d^3x\, [\, \{X, \phi_1(\mathbf{x}, t)\} \{\phi_2(\mathbf{x}, t), Y\} \\ &- \{X, \phi_2(\mathbf{x}, t)\} \{\phi_1(\mathbf{x}, t), Y\} \,] \,, \end{aligned} \tag{11.157}$$

the constraint equations (11.149) can be taken as strong equations and the total Hamiltonian (11.150) reduces to the canonical Hamiltonian

$$H = \int d^3x\, \frac{\hbar^2}{2m} \nabla \psi \cdot \nabla \psi^*. \tag{11.158}$$

The canonical momenta are eliminated from the theory by means of the constraint equations (11.149) and the fundamental Dirac brackets are

$$\{\psi(\mathbf{x}, t), \psi(\mathbf{y}, t)\}^D = \{\psi^*(\mathbf{x}, t), \psi^*(\mathbf{y}, t)\}^D = 0, \tag{11.159a}$$

$$\{\psi(\mathbf{x}, t), \psi^*(\mathbf{y}, t)\}^D = \frac{1}{i\hbar} \delta(\mathbf{x} - \mathbf{y}). \tag{11.159b}$$

Exercise 11.7.1   Using (11.157), verify Eqs. (11.159).

According to the quantisation rules for systems with second-class constraints, one must replace the Dirac bracket by $i\hbar$ times the commutator (for bosonic fields). Thus, the fundamental commutation relation between the field operators $\hat{\Psi}$ and $\hat{\Psi}^\dagger$ is

$$[\hat{\Psi}(\mathbf{x}, t), \hat{\Psi}^\dagger(\mathbf{y}, t)] = \delta(\mathbf{x} - \mathbf{y}), \tag{11.160}$$

where $\hat{\Psi}^\dagger$ is the adjoint (Hermitian conjugate) of $\hat{\Psi}$. It is the commutator (11.160) that lies at the basis of what is called the second quantisation of non-relativistic quantum mechanics (Schiff, 1968). The description of fermions requires the replacement of the commutator by the anticommutator in (11.160).

Electromagnetic Field

The free electromagnetic field is another system whose Hamiltonian formulation involves constraints. From Eq. (11.41) with $\mu = 0$ one finds that the canonical momenta are

$$\pi^\alpha = \frac{\partial \mathcal{L}}{\partial \dot{A}_\alpha} = F^{\alpha 0}, \tag{11.161}$$

or, componentwise,

$$\pi^0 = 0, \qquad \pi^i = F^{i0} = \partial^i A^0 - \dot{A}^i = E^i, \tag{11.162}$$

where E^i, $i = 1, 2, 3$, are the Cartesian components of the electric field. There is, therefore, one primary constraint:

$$\phi_1 \equiv \pi^0 \approx 0. \tag{11.163}$$

A straightforward computation shows that the canonical Hamiltonian takes the form

$$H = \int d^3x \left(\frac{1}{2} \pi^i \pi^i + \frac{1}{4} F^{ij} F_{ij} + \pi^i \partial_i A_0 \right) \tag{11.164}$$

in which all repeated indices are summed over. Therefore, the total Hamiltonian is

$$H = \int d^3x \left(\frac{1}{2} \pi^i \pi^i + \frac{1}{4} F^{ij} F_{ij} + \pi^i \partial_i A_0 + \lambda_1 \phi_1 \right). \tag{11.165}$$

With the use of the fundamental Poisson brackets

$$\{A_\mu(\mathbf{x}, t), \pi^\nu(\mathbf{y}, t)\} = \delta^\nu_\mu \, \delta(\mathbf{x} - \mathbf{y}), \tag{11.166}$$

the consistency condition $\dot{\phi}_1 \approx 0$ gives rise to the secondary constraint

$$\phi_2 \equiv -\partial_i \pi^i \approx 0. \tag{11.167}$$

This constraint is none other than $\nabla \cdot \mathbf{E} = 0$, hence it being known as the Gauss law constraint. There are no other constraints because ϕ_2 has zero Poisson bracket both with H and ϕ_1.

Exercise 11.7.2   Show that $\{\phi_2(\mathbf{x}, t), H\} = 0$ and $\{\phi_1(\mathbf{x}, t), \phi_2(\mathbf{y}, t)\} = 0$. Hint: use (11.166) and $\{\partial_i \pi^i(\mathbf{x}, t), X\} = \partial_i \{\pi^i(\mathbf{x}, t), X\}$.

By means of an an integration by parts with neglect of a surface term, which does not affect the field equations of motion, the total Hamiltonian can be cast in the form

$$H_T = \int d^3x \left(\frac{1}{2} \pi^i \pi^i + \frac{1}{4} F^{ij} F_{ij} - A_0 \, \partial_i \pi^i + \lambda_1 \phi_1 \right). \tag{11.168}$$

The constraints ϕ_1 and ϕ_2 are first class and the dynamics involve arbitrary functions such as A_0, reflecting the gauge invariance of the theory. The elimination of the arbitrary functions demands a gauge fixing. It should be noted that A_0 acts as a Lagrange multiplier for the Gauss law, with the consequence that the genuine canonical variables are the space components of the potential four-vector and their conjugate momenta. Other aspects of the Hamiltonian treatment of Maxwell's electrodynamics are left to the specialised treatises (Dirac, 1964; Sundermeyer, 1982; Henneaux & Teitelboim, 1992).

Problems

11.1 Prove that any two Lagrangian densities that differ by a four-divergence $\partial_\lambda \Phi^\lambda(\varphi_\alpha, x)$ yield the same field equations of motion. Compare this result with Theorem 2.3.1.

11.2 Several authors prefer the following direct definition of functional derivative:

$$\frac{\delta F}{\delta f(y)} = \lim_{\epsilon \to 0} \frac{F[f + \epsilon\, \delta_y] - F[f]}{\epsilon},$$

with $\delta_y(x) = \delta(x - y)$. Show that this definition is equivalent to that of Section 11.3.

11.3 Show that

$$\frac{\delta}{\delta f(x)} \int_{\mathbb{R}^n} d^n y\, f(y)^n = n f(x)^{n-1},$$

where n is a natural number. Using this result and (11.47), show that the funcional differential equation

$$\frac{\delta F}{\delta f(x)} + f(x)F = 0$$

admits the solution

$$F[f] = C\, \exp\left\{ -\frac{1}{2} \int_{\mathbb{R}^n} d^n x\, f(x)^2 \right\},$$

where C is an arbitrary constant (Rosen, 1969).

11.4 Given the functional power series

$$Z[f] = K_0 + \int d^n x_1\, K_1(x_1) f(x_1)$$

$$+ \frac{1}{2!} \int d^n x_1\, d^n x_2\, K_2(x_1, x_2) f(x_1) f(x_2)$$

$$+ \frac{1}{3!} \int d^n x_1\, d^n x_2\, d^n x_3\, K_3(x_1, x_2, x_3) f(x_1) f(x_2) f(x_3) + \cdots,$$

where K_0 is a constant and $K_n(x_1, \ldots, x_n)$ is symmetric under any permutation of its arguments, show that

$$K_n(x_1, \ldots, x_n) = \left. \frac{\delta^n Z}{\delta f(x_1) \cdots \delta f(x_n)} \right|_{f=0}.$$

This formula is of great utility in quantum field theory.

11.5 The (orbital) angular momentum current density $\mathcal{M}^{\alpha\beta\gamma}$ is the tensor field defined by

$$\mathcal{M}^{\alpha\beta\gamma} = T^{\alpha\beta} x^\gamma - T^{\alpha\gamma} x^\beta,$$

where $T^{\alpha\beta}$ is the energy-momentum tensor. (a) The angular momentum tensor $L^{\beta\gamma} = \int \mathcal{M}^{0\beta\gamma} d^3 x$ will be conserved if $\partial_\alpha \mathcal{M}^{\alpha\beta\gamma} = 0$. Prove that this last equation is satisfied if and only if the energy-momentum tensor is symmetric: $T^{\alpha\beta} = T^{\beta\alpha}$. (b) Another physical reason to demand that the energy-momentum tensor be symmetric is its acting as source of the gravitational field in Einstein's equations of general relativity, which requires that the sources of gravitation be represented by symmetric tensors. If the canonical energy-momentum tensor (11.118) is not symmetric, it can be shown (Barut, 1980; Soper, 1976) that it is always possible to symmetrise it by the addition of a four-divergence $\partial_\lambda \psi^{\lambda\mu\nu}$ where the tensor $\psi^{\lambda\mu\nu}$ is antisymmetric in its first two indices. Prove that if $\partial_\mu T^{\mu\nu} = 0$ then it is also true that $\partial_\mu \Theta^{\mu\nu} = 0$ with $\Theta^{\mu\nu} = T^{\mu\nu} + \partial_\lambda \psi^{\lambda\mu\nu}$. Hint: see part (b) of Problem 6.6. (c) Provided $\psi^{\lambda\mu\nu}$ vanishes sufficiently fast at infinity, prove that the conserved physical quantities associated with $\Theta^{\mu\nu}$ and $T^{\mu\nu}$ have the same value, which justifies taking $\Theta^{\mu\nu}$ as the physical energy-momentum tensor.

11.6 (a) Show that, according to (11.118), the canonical energy-momentum tensor of the free electromagnetic field is

$$T^{\mu\nu} = F^{\alpha\mu}\partial^\nu A_\alpha + \frac{1}{4} g^{\mu\nu} F^{\alpha\beta} F_{\alpha\beta} .$$

Note that this tensor is not symmetric. Another physical objection to this tensor is that it is not gauge invariant. (b) Substituting $\partial^\nu A_\alpha = \partial^\alpha A_\nu + F^\nu{}_\alpha$ into the above expression for $T^{\mu\nu}$ and using Maxwell's equations for the free electromagnetic field, show that the tensor

$$\Theta^{\mu\nu} = F^{\alpha\mu} F^\nu{}_\alpha + \frac{1}{4} g^{\mu\nu} F^{\alpha\beta} F_{\alpha\beta}$$

is symmetric, gauge invariant, and differs from $T^{\mu\nu}$ only by a term of the form described in part (b) of the previous problem. Except for being expressed in Heaviside-Lorentz units, $\Theta^{\mu\nu}$ coincides with the energy-momentum tensor introduced in Problem 6.18.

11.7 Consider a scalar field in one spatial dimension described by the Lagrangian $\mathcal{L} = \frac{1}{2}\partial_\mu\partial^\mu\phi - V(\phi) \equiv \frac{1}{2}(\dot\phi^2 - \phi'^2) - V(\phi)$. (a) If ϕ is a static solution of Lagrange's equation, show that

$$\frac{d}{dx}\left[-\frac{1}{2}\phi'^2 + V(\phi)\right] = 0 .$$

(b) Suppose $V(\phi) \geq 0$ and write $V(\phi)$ in terms of a function $W(\phi)$ in the form $V(\phi) = \frac{1}{2}\left(\frac{dW}{d\phi}\right)^2$ (explain why this is always possible). If ϕ is a static solution that converges to some solution of the algebraic equation $V(\phi) = 0$ as $x \to \pm\infty$, show that ϕ satisfies the first-order differential equation

$$\frac{d\phi}{dx} = \pm\frac{dW}{d\phi} .$$

(c) Show that the energy $E = \int_{-\infty}^\infty dx[\frac{1}{2}\phi'^2 + V(\phi)]$ for a static solution of the type described in part (b) can be written in the form

$$E = \frac{1}{2}\int_{-\infty}^\infty dx\left(\phi' \mp \frac{dW}{d\phi}\right)^2 \pm [W(\phi(\infty)) - W(\phi(-\infty))]$$
$$= \pm[W(\phi(\infty)) - W(\phi(-\infty))] .$$

(d) Apply this result to the example discussed in Section 11.6 and get the kink's energy without having to explicitly compute the integral in Eq. (11.136).

11.8 Repeat the analysis of Section 11.6 for the Lagrangian

$$\mathcal{L} = \frac{1}{2} \partial^\mu \phi \partial_\mu \phi - m^2 \left(1 - \cos \phi\right),$$

considering the static solution that connects the ground states $\phi = 0$ and $\phi = 2\pi$. The field equation corresponding to this Lagrangian is informally known as the sine-Gordon equation. (a) Show that the sine-Gordon equation is satisfied by the soliton ϕ_+ and by the antisoliton ϕ_- given by

$$\phi_\pm(x, t) = 4 \arctan \left[e^{\pm \gamma (x - vt)} \right],$$

where $\gamma = (1 - v^2)^{-1/2}$ with $c = 1$. (b) Find the energy and the momentum carried by a soliton and by an antisoliton. (c) Show that

$$\phi_{SS}(x, t) = 4 \arctan \left[v \sinh\left(\gamma x\right) \text{sech}(\gamma v t) \right],$$

also satisfies the sine-Gordon equation. It can be shown (José & Saletan, 1998) that this solution describes the collision of two solitons: as $t \to -\infty$ the field ϕ_{SS} behaves as two separate solitons coming from $x = \pm\infty$ that approach the origin moving in opposite directions; the solitons interact as they pass the origin and get deformed during the interaction; as $t \to \infty$ the solitons reassemble themselves and move away to $x = \mp\infty$.

11.9 A string with clamped ends at $x = 0$ and $x = \ell$ has constant linear mass density σ and tension τ. If $\varphi_1(x, t)$ and $\varphi_2(x, t)$ are small transversal displacements of the string along two mutually perpendicular directions, the Lagrangian is

$$\mathcal{L} = \frac{\sigma}{2} \left[\left(\frac{\partial \varphi_1}{\partial t}\right)^2 + \left(\frac{\partial \varphi_2}{\partial t}\right)^2 \right] - \frac{\tau}{2} \left[\left(\frac{\partial \varphi_1}{\partial x}\right)^2 + \left(\frac{\partial \varphi_2}{\partial x}\right)^2 \right].$$

(a) Obtain Lagrange's equations for $\varphi_1(x, t)$ and $\varphi_2(x, t)$. (b) Taking into account that $\varphi_1(x, t)$ and $\varphi_2(x, t)$ vanish at $x = 0$ and $x = \ell$, prove that the quantity

$$Q = \int_0^\ell \sigma(\varphi_1 \dot{\varphi}_2 - \dot{\varphi}_1 \varphi_2) dx$$

is conserved, that is, its value does not depend on time. (c) What physical quantity does Q represent? (d) For what value of ω is the stationary wave $\varphi_1(x, t) = A \cos \omega t \sin(\pi x/\ell)$, $\varphi_2 = 0$ a solution of Lagrange's equations? Compute the energy associated with this wave.

11.10 The system composed of a charged scalar field in interaction with the electromagnetic field is characterised by the Lagrangian

$$\mathcal{L} = \left(\partial^\mu \phi + ieA^\mu \phi\right) \left(\partial_\mu \phi^* - ieA_\mu \phi^*\right) - \frac{1}{4} F^{\mu\nu} F_{\mu\nu},$$

where e denotes the electric charge associated with the field ϕ. (a) Write down Lagrange's equations. (b) Show that the Lagrangian is invariant under the **local gauge transformation** (also known as **gauge transformation of the second kind**)

$$\phi(x) \;\to\; \tilde{\phi}(x) = e^{-ie\chi(x)}\phi(x)\,, \qquad \phi^*(x) \;\to\; \tilde{\phi}^*(x) = e^{ie\chi(x)}\phi^*(x)\,,$$

$$A_\mu(x) \;\to\; \tilde{A}_\mu(x) = A_\mu(x) + \partial_\mu\chi(x)\,,$$

where χ is an arbitrary real function. (c) Show that the Lagrangian can be written in the form

$$\mathcal{L} = \partial^\mu\phi\,\partial_\mu\phi^* - \frac{1}{4}\,F^{\mu\nu}F_{\mu\nu} - J^\mu A_\mu\,,$$

where J^μ is the Noether current (11.125) associated with the invariance under global gauge transformations, those with constant χ. Several unusual features are involved in the application of Noether's theorem to theories with local gauge invariance (Karatas & Kowalski, 1990).

11.11 Maxwell's equations imply that the photon has zero mass. If the Coulomb law is not exactly valid and the field of a point charge has finite range, this can be ascribed to the photon having a mass. Apparently, the first to examine the consequences of assigning a mass to the photon was Proca, who introduced the Lagrangian

$$\mathcal{L}_{\text{Proca}} = -\frac{1}{4}F_{\alpha\beta}F^{\alpha\beta} + \frac{\mu^2}{2}A_\alpha A^\alpha - \frac{1}{c}J_\alpha A^\alpha\,,$$

where the parameter μ has dimension of inverse length. (a) Show that Lagrange's equations for the Proca field A^α are

$$\partial_\beta F^{\beta\alpha} + \mu^2 A^\alpha = \frac{1}{c}J^\alpha\,,$$

which are *not* invariant under the gauge transformation $A_\alpha \to A_\alpha + \partial_\alpha\Lambda$. (b) Prove that the local conservation of electric charge requires that the Proca field satisfy the Lorentz condition $\partial_\alpha A^\alpha = 0$. Show that, as a consequence, the Proca field obeys the following equation:

$$\Box A^\alpha + \mu^2 A^\alpha = \frac{1}{c}J^\alpha\,.$$

In a static case with only a point charge at rest at the origin, show that the spherically symmetric solution for the Proca field is $A^\alpha = (\phi, 0, 0, 0)$ with

$$\phi(r) = \frac{q}{4\pi}\frac{e^{-\mu r}}{r}\,.$$

In Proca's theory the potential of a static charge decays exponentially with distance and μ^{-1} provides a measure of the the range of the potential. (c) Assuming plane wave solutions $A^\alpha(\mathbf{r}, t) = A^\alpha_{(0)}\exp(i\mathbf{k}\cdot\mathbf{r} - i\omega t)$ in empty space ($J^\alpha = 0$), show that

$$\omega^2 = c^2 k^2 + \mu^2 c^2\,.$$

In terms of $E = \hbar\omega$ and $\mathbf{p} = \hbar\mathbf{k}$, this represents the relativistic relation between energy and momentum for a particle of mass $\mu\hbar/c$.

11.12 In a space-time with only two spatial dimensions there is a variant of classical electrodynamics, known as the Chern-Simons theory, in which the photon is massive but gauge invariance is preserved (Deser, Jackiw & Templeton, 1982). The Chern-Simons Lagrangian is

$$\mathcal{L}_{CS} = -\frac{1}{4}F_{\alpha\beta}F^{\alpha\beta} + \frac{\mu}{4}\epsilon^{\alpha\beta\gamma}F_{\alpha\beta}A_{\gamma},$$

where $F_{\alpha\beta} = \partial_\alpha A_\beta - \partial_\beta A_\alpha$, μ is the Chern-Simons parameter and $\epsilon^{\alpha\beta\gamma}$ is the three-dimensional Levi-Civita symbol with $\epsilon^{012} = 1$; Greek indices take the values $0, 1, 2$. (a) Show that under the gauge transformation $A_\alpha \to A_\alpha + \partial_\alpha \Lambda$ the Lagrangian $\mathcal{L}_{CS}$ varies only by the divergence $\partial_\gamma \left[(\mu/4) \epsilon^{\alpha\beta\gamma} F_{\alpha\beta} \Lambda \right]$. (b) Show that Lagrange's equations for the Chern-Simons field A_α are

$$\partial_\beta F^{\beta\alpha} + \frac{\mu}{2}\epsilon^{\alpha\beta\gamma}F_{\beta\gamma} = 0,$$

which are gauge invariant. Explain this property in the light of Problem 11.1. (c) Show that

$$(F^{\alpha\beta}) = \begin{pmatrix} 0 & -E_x & -E_y \\ E_x & 0 & -B \\ E_y & B & 0 \end{pmatrix}.$$

In a two-dimensional world the magnetic field is a scalar under space rotations. How to explain this phenomenon? (d) Show that, in terms of $\mathbf{E}$ and B, the complete Chern-Simons equations, including the homogeneous equations $\partial^\alpha F^{\beta\gamma} + \partial^\beta F^{\gamma\alpha} + \partial^\gamma F^{\alpha\beta} = 0$, are

$$\nabla \cdot \mathbf{E} = \mu B, \qquad \frac{1}{c}\frac{\partial E_x}{\partial t} + \mu E_y = \frac{\partial B}{\partial y},$$

$$\frac{1}{c}\frac{\partial E_y}{\partial t} - \mu E_x = -\frac{\partial B}{\partial x}, \qquad \frac{\partial E_y}{\partial x} - \frac{\partial E_x}{\partial y} = -\frac{1}{c}\frac{\partial B}{\partial t}.$$

(e) Show that both $\mathbf{E}$ and B satisfy the Klein-Gordon equation $(\Box + \mu^2)\phi = 0$, so μ plays the role of photon mass. (f) Generalise the Chern-Simons theory for the case in which there are charges and currents described by the current density J^α. Prove that, just as Maxwell's equations, the field equations of the Chern-Simons theory automatically imply the local conservation of electric charge, that is, $\partial_\alpha J^\alpha = 0$.

11.13 The Navier-Stokes equation for a fluid is

$$\frac{\partial \mathbf{v}}{\partial t} + (\mathbf{v} \cdot \nabla)\mathbf{v} = -\frac{1}{\rho}\nabla p + \nu\nabla^2\mathbf{v},$$

where $\mathbf{v}(\mathbf{x}, t)$ is the velocity at each point of the fluid, $\rho(\mathbf{x}, t)$ is the mass density, $p(\mathbf{x}, t)$ is the pressure and ν is the kinematic viscosity. The one-dimensional version of this equation with omission of the pressure term is known as the Burgers equation:

$$v_t + vv_x = \nu v_{xx}.$$

(a) Show that the Burgers equation admits a solution of the form

$$v(x, t) = c\left[1 - \tanh\frac{c(x - ct)}{2\nu}\right],$$

where c is an arbitrary constant. (b) Show that the Hopf-Cole transformation

$$v = -2v\frac{u_x}{u}$$

yields solutions to the Burgers equation provided u obeys the heat equation:

$$u_t = vu_{xx}\,.$$

(c) The solution to the heat equation with initial value $u(x,0)$ is

$$u(x,t) = \int_{-\infty}^{\infty} e^{-\frac{(x-y)^2}{4vt}}\, u(y,0)\,dy\,,\qquad t > 0.$$

Prove that the inverse of the Hopf-Cole transformation is

$$u(x,t) = \exp\left[-\frac{1}{2v}\int^x v(y,t)dy\right].$$

Combine the last two equations with the Hopf-Cole transformation to show that the solution of the Burgers equation is

$$v(x,t) = \frac{1}{t}\frac{\int_{-\infty}^{\infty} dy\,(x-y)\,e^{-\frac{(x-y)^2}{4vt}}\,\exp\left[-\frac{1}{2v}\int^y v_0(z)dz\right]}{\int_{-\infty}^{\infty} dy\,e^{-\frac{(x-y)^2}{4vt}}\,\exp\left[-\frac{1}{2v}\int^y v_0(z)dz\right]},$$

where $v_0(x) = v(x,0)$ is the initial velocity field.

11.14 The Korteweg-de Vries equation

$$u_t = uu_x + u_{xxx}$$

describes the propagation of waves in shallow waters. (a) Consider the solutions $u(x,t)$ to the Korteweg-de Vries equation that vanish as $x \to \pm\infty$ together with their spatial derivatives of all orders. Prove that F_0, F_1 and F_2 given by

$$F_0 = \int_{-\infty}^{\infty} u(x,t)dx,\ \ F_1 = \int_{-\infty}^{\infty} u(x,t)^2 dx,\ \ F_2 = \int_{-\infty}^{\infty}\left(\frac{u(x,t)^3}{6} - \frac{u_x(x,t)^2}{2}\right)dx$$

are constants of the motion. In José and Saletan (1998) there is a derivation of the Korteweg-de Vries equation as well as an iterative procedure that allows one to construct an infinite sequence of constants of the motion. (b) In Appendix H it is shown that the Korteweg-de Vries equation can be put in Hamiltonian form by means of the Poisson bracket

$$\{F,G\} = \int_{-\infty}^{\infty} dx\,\frac{\delta F}{\delta u(x)}\frac{\partial}{\partial x}\left(\frac{\delta G}{\delta u(x)}\right).$$

Show that the constants of the motion F_0, F_1 and F_2 defined here are in involution with respect to this Poisson bracket. The Korteweg-de Vries equation defines an integrable system with infinitely many degrees of freedom in the following sense: there exists an infinite sequence of constants of the motion in involution.

$$c_i = \sum_{j,k=1}^{3} \epsilon_{ijk}\, a_j\, b_k \,. \qquad (A.23)$$

A useful identity is

$$\sum_{k=1}^{3} \epsilon_{ijk}\, \epsilon_{klm} = \delta_{il}\, \delta_{jm} - \delta_{im}\, \delta_{jl} \,, \qquad (A.24)$$

the proof of which consists in some simple observations. The left-hand side of the identity differs from zero only if $i = l$ and $j = m$ or if $i = m$ and $j = l$, and it is antisymmetric in the indices i, j as well as in the indices l, m. So, it can only be a constant multiple of the combination of products of Kronecker deltas that appears on the right-hand side of the identity. A particular choice of values for the indices determines the multiplicative constant.

Substitution and Factorisation

There are certain precautions that must be taken when algebraically manipulating quantities expressed in indicial notation. Let us first consider the process of *substitution*. If

$$a_i = \sum_{k=1}^{N} U_{ik}\, b_k \qquad (A.25)$$

and

$$b_i = \sum_{k=1}^{N} V_{ik}\, c_k \,, \qquad (A.26)$$

in order to substitute into (A.25) the bs given by (A.26) we first change the free index in (A.26) from i to k and the dummy index k to some other letter not already in use, say l, so that

$$b_k = \sum_{l=1}^{N} V_{kl}\, c_l \,. \qquad (A.27)$$

Now, the substitution of (A.27) into (A.25) yields

$$a_i = \sum_{k,l=1}^{N} U_{ik}\, V_{kl}\, c_l \,. \qquad (A.28)$$

Had the change of summation index not been made in the transition from (A.26) to (A.27), the mere replacement of i by k in (A.26) followed by its insertion into (A.25) would have given

$$a_i = \sum_{k=1}^{N} \sum_{k=1}^{N} U_{ik}\, V_{kk}\, c_k \,, \qquad (A.29)$$

which is glaringly wrong because this summation excludes non-diagonal terms such as V_{12}, V_{23}, which have to appear. In order that the substitution lead to a correct equation, it is *mandatory* that the summation indices in (A.25) and (A.26) be *distinct*, since the respective summations are mutually independent. The same precaution must be taken when multiplying sums. For instance, if

$$p = \sum_{k=1}^{N} a_k \tag{A.30}$$

and

$$q = \sum_{k=1}^{N} b_k , \tag{A.31}$$

the correct expression for the product pq is

$$pq = \sum_{k,l=1}^{N} a_k b_l . \tag{A.32}$$

We insist that

$$pq \neq \sum_{k=1}^{N} \sum_{k=1}^{N} a_k b_k = N(a_1 b_1 + a_2 b_2 + \cdots + a_N b_N) . \tag{A.33}$$

The product of two sums can only be expressed correctly as a double sum if distinct dummy indices are employed to indicate the independent summations involved, as in Eq. (A.32).

Finally, regarding the process of *factorisation*, the correct procedure is pointed out by means of an example. Consider the equation

$$\sum_{j=1}^{N} T_{ij} n_j - \lambda n_i = 0 , \tag{A.34}$$

in which we would like to factorise n_j. With the help of the Kronecker delta we write

$$n_i = \sum_{j=1}^{N} \delta_{ij} n_j , \tag{A.35}$$

which once inserted into (A.34) yields

$$\sum_{j=1}^{N} T_{ij} n_j - \lambda \sum_{j=1}^{N} \delta_{ij} n_j = 0 , \tag{A.36}$$

or

$$\sum_{j=1}^{N} \left(T_{ij} - \lambda \delta_{ij} \right) n_j = 0 , \tag{A.37}$$

which is the intended result.

Convention of Sum over Repeated Indices

In most physical contexts the sums are performed over repeated indices, as in (A.11), (A.23), (A.28) and (A.37). In the general theory of relativity triple and quadruple sums occur all the time. Einstein introduced the following expedient convention: whenever an index appears repeated, a summation over the repeated index is understood without need of inserting the summation symbol. This simplifies enormously the writing of sums, especially of multiple sums. For instance, with this convention the scalar product of the vectors $\mathbf{a} = (a_1, a_2, a_3)$ and $\mathbf{b} = (b_1, b_2, b_3)$ is simply written as $\mathbf{a} \cdot \mathbf{b} = a_i b_i$.

With the convention of sum over repeated indices the vector product of two vectors is expressed as

$$(\mathbf{a} \times \mathbf{b})_i = \epsilon_{ijk} a_j b_k ,\qquad (A.38)$$

the identity (A.24) takes the simpler form

$$\epsilon_{ijk}\, \epsilon_{klm} = \delta_{il}\, \delta_{jm} - \delta_{im}\, \delta_{jl} \qquad (A.39)$$

and the scalar triple product $\mathbf{a} \cdot (\mathbf{b} \times \mathbf{c})$ is written

$$\mathbf{a} \cdot (\mathbf{b} \times \mathbf{c}) = \epsilon_{ijk} a_i b_j c_k . \qquad (A.40)$$

Exercise A.1   One says that a_{ij} is symmetric and b_{ij} is antisymmetric in the indices i, j if $a_{ji} = a_{ij}$ and $b_{ji} = -b_{ij}$. Exploring the fact that the names chosen for the summation indices do not affect the sum, prove that if a_{ij} is symmetric and b_{ij} is antisymmetric then one necessarily has $a_{ij}b_{ij} = 0$.

Exercise A.2   Show that the well-known vector identity

$$\mathbf{a} \times (\mathbf{b} \times \mathbf{c}) = (\mathbf{a} \cdot \mathbf{c})\,\mathbf{b} - (\mathbf{a} \cdot \mathbf{b})\,\mathbf{c} \qquad (A.41)$$

is an immediate consequence of the identity (A.39).

Exercise A.3   Prove the vector identity

$$(\mathbf{a} \times \mathbf{b}) \cdot (\mathbf{c} \times \mathbf{d}) = (\mathbf{a} \cdot \mathbf{c})(\mathbf{b} \cdot \mathbf{d}) - (\mathbf{a} \cdot \mathbf{d})(\mathbf{b} \cdot \mathbf{c}) \qquad (A.42)$$

by means of (A.39).

Let $A = (A_{ij})$ be a 3×3 matrix. An important identity involving the determinant of A is

$$\epsilon_{ijk} A_{il} A_{jm} A_{kn} = (\det A)\, \epsilon_{lmn} , \qquad (A.43)$$

where sums over the repeated indices are understood. This identity extends to $n \times n$ matrices in the form

$$\epsilon_{i_1 i_2 \cdots i_n} A_{i_1 j_1} A_{i_2 j_2} \cdots A_{i_n j_n} = (\det A)\, \epsilon_{j_1 j_2 \cdots j_n} , \qquad (A.44)$$

where $\epsilon_{i_1 i_2 \cdots i_n}$ is the n-dimensionl Levi-Civita symbol, defined by a straightforward generalisation of (A.22).

Appendix B **Frobenius Integrability Condition**

The most common velocity-dependent constraints are those in which the velocities occur linearly, as in Example 1.6. Constraints of this nature are first-order differential equations which, once cast in the equivalent form (2.66), assert that certain differential forms vanish. If constraints of this type are integrable the system is holonomic: there exist generalised coordinates in terms of which the constraints are identically satisfied and the dimension of configuration space (the number of degrees of freedom) is smaller that the number of originally chosen coordinates. Otherwise, the system is non-holonomic and the constraints restrict the velocities alone: the totality of the original configuration space is accessible to the system. The theory of differential forms is the milieu in which the integrability condition for constraints of the form (2.66) is expressed in the most crystal-clear fashion. Space limitations compel this appendix to a succinct, pragmatic and informal exposition. Precise definitions and rigorous proofs can be found in the beautiful book by Flanders (1989), whose study is highly recommended and upon which the following text is largely based.

Differential Forms: Mathematical and Physical Motivations

Differential forms arise naturally in physics. Given a vector field $\mathbf{F} = (A, B, C)$ in Cartesian coordinates, its line integral along an oriented curve $\mathcal{C}$ is

$$\int_{\mathcal{C}} \mathbf{F} \cdot d\mathbf{r} = \int_{\mathcal{C}} A\,dx + B\,dy + C\,dz, \tag{B.1}$$

and its surface integral over an oriented surface $\mathcal{S}$ is

$$\int_{\mathcal{S}} \mathbf{F} \cdot d\mathbf{S} = \int \int_{\mathcal{S}} A\,dy\,dz + B\,dz\,dx + C\,dx\,dy. \tag{B.2}$$

If $\mathbf{F}$ stands for the force on a particle, the integral (B.1) is the work done by the force during a displacement of the particle along the curve $\mathcal{C}$; if $\mathbf{F}$ is the current density, the integral (B.2) is the electric current through $\mathcal{S}$; if $\mathbf{F}$ is an electric field and $\mathcal{S}$ is a closed surface, the integral (B.2) is the flux of the electric field through $\mathcal{S}$ that is involved in Gauss's law of electromagnetism.

Thus, we are naturally led to consider the 1-form (or one-form)

$$\omega = A\,dx + B\,dy + C\,dz \tag{B.3}$$

and the 2-form (or two-form)

$$\lambda = Pdydz + Qdzdx + Rdxdy. \tag{B.4}$$

Intuitively, therefore, differential forms are things that can be integrated.

The absence of $dzdy$, $dxdz$ and $dydx$ from λ suggests symmetry or antisymmetry of the product of differentials; the absence of $dxdx$, $dydy$ and $dzdz$ favours antisymmetry. Antisymmetry is reinforced by making immediate the rule for change of variables in a multiple integral. Under the change of variables $x = x(u, v)$, $y = y(u, v)$, the use of $dudu = dvdv = 0$ and $dvdu = -dudv$ yields

$$
\begin{aligned}
\int \int f(x, y) dx dy &= \int \int f(x(u, v), y(u, v)) \left(\frac{\partial x}{\partial u} du + \frac{\partial x}{\partial v} dv \right) \left(\frac{\partial y}{\partial u} du + \frac{\partial y}{\partial v} dv \right) \\
&= \int \int f(x(u, v), y(u, v)) \left(\frac{\partial x}{\partial u} \frac{\partial y}{\partial v} du dv - \frac{\partial x}{\partial v} \frac{\partial y}{\partial u} du dv \right) \\
&= \int \int f(x(u, v), y(u, v)) \frac{\partial(x, y)}{\partial(u, v)} du dv,
\end{aligned}
\tag{B.5}
$$

in which there automatically appears the Jacobian of the transformation:

$$
\frac{\partial(x, y)}{\partial(u, v)} =
\begin{vmatrix}
\dfrac{\partial x}{\partial u} & \dfrac{\partial x}{\partial v} \\[2ex]
\dfrac{\partial y}{\partial u} & \dfrac{\partial y}{\partial v}
\end{vmatrix}.
\tag{B.6}
$$

Exterior Product

In $\mathbb{R}^n$, with coordinates $(x^1, \ldots, x^n)$, the differential forms are built from the basic differentials $dx^1, \ldots, dx^n$ by means of an operation known as *exterior product* or *wedge product*.[1]

Definition B.1 *The **exterior product** (denoted by $\wedge$) of the basic differentials is defined by*

$$dx^i \wedge dx^i = 0, \quad dx^i \wedge dx^j = -dx^j \wedge dx^i \ \text{if} \ i \neq j \ (i, j = 1, \ldots, n). \tag{B.7}$$

In the case of the exterior product of more than two basic differentials associativity holds:

$$dx^i \wedge (dx^j \wedge dx^k) = (dx^i \wedge dx^j) \wedge dx^k = dx^i \wedge dx^j \wedge dx^k. \tag{B.8}$$

With $x = (x^1, \ldots, x^n)$, a 1-form is defined by

$$\omega = \sum_{k=1}^{n} a_k(x) dx^k; \tag{B.9}$$

---

[1] The basic differentials are elements of the dual space of the tangent space $T_p(\mathcal{M})$ of a differentiable manifold $\mathcal{M}$ at each point $p \in \mathcal{M}$ – that is, they are linear applications from $T_p(\mathcal{M})$ into $\mathbb{R}$. See Flanders (1989), the second chapter of Hawking and Ellis (1973) or the seventh chapter of Arnold (1989). The exterior product is the fully antisymmetrised tensor product.

a p-form (with $1 < p < n$) is defined by

$$\lambda = \frac{1}{p!} \sum_{k_1 \ldots k_p} a_{k_1 \ldots k_p}(x) dx^{k_1} \wedge \cdots \wedge dx^{k_p}, \tag{B.10}$$

where $a_{k_1 \ldots k_p}$ is totally antisymmetric in its p indices. An n-form is

$$\epsilon = g(x) dx^1 \wedge \cdots \wedge dx^n \tag{B.11}$$

and a 0-form is just a function f. All functions a_k, $a_{k_1 \ldots k_p}$, g and f are applications from $\mathbb{R}^n$ into $\mathbb{R}$ of class C^∞ – that is, they are infinitely differentiable. The purpose of the denominator $p!$ in the expression of a p-form with $1 < p < n$ is to cancel the factor $p!$ that arises from the equal repeated terms associated with the permutations of the summation indices in Eq. (B.10). The factor $1/p!$ can be omitted by conventioning that the summation is restricted to the indices $k_1 < k_2 < \cdots < k_p$.

A p-form is also called a form of degree p. A differential form is non-zero if at least one of the coefficients of its expression in terms of the basic differentials does not vanish. Note that in a space of dimension n any p-form with $p > n$ is zero.

From Definition B.1 the exterior product of two arbitrary differential forms is naturally determined by linearity. Examples in $\mathbb{R}^3$ illustrate not only the calculation of exterior products but also how elementary operations on vectors emerge naturally from the exterior algebra of differential forms. For instance, the exterior product

$$(Adx + Bdy + Cdz) \wedge (Edx + Fdy + Gdz)$$
$$= (BG - CF)dy \wedge dz + (CE - AG)dz \wedge dx + (AF - BE)dx \wedge dy \tag{B.12}$$

brings forth the components of the vector product of the vectors (A, B, C) and (E, F, G). On the other hand, from the exterior product

$$(Adx + Bdy + Cdz) \wedge (Pdy \wedge dz + Qdz \wedge dx + Rdx \wedge dy) = (AP + BQ + CR)dx \wedge dy \wedge dz \tag{B.13}$$

the scalar product of the vectors (A, B, C) and (P, Q, R) emerges automatically.

The exterior or wedge product has the following properties:

(EP1) $\lambda \wedge (\mu + \nu) = \lambda \wedge \mu + \lambda \wedge \nu$ (distributivity);
(EP2) $\lambda \wedge (\mu \wedge \nu) = (\lambda \wedge \mu) \wedge \nu$ (associativity);
(EP3) $\lambda \wedge \mu = (-1)^{pq} \mu \wedge \lambda$ (p is the degree of λ and q is the degree of μ).

In order to verify the third property it is enough to consider the product of monomials:

$$dx^{i_1} \wedge dx^{i_2} \wedge \cdots \wedge dx^{i_p} \wedge dx^{j_1} \wedge dx^{j_2} \wedge \cdots \wedge dx^{j_q}.$$

The displacement of dx^{j_1} to the left of dx^{i_1} causes p changes of sign; the same goes for dx^{j_2} and so on:

$$dx^{i_1} \wedge dx^{i_2} \wedge \cdots \wedge dx^{i_p} \wedge dx^{j_1} \wedge dx^{j_2} \wedge \cdots \wedge dx^{j_q}$$
$$= (-1)^p dx^{j_1} \wedge dx^{i_1} \wedge dx^{i_2} \wedge \cdots \wedge dx^{i_p} \wedge dx^{j_2} \wedge \cdots \wedge dx^{j_q}$$
$$= (-1)^{2p} dx^{j_1} \wedge dx^{j_2} \wedge dx^{i_1} \wedge dx^{i_2} \wedge \cdots \wedge dx^{i_p} \wedge dx^{j_3} \wedge \cdots \wedge dx^{j_q}$$
$$= \cdots = (-1)^{qp} dx^{j_1} \wedge dx^{j_2} \wedge \cdots dx^{j_q} \wedge dx^{i_1} \wedge dx^{i_2} \wedge \cdots \wedge dx^{i_p}. \tag{B.14}$$

Remark   From property **EP3** of the exterior product it follows that $\lambda \wedge \lambda = 0$ for any form λ of odd degree.

Definition B.2   *Given the non-zero 1-form* $\omega = a_1(x)dx^1 + \cdots + a_n(x)dx^n$, *the equation* $\omega = 0$ *determines all parameterised curves* $x(t) = (x^1(t), \cdots, x^n(t))$ *that satisfy the ordinary differential equation* $a_1(x(t))\dot{x}^1(t) + \cdots + a_n(x(t))\dot{x}^n(t) = 0$ – *that is:*

$$a_1(x)dx^1 + \cdots + a_n(x)dx^n = 0 \implies a_1(x(t))\frac{dx^1(t)}{dt} + \cdots + a_n(x(t))\frac{dx^n(t)}{dt} = 0. \quad (B.15)$$

It is worth emphasising that the equation $\omega = 0$ *does not* assert that the 1-form ω is identically zero, but instead *specifies the trajectories along which it vanishes*.

Example B.1   The 1-form $\omega = \sin\theta\, dx - \cos\theta\, dy$ defined on $\mathbb{R}^3$ with coordinates (x, y, θ) never vanishes because $\sin\theta$ and $\cos\theta$ do not vanish simultaneously. The equation $\omega = 0$ expresses the constraint $\dot{x}\sin\theta - \dot{y}\cos\theta = 0$ for the skate in Example 2.8. The curve $x(t) = \sin t$, $y(t) = -\cos t$, $\theta(t) = t$ satisfies $\omega = 0$.

Exterior Derivative

The notion of exterior derivative plays a crucial role in the theory of differential forms.

Definition B.3   *The **exterior derivative** of a p-form*

$$\lambda = \frac{1}{p!}\sum_{k_1...k_p} a_{k_1...k_p}(x)dx^{k_1} \wedge \cdots \wedge dx^{k_p} \quad (B.16)$$

is the $(p+1)$-*form defined by*

$$d\lambda = \frac{1}{p!}\sum_k \sum_{k_1...k_p} \frac{\partial a_{k_1...k_p}}{\partial x^k}(x)dx^k \wedge dx^{k_1} \wedge \cdots \wedge dx^{k_p}. \quad (B.17)$$

The exterior derivative has the following properties:
   (**ED1**) $d(\mu + \nu) = d\mu + d\nu$;
   (**ED2**) $d(\mu \wedge \nu) = d\mu \wedge \nu + (-1)^p \mu \wedge d\nu$ where p is the degree of μ;
   (**ED3**) $d(d\lambda) = 0$ for any differential form λ;
   (**ED4**)   if f is a 0-form, $df = \sum_k \frac{\partial f}{\partial x^k}dx^k$.

Property **ED4** is, in fact, a definition. As to the remaining ones, we will prove only **ED3**, which is of utmost importance. Taking into account the linearity of the exterior derivative, it suffices to consider the monomial

$$\lambda = a(x)dx^{k_1} \wedge \cdots \wedge dx^{k_p}. \quad (B.18)$$

We have

$$d\lambda = \sum_j \frac{\partial a}{\partial x^j} dx^j \wedge dx^{k_1} \wedge \cdots \wedge dx^{k_p} \tag{B.19}$$

and

$$d(d\lambda) = \sum_i \sum_j \frac{\partial^2 a}{\partial x^i \partial x^j} dx^i \wedge dx^j \wedge dx^{k_1} \wedge \cdots \wedge dx^{k_p}. \tag{B.20}$$

The exchange of indices $i \leftrightarrow j$ does not alter the sum. Therefore, using Eq. (A.9) of Appendix A as well as $dx^i \wedge dx^j = -dx^j \wedge dx^i$, there results

$$d(d\lambda) = \frac{1}{2} \sum_{i,j} \left(\frac{\partial^2 a}{\partial x^i \partial x^j} - \frac{\partial^2 a}{\partial x^j \partial x^i} \right) dx^i \wedge dx^j \wedge dx^{k_1} \wedge \cdots \wedge dx^{k_p} = 0, \tag{B.21}$$

owing to the equality of the mixed second partial derivatives.

Applied to differential forms defined on $\mathbb{R}^3$ the exterior derivative engenders the most important operations of vector analysis.

Example B.2 Given the 1-form $\omega = A dx + B dy + C dz$, we have

$$d\omega = \left(\frac{\partial A}{\partial x} dx + \frac{\partial A}{\partial y} dy + \frac{\partial A}{\partial z} dz \right) \wedge dx + \left(\frac{\partial B}{\partial x} dx + \frac{\partial B}{\partial y} dy + \frac{\partial B}{\partial z} dz \right) \wedge dy$$

$$+ \left(\frac{\partial C}{\partial x} dx + \frac{\partial C}{\partial y} dy + \frac{\partial C}{\partial z} dz \right) \wedge dz.$$

With the use of the properties of the exterior product we get

$$d\omega = \left(\frac{\partial C}{\partial y} - \frac{\partial B}{\partial z} \right) dy \wedge dz + \left(\frac{\partial A}{\partial z} - \frac{\partial C}{\partial x} \right) dz \wedge dx + \left(\frac{\partial B}{\partial x} - \frac{\partial A}{\partial y} \right) dx \wedge dy, \tag{B.22}$$

where we can see the components of the curl of the vector field (A, B, C). In the case of the 2-form $\lambda = P dy \wedge dz + Q dz \wedge dx + R dx \wedge dy$, we have

$$d\lambda = \left(\frac{\partial P}{\partial x} dx + \frac{\partial P}{\partial y} dy + \frac{\partial P}{\partial z} dz \right) \wedge dy \wedge dz + \left(\frac{\partial Q}{\partial x} dx + \frac{\partial Q}{\partial y} dy + \frac{\partial Q}{\partial z} dz \right) \wedge dz \wedge dx$$

$$+ \left(\frac{\partial R}{\partial x} dx + \frac{\partial R}{\partial y} dy + \frac{\partial R}{\partial z} dz \right) \wedge dx \wedge dy.$$

It follows that

$$d\lambda = \left(\frac{\partial P}{\partial x} + \frac{\partial Q}{\partial y} + \frac{\partial R}{\partial z} \right) dx \wedge dy \wedge dz, \tag{B.23}$$

which contains the divergence of the vector field (P, Q, R).

Exercise B.3 From $d^2 = 0$ prove the following well-known identities of vector analysis: $\mathrm{curl}(\mathrm{grad} f) = 0$; $\mathrm{div}(\mathrm{curl} \mathbf{V}) = 0$.

with $x_1, \ldots, x_n$ fixed. Then, again with $u_1 = \lambda x_1, \ldots, u_n = \lambda x_n$, we have

$$\frac{dg}{d\lambda} = -p\,\lambda^{-p-1}\,F(u_1, \ldots, u_n) + \lambda^{-p}\sum_{k=1}^{n}\frac{\partial F}{\partial u_k}\,x_k$$

$$= \lambda^{-p-1}\left[-p\,F(u_1, \ldots u_n) + \sum_{k=1}^{n}\frac{\partial F}{\partial u_k}\,u_k\right] = 0 \qquad (C.6)$$

in virtue of (C.2). Therefore, $g(\lambda) = $ constant. But from (C.5) it is immediate that $g(1) = F(x_1, \ldots, x_n)$. Therefore, $g(\lambda) = g(1) = F(x_1, \ldots, x_n)$, which proves (C.1) and establishes that the function F is homogeneous of degree p. $\qquad\square$

Example C.2　By Euler's theorem, the general solution of the first-order partial differential equation

$$x\frac{\partial f}{\partial x} + y\frac{\partial f}{\partial y} = pf \qquad (C.7)$$

for $f(x, y)$ can be written in the form

$$f(x, y) = x^p g\left(\frac{x}{y}\right), \qquad (C.8)$$

where $g : \mathbb{R} \to \mathbb{R}$ is an arbitrary differentiable function.

Appendix D Vector Spaces and Linear Operators

In this appendix we outline the theory of linear transformations on finite-dimensional vector spaces. The reader is assumed to be familiar with the definition of vector space and with the notions of linear dependence and independence, basis, matrix, trace and determinant (Halmos, 1974; Hoffman & Kunze, 1971). We designate by $\mathcal{F}$ the field of real or complex numbers, which are generically called **scalars**. We denote the scalars by Latin letters, such as a, b, c, x, y, z, and the vectors by Greek letters, such as $\alpha, \beta, \gamma, \xi, \eta$.

Linear Operators

Linear operators on vector spaces are important in classical mechanics and play a crucial role in quantum mechanics.

Definition D.1 *Let $\mathcal{V}$ be a vector space over the field $\mathcal{F}$. A **linear operator** or **linear transformation** A on $\mathcal{V}$ is a rule that assigns to each vector α in $\mathcal{V}$ a vector $A\alpha$ in $\mathcal{V}$ in such a way that*

$$A(c_1\alpha + c_2\beta) = c_1 A\alpha + c_2 A\beta \tag{D.1}$$

for all α, β in $\mathcal{V}$ and all scalars c_1, c_2 in $\mathcal{F}$.

For a linear operator A it is always true that $A0 = 0$. The **identity operator** I, defined by $I\alpha = \alpha$, is a linear transformation on $\mathcal{V}$, and the same holds for the **zero operator** 0, defined by $0\alpha = 0$.

Given two linear operators A and B, their *sum* $S = A + B$ is the linear operator defined by $S\alpha = A\alpha + B\alpha$, while their *product* $P = AB$ is defined by $P\alpha = A(B\alpha)$. In general, the product of linear operators is not commutative. The algebraic properties of sum and product of real numbers remain valid for linear operators, with the notable exception of commutativity.

Definition D.2 *An operator A, not necessarily linear, is said to be **invertible** if it has the two following properties:*
  (a) for each vector β there is at least one vector α such that $A\alpha = \beta$;
  (b) if $\alpha_1 \neq \alpha_2$ then $A\alpha_1 \neq A\alpha_2$.
These properties characterise A as an application that is both surjective and injective – that is, as a bijection from $\mathcal{V}$ to $\mathcal{V}$.

If A is an invertible linear operator one can define a linear operator A^{-1}, called the **inverse** of A, as we proceed to describe. If β is any vector, we can, by (a), find a vector α such that $A\alpha = \beta$. In virtue of (b), α is uniquely determined. By definition, $A^{-1}\beta = \alpha$. In order to prove that A^{-1} is linear, let us compute $A^{-1}(c_1\beta_1 + c_2\beta_2)$ with $A^{-1}\beta_1 = \alpha_1$ and $A^{-1}\beta_2 = \alpha_2$. From the linearity of A it follows that $A(c_1\alpha_1 + c_2\alpha_2) = c_1\beta_1 + c_2\beta_2$, so that $A^{-1}(c_1\beta_1 + c_2\beta_2) = c_1\alpha_1 + c_2\alpha_2 = c_1A^{-1}\beta_1 + c_2A^{-1}\beta_2$ and A^{-1} is linear. From the definition of inverse we immediately have

$$AA^{-1} = A^{-1}A = I. \tag{D.2}$$

Theorem D.1 *A linear operator A on a finite-dimensional vector space $\mathcal{V}$ is invertible if and only if $A\alpha = 0$ implies $\alpha = 0$.*

Proof If A is invertible, combining $A0 = 0$ with property (b) one concludes that the only solution of $A\alpha = 0$ is $\alpha = 0$. Suppose, now, that $A\alpha = 0$ implies $\alpha = 0$. In this case, $\alpha_1 \neq \alpha_2$ or, equivalently, $\alpha_1 - \alpha_2 \neq 0$ implies $A(\alpha_1 - \alpha_2) \neq 0$ – that is, $A\alpha_1 \neq A\alpha_2$, which proves (b). In order to prove (a), let $\{\alpha_1, \ldots, \alpha_n\}$ be a basis of $\mathcal{V}$. If $\sum_i c_iA\alpha_i = 0$ then $A(\sum_i c_i\alpha_i) = 0$. Hence, according to our hypothesis, $\sum_i c_i\alpha_i = 0$ and the linear independence of the α_i leads to $c_1 = \cdots = c_n = 0$. Therefore, $\{A\alpha_1, \ldots, A\alpha_n\}$ is a set of linearly independent vectors, also constituting a basis of $\mathcal{V}$. Therefore, any vector β can be written in the form $\beta = \sum_i c_iA\alpha_i = A(\sum_i c_i\alpha_i) \equiv A\alpha$, which completes the proof. $\qquad\square$

It easily verified that: (1) if A and B are invertible then the product AB is also invertible and $(AB)^{-1} = B^{-1}A^{-1}$; (2) if A is invertible and $c \neq 0$, then cA is invertible and $(cA)^{-1} = \dfrac{1}{c}A^{-1}$; (3) if A is invertible, then A^{-1} is also invertible and $(A^{-1})^{-1} = A$.

Matrix Representation of Operators

Any linear operator can be represented by a matrix once a basis is chosen for the vector space on which it acts.

Definition D.3 *Let $\mathcal{V}$ be a vector space of dimension n. Let $\mathscr{A} = \{\alpha_1, \ldots, \alpha_n\}$ be a basis of $\mathcal{V}$ and A be a linear operator on $\mathcal{V}$. Since any vector is a linear combination of the α_i, we have*

$$A\alpha_j = \sum_{i=1}^{n} a_{ij}\alpha_i, \quad j = 1, \ldots, n. \tag{D.3}$$

The set (a_{ij}) of n^2 scalars is the matrix of A in basis (or coordinate system) $\mathscr{A}$, which we denote by $[A]$. When it is necessary to specify the basis under consideration we write $[A]_{\mathscr{A}}$.

Let $\xi = \sum_i x_i\alpha_i$ and $\eta = \sum_i y_i\alpha_i$ be vectors such that $\eta = A\xi$. Let $[\xi]$ and $[\eta]$ be the column matrices

$$[\xi] = \begin{pmatrix} x_1 \\ \vdots \\ x_n \end{pmatrix}, \qquad [\eta] = \begin{pmatrix} y_1 \\ \vdots \\ y_n \end{pmatrix}. \tag{D.4}$$

Then

$$\eta = A\xi = \sum_{j=1}^{n} x_j A\alpha_j = \sum_{j=1}^{n}\sum_{i=1}^{n} x_j a_{ij}\alpha_i, \tag{D.5}$$

whence

$$y_i = \sum_{j=1}^{n} a_{ij} x_j, \tag{D.6}$$

or, in matrix language,

$$[\eta] = [A][\xi]. \tag{D.7}$$

It is to ensure the validity of this equation that a seemingly perverse choice of indices was made in Eq. (D.3). Algebraic operations with linear operators transfer unchanged to the corresponding matrices. For instance, it is left as an exercise to the reader the verification that if $C = AB$ then $[C] = [A][B]$. In particular, if S is an invertible linear operator, from $SS^{-1} = S^{-1}S = I$ it follows that $[S][S^{-1}] = [S^{-1}][S] = I$, where I is the identity matrix, and, as a consequence, $[S^{-1}] = [S]^{-1}$. A linear operator is invertible if and only if its associated matrix is non-singular – that is, its determinant is non-zero in some basis (therefore, in all bases). It can be shown (Hoffman & Kunze, 1971) that the elements of the inverse of a non-singular matrix M are given by

$$(M^{-1})_{ij} = (-1)^{i+j}\,\frac{\det M(j|i)}{\det M}, \tag{D.8}$$

where $M(i|j)$ is the matrix obtained by erasing the ith row and the jth column of M. On the right-hand side of (D.8) attention should be paid to the inverted order of the indices.

How is the matrix associated with a linear operator A affected by a change of basis? In order to answer this question, let $\mathscr{X} = \{\xi_1, \ldots, \xi_n\}$ and $\mathscr{Y} = \{\eta_1, \ldots, \eta_n\}$ be two bases of V and consider the change-of-basis linear operator S defined by $S\xi_i = \eta_i$ or, more explicitly,

$$S\left(\sum_i x_i \xi_i\right) = \sum_i x_i \eta_i. \tag{D.9}$$

The operator S is clearly invertible, since $S(\sum_i x_i \xi_i) = 0$ is equivalent to $\sum_i x_i \eta_i = 0$, which implies $x_1 = \cdots = x_n = 0$. Set $[A]_{\mathscr{X}} = (a_{ij})$, $[A]_{\mathscr{Y}} = (b_{ij})$ and $[S]_{\mathscr{X}} = (s_{ij})$. We can write

$$A\xi_j = \sum_i a_{ij}\xi_i \tag{D.10}$$

and

$$A\eta_j = \sum_i b_{ij}\eta_i. \tag{D.11}$$

On the one hand,

$$A\eta_j = AS\xi_j = A\left(\sum_k s_{kj}\xi_k\right) = \sum_k s_{kj}A\xi_k = \sum_k s_{kj}\left(\sum_i a_{ik}\xi_i\right) = \sum_i\left(\sum_k a_{ik}s_{kj}\right)\xi_i.$$

(D.12)

On the other hand,

$$A\eta_j = \sum_k b_{kj}\eta_k = \sum_k b_{kj}S\xi_k = \sum_k b_{kj}\left(\sum_i s_{ik}\xi_i\right) = \sum_i\left(\sum_k s_{ik}b_{kj}\right)\xi_i.$$ (D.13)

Comparing (D.12) with (D.13) we see that

$$\sum_k a_{ik}s_{kj} = \sum_k s_{ik}b_{kj},$$ (D.14)

that is,

$$[A]_{\mathscr{X}}[S]_{\mathscr{X}} = [S]_{\mathscr{X}}[A]_{\mathscr{Y}}.$$ (D.15)

The most usual form of writing this equation is

$$[A]_{\mathscr{Y}} = [S]_{\mathscr{X}}^{-1}[A]_{\mathscr{X}}[S]_{\mathscr{X}}.$$ (D.16)

Two matrices M and N are **similar** if they are related by a **similarity transformation** – that is, if there exists an invertible matrix P such that $M = P^{-1}NP$. Equation (D.16) shows that, under a change of basis, the matrix associated with a linear operator undergoes a similarity transformation. Since $\mathrm{tr}\,(MN) = \mathrm{tr}\,(NM)$ and $\det(MN) = (\det M)(\det N)$, it is immediate that the trace and the determinant of a matrix are invariant under a similarity transformation. Thus, we can speak of **trace of an operator** and **determinant of an operator** as basis independent quantities.

Eigenvalues and Eigenvectors

Virtually all linear operators that occur in physical contexts can be completely characterised by their eigenvalues and eigenvectors.

Definition D.4   *Let A be a linear operator on a vector space $\mathcal{V}$ over the field $\mathcal{F}$. An **eigenvalue** of A is a scalar c such that there is a non-zero vector α in $\mathcal{V}$ that satisfies $A\alpha = c\alpha$. We say that α is an **eigenvector** of A associated with the eigenvalue c. The **spectrum** of A on a finite-dimensional vector space is the set of its eigenvalues.*

Theorem D.2   *Every linear operator on a finite-dimensional complex vector space has an eigenvector.*

Proof   A linear operator A on a complex vector space $\mathcal{V}$ of dimension n has an eigenvector if there exists $\alpha \neq 0$ such that $(A - cI)\alpha = 0$ for some complex number c. So, in order for there to be an eigenvector, it is necessary and sufficient that the linear operator $A - cI$ not be invertible. This means that in some basis $\mathcal{V}$ we must have $\det([A] - c\boldsymbol{I}) = 0$

for some complex number c. But $\det([A] - c\boldsymbol{I})$ is a polynomial of degree n in c, and the fundamental theorem of algebra ensures that any polynomial of positive degree with complex coefficients has a complex root. Therefore, there exists a complex number c such that $\det([A] - c\boldsymbol{I}) = 0$ and the proof is complete.   □

Inner Product and Orthonormal Bases

The introduction of an inner product on a vector space gives rise to the essential notions of orthogonality and norm.

Definition D.5   *Let $\mathcal{F}$ be the field of real or complex numbers and let $\mathcal{V}$ be a vector space over $\mathcal{F}$. An **inner product** or **scalar product** on $\mathcal{V}$ is a rule that to each ordered pair of vectors α, β in $\mathcal{V}$ assigns the scalar (α, β) in such a way that:*

(IP1) $(\alpha + \beta, \gamma) = (\alpha, \gamma) + (\beta, \gamma)$;
(IP2) $(\alpha, c\beta) = c(\alpha, \beta)$ *for any* $c \in \mathcal{F}$;
(IP3) $(\alpha, \beta) = (\beta, \alpha)^*$ *where the asterisk indicates complex conjugation;*
(IP4) $(\alpha, \alpha) \geq 0$ *and* $(\alpha, \alpha) = 0$ *if and only if* $\alpha = 0$.

Properties **IP1** to **IP3** have the following immediate consequences:

(IP5) $(c\alpha, \beta) = c^*(\alpha, \beta)$;   **(IP6)** $(\alpha, \beta + \gamma) = (\alpha, \beta) + (\alpha, \gamma)$.

When $\mathcal{F}$ is the field of real numbers, the complex conjugation in **IP3** and **IP5** is superfluous. Taking $c = 0$ in **IP2** one concludes that $(\alpha, 0) = 0$. It is worth mentioning that mathematicians prefer to replace property **IP2** by **IP2′**: $(c\alpha, \beta) = c(\alpha, \beta)$.

A vector space on which an inner product is defined is called an **inner product space**. On an inner product space the non-negative real number $||\alpha||$ defined by

$$||\alpha|| = \sqrt{(\alpha, \alpha)} \tag{D.17}$$

is called **norm** or **modulus** of the vector α. The following properties of the norm are easily verified: **(N1)** $||c\alpha|| = |c| \, ||\alpha||$; **(N2)** $||\alpha|| \geq 0$ and $||\alpha|| = 0$ if and only if $\alpha = 0$. The property **(N3)** $||\alpha + \beta|| \leq ||\alpha|| + ||\beta||$ will be proved below.

Let $\alpha = (x_1, \ldots, x_n)$ and $\beta = (y_1, \ldots, y_n)$ be vectors of $\mathcal{F}^n$. The **canonical inner product** on $\mathcal{F}^n$ is defined by

$$(\alpha, \beta) = x_1^* y_1 + x_2^* y_2 + \cdots + x_n^* y_n \,. \tag{D.18}$$

The **canonical norm** on $\mathcal{F}^n$ is given by

$$||\alpha|| = \sqrt{|x_1|^2 + |x_2|^2 + \cdots + |x_n|^2} \,. \tag{D.19}$$

Theorem D.3   *If $\mathcal{V}$ is an inner product space, then for any vectors $\alpha, \beta \in \mathcal{V}$ the following inequalities hold:*

$$|(\alpha, \beta)| \leq ||\alpha|| \, ||\beta|| \,, \tag{D.20}$$

called the **Schwarz inequality**,[1] *and*

$$||\alpha + \beta|| \leq ||\alpha|| + ||\beta||, \tag{D.21}$$

known as the **triangle inequality**.

Proof   If $\alpha = 0$ the inequality (D.20) is obviously true. If $\alpha \neq 0$, consider $\gamma = \beta - c\alpha$. Using the properties of the inner product we can write

$$0 \leq (\gamma, \gamma) = ||\beta||^2 - c^*(\alpha, \beta) - c(\beta, \alpha) + |c|^2||\alpha||^2. \tag{D.22}$$

The choice $c = (\alpha, \beta)/||\alpha||^2$ leads to

$$||\beta||^2 - \frac{|(\alpha, \beta)|^2}{||\alpha||^2} \geq 0, \tag{D.23}$$

from which the Schwarz inequality follows immediately. Now using the obvious inequality $\text{Re}\, z \leq |z|$ together with (D.20), we have

$$\begin{aligned}||\alpha + \beta||^2 &= ||\alpha||^2 + (\alpha, \beta) + (\beta, \alpha) + ||\beta||^2 \\ &= ||\alpha||^2 + 2\text{Re}\,(\alpha, \beta) + ||\beta||^2 \\ &\leq ||\alpha||^2 + 2||\alpha||\,||\beta|| + ||\beta||^2 = (||\alpha|| + ||\beta||)^2. \end{aligned} \tag{D.24}$$

By extraction of the positive square root the triangle inequality follows.   □

Definition D.6   *If V is an inner product space, two vectors α, β in V are said to be* **orthogonal** *if $(\alpha, \beta) = 0$, and we write $\alpha \perp \beta$. Two subsets R and S of V are said to be orthogonal, and we write $R \perp S$, if each vector in R is orthogonal to all vectors in S. A set of vectors such that any two of them are orthogonal is said to be an* **orthogonal system** *of vectors. A vector α is said to be* **normalised** *if $||\alpha|| = 1$. An orthogonal system of vectors is said to be an* **orthonormal system** *if each vector of the system is normalised.*

For an orthonormal system of vectors $\{\xi_1, \ldots, \xi_n\}$ we have

$$(\xi_k, \xi_l) = \delta_{kl}, \qquad k, l = 1, \ldots, n. \tag{D.25}$$

Any orthonormal system of vectors is linearly independent because if $\sum_k c_k \xi_k = 0$ then $0 = (\xi_l, \sum_k c_k \xi_k) = \sum_k c_k (\xi_l, \xi_k) = \sum_k c_k \delta_{kl} = c_l$. Thus, any orthonormal system of n vectors in an inner product space V of dimension n is a basis of V. In an orthonormal basis $\{\xi_1, \ldots, \xi_n\}$ a vector α is written as $\alpha = \sum_j a_j \xi_j$, whence

$$(\xi_k, \alpha) = \sum_j (\xi_k, a_j \xi_j) = \sum_j a_j (\xi_k, \xi_j) = \sum_j a_j \delta_{kj} = a_k. \tag{D.26}$$

Therefore,

$$\alpha = \sum_k (\xi_k, \alpha)\, \xi_k. \tag{D.27}$$

[1] Also known as Cauchy-Schwarz or Cauchy-Schwarz-Bunyakovsky inequality.

An analogous calculation shows that, in terms of components relative to an orthonormal basis, the inner product takes the canonical form:

$$(\alpha, \beta) = \sum_k (\xi_k, \alpha)^* (\xi_k, \beta).$$
(D.28)

If $[A] = (a_{ij})$ is the matrix associated with an operator A in an orthonormal basis $\{\xi_1, \ldots, \xi_n\}$, Eqs. (D.10) and (D.25) combined yield at once

$$a_{ij} = (\xi_i, A\xi_j).$$
(D.29)

Theorem D.4   *Every finite-dimensional inner product space has an orthonormal basis.*

Proof   Let V be a finite-dimensional inner product space and let $\{\alpha_1, \ldots, \alpha_n\}$ be a basis of V. From this basis one can construct an orthogonal basis by the **Gram-Schmidt orthogonalisation process** (Hoffman & Kunze, 1971). Dividing each vector of this orthogonal basis by its norm one obtains the desired orthonormal basis.   □

Orthogonal Complement and Direct Sum

Any inner product space can be expressed as a sort of union of mutually orthogonal subspaces.

Definition D.7   *Let V be an inner product space and S be an arbitrary set of vectors of V. The **orthogonal complement** of S consists of the set $S^\perp$ of vectors of V that are orthogonal to every vector of S.*

Definition D.8   *The vector space V is the **direct sum** of the vector spaces U and W, and we write $V = U \oplus W$, if each element α of V can be uniquely expressed as $\alpha = \beta + \gamma$ where $\beta \in U$ and $\gamma \in W$.*

Theorem D.5   *If W is a finite-dimensional vector subspace of the inner product space V then $V = W \oplus W^\perp$.*

Proof   Let $\{\xi_1, \ldots, \xi_n\}$ be an orthonormal basis of W, whose existence is guaranteed by Theorem D.4. Any vector α of V can be written as

$$\alpha = \sum_{k=1}^{n} (\xi_k, \alpha) \xi_k + \gamma$$
(D.30)

by simply choosing

$$\gamma = \alpha - \sum_{k=1}^{n} (\xi_k, \alpha) \xi_k.$$
(D.31)

With the use of (D.25) one easily verifies that $(\xi_i, \gamma) = 0$ for all ξ_i. It follows that γ is orthogonal to any linear combination of $\{\xi_1, \ldots, \xi_n\}$, therefore it is an element of $W^\perp$. According to (D.30) we have $\alpha = \beta + \gamma$ with β in W and γ in $W^\perp$. It remains to prove that this decomposition is unique. If $\alpha = \beta' + \gamma'$ with $\beta' \in \mathcal{W}$ and $\gamma' \in \mathcal{W}^\perp$, we have $\beta - \beta' = \gamma' - \gamma$. Since $\mathcal{W}^\perp$ is also a vector subspace of $\mathcal{V}$ (check it!), the vector $\beta'' = \beta - \beta'$ belongs at the same time to $\mathcal{W}$ and to $\mathcal{W}^\perp$ – that is, $(\beta'', \beta'') = 0$, which implies $\beta'' = 0$. So $\beta' = \beta$ and $\gamma' = \gamma$, which establishes uniqueness and completes the proof of the theorem. $\square$

Adjoint of a Linear Operator

A most important notion is that of the adjoint $A^\dagger$ of a linear operator A. In order to define the adjoint and establish its main properties, it is first necessary to introduce the concept of *linear functional*.

Definition D.9   *A **linear functional** Φ on a vector space $\mathcal{V}$ is a rule that assigns a scalar $\Phi(\alpha)$ to each vector $\alpha \in \mathcal{V}$ in such a way that*

$$\Phi(c_1\alpha_1 + c_2\alpha_2) = c_1\Phi(\alpha_1) + c_2\Phi(\alpha_2) \tag{D.32}$$

for all vectors α_1, α_2 in $\mathcal{V}$ and all scalars c_1, c_2.

For example, if $\mathcal{V}$ is an inner product space and $\beta \in \mathcal{V}$ is a fixed vetor, the application Φ_β defined by

$$\Phi_\beta(\alpha) = (\beta, \alpha) \tag{D.33}$$

is a linear functional in virtue of the properties of the inner product. A notable result is that if $\mathcal{V}$ is finite-dimensional then any linear functional is of this form.

Theorem D.6   *Let $\mathcal{V}$ be a finite-dimensional inner product space and Φ be a linear functional on $\mathcal{V}$. Then there exists a unique vector $\beta \in \mathcal{V}$ such that $\Phi(\alpha) = (\beta, \alpha)$ for all $\alpha \in \mathcal{V}$.*

Proof   Given a linear functional Φ, let $\{\alpha_1, \ldots, \alpha_n\}$ be an orthonormal basis of $\mathcal{V}$ and define the vector

$$\beta = \sum_{j=1}^{n} \Phi(\alpha_j)^* \alpha_j. \tag{D.34}$$

Then

$$(\beta, \alpha_k) = \sum_{j=1}^{n} (\Phi(\alpha_j)^* \alpha_j, \alpha_k) = \sum_{j=1}^{n} \Phi(\alpha_j)(\alpha_j, \alpha_k) = \sum_{j=1}^{n} \Phi(\alpha_j)\delta_{jk} = \Phi(\alpha_k), \tag{D.35}$$

and, for a generic vector α with $\alpha = \sum_k c_k\alpha_k$,

$$\Phi(\alpha) = \sum_{k=1}^{n} c_k\Phi(\alpha_k) = \sum_{k=1}^{n} c_k(\beta, \alpha_k) = \left(\beta, \sum_{k=1}^{n} c_k\alpha_k\right) = (\beta, \alpha). \tag{D.36}$$

In order to prove uniqueness, suppose γ is another vector such that $\Phi(\alpha) = (\beta, \alpha) = (\gamma, \alpha)$ for all $\alpha \in \mathcal{V}$. Then the vector $\rho = \gamma - \beta$ is orthogonal to all vectors in $\mathcal{V}$, therefore it is orthogonal to itself. Thus, $(\rho, \rho) = 0$ whence $\rho = 0$ and $\gamma = \beta$. □

Theorem D.7　*If A is a linear operator on a finite-dimensional inner product space $\mathcal{V}$ then there is a unique linear operator $A^\dagger$ such that[2]*

$$(\beta, A\alpha) = (A^\dagger \beta, \alpha) \tag{D.37}$$

for all $\alpha, \beta \in \mathcal{V}$. The operator $A^\dagger$ is called the **adjoint** *or* **Hermitian conjugate** *of A.*

Proof　Since $\Phi(\alpha) = (\beta, A\alpha)$ is a linear functional for fixed β, Theorem D.6 tells us that there is a unique vector $\beta' \in \mathcal{V}$ such that $(\beta, A\alpha) = (\beta', \alpha)$. The operator $A^\dagger$ is defined as the rule that assigns β' to β:

$$A^\dagger \beta = \beta'. \tag{D.38}$$

It remains to verify that $A^\dagger$ is a linear operator. Let β, γ be elements of $\mathcal{V}$ and c_1, c_2 be scalars. For any $\alpha \in \mathcal{V}$,

$$\begin{aligned}
(A^\dagger(c_1\beta + c_2\gamma), \alpha) &= (c_1\beta + c_2\gamma, A\alpha) = c_1^*(\beta, A\alpha) + c_2^*(\gamma, A\alpha) \\
&= c_1^*(A^\dagger\beta, \alpha) + c_2^*(A^\dagger\gamma, \alpha) = (c_1 A^\dagger\beta, \alpha) + (c_2 A^\dagger\gamma, \alpha) \\
&= (c_1 A^\dagger\beta + c_2 A^\dagger\gamma, \alpha).
\end{aligned} \tag{D.39}$$

So $A^\dagger(c_1\beta + c_2\gamma) = c_1 A^\dagger\beta + c_2 A^\dagger\gamma$ and $A^\dagger$ is linear. □

The following properties hold on a finite-dimensional inner product space and are easily verified:

(ADJ1)　$(A + B)^\dagger = A^\dagger + B^\dagger$;

(ADJ2)　$(cA)^\dagger = c^* A^\dagger$ for any $c \in \mathcal{F}$;

(ADJ3)　$(AB)^\dagger = B^\dagger A^\dagger$;

(ADJ4)　$(A^\dagger)^\dagger = A$.

Unitary and Self-Adjoint Operators

For physical applications the self-adjoint and unitary operators constitute the most important class of linear operators.

Definition D.10　*A linear operator U on an inner product space $\mathcal{V}$ is said to be* **unitary** *if its adjoint $U^\dagger$ equals its inverse U^{-1} – that is,*

$$UU^\dagger = U^\dagger U = I. \tag{D.40}$$

---

[2] Mathematicians prefer to denote the complex conjugate of a complex number z by $\bar{z}$ and the adjoint of a linear operator A by A^*.

Theorem D.8 *If U is a linear operator on a finite-dimensional inner product space any of the following conditions is necessary and sufficient for U to be unitary: (1) $UU^\dagger = I$ or $U^\dagger U = I$; (2) $(U\xi, U\eta) = (\xi, \eta)$ for any vectors $\xi, \eta \in \mathcal{V}$.*

Proof Regarding part (1), if U is unitary we have both $UU^\dagger = I$ and $U^\dagger U = I$ and there is nothing to prove. Suppose now that $U^\dagger U = I$. By Theorem D.1 the operator U is necessarily invertible because $U\alpha = 0$ implies $\alpha = I\alpha = U^\dagger U\alpha = U^\dagger 0 = 0$. Multiplying $U^\dagger U = I$ on the left by U and on the right by U^{-1} there results $UU^\dagger = I$ and U is unitary. The same argument applies if $UU^\dagger = I$ by just exchanging the roles played by U and $U^\dagger$, and the proof concerning part (1) is complete. As to part (2), if U is unitary then $(\xi, \eta) = (U^\dagger U\xi, \eta) = (U\xi, U\eta)$ by the definition of adjoint. Conversely, if $(U\xi, U\eta) = (\xi, \eta)$ for all vectors $\xi, \eta \in \mathcal{V}$, it follows from the definition of adjoint that $(U^\dagger U\xi, \eta) = (\xi, \eta)$ or, equivalently, $((U^\dagger U - I)\xi, \eta) = 0$. Inasmuch as ξ and η are arbitrary vectors, this means that $U^\dagger U = I$ and, by part (1), U is unitary. ∎

Unitary operators can be characterised by the property of mapping orthonormal bases to orthonormal bases.

Theorem D.9 *If U is a unitary operator on a finite-dimensional inner product space $\mathcal{V}$ and $\{\alpha_1, \ldots, \alpha_n\}$ is an orthonormal basis of $\mathcal{V}$, then $\{U\alpha_1, \ldots, U\alpha_n\}$ is also an orthonormal basis of $\mathcal{V}$. Conversely, if $\{\alpha_1, \ldots, \alpha_n\}$ and $\{\beta_1, \ldots, \beta_n\}$ are orthonormal bases of $\mathcal{V}$ then there exists a unitary operator U such that $\beta_k = U\alpha_k$, $k = 1, \ldots, n$.*

Proof If U is unitary, $(U\alpha_k, U\alpha_l) = (U^\dagger U\alpha_k, \alpha_l) = (\alpha_k, \alpha_l) = \delta_{kl}$ and it follows that $\{U\alpha_1, \ldots, U\alpha_n\}$ is also an orthonormal basis of $\mathcal{V}$. If $\{\alpha_1, \ldots, \alpha_n\}$ and $\{\beta_1, \ldots, \beta_n\}$ are orthonormal bases of $\mathcal{V}$, define the linear operator U by $U\alpha_k = \beta_k$, $k = 1, \ldots, n$. For any two vectors $\xi = \sum_k x_k \alpha_k$ and $\eta = \sum_k y_k \alpha_k$ in $\mathcal{V}$ we have

$$(U\xi, U\eta) = \left(\sum_{k=1}^n x_k U\alpha_k, \sum_{l=1}^n y_l U\alpha_l\right) = \sum_{k,l=1}^n x_k^* y_l (\beta_k, \beta_l) = \sum_{k,l=1}^n x_k^* y_l \delta_{kl}$$

$$= \sum_{k,l=1}^n x_k^* y_l (\alpha_k, \alpha_l) = \left(\sum_{k=1}^n x_k \alpha_k, \sum_{l=1}^n y_l \alpha_l\right) = (\xi, \eta) \tag{D.41}$$

and U is unitary by Theorem D.8. ∎

Definition D.11 *A linear operator A on an inner poduct space is said to be* **self-adjoint** *or* **Hermitian** *if $A^\dagger = A$.*

Unitary and self-adjoint operators are of fundamental importance to physics, and their eigenvalues and eigenvectors enjoy simple properties.

Theorem D.10 *If c is an eigenvalue of a unitary operator then $|c| = 1$. The eigenvalues of a self-adjoint operator are real and the eigenvectors of a self-adjoint operator associated with distinct eigenvalues are orthogonal.*

Proof If U is unitary and $U\alpha = c\alpha$ with $\alpha \neq 0$ then

$$(\alpha, \alpha) = (U\alpha, U\alpha) = (c\alpha, c\alpha) = |c|^2 (\alpha, \alpha). \tag{D.42}$$

So, $|c|^2 = 1$ whence $|c| = 1$. If A is self-adjoint and $A\alpha = c\alpha$ with $\alpha \neq 0$ then

$$c(\alpha, \alpha) = (\alpha, c\alpha) = (\alpha, A\alpha) = (A^\dagger \alpha, \alpha) = (A\alpha, \alpha) = (c\alpha, \alpha) = c^*(\alpha, \alpha). \quad \text{(D.43)}$$

Since $(\alpha, \alpha) \neq 0$, it follows that $c = c^*$ and c is real. Finally, let α_1, α_2 be non-zero vectors such that $A\alpha_1 = c_1\alpha_1$ and $A\alpha_2 = c_2\alpha_2$. Then

$$\begin{aligned} c_2(\alpha_1, \alpha_2) &= (\alpha_1, c_2\alpha_2) = (\alpha_1, A\alpha_2) = (A^\dagger \alpha_1, \alpha_2) \\ &= (A\alpha_1, \alpha_2) = (c_1\alpha_1, \alpha_2) = c_1(\alpha_1, \alpha_2), \end{aligned} \quad \text{(D.44)}$$

where we have used the fact that c_1 is real. Thus $(c_2 - c_1)(\alpha_1, \alpha_2) = 0$, with the consequence that $(\alpha_1, \alpha_2) = 0$ if $c_1 \neq c_2$. $\square$

Theorem D.11   *Every self-adjoint operator on a finite-dimensional inner product space has an eigenvector.*

Proof   If the inner product space is complex the result is contained in Theorem D.2. If the inner product space is real the fundamental theorem of algebra guarantees the existence of a complex number c such that $\det([A] - c\boldsymbol{I}) = 0$ where $[A]$ is the matrix of the operator A in some basis. Thus, there is a non-zero column matrix $[\alpha]$ such that $[A][\alpha] = c[\alpha]$. But inasmuch as A is self-adjoint, c is real. Taking into account that $[A]$ is a real matrix, we can choose $[\alpha]$ in such a way that all its elements are real. Therefore, there exists a non-zero real vector α such that $A\alpha = c\alpha$. $\square$

A crucial result for applications to physics asserts that, given a self-adjoint operator A on a finite-dimensional inner product space $\mathcal{V}$, there exists a basis of $\mathcal{V}$ composed by eigenvectores of A. As will be seen, this is tantamount to saying that any Hermitian matrix can be diagonalised by a similarity transformation effected by a unitary matrix. In order to prove this result in full generality, we need a definition and a preliminary theorem.

Definition D.12   *Let $\mathcal{V}$ be a vector space and A be a linear operator on $\mathcal{V}$. Let $\mathcal{W}$ be a vector subspace of $\mathcal{V}$. We say that $\mathcal{W}$ is **invariant under** A if for any vector $\alpha \in \mathcal{W}$ the vector $A\alpha$ also belongs to $\mathcal{W}$.*

Theorem D.12   *Let $\mathcal{V}$ be a finite-dimensional inner product space and let A be a linear operator on $\mathcal{V}$. If $\mathcal{W}$ is a vector subspace of $\mathcal{V}$ that is invariant under A, then the orthogonal complement of $\mathcal{W}$ is invariant under $A^\dagger$.*

Proof   Take an arbitrary element $\beta \in \mathcal{W}^\perp$. We need to prove that $A^\dagger \beta$ is in $\mathcal{W}^\perp$, which is to say that $(\alpha, A^\dagger \beta) = 0$ for all $\alpha \in \mathcal{W}$. But, by hypothesis, if α is in $\mathcal{W}$ then $A\alpha$ is also in $\mathcal{W}$ and, therefore, $(A\alpha, \beta) = 0$. From the definition of adjoint we conclude that $(\alpha, A^\dagger \beta) = (A\alpha, \beta) = 0$. $\square$

We are now ready to state and prove the result that is the culmination of this appendix: the spectral theorem for self-adjoint operators.

Theorem D.13   *Let $\mathcal{V}$ be a finite-dimensional inner product space and let A be a self-adjoint operator on $\mathcal{V}$. Then there exists an orthonormal basis of $\mathcal{V}$ composed by eigenvectors of A.*

Proof   Let $n > 0$ be the dimension of $\mathcal{V}$. By Theorem D.11, A has an eigenvector α, so $\alpha_1 = \alpha/||\alpha||$ is a normalised eigenvector of A. If $n = 1$ the proof is finished. Let us now proceed by induction over n. Suppose the theorem holds for spaces with dimension n and let $\mathcal{V}$ have dimension $n + 1$. If $\mathcal{W}$ is the one-dimensional subspace of $\mathcal{V}$ generated by α_1, saying that α_1 is an eigenvector of A means just that $\mathcal{W}$ is invariant under A. By Theorem D.12, the orthogonal complement $\mathcal{W}^\perp$ is invariant under $A^\dagger = A$. But $\mathcal{W}^\perp$, with the inner product of $\mathcal{V}$, is an inner product space of dimension n. Let T be the linear operator induced on $\mathcal{W}^\perp$ by A – that is, the restriction of A to $\mathcal{W}^\perp$. The operator T is well defined on $\mathcal{W}^\perp$ because $\mathcal{W}^\perp$ is invariant under A. Clearly, T is self-adjoint and, by the induction hypothesis, $\mathcal{W}^\perp$ has an orthonormal basis $\{\alpha_2, \ldots, \alpha_{n+1}\}$ composed by eigenvectors of T, therefore of A. Since $\mathcal{V} = \mathcal{W} \oplus \mathcal{W}^\perp$, we conclude that $\{\alpha_1, \ldots, \alpha_{n+1}\}$ is the desired basis. $\square$

If A is a self-adjoint operator, from (D.29) one infers that its matrix $[A]$ in an orthonormal basis $\{\alpha_1, \ldots, \alpha_n\}$ has elements

$$a_{ij} = (\alpha_i, A\alpha_j) = (A^\dagger \alpha_i, \alpha_j) = (A\alpha_i, \alpha_j) = (\alpha_j, A\alpha_i)^* = a_{ji}^* , \tag{D.45}$$

so that

$$[A]^\dagger = [A] \tag{D.46}$$

and $[A]$ is a Hermitian matrix. It follows from this result that a linear operator is self-adjoint if and only if its matrix in some orthonormal basis is Hermitian. On the other hand, if U is a unitary operator whose matrix in the basis $\{\alpha_1, \ldots, \alpha_n\}$ has elements u_{ij}, we have

$$\delta_{ij} = (\alpha_i, \alpha_j) = (U\alpha_i, U\alpha_j) = \sum_{k,l} u_{ki}^* u_{lj}(\alpha_k, \alpha_l) = \sum_{k,l} u_{ki}^* u_{lj}\delta_{kl} = \sum_k u_{ki}^* u_{kj} , \tag{D.47}$$

whence

$$[U]^\dagger [U] = \boldsymbol{I} . \tag{D.48}$$

Thus, $\det [U] \neq 0$ and $[U]$ is an invertible matrix. Therefore

$$[U]^\dagger = [U]^{-1} \tag{D.49}$$

and $[U]$ is a unitary matrix. It follows that a linear operator is unitary if and only if its matrix in some orthonormal basis is unitary. It should be noted that on a real vector space a Hermitian matrix is the same as a real *symmetric* matrix – that is, a matrix that equals its transpose. Similarly, on a real vector space a unitary matrix reduces to a real *orthogonal* matrix – that is, a matrix whose transpose equals its inverse.

By (D.16) a change of orthonormal basis corresponds to a similarity transformation of the matrix associated with an operator A effected by a unitary matrix $[S]_{\mathcal{X}}$:

$$[A]_{\mathcal{Y}} = [S]_{\mathcal{X}}^\dagger [A]_{\mathcal{X}} [S]_{\mathcal{X}} . \tag{D.50}$$

If $[A]$ is Hermitian in the orthonormal basis $\mathcal{X}$ we have

$$[A]_{\mathcal{Y}}^\dagger = \left([S]_{\mathcal{X}}^\dagger [A]_{\mathcal{X}} [S]_{\mathcal{X}} \right)^\dagger = [S]_{\mathcal{X}}^\dagger [A]_{\mathcal{X}}^\dagger [S]_{\mathcal{X}}^{\dagger\dagger}$$
$$= [S]_{\mathcal{X}}^\dagger [A]_{\mathcal{X}} [S]_{\mathcal{X}} = [A]_{\mathcal{Y}} . \tag{D.51}$$

Therefore $[A]$ is also Hermitian in any other orthonormal basis $\mathscr{Y}$. Similarly, if the matrix $[U]$ is unitary in the orthonormal basis $\mathscr{X}$ then

$$
\begin{aligned}
[U]_{\mathscr{Y}}^{\dagger}[U]_{\mathscr{Y}} &= [S]_{\mathscr{X}}^{\dagger}\,[U]_{\mathscr{X}}^{\dagger}\,[S]_{\mathscr{X}}\,[S]_{\mathscr{X}}^{\dagger}\,[U]_{\mathscr{X}}\,[S]_{\mathscr{X}} \\
&= [S]_{\mathscr{X}}^{\dagger}\,[U]_{\mathscr{X}}^{\dagger}\,[U]_{\mathscr{X}}\,[S]_{\mathscr{X}} = [S]_{\mathscr{X}}^{\dagger}\,[S]_{\mathscr{X}} = I
\end{aligned}
\tag{D.52}
$$

and $[U]$ is also unitary in any other orthonormal basis $\mathscr{Y}$. Therefore, a change of orthonormal basis maps Hermitian matrices to Hermitian matrices and unitary matrices to unitary matrices.

Noting that in an orthonormal basis composed by eigenvectors of a linear operator A its associated matrix $[A]$ is diagonal, this discusssion permits a reformulation of Theorem D.13 in terms of matrices.

Theorem D.14 *Any Hermitian matrix can be diagonalised by a similarity transformation effected by a unitary matrix. In particular, any real symmetric matrix can be diagonalised by a similarity transformation effected by an orthogonal matrix.*

This appendix, based on Wiggins (2003), extends the elementary discussion of stability of Section 5.1 to a more general setting.

A system of ordinary differential equations of the form

$$\dot{x} = f(x, t), \tag{E.1}$$

where $x \in \mathbb{R}^n$ and $f : \mathbb{R}^{n+1} \to \mathbb{R}^n$ is an infinitely differentiable function, constitutes a **dynamical system**. If f does not explicitly depend on time the dynamical system is said to be **autonomous**. Our attention will be restricted to autonomous dynamical systems:

$$\dot{x} = f(x). \tag{E.2}$$

Explicitly, this system of n coupled ordinary differential equations is written

$$\dot{x}_1 = f_1(x_1, \ldots, x_n),$$

$$\vdots \tag{E.3}$$

$$\dot{x}_n = f_n(x_1, \ldots, x_n).$$

It is worth noting that any system of second-order ordinary differential equations, such as Lagrange's equations, can be cast in the the form (E.3) by doubling the number of variables. For example, the one-dimensional equation of motion

$$\ddot{x} = f(x, \dot{x}) \tag{E.4}$$

is equivalent to the dynamical system

$$\dot{x} = y, \tag{E.5a}$$

$$\dot{y} = f(x, y). \tag{E.5b}$$

Even if the equations of motion of a mechanical system depend explicitly on time, we can express them in the form of an autonomous dynamical system: it suffices to introduce an additional variable τ in place of t and add the differential equation $\dot{\tau} = 1$ to the set of equations of motion.

Definition E.1   *An **equilibrium point** of (E.2) is a point $\bar{x} \in \mathbb{R}^n$ such that*

$$f(\bar{x}) = 0, \tag{E.6}$$

that is, $\bar{x}$ is a solution that does not vary with time. Other terms used as synonymous to equilibrium point are stationary point, fixed point or critical point.

For dynamical systems the notion of stability does not apply only to equilibrium points, but can be defined for any dynamical path – that is, to any solution of the system of differential equations (E.2). Intuitively, a solution $\bar{x}(t)$ of (E.2) is *stable* if any other solution $x(t)$ that starts sufficiently close to $\bar{x}(t)$ at a given time remains close to $\bar{x}(t)$ at all later times. If, besides, $x(t)$ converges to $\bar{x}(t)$ as $t \to \infty$, the solution $\bar{x}(t)$ is said to be *asymptotically stable*. In order to make precise the notion of closeness we use the Euclidean distance between any two points $x = (x_1, \ldots, x_n)$ and $\tilde{x} = (\tilde{x}_1, \ldots, \tilde{x}_n)$ of $\mathbb{R}^n$, namely

$$|x - \tilde{x}| = \sqrt{(x_1 - \tilde{x}_1)^2 + \cdots + (x_n - \tilde{x}_n)^2}. \tag{E.7}$$

Definition E.2 (Lyapunov Stability) *A solution $\bar{x}(t)$ of (E.2) is said to be* **stable** *if, given $\epsilon > 0$, there exists $\delta > 0$ such that if $x(t)$ is any other solution that satisfies $|x(t_0) - \bar{x}(t_0)| < \delta$ then $|x(t) - \bar{x}(t)| < \epsilon$ for all $t > t_0$. If, furthermore, $\lim_{t \to \infty} |x(t) - \bar{x}(t)| = 0$, then $\bar{x}(t)$ is said to be* **asymptotically stable**.

Stability and Lyapunov Functions

The idea of using a scalar function, such as the energy, to prove stability of an equilibrium point was generalised by Lyapunov near the end of the nineteenth century. Although Lyapunov's method works in $\mathbb{R}^n$, its basic geometric motivations are more easily visualised in the plane (Wiggins, 2003).

Given the planar autonomous dynamical system

$$\dot{x} = f(x, y), \tag{E.8a}$$

$$\dot{y} = g(x, y), \tag{E.8b}$$

let $(\bar{x}, \bar{y}) \in \mathbb{R}^2$ be an equilibrium point. In order to establish stability in the sense of Lyapunov it appears to be enough to find a neighbourhood U of the equilibrium point such that all trajectories that start in U at $t = 0$ remain in U for all $t > 0$. In its turn, this condition is guaranteed if the velocity vector $(\dot{x}, \dot{y})$ at each point of the boundary of U is either tangent to the boundary or points toward the interior of U (see Fig. E.1). This state of affairs should persist as U is shrunk about the point $(\bar{x}, \bar{y})$.

Suppose there exists a differentiable function $V : \mathbb{R}^2 \to \mathbb{R}$ such that: (a) $V(\bar{x}, \bar{y}) = 0$; (b) $V(x, y) > 0$ in a neighbourhood of $(\bar{x}, \bar{y})$ and V increases as one moves further away from $(\bar{x}, \bar{y})$. Since the gradient of V is a vector perpendicular to the level curve $V = $ constant, which points in the direction of increasing V, if the velocity vector at the boundary of U is as depicted in Fig. E.1 we have

$$\dot{V} = \nabla V \cdot (\dot{x}, \dot{y}) = \frac{\partial V}{\partial x}\dot{x} + \frac{\partial V}{\partial y}\dot{y} = \frac{\partial V}{\partial x}f(x, y) + \frac{\partial V}{\partial y}g(x, y) \le 0. \tag{E.9}$$

This discussion suggests that if there is a function with the described properties then the equilibrium point $(\bar{x}, \bar{y})$ is stable.

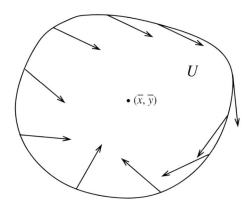

Velocity field at the boundary points of U.

Theorem E.1 (Lyapunov's Stability Theorem)   *Let $\bar{x}$ be an equilibrium point of the autonomous dynamical system*

$$\dot{x} = f(x), \qquad x \in \mathbb{R}^n. \tag{E.10}$$

Let $U \subset \mathbb{R}^n$ be some neighbourhood of $\bar{x}$ and assume there exists a differentiable function $V : U \to \mathbb{R}$ such that:

(L1) $V(\bar{x}) = 0$ and $V(x) > 0$ if $x \neq \bar{x}$;
(L2) $\dot{V}(x) \leq 0$ in $U \setminus \{\bar{x}\}$ (U without the point $\bar{x}$).

Then $\bar{x}$ is stable. If, moreover,

(L3) $\dot{V}(x) < 0$ in $U \setminus \{\bar{x}\}$,

then $\bar{x}$ is asymptotically stable.

Proof   Let $B_\sigma(\bar{x})$ be an open ball centred at $\bar{x}$ with radius σ – that is,

$$B_\sigma(\bar{x}) = \left\{ x \in \mathbb{R}^n \,\middle|\, |x - \bar{x}| < \sigma \right\}, \tag{E.11}$$

where σ is so small that $B_\sigma(\bar{x}) \subset U$. Let m be the minimum value of V on the boundary of $B_\sigma(\bar{x})$. The existence of m follows from the fact that a continuous function attains a maximum as well as a minimum value on a closed and bounded subset of $\mathbb{R}^n$. Because of (L1), $m > 0$. Let $\tilde{U} \subset B_\sigma(\bar{x})$ be the open set defined by

$$\tilde{U} = \left\{ x \in B_\sigma(\bar{x}) \,\middle|\, V(x) < m \right\}, \tag{E.12}$$

as shown in Fig. E.2. Along any dynamical trajectory that starts in $\tilde{U}$ the function V does not increase owing to condition (L2). It follows that, by construction, any such trajectory cannot leave $\tilde{U}$. Since $\sigma > 0$ is arbitrary, given $\epsilon > 0$ it is enough to take $\sigma < \epsilon$ in order to have $\tilde{U} \subset B_\epsilon(\bar{x})$. Finally, picking $\delta > 0$ such that $B_\delta(\bar{x}) \subset \tilde{U}$, it follows that any dynamical trajectory that starts in $B_\delta(\bar{x})$ never leaves $B_\epsilon(\bar{x})$ because it never leaves $\tilde{U}$. Therefore, the equilibrium point $\bar{x}$ is stable in the sense of Lyapunov. For a proof that the additional condition (L3) implies asymptotic stability, the reader is referred to Wiggins (2003). $\square$

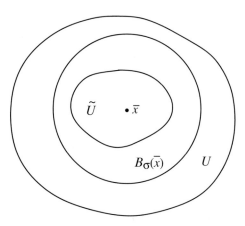

Neighbourhood $\tilde{U}$ defined by (E.12).

A function V that satisfies conditions (L1) and (L2) is called a **Lyapunov function**. If condition (L3) is also satisfied V is called a **strict Lyapunov function**.

Example E.1 Consider a particle subject to a conservative force $\mathbf{F}(\mathbf{r}) = -\nabla\phi(\mathbf{r})$ and suppose the potential energy ϕ has a local isolated minimum at $\bar{\mathbf{r}}$ – that is, $\phi(\mathbf{r}) > \phi(\bar{\mathbf{r}})$ for $\mathbf{r} \neq \bar{\mathbf{r}}$ in a neighbourhood of $\bar{\mathbf{r}}$. Show that $\bar{\mathbf{r}}$ is a stable equilibrium position.

Solution

The equation of motion of the particle is $m\ddot{\mathbf{r}} = -\nabla\phi(\mathbf{r})$. Let us double the number of variables – and write $\boldsymbol{\xi} = (\mathbf{r}, \mathbf{v}) \in \mathbb{R}^6$ – in such a way as to put the equation of motion in the standard form of a dynamical system:

$$\dot{\mathbf{r}} = \mathbf{v}, \qquad \dot{\mathbf{v}} = -\frac{1}{m}\nabla\phi(\mathbf{r}). \tag{E.13}$$

Inasmuch as $\nabla\phi(\bar{\mathbf{r}}) = 0$, this dynamical system has an equilibrium point $\bar{\boldsymbol{\xi}} = (\bar{\mathbf{r}}, \mathbf{0}) \in \mathbb{R}^6$. Set

$$V(\boldsymbol{\xi}) = \frac{m}{2}v^2 + \phi(\mathbf{r}) - \phi(\bar{\mathbf{r}}), \tag{E.14}$$

where $v^2 = \mathbf{v}\cdot\mathbf{v}$. This function satisfies condition (L1) in a neighbourhood of $\bar{\boldsymbol{\xi}}$. Besides, the conservation of energy secures the validity of condition (L2):

$$\dot{V} = m\mathbf{v}\cdot\dot{\mathbf{v}} + \mathbf{v}\cdot\nabla\phi(\mathbf{r}) = -\mathbf{v}\cdot\nabla\phi + \mathbf{v}\cdot\nabla\phi = 0. \tag{E.15}$$

It follows that $\bar{\mathbf{r}}$ is a point of stable equilibrium.

Example E.2 Consider the dynamical system

$$\dot{x} = -x - 2y^3, \qquad \dot{y} = x - y, \tag{E.16}$$

where $(x, y) \in \mathbb{R}^2$. It is easy to show that the only equilibrium point is $(0, 0)$. The function V defined by

$$V(x, y) = x^2 + y^4 \tag{E.17}$$

satisfies condition (L1). Besides,

$$\dot{V} = 2x\dot{x} + 4y^3\dot{y} = 2x(-x - 2y^3) + 4y^3(x - y) = -2(x^2 + 2y^4) < 0 \tag{E.18}$$

for $(x, y) \neq (0, 0)$. Thus condition (L3) is verified and the equilibrium point $(0, 0)$ is asymptotically stable.

Exercise E.3 The Duffing oscillator consists in the dynamical system

$$\dot{x} = y, \qquad \dot{y} = x - x^3 - \delta y, \tag{E.19}$$

where $(x, y) \in \mathbb{R}^2$ and $\delta > 0$. Show that

$$V(x, y) = \frac{y^2}{2} + \frac{(x^2 - 1)^2}{4} \tag{E.20}$$

is a Lyapunov function for the equilibrium points $(\bar{x}, \bar{y})_\pm = (\pm 1, 0)$ and conclude that they are both stable.

Unfortunately, there is no systematic method to construct Lyapunov functions, which are sort of a generalisation of the energy. Except for the simplest cases, finding a Lyapunov function requires ingenuity and art.

Appendix F **Exact Differentials**

In this appendix the Poincaré lemma is proved in the special case of 1-forms: we state necessary and sufficient conditions for a 1-form to be exact.

Let $f_1, f_2, \ldots, f_n$ be real functions of n-real variables $x_1, x_2, \ldots, x_n$ and suppose each of the f_i as well as its partial derivatives $\partial f_i / \partial x_j$ are continuous in an open set $\mathcal{O} \subset \mathbb{R}^n$. Additional restrictions on the region $\mathcal{O}$ will naturally appear in the course of the discussion that follows. In order to make the equations as simple as possible we use the short notation $x = (x_1, \ldots, x_n)$. For the sake of convenience, we repeat here the general definition of exact differential form, given in Appendix B, for the special case of 1-forms.

Definition F.1   *The 1-form on R^n*

$$\omega = \sum_{i=1}^{n} f_i \, dx_i \tag{F.1}$$

is said to be an **exact differential** *in $\mathcal{O}$ if there exists a function $F : \mathbb{R}^n \to \mathbb{R}$ such that*

$$\omega = dF \tag{F.2}$$

in $\mathcal{O}$ or, equivalently, if there exists a function $F : \mathbb{R}^n \to \mathbb{R}$ such that

$$f_i = \frac{\partial F}{\partial x_i} \tag{F.3}$$

at each point of the region $\mathcal{O}$.

Theorem F.1   *The 1-form (F.1) is an exact differential in $\mathcal{O}$ if and only if*

$$\frac{\partial f_i}{\partial x_j} = \frac{\partial f_j}{\partial x_i}, \qquad i, j = 1, \ldots, n \tag{F.4}$$

at all points of $\mathcal{O}$.

Proof   Let us prove the necessity of (F.4). If $\omega = \sum_i f_i \, dx_i$ is an exact differential in $\mathcal{O}$ there is a function F such that $f_i = \partial F / \partial x_i$. From the continuity of the partial derivatives of the functions f_i it follows that

$$\frac{\partial f_i}{\partial x_j} = \frac{\partial^2 F}{\partial x_j \partial x_i} = \frac{\partial^2 F}{\partial x_i \partial x_j} = \frac{\partial f_j}{\partial x_i}, \qquad i, j = 1, \ldots, n \tag{F.5}$$

at each point of $\mathcal{O}$. In order to prove the sufficiency, assume conditions (F.4) hold and consider the function $F(x_1, \ldots, x_n)$ defined as the line integral of $\sum_i f_i dx_i$ along the straight line segment connecting the fixed point P_0 to a generic point $P = (x_1, \ldots, x_n)$ of $\mathcal{O}$.

Note that $(\Delta s)_{dyn} = \dot{s}_0 T$ is the displacement of dynamical origin that coincides with the displacement for a motionless hoop, in which case Eq. (G.8) reduces to $\ddot{s} = 0$. If the hoop rotates very slowly, at the end of a full turn there appears, on average, the additional displacement $(\Delta s)_{geom} = -4\pi A/\ell$ that depends only on the geometry of the hoop.

> **Exercise G.3** If the hoop is circular, show that formula (G.14) is exact – that is, it holds without the need of taking averages or appealing to the adiabatic approximation. Is the result in harmony with what intuition suggests?

Freely Rotating Rigid Body

A free rigid body also displays a geometric phase when one considers a full turn of the angular momentum vector as observed from the reference frame attached to the body.

As noted in Section 4.8, the angular momentum vector $\mathbf{L} = (L_x, L_y, L_z)$ of a freely rotating rigid body relative to an inertial reference frame is a constant of the motion. However, viewed from the body frame, the angular momentum vector is typically a periodic function of time, so its tip traces out a closed curve during one period of the motion. If $\mathbf{L}_b = (L_1, L_2, L_3)$ denotes the angular momentum relative to the frame fixed in the body, $\mathbf{L} = \mathcal{R}(t)\mathbf{L}_b$ where $\mathcal{R}(t)$ is a rotation matrix (an orthogonal matrix with unit determinant). Inasmuch as both the kinetic energy E and the angular momentum squared $\mathbf{L} \cdot \mathbf{L} = \mathbf{L}_b \cdot \mathbf{L}_b$ are constants of the motion, we have

$$L_1^2 + L_2^2 + L_3^2 = L^2, \qquad \frac{L_1^2}{I_1} + \frac{L_2^2}{I_2} + \frac{L_3^2}{I_3} = 2E \tag{G.15}$$

with L and E constants, as well as the principal moments of inertia I_1, I_2, I_3. Therefore, as discussed in Section 4.8, the tip of the vector $\mathbf{L}_b$ moves along a curve which is the intersection of a sphere with an ellipsoid. For typical initial conditions these curves are closed.

During a period T of the motion of the tip of $\mathbf{L}_b$, from the point of view of the external inertial frame the rigid body rotates by an angle $\Delta\theta$ about the angular momentum vector $\mathbf{L}$. The value of this angle is

$$\Delta\theta = \frac{2ET}{L} - \Omega, \tag{G.16}$$

where Ω is the solid angle subtended by the closed curve traced out by L_b (see Fig. G.2). Note that $\Delta\theta$ = dynamical phase + geometric phase. The geometric phase $-\Omega$ depends only on the geometry of the closed curve described by the tip of $\mathbf{L}_b$. Note, still, that this result is exact, it does not require the adiabatic approximation. The reader is referred to Montgomery (1991) for a proof of formula (G.16).

In spite of springing from a purely theoretical curiosity about highly idealised mechanical systems, geometric phases are relevant to practical problems in the field of robotics, in the search for efficient means of stabilising and controlling the orientation of satellites

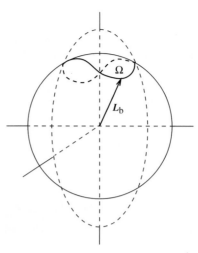

Fig. G.2   Solid angle swept out by the tip of $\mathbf{L}_b(t)$.

in their orbits and even to explain the locomotion of swimming microorganisms. These and several other fascinating applications of geometric phases (also known by the name of **holonomies**) are discussed in the review paper by Marsden *et al.* (1991) and in the references therein.

Appendix H Poisson Manifolds

In this appendix, largely based on Marsden and Ratiu (1999), we describe how equations of motion in Hamiltonian form can be formulated over differentiable manifolds more general than the symplectic manifolds of the conventional Hamiltonian formalism. This is feasible as long as the emphasis is put on the Poisson bracket and it is admitted that it does not necessarily have the standard form.

Basic Definitions

Given a differentiable manifold P, let $\mathcal{F}(P)$ be the set of the infinitely differentiable functions from P to $\mathbb{R}$.

Definition H.1 *A **Poisson bracket** or a **Poisson structure** on $\mathcal{F}(P)$ is an application $\{\cdot, \cdot\}$ that associates with each ordered pair of elements of $\mathcal{F}(P)$ an element of $\mathcal{F}(P)$ and enjoys the following properties:*

    (PB1) $\{F, G\}$ *depends linearly on both F and G;*
    (PB2) $\{F, G\} = -\{G, F\}$ *(antisymmetry);*
    (PB3) $\{\{E, F\}, G\}\} + \{\{G, E\}, F\}\} + \{\{F, G\}, E\}\} = 0$ *(Jacobi identity);*
    (PB4) $\{EF, G\} = E\{F, G\} + \{E, G\}F$ *(Leibniz rule).*

Definition H.2 *A **Poisson manifold** is a differentiable manifold P endowed with a Poisson bracket on $\mathcal{F}(P)$. In more precise mathematical language: a Poisson manifold is an ordered pair $(P, \{\cdot, \cdot\})$ where P is a differentiable manifold and $\{\cdot, \cdot\}$ is a Poisson bracket on $\mathcal{F}(P)$. For simplicity, when there is only one Poisson bracket involved one denotes the Poisson manifold just by P.*

Definition H.3 *On a Poisson manifold P the **Hamiltonian** is a function $H \in \mathcal{F}(P)$ such that the time evolution of any dynamical variable $F \in \mathcal{F}(P)$ is given by*

$$\dot{F} = \{F, H\}. \tag{H.1}$$

Differently from a symplectic manifold, a Poisson manifold does not have to be even-dimensional. Consequently, on a generic Poisson manifold the Poisson bracket does not take the standard form (8.79). Let us examine a few representative examples.

Free Rigid Body

As discussed in Section 4.8, the free rotational motion of a rigid body about its centre of mass is governed by the Euler equations

$$
\begin{aligned}
I_1\dot{\omega}_1 - (I_2 - I_3)\omega_2\omega_3 &= 0, \\
I_2\dot{\omega}_2 - (I_3 - I_1)\omega_3\omega_1 &= 0, \\
I_3\dot{\omega}_3 - (I_1 - I_2)\omega_1\omega_2 &= 0,
\end{aligned}
\tag{H.2}
$$

where I_1, I_2, I_3 are principal moments of inertia and $\omega_1, \omega_2, \omega_3$ are the components of the angular velocity along principal axes fixed in the body with origin at the centre of mass.

Let $\mathbf{\Pi} = (\Pi_1, \Pi_2, \Pi_3)$ be the angular momentum with respect to the body frame:

$$
\Pi_1 = I_1\omega_1, \quad \Pi_2 = I_2\omega_2, \quad \Pi_3 = I_3\omega_3.
\tag{H.3}
$$

In terms of the Πs the Euler equations become

$$
\begin{aligned}
\dot{\Pi}_1 &= \frac{I_2 - I_3}{I_2 I_3}\Pi_2\Pi_3, \\
\dot{\Pi}_2 &= \frac{I_3 - I_1}{I_3 I_1}\Pi_3\Pi_1, \\
\dot{\Pi}_3 &= \frac{I_1 - I_2}{I_1 I_2}\Pi_1\Pi_2.
\end{aligned}
\tag{H.4}
$$

The differentiable manifold $P = \mathbb{R}^3$ with coordinates Π_1, Π_2, Π_3 becomes a Poisson manifold upon being endowed with the Poisson bracket

$$
\{F, G\} = -\mathbf{\Pi} \cdot (\nabla F \times \nabla G),
\tag{H.5}
$$

where F and G are elements of $\mathcal{F}(\mathbb{R}^3)$ and the gradient is taken with respect to the components of the angular momentum $\mathbf{\Pi}$.

In order to establish that the application (H.5) is a Poisson bracket it is enough to prove the Jacobi identity because the remaining properties demanded by Definition H.1 are trivially satisfied. With the indicial notation

$$
\partial_i F = \frac{\partial F}{\partial \Pi_i} = F_i, \quad \partial_i \partial_j F = \frac{\partial^2 F}{\partial \Pi_i \partial \Pi_j} = F_{ij} = F_{ji}
\tag{H.6}
$$

and the convention of sum over repeated indices, we can write Eq. (H.5) in the following form:

$$
\{F, G\} = -\Pi_i \epsilon_{ijk} F_j G_k.
\tag{H.7}
$$

Consequently,

$$
\begin{aligned}
\{\{E, F\}, G\} &= \Pi_i \epsilon_{ijk} \partial_j \left(\Pi_l \epsilon_{lmn} E_m F_n \right) G_k \\
&= \Pi_i \epsilon_{ijk} \epsilon_{lmn} \left(\delta_{jl} E_m F_n + \Pi_l E_{mj} F_n + \Pi_l E_m F_{nj} \right) G_k \\
&= \underbrace{\epsilon_{ijk} \epsilon_{jmn} \Pi_i E_m F_n G_k}_{(\mathrm{I})} + \epsilon_{ijk}\epsilon_{lmn}\Pi_i\Pi_l E_{mj}F_n G_k + \epsilon_{ijk}\epsilon_{lmn}\Pi_i\Pi_l E_m F_{nj}G_k.
\end{aligned}
$$

$$
\tag{H.8}
$$

With the use of $\epsilon_{ijk}\epsilon_{jmn} = \delta_{in}\delta_{km} - \delta_{im}\delta_{kn}$ [identity (A.39) in Appendix A] the term (I) of this last equation is written

$$(I) = -(\mathbf{\Pi} \cdot \nabla E)(\nabla F \cdot \nabla G) + (\mathbf{\Pi} \cdot \nabla F)(\nabla E \cdot \nabla G). \tag{H.9}$$

Therefore,

$$\{\{E, F\}, G\}\} + \{\{G, E\}, F\}\} + \{\{F, G\}, E\}\}$$

$$= \underbrace{-(\mathbf{\Pi} \cdot \nabla E)(\nabla F \cdot \nabla G) + (\mathbf{\Pi} \cdot \nabla F)(\nabla E \cdot \nabla G)}_{(I)} + \text{(cyclic permutations of } EFG)$$

$$+ \underbrace{\epsilon_{ijk}\epsilon_{lmn}\Pi_i\Pi_l E_{mj}F_n G_k}_{(II)} + \text{(cyclic permutations of } EFG)$$

$$+ \underbrace{\epsilon_{ijk}\epsilon_{lmn}\Pi_i\Pi_l E_m F_{nj} G_k}_{(III)} + \text{(cyclic permutations of } EFG). \tag{H.10}$$

It is immediate that the sum of the term (I) in Eq. (H.10) with the other two terms obtained by cyclic permutations is zero. As to the sum of the terms (II) and (III), there is a pairwise cancellation: each term of (II) is cancelled by a corresponding term of (III). For instance, the term $\epsilon_{ijk}\epsilon_{lmn}\Pi_i\Pi_l E_{mj}F_n G_k$ is cancelled by the corresponding term of (III) that contains second derivatives of E, namely $\epsilon_{ijk}\epsilon_{lmn}\Pi_i\Pi_l G_m E_{nj}F_k$. Indeed, making in this last expression the consecutive exchanges of dummy indices $i \leftrightarrow l$, $m \leftrightarrow k$, $m \leftrightarrow n$, $m \leftrightarrow j$, we obtain $\epsilon_{lmn}\epsilon_{ikj}\Pi_i\Pi_l G_k E_{jm}F_n$, which is equal to $-\epsilon_{ijk}\epsilon_{lmn}\Pi_i\Pi_l E_{mj}F_n G_k$. It is left to the reader the verification that the remaining pairs of terms associated with (II) and (III) also cancel each other, completing the proof.

As it is to be expected, the Hamiltonian for the free rigid body is the kinetic energy:

$$H = \frac{\Pi_1^2}{2I_1} + \frac{\Pi_2^2}{2I_2} + \frac{\Pi_3^2}{2I_3}. \tag{H.11}$$

The Euler equations (H.4) take the form

$$\dot{\Pi}_i = \{\Pi_i, H\}. \tag{H.12}$$

Exercise H.1   Show that these last equations exactly reproduce the Euler equations of motion (H.4).

Exercise H.2   Prove that

$$\dot{F} = \{F, H\} \tag{H.13}$$

for any function $F(\mathbf{\Pi})$.

Definition H.4   *A function C such that $\{C, F\} = 0$ for any F is called a* **Casimir function**.

The function C defined by

$$C(\mathbf{\Pi}) = \mathbf{\Pi}^2 = \Pi_1^2 + \Pi_2^2 + \Pi_3^2 \tag{H.14}$$

is a Casimir function for the free rigid body. In fact, since $\nabla C = 2\mathbf{\Pi}$, we have

$$\{C, F\} = -2\mathbf{\Pi} \cdot (\mathbf{\Pi} \times \nabla F) = 0 \qquad (H.15)$$

for all F. In particular, C is a constant of the motion, as every Casimir function is a constant of the motion. More generally, any infinitely differentiable function $\Phi(\mathbf{\Pi}^2)$ is a Casimir function.

For a description of the important role played by Casimir functions in the study of stability of mechanical systems, the reader is referred to Marsden and Ratiu (1999).

Exercise H.3 In the conventional Hamiltonian formalism with the Poisson bracket (8.79), show that the Casimir functions are trivial: every Casimir function is identically constant.

Gyroscopic Systems

Given $\mathbf{q} = (q_1, \cdots, q_n) \in \mathbb{R}^n$, consider the linear gyroscopic system characterised by the equations of motion

$$M\ddot{\mathbf{q}} + S\dot{\mathbf{q}} + V\mathbf{q} = 0, \qquad (H.16)$$

where $\mathbf{q}$ is treated as a column matrix and M, S, V are $n \times n$ constant matrices, with M symmetric and positive, S antisymmetric and V symmetric. This system does not admit a conventional Lagrangian formulation: no Lagrangian $L(\mathbf{q}, \dot{\mathbf{q}}, t)$ yields Lagrange equations identical to (H.16). As a consequence, these equations of motion do not admit a conventional Hamiltonian formulation either.

However, with $\mathbf{p} = M\dot{\mathbf{q}}$ and the Poisson bracket

$$\{F, G\} = \frac{\partial F}{\partial q_i}\frac{\partial G}{\partial p_i} - \frac{\partial G}{\partial q_i}\frac{\partial F}{\partial p_i} - S_{ij}\frac{\partial F}{\partial p_i}\frac{\partial G}{\partial p_j}, \qquad (H.17)$$

where there is a sum over repeated indices, the system is Hamiltonian with Hamilton's function equal to the mechanical energy:

$$H = \frac{1}{2}\mathbf{p} \cdot M^{-1}\mathbf{p} + \frac{1}{2}\mathbf{q} \cdot V\mathbf{q}. \qquad (H.18)$$

Although straightforward, the proof that the Poisson bracket (H.17) obeys the Jacobi identity is too long and tedious, and will be omitted.

Writing $M = (M_{ij})$, $M^{-1} = (M^{ij})$ and $V = (V_{ij})$, from

$$H = \frac{1}{2}p_i M^{ij} p_j + \frac{1}{2}q_i V_{ij} q_j \qquad (H.19)$$

one gets

$$\frac{\partial H}{\partial p_k} = \frac{1}{2}[\delta_{ik}M^{ij}p_j + p_i M^{ij}\delta_{jk}] = M^{kj}p_j, \qquad \frac{\partial H}{\partial q_k} = V_{kj}p_j, \qquad (H.20)$$

where the fact has been used that the inverse of a symmetric matrix is likewise symmetric. It follows that

$$\dot{q}_k = \{q_k, H\} = \frac{\partial H}{\partial p_k} = M^{kj} p_j \implies \dot{\mathbf{q}} = M^{-1}\mathbf{p} \implies \mathbf{p} = M\dot{\mathbf{q}} \qquad \text{(H.21)}$$

and, similarly,

$$\dot{p}_k = \{p_k, H\} = -\frac{\partial H}{\partial q_k} - S_{ij}\delta_{ik}\frac{\partial H}{\partial p_j} = -V_{kj}q_j - S_{kj}M^{jl}p_l, \qquad \text{(H.22)}$$

which is the same as

$$\dot{\mathbf{p}} = -V\mathbf{q} - SM^{-1}\mathbf{p}. \qquad \text{(H.23)}$$

The equation of motion (H.16) is an immediate consequence of Eqs. (H.21) and (H.23). Therefore, the system is Hamiltonian with Poisson bracket (H.17) and Hamilton function (H.18).

Exercise H.4   Prove that $\dot{F} = \{F, H\}$ for any function $F(\mathbf{q}, \mathbf{p})$.

Exercise H.5   A non-conservative force is said to be **gyroscopic** if it does no work during the system's motion. For instance, the magnetic force is gyroscopic. Prove that the force $\mathbf{F} = -S\dot{\mathbf{q}}$ is gyroscopic.

It is the result of this last exercise that justifies the possibility of giving Hamiltonian form to the equations of motion (H.16) with Hamilton's function equal to the mechanical (kinetic plus potential) energy of the system.

Nonlinear Schrödinger Equation

As an infinite-dimensional example, consider the nonlinear Schrödinger equation

$$i\frac{\partial\psi}{\partial t} = -\frac{\partial^2\psi}{\partial x^2} + 2\chi|\psi|^2\psi, \qquad \text{(H.24)}$$

where χ is a real parameter. This equation is used in mathematical biology for the modelling of muscle contractions (Davydov, 1973; Ivancevic & Pearce, 2001).

Writing $\psi_t(x) = \psi(x, t)$, for each fixed t the function $\psi_t : \mathbb{R} \to \mathbb{C}$ is an element of the Schwartz space $\mathcal{S}(\mathbb{R})$ – that is, it is a complex infinitely differentiable function of the real variable x such that, as $|x| \to \infty$, the function itself as well as its derivatives of all orders tend to zero faster that any inverse power of $|x|$. In the case of the infinite-dimensional manifold $\mathcal{P} = (\psi, \bar{\psi})$ with the functions $\psi, \bar{\psi} \in \mathcal{S}(\mathbb{R})$ treated as mutually independent, a dynamical variable F is a functional $F[\psi, \bar{\psi}]$ that associates with each element of $\mathcal{P}$ a real number (for notational convenience, in this appendix $\bar{z}$ denotes the complex conjugate of

z). The Poisson bracket

$$\{F, G\} = i \int_{-\infty}^{\infty} \left(\frac{\delta F}{\delta \bar{\psi}(x)} \frac{\delta G}{\delta \psi(x)} - \frac{\delta G}{\delta \bar{\psi}(x)} \frac{\delta F}{\delta \psi(x)} \right) dx \qquad (\text{H.25})$$

makes $\mathcal{P}$ a Poisson manifold. With the Hamiltonian

$$H[\psi, \bar{\psi}] = \int_{-\infty}^{\infty} \left(\left| \frac{\partial \psi}{\partial x} \right|^2 + \chi |\psi|^4 \right) dx, \qquad (\text{H.26})$$

the time evolution of the system is governed by

$$\dot{\psi} = \{\psi, H\}, \qquad \dot{\bar{\psi}} = \{\bar{\psi}, H\}. \qquad (\text{H.27})$$

Exercise H.6 Using the definition of functional derivative given in Section 11.3, prove that Eqs. (H.27) coincide with the non-linear Schrödinger equation (H.24) and its complex conjugate. By way of clarification, an equation of motion of the form $\dot{\psi} = \{\psi, H\}$ is to be interpreted as $(\dot{\psi}_t)(x) = \{\psi_t(x), H[\psi_t, \bar{\psi}_t]\}$ or, still, $\partial \psi(x, t)/\partial t = \{\psi_t(x), H[\psi_t, \bar{\psi}_t]\}$ where $\psi_t(x) = \psi(x, t)$.

The Korteweg-de Vries Equation

The infinite-dimensional system governed by the Korteweg-de Vries equation

$$u_t = u u_x + u_{xxx} \qquad (\text{H.28})$$

will serve as our last example. This nonlinear equation describes shallow water waves and has soliton solutions (José & Saletan, 1998). Suppose that for each $t \in \mathbb{R}$ the function $u(x, t)$ is a real element of the Schwartz space $\mathcal{S}(\mathbb{R})$. Let F and G be functionals of u. Define the Poisson bracket $\{F, G\}$ by

$$\{F, G\} = \int_{-\infty}^{\infty} dx \, \frac{\delta F}{\delta u(x)} \frac{\partial}{\partial x} \left(\frac{\delta G}{\delta u(x)} \right). \qquad (\text{H.29})$$

Under the assumption that both $\delta F/\delta u$ and $\delta G/\delta u$ belong to $\mathcal{S}(\mathbb{R})$, an integration by parts shows that $\{F, G\} = -\{G, F\}$. It can be checked that $\{F, G\}$ defined by (H.29) has all the required properties of a Poisson bracket.

Exercise H.7 Under mild conditions the mixed second functional derivatives are equal:

$$\frac{\delta^2 F}{\delta u(x) \delta u(y)} = \frac{\delta^2 F}{\delta u(y) \delta u(x)}. \qquad (\text{H.30})$$

Use this to prove the Jacobi identity. Show that the Leibniz rule is also satisfied, establishing that (H.29) defines a Poisson structure.

The Hamiltonian for the Korteweg-de Vries system is

$$H[u] = \int_{-\infty}^{\infty} \left(\frac{1}{6} u^3 - \frac{1}{2} u_x^2 \right) dx. \tag{H.31}$$

Exercise H.8   Using the definition of funcional derivative of Section 11.3, prove that

$$\frac{\delta H}{\delta u(x)} = \frac{1}{2} u(x)^2 + u_{xx}(x). \tag{H.32}$$

Use this result to show that the Hamiltonian equation of motion

$$\dot{u}(x) = \{u(x), H\} \tag{H.33}$$

is equivalent to the Korteweg-de Vries equation (H.28).

Canonical Transformations

A transformation between Poisson manifolds is said to be canonical if it preserves Poisson brackets. In order to formulate a precise definition, let $(P_1, \{\cdot, \cdot\}_1)$ and $(P_2, \{\cdot, \cdot\}_2)$ be two Poisson manifolds and $f : P_1 \to P_2$ be an infinitely differentiable application. Note that if $F \in \mathcal{F}(P_2)$ then the composite function $F \circ f$ is an element of $\mathcal{F}(P_1)$.

Definition H.5   *If $(P_1, \{\cdot, \cdot\}_1)$ and $(P_2, \{\cdot, \cdot\}_2)$ are Poisson manifolds, the infinitely differentiable application $f : P_1 \to P_2$ is canonical if*

$$\{F, G\}_2 \circ f = \{F \circ f, G \circ f\}_1 \tag{H.34}$$

for all $F, G \in \mathcal{F}(P_2)$.

For example, on the Poisson manifold $P = \mathbb{R}^3$ associated with the free rigid body with the Poisson bracket (H.5), any rotation $\mathcal{R} : \mathbb{R}^3 \to \mathbb{R}^3$ is a canonical transformation because both the scalar product and the vector product are invariant under rotations (see Example 3.2).

Quantisation

Canonical quantisation consists in a set of rules that permit the transition from a traditional classical Hamiltonian system – formulated on a symplectic manifold equipped with the standard Poisson bracket – to the corresponding quantum system. It is remarkable that the traditional procedure for quantising a classical system can be generalised to Hamiltonian systems on Poisson manifolds. This generalised procedure is known as **deformation quantisation**. An introduction to this subject, which by itself is an extensive research area, can be found in a review paper by Bordemann (2008).

This appendix provides rigorous estimates of the magnitude of the Fourier coefficients of a sufficiently regular function on $\mathbb{R}^n$.

Definition I.1   *A function $f : \mathbb{R}^n \to \mathbb{R}$ is said to be periodic with period 2π in each of its arguments if*

$$f(x_1 + 2\pi, x_2, \ldots, x_n) = f(x_1, x_2 + 2\pi, \ldots, x_n) = \cdots$$
$$= f(x_1, x_2, \ldots, x_n + 2\pi) = f(x_1, x_2, \ldots, x_n). \tag{I.1}$$

Let $f : \mathbb{R}^n \to \mathbb{R}$ be a periodic function with period 2π in each of its arguments. One says that f admits a **Fourier series expansion** if there exist **Fourier coefficients** $\hat{f}_m \in \mathbb{C}$ such that

$$f(x) = \sum_{m \in \mathbb{Z}^n} \hat{f}_m \, e^{im \cdot x} \tag{I.2}$$

where $m \cdot x = m_1 x_1 + \cdots + m_n x_n$ and

$$\hat{f}_m = \frac{1}{(2\pi)^n} \int_0^{2\pi} \cdots \int_0^{2\pi} f(x) \, e^{-im \cdot x} \, dx_1 \cdots dx_n \equiv \frac{1}{(2\pi)^n} \oint_{T^n} f(x) \, e^{-im \cdot x} \, d^n x. \tag{I.3}$$

Theorem I.1   *Let $f : \mathbb{R}^n \to \mathbb{R}$ be an infinitely differentiable function which is periodic with period 2π in each of its arguments. Then, for $m \neq \mathbf{0}$ and any natural number s there exists a constant $C > 0$ such that*

$$|\hat{f}_m| \leq \frac{C}{|m|^s}, \tag{I.4}$$

where

$$|m| = |m_1| + \cdots + |m_n|. \tag{I.5}$$

Proof   Taking advantage of the periodicity and differentiability of f, for $m_k \neq 0$ we have

$$\int_0^{2\pi} f(x) \, e^{-im \cdot x} \, dx_k = \int_0^{2\pi} f(x) \frac{1}{-im_k} \frac{\partial}{\partial x_k} \left(e^{-im \cdot x} \right) dx_k$$
$$= \frac{1}{-im_k} f(x) \, e^{-im \cdot x} \Big|_0^{2\pi} + \frac{1}{im_k} \int_0^{2\pi} \frac{\partial f}{\partial x_k}(x) \, e^{-im \cdot x} \, dx_k$$
$$= \frac{1}{im_k} \int_0^{2\pi} \frac{\partial f}{\partial x_k}(x) \, e^{-im \cdot x} \, dx_k. \tag{I.6}$$

Repeating this procedure s times we are left with

$$\hat{f}_m = \frac{1}{(2\pi)^n} \frac{1}{(im_k)^s} \oint_{T^n} \frac{\partial^s f}{\partial x_k^s}(x)\, e^{-im\cdot x}\, d^n x, \tag{I.7}$$

since all partial derivatives of f are equally periodic. It follows that

$$|\hat{f}_m| \leq \frac{1}{(2\pi)^n} \frac{1}{|m_k|^s} \oint_{T^n} \left| \frac{\partial^s f}{\partial x_k^s}(x) \right| d^n x \leq \frac{M_k}{|m_k|^s}, \tag{I.8}$$

where

$$M_k = \max_{x \in [0, 2\pi]^n} \left| \frac{\partial^s f}{\partial x_k^s}(x) \right|. \tag{I.9}$$

From the validity of (I.8) for any k such that $m_k \neq 0$, and taking $M = \max\limits_{1 \leq k \leq n} M_k$, one infers that there exists $M > 0$ such that

$$|\hat{f}_m| \leq \frac{M}{\left(\max\limits_{1 \leq k \leq n} |m_k| \right)^s}. \tag{I.10}$$

Finally, by means of the obvious inequality

$$|m| = |m_1| + \cdots + |m_n| \leq n \max_{1 \leq k \leq n} |m_k|, \tag{I.11}$$

it is immediate to rewrite (I.10) in the form (I.4). $\qquad\square$

If f is an analytic function, it can be shown (Siegel & Moser, 1971) that there exist positive constants C and c such that

$$|\hat{f}_m| \leq C e^{-c|m|}. \tag{I.12}$$

References

Abraham, R. & Marsden, J. E. (1978). *Foundations of Mechanics*, 2nd edn. London: Benjamin/Cummmings.

Almeida, M. A., Moreira, I. C. & Santos, F. C. (1998). On the Ziglin-Yoshida analysis for some classes of homogeneous Hamiltonian systems. *Brazilian Journal of Physics*, **28**, 470–480.

Almeida, M. A., Moreira, I. C. & Yoshida, H. (1992). On the non-integrability of the Störmer problem. *Journal of Physics A: Mathematical and General*, **25**, L227–L230.

Anderson, J. L. (1990). Newton's first two laws of motion are not definitions. *American Journal of Physics*, **58**, 1192–1195.

Andersson, K. G. (1994). Poincaré's discovery of homoclinic points. *Archive for History of Exact Sciences*, **48**, 133–147.

Aravind, P. K. (1989). Geometrical interpretation of the simultaneous diagonalization of two quadratic forms. *American Journal of Physics*, **57**, 309–311.

Arfken, G. B. & Weber, H. J. (1995). *Mathematical Methods for Physicists*, 4th edn. New York: Academic Press.

Armitage, J. V. & Eberlein, W. F. (2006). *Elliptic Functions*. Cambridge: Cambridge University Press.

Arnold, V. I. (1963). Proof of a theorem of A. N. Kolmogorov on the preservation of conditionally periodic motions under a small perturbation of the Hamiltonian. *Russian Mathematical Surveys*, **18**(5), 9–36.

Arnold, V. I. (1989). *Mathematical Methods of Classical Mechanics*, 2nd edn. New York: Springer.

Arnold, V. I., Kozlov, V. V. & Neishtadt, A. I. (2006). *Mathematical Aspects of Classical and Celestial Mechanics*, 3rd edn. Berlin: Springer.

Arnowitt, R., Deser, S. & Misner, C. W. (1962). The Dynamics of General Relativity. In L. Witten, ed., *Gravitation: An Introduction to Current Research*. New York: Wiley, pp. 227–264. arXiv:gr-qc/0405109v1.

Barrow-Green, J. (1994). Oscar II's prize competition and the error in Poincaré's memoir on the three body problem. *Archive for History of Exact Sciences*, **48**, 107–131.

Barrow-Green, J. (1997). *Poincaré and the Three Body Problem*. Providence, RI: American Mathematical Society.

Barrow-Green, J. (2010). The dramatic episode of Sundman. *Historia Mathematica*, **37**, 164–203. www.sciencedirect.com/science/article/pii/S0315086009001360.

Bartucelli, M. & Gentile, G. (2002). Lindstedt series for perturbations of isochronous systems: A review of the general theory. *Reviews in Mathematical Physics*, **14**, 121–171.

Barut, A. O. (1980). *Electrodynamics and Classical Theory of Fields and Particles*. New York: Dover.

Bateman, H. (1931). On dissipative systems and related variational principles. *Physical Review*, **38**, 815–819.

Benettin, G., Galgani, L., Giorgilli, A. & Strelcyn, J.-M. (1984). A proof of Kolmogorov's theorem on invariant tori using canonical transformations defined by the Lie method. *Nuovo Cimento B*, **79**, 201–223.

Bergmann, P. G. (1976). *Introduction to the Theory of Relativity*. New York: Dover.

Berry, M. V. (1978). Regular and Irregular Motion. In S. Jorna, ed., *Topics in Nonlinear Dynamics: A Tribute to Sir Edward Bullard*. New York: American Institute of Physics, pp. 16–120.

Berry, M. V. (1985). Classical adiabatic angles and quantal adiabatic phase. *Journal of Physics A: Mathematical and General*, **18**, 15–27.

Bertrand, J. (1873). Théorème relatif au mouvement d'un point attire vers un centre fixe. *Comptes Rendus de l'Académie des Sciences de Paris*, **77**, 849–853. English translation by Santos, F. C., Soares, V. and Tort, A. C. (2011). An English translation of Bertrand's theorem. *Latin American Journal of Physics Education*, **5**, 694–696. arXiv:0704.2396.

Boas, Jr., R. P. (1996). *A Primer of Real Functions*, 4th edn. Washington, DC: The Mathematical Association of America.

Bordemann, M. (2008). Deformation quantization: A survey. *Journal of Physics: Conference Series*, **103**, 1–31.

Bottazzini, U. & Gray, J. (2013). *Hidden Harmony-Geometric Fantasies: The Rise of Complex Function Theory*. New York: Springer.

Braun, M. (1981). Mathematical remarks on the Van Allen radiation belt: A survey of old and new results. *SIAM Review*, **23**, 61–93.

Bricmont, J. (1996). Science of chaos or chaos in science? arXiv:chao-dyn/9603009.

Cadoni, M., Carta, P. & Mignemi, S. (2000). Realization of the infinite-dimensional symmetries of conformal mechanics. *Physical Review D*, **62**, 086002-1-4.

Callen, H. (1960). *Thermodynamics*. New York: Wiley.

Cary, J. R. (1981). Lie transform perturbation theory for Hamiltonian systems. *Physics Reports*, **79**, 129–159.

Casasayas, J., Nunes, A. & Tufillaro, N. (1990). Swinging Atwood's machine: Integrability and dynamics. *Journal de Physique*, **51**, 1693–1702.

Casey, J. (2014). Applying the principle of angular momentum to constrained systems of point masses. *American Journal of Physics*, **82**, 165–168. Casey, J. (2015). Erratum. *American Journal of Physics*, **83**, 185.

Cercignani, C. (1998). *Ludwig Boltzmann: The Man Who Trusted Atoms*. Oxford: Oxford University Press.

Chagas, E. F. das & Lemos, N. A. (1981). Um exemplo de como não usar teoremas matemáticos em problemas físicos. *Revista Brasileira de Física*, **11**, 481–488.

Chandre, C. & Jauslin, H. R. (1998). A version of Thirring's approach to the Kolmogorov-Arnold-Moser theorem for quadratic Hamiltonians with degenerate twist. *Journal of Mathematical Physics*, **39**, 5856–5865.

Chenciner, A. (2007).   Three Body Problem. www.scholarpedia.org/article/Three_body_problem.

Christ, N. H. & Lee, T. D. (1980). Operator ordering and Feynman rules in gauge theories. *Physical Review D*, **22**, 939–958.

Corben, H. C. & Stehle, P. (1960). *Classical Mechanics*. New York: Dover.

Courant, R. & Hilbert, D. (1953). *Methods of Mathematical Physics*. Vol. I. New York: Interscience Publishers.

Crawford, F. S. (1990). Elementary examples of adiabatic invariance. *American Journal of Physics*, **58**, 337–344.

Davydov, A. S. (1973). The theory of contraction of proteins under their excitation. *Journal of Theoretical Biology*, **38**, 559–569.

Deser, S., Jackiw, R. & Templeton, S. (1982). Topologically massive gauge theories. *Annals of Physics*, **140**, 372–411.

Desloge, E. A. (1982). *Classical Mechanics*, 2 vols. Malabar, FL: Robert E. Krieger.

Dettman, J. W. (1986). *Introduction to Linear Algebra and Differential Equations*. New York: Dover.

Dhar, A. (1993). Nonuniqueness in the solutions of Newton's equations of motion. *American Journal of Physics*, **61**, 58–61.

Diacu, F. (1996). The Solution of the *n*-body Problem. *The Mathematical Intelligencer*, **18**(3), 66–70.

Diacu, F. & Holmes, P. (1996). *Celestial Encounters: The Origins of Chaos and Stability*. Princeton, NJ: Princeton University Press.

Dilão, R. & Alves-Pires, R. (2007). Chaos in the Störmer problem. arXiv:0704.3250v1.

Dirac, P. A. M. (1964). *Lectures on Quantum Mechanics*. New York: Yeshiva University.

Dugas, R. (1988). *A History of Mechanics*. New York: Dover.

Dumas, H. S. (2014). *The KAM Story*. New Jersey: World Scientific.

Eisenbud, L. (1958). On the classical laws of motion. *American Journal of Physics*, **26**, 144–159.

Eliasson, L. H. (1996). Absolutely convergent series expansions for quasi periodic motions. *Mathematical Physics Electronic Journal*, **2**(4), 1–33.

Elsgoltz, L. (1969). *Ecuaciones Diferenciales y Cálculo Variacional*. Moscow: Mir Publishers.

Epstein, S. T. (1982). The angular velocity of a rotating rigid body. *American Journal of Physics*, **50**, 948.

Farina de Souza, C. and Gandelman, M. M. (1990). An algebraic approach for solving mechanical problems. *American Journal of Physics*, **58**, 491–495.

Fasano, A. & Marmi, S. (2006). *Analytical Mechanics*. Oxford: Oxford University Press.

Fetter, A. L. & Walecka, J. D. (1980). *Theoretical Mechanics of Particles and Continua*. New York: McGraw-Hill.

Feynman, R. P. (1985). *Surely You're Joking, Mr. Feynman!* New York: Bantam Books.

Feynman, R. P. & Hibbs, A. (1965). *Quantum Mechanics and Path Integrals*. New York: McGraw-Hill.

Feynman, R. P., Leighton, R. B. & Sands, M. (1963). *The Feynman Lectures on Physics*, 3 vols. Reading, MA: Addison-Wesley.

Flanders, H. (1989). *Differential Forms with Applications to the Physical Sciences*. New York: Dover.

Fox, C. (1987). *An Introduction to the Calculus of Variations*. New York: Dover.

Gallavotti, G. (1983). *The Elements of Mechanics*. New York: Springer.

Gallavotti, G. (1999). *Statistical Mechanics: A Short Treatise*. Berlin: Springer.

Gangopadhyaya, A. & Ramsey, G. (2013). Unintended consequences of imprecise notation: An example from mechanics. *American Journal of Physics*, **81**, 313–315.

Gantmacher, F. (1970). *Lectures in Analytical Mechanics*. Moscow: Mir Publishers.

Gelfand, I. M. & Fomin, S. V. (1963). *Calculus of Variations*. Englewood Cliffs, NJ: Prentice-Hall.

Giorgilli, A., Locatelli, U. & Sansottera, M. (2013). Kolmogorov and Nekhoroshev theory for the problem of three bodies. arXiv:1303.7395v1.

Gleiser, R. J. & Kozameh, C. N. (1980). A simple application of adiabatic invariance. *American Journal of Physics*, **48**, 756–759.

Goldstein, H. (1980). *Classical Mechanics*, 2nd edn. Reading, MA: Addison-Wesley.

Goriely, A. (2001). *Integrability and Nonintegrability of Dynamical Systems*. New Jersey: World Scientific.

Gradshteyn, I. S. & Ryzhik, I. M. (1980). *Tables of Integrals, Series and Products*. New York: Academic Press.

Gray, C. G. & Taylor, E. F. (2007). When action is not least. *American Journal of Physics*, **75**, 434–458.

Gutiérrez-López, S., Castellanos-Moreno, A. & Rosas-Burgos, R. A. (2008). A new constant of motion for an electric charge acted on by a point electric dipole. *American Journal of Physics*, **76**, 1141–1145.

Halmos, P. R. (1974). *Finite-Dimensional Vector Spaces*. New York: Springer.

Hamermesh, M. (1962). *Group Theory and Its Application to Physical Problems*. New York: Dover.

Hamilton, W. R. (1834). On a General Method in Dynamics. In *Philosophical Transactions of the Royal Society*, Part II, 247–308.

Hamilton, W. R. (1835). Second Essay on a General Method in Dynamics. In *Philosophical Transactions of the Royal Society*, Part I, 95–144.

Hand, L. N. & Finch, J. D. (1998). *Analytical Mechanics*. Cambridge: Cambridge University Press.

Hartmann, H. (1972). Die Bewegung eines Körpers in einem ringförmigen Potentialfeld. *Theoretica Chimica Acta*, **24**, 201–206.

Havas, P. (1957). The range of application of the Lagrange formalism - I. *Nuovo Cimento Supplement*, **5**, 363–388.

Hawking, S. W. & Ellis, G. F. R. (1973). *The Large Scale Structure of Space-Time*. Cambridge: Cambridge University Press.

Henneaux, M. & Shepley, L. C. (1982). Lagrangians for spherically symmetric potentials. *Journal of Mathematical Physics*, **23**, 2101–2107.

Henneaux, M. & Teitelboim, C. (1992). *Quantization of Gauge Systems*. New Jersey: Princeton University Press.

Hietarinta, J. (1984). New integrable Hamiltonians with transcendental invariants. *Physical Review Letters*, **52**, 1057–1060.

Hoffman, K. & Kunze, R. (1971). *Linear Algebra*, 2nd edn. Englewood Cliffs, NJ: Prentice-Hall.

Huang, K. (1963). *Statistical Mechanics*. New York: Wiley.

Ichtiaroglou, S. (1997). Non-integrability in Hamiltonian mechanics. *Celestial Mechanics and Dynamical Astronomy*, **65**, 21–31.

Ivancevic, S. & Pearce, C. E. M. (2001). Poisson manifolds in generalised Hamiltonian biomechanics. *Bulletin of the Australian Mathematical Society*, **64**, 515–526.

Jackson, J. D. (1999). *Classical Electrodynamics*, 3rd edn. New York: Wiley.

Jacobi, C. G. J. (1837). Über die reduction der partiellen Differentialgleichungen erster Ordnung zwischen irgend einer Zahl Variabeln auf die Integration eines einzigen Systems gewöhnlicher Differentialgleichungen. *Crelle Journal fur die Reine und Angewandte Mathematik*, **17**, 97–162. Vol. IV of *Oeuvres Complètes*. http://sites.mathdoc.fr/OEUVRES/.

Jones, R. S. (1995). Circular motion of a charged particle in an electric dipole field. *American Journal of Physics*, **63**, 1042–1043.

José, J. V. & Saletan, E. J. (1998). *Classical Dynamics: A Contemporary Approach*. Cambridge: Cambridge University Press.

Karatas, D. L. & Kowalski, K. L. (1990). Noether's theorem for local gauge transformations. *American Journal of Physics*, **58**, 123–131.

Kolmogorov, A. N. (1954). Preservation of conditionally periodic movements with small change in the Hamiltonian function. *Doklady Akademii Nauk SSSR*, **98**, 527–530 (in Russian). English translation in Appendix A of Dumas (2014).

Kolmogorov, A. N. (1957). The general theory of dynamical systems and classical mechanics. In *Proceedings of the International Congress of Mathematicians 1954* (in Russian). Amsterdam: North-Holland. English translation in the Appendix of Abraham & Marsden (1978).

Konopinski, E. J. (1969). *Classical Descriptions of Motion*. San Francisco: W. H. Freeman.

Kot, M. (2014). *A First Course in the Calculus of Variations*. Providence, RI: American Mathematical Society.

Kotkin, G. L. & Serbo, V. G. (1971). *Collection of Problems in Classical Mechanics*. Oxford: Pergamon Press.

Kozlov, V. V. (1983). Integrability and non-integrability in Hamiltonian mechanics. *Russian Mathematical Surveys*, **38**, 1–76.

Lagrange, J. L. (1888). *Mécanique Analytique*. Vol. XI of *Oeuvres Complètes*. http://sites.mathdoc.fr/OEUVRES/.

Lam, K. S. (2014). *Fundamental Principles of Classical Mechanics: A Geometrical Perspective*. New Jersey: World Scientific.

Lanczos, C. (1970). *The Variational Principles of Mechanics*. New York: Dover.

Landau, L. D. & Lifshitz, E. (1976). *Mechanics*, 3rd edn. Oxford: Butterworth-Heinemann.

Laplace, P. S. (1840). *Essai Philosophique sur les Probabilités*. Paris: Bachelier. https://archive.org/stream/essaiphilosophiq00lapluoft#page/n5/mode/2up.

Laskar, J. (2013). Is the Solar System stable? *Progress in Mathematical Physics*, **66**, 239–270.

Lawden, D. F. (1989). *Elliptic Functions and Applications*. New York: Springer.

Lee, T. D. (1981). *Particle Physics and Introduction to Field Theory*. New York: Harwood Academic Publishers.

Lemos, N. A. (1979). Canonical approach to the damped harmonic oscillator. *American Journal of Physics*, **47**, 857–858.

Lemos, N. A. (1991). Remark on Rayleigh's dissipation function. *American Journal of Physics*, **59**, 660–661.

Lemos, N. A. (1993). Symmetries, Noether's theorem and inequivalent Lagrangians applied to nonconservative systems. *Revista Mexicana de Física*, **39**, 304–313.

Lemos, N. A. (1996). Singularities in a scalar field quantum cosmology. *Physical Review D*, **53**, 4275–4279.

Lemos, N. A. (2000a). Short proof of Jacobi's identity for Poisson brackets. *American Journal of Physics*, **68**, 88.

Lemos, N. A. (2000b). Uniqueness of the angular velocity of a rigid body: Correction of two faulty proofs. *American Journal of Physics*, **68**, 668–669.

Lemos, N. A. (2003). Sutilezas dos vínculos não-holônomos. *Revista Brasileira de Ensino de Física*, **25**, 28–34.

Lemos, N. A. (2005). Formulação geométrica do princípio de d'Alembert. *Revista Brasileira de Ensino de Física*, **27**, 483–485.

Lemos, N. A. (2014a). On what does not expand in an expanding universe: A very simple model. *Brazilian Journal of Physics*, **44**, 91–94.

Lemos, N. A. (2014b). Comment on "Unintended consequences of imprecise notation: An example from mechanics", [Am. J. Phys. **81**, 313–315 (2013)]. *American Journal of Physics*, **82**, 164–165.

Lemos, N. A. (2014c). Incompleteness of the Hamilton-Jacobi theory. *American Journal of Physics*, **82**, 848–852.

Lemos, N. A. (2015). Vínculos dependentes de velocidades e condição de integrabilidade de Frobenius. *Revista Brasileira de Ensino de Física*, **37**, 4307-1–8.

Lemos, N. A. (2017a). Oscilador quártico e funções elípticas de Jacobi. *Revista Brasileira de Ensino de Física*, **39**, e1305-1–8.

Lemos, N. A. (2017b). Atwood's machine with a massive string. *European Journal of Physics*, **38**, 065001.

Lemos, N. A. & Natividade, C. P. (1987). Harmonic oscillator in expanding universes. *Nuovo Cimento B*, **99**, 211–225.

Leubner, C. (1979). Coordinate-free rotation operator. *American Journal of Physics*, **47**, 727–729.

Lewis Jr., H. R. & Riesenfeld, W. B. (1969). An exact quantum theory of the time-dependent harmonic oscillator and of a charged particle in a time-dependent electromagnetic field. *Journal of Mathematical Physics*, **10**, 1458–1473.

Lissauer, J. L. (1999). Chaotic motion in the Solar System. *Reviews of Modern Physics*, **71**, 835–845.

Lorentz, H. A., Einstein, A., Minkowski, H. & Weyl, H. (1952). *The Principle of Relativity*. New York: Dover.

Lutzky, J. L. (1978). Symmetry groups and conserved quantities for the harmonic oscillator. *Journal of Physics A: Mathematical and General*, **11**, 249–258.

Mann, R. A. (1974). *The Classical Dynamics of Particles*. New York: Academic Press.

Marion, J. B. & Heald, M. A. (1980). *Classical Electromagnetic Radiation*, 2nd edn. New York: Academic Press.

Marion, J. B. & Thornton, S. T. (1995). *Classical Dynamics of Particles and Systems*, 4th edn. Fort Worth: Saunders College Publishing.

Markus, L. & Meyer, K. R. (1974). Generic Hamiltonian dynamical systems are neither integrable nor ergodic. *Memoirs of the American Mathematical Society*, no. 144. Providence, RI: American Mathematical Society.

Marsden, J. E. & Ratiu, T. S. (1999). *Introduction to Mechanics and Symmetry*, 2nd edn. New York: Springer.

Marsden, J. E., O'Reilly, O. M., Wicklin, F. J. & Zombro, B. W. (1991). Symmetry, stability, geometric phases, and mechanical integrators (Part I). *Nonlinear Sciences Today* **1**(1), 4–21; Symmetry, stability, geometric phases, and mechanical integrators (Part II). *Nonlinear Sciences Today* **1**(2), 14–21.

Mathews, P. M. & Lakshmanan, M. (1974). On a unique non-linear oscillator. *Quarterly of Applied Mathematics*, **32**, 215–218.

Maxwell, J. C. (1891). *A Treatise on Electricity and Magnetism*, 3rd edn. 2 vols. New York: Dover (1954 reprint).

Mehra, J. (1994). *The Beat of a Different Drum*. Oxford: Oxford University Press.

Montgomery, R. (1991). How much does a rigid body rotate? A Berry's phase from the 18th century. *American Journal of Physics*, **59**, 394–398.

Morales-Ruiz, J. J. & Ramis, J. P. (2001). A note on the non-integrability of some Hamiltonian systems with a homogeneous potential. *Methods and Applications of Analysis*, **8**, 113–120.

Moriconi, M. (2017). Condition for minimal harmonic oscillator action. *American Journal of Physics*, **85**, 633–634.

Moser, J. (1962). On invariant curves of area-preserving mappings of an annulus. *Nachrichten von der Akademie der Wissenschaften in Göttingen, Mathematisch-Physikalische Klasse II*, 1–20.

Moser, J. (1967). Convergent series expansions for quasi-periodic motions. *Mathematische Annalen*, **169**, 136–176.

Moser, J. (1973). *Stable and Random Motions in Dynamical Systems*. Princeton, NJ: Princeton University Press.

Nakagawa, K. & Yoshida, H. (2001). A necessary condition for the integrability of homogeneous Hamiltonian systems with two degrees of freedom. *Journal of Physics A: Mathematical and General*, **34**, 2137–2148.

Nakane, M. & Fraser, C. G. (2002). The early history of Hamilton-Jacobi dynamics 1834–1837. *Centaurus*, **44**, 161–227.

Neĭmark, J. I. & Fufaev, N. A. (1972). *Dynamics of Nonholonomic Systems*. Providence, RI: American Mathematical Society.

Nekhoroshev, N. N. (1977). An exponential estimate of the time of stability of nearly-integrable Hamiltonian systems. *Russian Mathematical Surveys*, **32**(6), 1–65.

Noether, E. (1918). Invariante Variationsprobleme. *Nachrichten von der Gesellschaft der Wissenschaften zu Göttingen, Mathematisch-Physikalische Klasse*, **2**, 235–257. English translation by M. A. Tavel: arXiv:physics/0503066.

Núñes-Yepes, H. N., Delgado, J. & Salas-Brito, A. L. (2001). Variational equations of Lagrangian systems and Hamilton's principle. arXiv:math-ph/0107006.

Ohanian, H. C. (1976). *Gravitation and Spacetime*. New York: W. W. Norton & Company.

Ott, E. (1993). *Chaos in Dynamical Systems*. Cambridge: Cambridge University Press.

Pais, A. (1982) *"Subtle is the Lord ..." The Science and the Life of Albert Einstein*. Oxford: Oxford University Press.

Pars, L. (1965). *A Treatise on Analytical Dynamics*. Woodbridge, CT: Ox Bow Press.

Pathria, R. K. (1972). *Statistical Mechanics*. Oxford: Pergamon Press.

Pauli, W. (1958). *Theory of Relativity*. New York: Dover.

Pearlman, N. (1967). Vector representation of rigid body rotation. *American Journal of Physics*, **35**, 1164.

Peterson, I. (1993). *Newton's Clock: Chaos in the Solar System*. New York: W. H. Freeman.

Protter, M. H. & Morrey Jr., C. B. (1985). *Intermediate Calculus*. New York: Springer.

Pullen, R. A. & Edmonds, A. R. (1981). Comparison of classical and quantum spectra for a totally bound potential. *Journal of Physics A: Mathematical and General*, **14**, L477–L484.

Reid, C. (1986). *Hilbert-Courant*. New York: Springer.

Rindler, W. (1982). *Introduction to Special Relativity*. Oxford: Clarendon Press.

Romer, R. H. (1978). Demonstration of the intermediate-axis theorem. *American Journal of Physics*, **46**, 575–576.

Rosen, G. (1969). *Formulations of Classical and Quantum Dynamical Theory*. New York: Academic Press.

Rund, H. (1966). *The Hamilton-Jacobi Theory in the Calculus of Variations*. London: D. Van Nostrand.

Russcll, J. S. (1844). Report on Waves. *14th Meeting of the British Association for the Advancement of Science*, 311–390.

Saari, D. G. (1990). A visit to the Newtonian N-body problem via elementary complex variables. *American Mathematical Monthly*, **97**, 105–119.

Saletan, E. J. & Cromer, A. H. (1970). A variational principle for nonholonomic systems. *American Journal of Physics*, **38**, 892–897.

Saletan, E. J. & Cromer, A. H. (1971). *Theoretical Mechanics*. New York: Wiley.

Scheck, F. (1994). *Mechanics – From Newton's Laws to Deterministic Chaos*. Berlin: Springer.

Schiff, L. I. (1968). *Quantum Mechanics*, 3rd edn. New York: McGraw-Hill.

Schrödinger, E. (1982). *Collected Papers on Wave Mechanics*. New York: Chelsea Publishing Company.

Schwartz, L. (2008). *Mathematics for the Physical Sciences*. New York: Dover.

Sharan, P. (1982). Two theorems in classical mechanics. *American Journal of Physics*, **50**, 351–354.

Shinbrot, T., Grebogi, G., Wisdom, J. & Yorke, J. A. (1992). Chaos in a double pendulum. *American Journal of Physics*, **60**, 491–499.

Siegel, C. L. & Moser, J. (1971) *Lectures on Celestial Mechanics*. Berlin: Springer.

Sivardière, J. (1983). A simple mechanical model exhibiting a spontaneous symmetry breaking. *American Journal of Physics*, **51**, 1016–1018.

Sivardière, J. (1986). Using the virial theorem. *American Journal of Physics*, **54**, 1100–1103.

Sommerfeld, A. (1952). *Mechanics*. New York: Academic Press.

Soper, D. E. (1976). *Classical Field Theory*. New York: Wiley.

Spiegel, M. R. (1963). *Advanced Calculus*. New York: McGraw-Hill.

Spivak, M. (1965). *Calculus on Manifolds*. Menlo Park, CA: Benjamin.

Spivak, M. (1994). *Calculus*, 3rd edn. Houston, TX: Publish or Perish.

Stadler, W. (1982). Inadequacy of the usual Newtonian formulation for certain problems in particle mechanics. *American Journal of Physics*, **50**, 595–598.

Sternberg, S. (1994). *Group Theory and Physics*. Cambridge: Cambridge University Press.

Stillwell, J. (2010). *Mathematics and Its History*, 3rd edn. New York: Springer.

Störmer, C. (1907). Sur les trajectoires des corpuscules électrisés dans le espace sous l'action du magnétisme terrestre avec application aux aurores boréales. *Archives des Sciences Physiques et Naturelles*, **24**, 5–18, 113–158, 221–247, 317–364. www.biodiversitylibrary.org/item/93687#page/5/mode/1up.

Sudarshan, E. C. G. & Mukunda, N. (1983). *Classical Dynamics: A Modern Perspective*. Malabar, FL: Robert E. Krieger.

Sundermeyer, K. (1982). *Constrained Dynamics*. New York: Springer.

Sussman, G. J. & Wisdom, J. with Mayer, M. E. (2001). *Structure and Interpretation of Classical Mechanics*. Cambridge, MA: MIT Press.

Symon, K. R. (1971). *Mechanics*, 3rd edn. Reading, MA: Addison-Wesley.

Synge, J. L. & Griffith, B. A. (1959). *Principles of Mechanics*, 3rd edn. New York: McGraw-Hill.

Tabor, M. (1989). *Chaos and Integrability in Nonlinear Dynamics*. New York: Wiley.

Taylor, J. R. (2005). *Classical Mechanics*. Mill Valley, CA: University Science Books.

Terra, M. S., Souza, R. M., and Farina, C. (2016). Is the tautochrone curve unique? *American Journal of Physics*, **84**, 917–923.

Thirring, W. (1997). *Classical Mathematical Physics*. Berlin: Springer.

Tiersten, M. S. (1991). Moments not to forget – The conditions for equating torque and rate of change of angular momentum around the instantaneous center. *American Journal of Physics*, **59**, 733–738.

Tiersten, M. S. (1992). Erratum. *American Journal of Physics*, **60**, 187.

Tufillaro, N. (1986). Integrable motion of a swinging Atwood's machine. *American Journal of Physics*, **54**, 142–143.

Tufillaro, N. B., Abbott, T. A. & Griffiths, D. J. (1984). Swinging Atwood's machine. *American Journal of Physics*, **52**, 895–903.

Tung, W. (1985). *Group Theory in Physics*. Singapore: World Scientific.

Van Dam, N. & Wigner, E. (1966). Instantaneous and asymptotic conservation laws for classical relativistic mechanics of interacting point particles. *Physical Review*, **142**, 838–843.

van der Waerden, B. L. (ed.) (1967). *Sources of Quantum Mechanics*. New York: Dover.

van Kampen, N. G. & Lodder, J. J. (1984). Constraints. *American Journal of Physics*, **52**, 419–424.

Weinstock, R. (1961). Laws of classical motion: What's **F**? What's *m*? What's **a**? *American Journal of Physics*, **29**, 698–702.

Weinstock, R. (1974). *Calculus of Variations*. New York: Dover.

Westfall, R. (1983). *Never at Rest: A Biography of Isaac Newton*. Cambridge: Cambridge University Press.

Whittaker, E. T. (1944). *A Treatise on the Analytical Dynamics of Particles and Rigid Bodies*. New York: Dover.

Whittaker, E. T. (1951). *A History of the Theories of Aether and Electricity, Vol. I: The Classical Theories*. London: Thomas Nelson and Sons.

Whittaker, E. T. & Watson, G. N. (1927). *A Course of Modern Analysis*. Cambridge: Cambridge University Press.

Wiggins, S. (2003). *Introduction to Applied Nonlinear Dynamical Systems and Chaos*, 2nd edn. New York: Springer.

Wisdom, J. (1987). Chaotic behaviour in the Solar System. *Proceedings of the Royal Society of London. Series A, Mathematical and Physical*, **413**, 109–129.

Yandell, B. H. (2002). *The Honors Class: Hilbert's Problems and Their Solvers*. Natick, MA: A K Peters.

Yoshida, H. (1987). A criterion for the nonexistence of an additional integral in Hamiltonian systems with a homogeneous potential. *Physica D*, **29**, 128–142.

Yoshida, H. (1989). A criterion for the non-existence of an additional analytic integral in Hamiltonian systems with n degrees of freedom. *Physics Letters A*, **141**, 108–112.

Yourgrau, W. & Mandelstam, S. (1968). *Variational Principles in Dynamics and Quantum Theory*. New York: Dover.

Zia, R. K. P., Redish, E. F. & McKay, S. R. (2009). Making sense of the Legendre transform. *American Journal of Physics*, **77**, 614–622.

Ziglin, S. L. (1983). Branching of solutions and nonexistence of first integrals in Hamiltonian mechanics I. *Journal of Functional Analysis and Applications*, **16**, 181–189; Branching of solutions and nonexistence of first integrals in Hamiltonian mechanics II. *Journal of Functional Analysis and Applications*, **17**, 6–17.

Index

Action, 50
  abbreviated, 234
  as a function of the coordinates, 304
  invariant, 74, 375
  least, 54
  never a maximum, 55
  as phase of a wave process, 330
  quasi-invariant, 75
  reduced, 230
  stationary, 52
Action at a distance in special relativity, 209
Action variables, 309
  isotropic oscillator (Problem 9.23), 337
  Kepler problem, 323
Action-angle variables, 306
  in perturbation theory, 340, 343
Active point-of-view, 90
Adiabatic
  hypothesis, 329
  invariant, 325
  transformation, 325, 428
Adjoint of linear operator, 416
ADM formalism, 232
Analogy between mechanics and geometric optics, 330
Angle variables, 309
Angular displacement vector, non-existence of (Problem 3.6), 110
Angular momentum, 3
  conservation, 4
  decomposition, 5
  rigid body, 112
  tensor (Problem 6.8), 212
  tensor in field theory (Problem 11.5), 385
Angular velocity, 98
  analytic definition (Problem 3.7), 110
  in terms of Euler angles, 101
  uniqueness, 100
Antikink, 379
Atwood's machine, 15
  with massive string (Problem 1.18), 41
  swinging, 37
    constant of the motion (Problem 2.8), 80
    Hamilton-Jacobi theory (Problem 9.12), 334
    non-integrability (Problem 9.12), 335
Averaging principle, 327

Basic differentials, 399
Bead in rotating hoop, 428
  geometric phase, 430
Berry phase, 136, 329, 428
Bertrand's theorem, 34
  and precession of Mercury's perihelion, 36
Bicycle
  kinematic model, 9
  non-holonomic constraints, 405
Body cone, 138
Bohr-Sommerfeld quantisation rules, 329
Brachistochrone, 48
Burgers equation (Problem 11.13), 389

Calculus of variations, 42
  Euler's equation, 45
  fundamental lemma, 44
  isoperimetric problem (Problem 2.5), 79
  simplest problem, 43
Canonical conjugate momentum, 66
Canonical quantisation, 258
Canonical transformation, 243
  between Poisson manifolds, 439
  generating function, 243
  infinitesimal, 258
    generator, 259
Casimir function, 435
  free rigid body, 435
Catenary, 48
Central forces, 6, 31
Centrifugal force, 104
Chain rule of functional differentiation, 368
Change of basis, 410
Chaos, 319
  in the Solar System, 322
Chaotic motion, 318
Characteristic frequencies, 166
  positivity, 176
Characteristic vectors, 166
  orthogonality, 177
Charge in electric dipole field, 299
  Hamilton-Jacobi theory, 301
  Lagrange's equations, 300
  pendular motion, 300
Chasles's theorem, 94

Chern-Simons theory (Problem 11.12), 388
Complete integral, 289
Complete set of commuting observables (Problem 9.18), 335
Compton effect, 206
Cone, 78, 85
   body, 138
   charge in the field of a magnetic monopole (Problem 2.29), 85
   geodesics (Problem 2.1), 78
   light, 189
   space, 138
Configuration manifold, 18
Configuration space, 18
Conformal mechanics (Problem 8.14), 283
Conjugate momentum, 66
Conservation of energy, 71
Consistency conditions, 274
Constant of the motion, 65
Constrained Hamiltonian system, 272
Constraint force, 14
Constraints, 7, 65
   first class, 277
   holonomic, 8
     included in the Lagrangian, 64
   ideal, 14
   non-holonomic, 8
   non-integrable, 10
   primary, 272
   second class, 278
   secondary, 275
Continuous symmetry, 74
Continuous system as limit of a discrete system, 362
Contraction of a tensor, 194
Coordinate
   cyclic, 66
   generalised, 18
   ignorable, 66
Coriolis force, 103
Correspondence between anti-symmetric matrices and vectors, 98
Cotangent bundle, 217
Cotangent manifold, 217
Coupled pendulums, 167
   normal modes, 168
Covariant Hamilton's equations, 225
Covariant vector, 193
Curl of a vector field in Minkowski space, 196
Cycloid, 49
   brachistochrone, 49
   tautochrone, 49
     non-uniqueness, 49
Cycloidal pendulum (Problem 1.6), 38

d'Alembert operator, 196
d'Alembert's principle, 15

d'Alembertian, 196
Damped harmonic oscillator
   Hamilton-Jacobi theory, 292
   Problem 1.16, 40
   Problem 2.14, 82
Darboux's theorem, 254
Deformation of a torus in invariant tori, 348
Deformation quantisation, 439
Degrees of freedom, 18
Diagonalisation of the inertia tensor, 123
Diffeomorphism, 23
Differentiable manifold, 18
Differential form, 399
   closed, 403
   exact, 403, 426
Diophantine condition, 350
Dirac delta "function", 369
Dirac bracket, 275
Dirac-Bergmann algorithm, 276
Direct sum, 414
Direction cosines, 87
Divergence of a vector field in Minkowski space, 195
Duffing oscillator, 425
Dummy index, 392
Dyad, 114
Dyadic, 115
Dyadic product, 114
Dynamical system, 421
   asymptotically stable solution, 422
   autonomous, 421
     equilibrium point of, 421
   stable solution, 422
Dynamical variable, 254

Eccentric anomaly (Problem 1.20), 41
Eccentricity, 33
Effective acceleration of gravity, 105
Effective potential, 33, 142, 162
Eigenvalue, 411
Eigenvector, 411
Einstein summation convention, 185, 397
Electromagnetic field as a constrained system, 384
Electromagnetic field tensor, 198
   dual (Problem 6.17), 213
Elliptic coordinates (Problem 9.14), 335
Elliptic functions, 156
Elliptic integral
   complete, 158
   of the first kind, 156, 160
   of the second kind, 160
   of the third kind, 160
Energy conservation, 7
   and time-displacement invariance, 72
Energy function, 71

Energy-momentum tensor, 377
  canonical, 377
    electromagnetic field (Problem 11.6),
      386
  electromagnetic field
    Problem 11.6, 386
    Problem 6.18, 213
  Klein-Gordon field, 377
Equilibrium configuration, 150
  stable, 150
Equivalent Lagrangians, 52
Ergodic hypothesis, 319
Euler angles, 95
Euler force, 104
Euler's equation, 45
Euler's equations of motion for a rigid
    body, 131
Euler's theorem on homogeneous
    functions, 406
Euler's theorem on the possible
    displacements of a rigid body, 93
Event, 183
Exact differential, 426
Exterior derivative, 401
  properties, 401
Exterior product, 399
  properties, 400

Fermat's principle, 236
First class function, 278
First integral, 65
Force, 2
  applied, 14
  central, 6
  centrifugal, 104
  conservative, 6
  constraint, 14
  Coriolis, 103
  Euler, 104
  generalised, 19
  gyroscopic, 437
Foucault pendulum, 107
Four-acceleration, 200
Four-current, 197
Four-force, 201
Fourier series, 345, 440
Fourier coefficients, 440
  decay rate, 440
    analytic function, 441
Four-momentum, 200
Four-vector
  contravariant, 191
  covariant, 193
Four-velocity, 199
Free index, 393
Frequencies, 308
  commensurate, 308

fundamental, 310
non-resonant, 320, 346
rationally independent, 320
resonant, 346
sufficiently irrational, 350
Frobenius theorem, 403
Functional, 42
  linear, 415
Functional derivative, 367
  chain rule, 368

Galilean transformation, 184
  Problem 2.15, 82
Gauge fixing, 277, 281
Gauge transformation
  effect on Lagrangian (Problem 2.11), 81
  first kind, 377
  global, 377
  local (Problem 11.10), 387
  second kind (Problem 11.10), 387
Gauss law constraint, 384
Generalised coordinates, 17
Generalised mechanics (Problem 2.13), 81
Generalised momentum, 66
Generalised potential, 28
Geodesic, 46
  cone (Problem 2.1), 78
  configuration space, 237
  sphere, 46
Geometric phase, 136, 329, 428
  bead in rotating hoop, 430
  free rigid body, 431
Global gauge transformation, 377
Gradient in Minkowski space, 193
Gram-Schmidt orthogonalisation process, 414
Group, 92
  composition law, 92
  Lie, 103
  Lorentz, 187
    restricted, 188
    sectors, 188
  of canonical transformations, 253
  of orthogonal matrices, 92
  real symplectic, 253
  rotation, 93
    infinitesimal generators, 102
    Lie algebra, 103
    structure constants, 103
Group velocity (Problem 9.17), 335
Gyroscopic force, 437
Gyroscopic system, 436

Hamilton's characteristic function, 295
Hamilton's equations, 217
  covariant, 225
  for fields, 371
  symplectic form, 251

Hamilton's partial differential equation, 289
Hamilton's principal function, 304
Hamilton's principle, 52
    in phase space, 226
Hamilton-Jacobi equation, 289
    charge in electric dipole field, 301
    complete integral, 289
    complete solution, 289
    separation of variables, 294
    time-independent, 295
Hamilton-Jacobi theory, 288
    and Schrödinger's equation, 331
    incompleteness of, 299, 302
Hamiltonian, 216
    conservation and connexion with total
        energy, 220
    constrained system, 272
    as a generator of time evolution, 259
    as a momentum conjugate to time, 228
    particle in non-inertial frame (Poblem 7.20),
        241
    particle in electromagnetic field, 219
    quasi-integrable system, 340
    relativistic, 224
        particle in electromagnetic field, 225
    total, 276
    and total energy, 218
Hamiltonian perturbation theory, 341
    fundamental equation, 342
    generating function method, 341
    Lie series method, 354
Hartmann's potential, 297
Herpolhode, 138
Hessian matrix, 216
Holonomic system, 8
Holonomy, 432
Homoclinic tangle, 321
Homogeneous function, 406
    Euler's thorem, 406
Hyperbolic motion, 203

Implicit function theorem, 216
Indicial notation, 392
    Einstein summation convention, 397
    factorisation, 396
    substitution, 395
Inertia ellipsoid, 137
Inertia matrix, 119
Inertia tensor, 113
    diagonalisation, 123
    in dyadic notation, 119
Inertial reference frame, 2
Infinitesimal rotations, 68
    commutativity, 97
    matrix form, 97
Infinitesimal transformation, 67
    and Noether's theorem, 73, 374

Infinitesimal translation, 68
Inner product, 412
    canonical, 412
Inner product space, 412
Instantaneous axis of rotation, 98
Integrability, 312
Integrability coefficient, 316
Integrable 1-form, 403
Integrable collection of 1-forms, 403
Integrable system, 314
Integral, 65
Integrating factor, 403, 404
Interval
    lightlike, 188
    spacelike, 188
    timelike, 188
Invariable plane, 138
Invariance under rotations and angular
        momentum conservation, 70
Invariance under time displacements and
        energy conservation, 72
Invariance under translations and linear
        momentum conservation, 69
Invariant, 65
Invariant interval, 185
Invariant subspace, 418
Inverse of a matrix, 410
Involution, 312
    constants of the motion in, 312
Isolated particle, 2
Isoperimetric problem (Problem 2.5),
        79

Jacobi elliptic functions, 156
    modulus, 156
Jacobi identity, 257
    proof, 260
    proof for free rigid body, 434
Jacobi integral, 71
Jacobi's principle, 236
Jacobi's theorem, 290

KAM theorem, 321, 352
Kepler problem, 33
    relativistic, 240
Kepler's equation (Problem 1.20), 41
Kepler's law of areas, 32
Kepler's third law, 34
Kinetic energy, 5
    decomposition, 5
    rigid body, 120
King Oscar II, 322
    prize, 322
Kink, 379
Kirkwood gaps, 321
Klein-Gordon equation, 366
    Hamiltonian form, 372

Korteweg-de Vries equation, 438
   Hamiltonian form, 439
   Problem 11.14, 390
Kronecker delta, 394

Lagrange bracket
   canonicity criterion, 249
   invariance under canonical transformations, 253
Lagrange multipliers, 57
   and generalised constraint force, 58
   as generalised coordinates, 64
Lagrange's equations, 22
   for fields, 361, 364
   invariance, 23
Lagrangian, 22
   Bateman's (Problem 1.16), 40
   charged particle in external electromagnetic
      field, 29
   Chern-Simons, 389
   covariant, 208
   electromagnetic field, 366
   in field theory, 360
   Klein-Gordon field, 366
   particle in central potential, 32
   phase space, 227
   plane double pendulum, 27
   Proca, 388
   relativistic, 208
      particle in electromagnetic field, 208
   sine-Gordon field (Problem 11.8), 387
   symmetric top, 140
Lagrangian density, 360
   Chern-Simons, 389
   electromagnetic field, 366
   Klein-Gordon field, 366
   Schrödinger field, 365
   sine-Gordon field (Problem 11.8), 387
   vibrating string, 362
Laplace-Runge-Lenz vector (Problem 2.22), 83
Law of local conservation of electric charge, 197
Legendre transformation, 216
Levi-Civita totally antisymmetric symbol, 394
Libration, 307
Lie algebra, 257
   infinitely differentiable functions on phase
      space, 257
Lie operator, 264
Lie series, 264
   perturbation theory, 355
Light cone, 189
Lindstedt series, 349
Linear functional, 415
Linear momentum, 3
   conservation, 3
Linear operator, 408
   adjoint of, 416
   Hermitian, 417

inverse of, 409
invertible, 408
self-adjoint, 417
spectrum of, 411
unitary, 416
Linear transformation, 408
Linear triatomic molecule, 173
   normal modes, 175
Lorentz force, 28, 201
Lorentz group
   restricted, 188
   sectors, 188
Lorentz scalar, 185
Lorentz transformation, 183
   improper, 187
   infinitesimal (Problem 6.5), 212
   non-orthocronous, 187
   orthocronous, 187
   proper, 187
   pure, 187
   restricted, 188
Lyapunov function, 424
   strict, 424
Lyapunov's stability theorem, 423

Magnetic monopole
   Problem 1.17, 41
   Problem 2.29, 84
Manifold
   compact, 315
   configuration, 18
   cotangent, 217
   differentiable, 18, 51, 217, 399
   Poisson, 257, 433
   symplectic, 254
Mathieu transformation (Problem 8.21),
      285
Maupertuis principle, 234
   and Fermat principle, 236
   for non-conservative systems, 240
   theological interpretation, 234
Maxwell's equations, 196
   in covariant form, 198
   as a Hamiltonian constrained system, 384
Metric, 186
Metric tensor, 186
Minimal coupling, 225
Minimal surface of revolution, 47
Minkowski diagram, 190
Minkowski force, 201
Minkowski space, 185
Modal matrix, 171
Moments of inertia, 119
   principal, 123
Momentum, 3
   angular, 3
      conservation, 4

canonical conjugate, 66
   generalised, 66
   linear, 3
      conservation, 3
   relativistic, 200

n-body problem, 322
   Problem 7.6, 238
   Wang's exact solution, 319
Navier-Stokes equation (Problem 11.13), 389
Nekhoroshev's theorem, 354
Neumann problem (Problem 2.31), 85
Newton's laws, 1
   First, 2
   Second, 2
   Third, 2
      strong form, 2
      weak form, 2
Noether charge, 376
Noether condition, 74
   generalised, 75
Noether's theorem, 74
   in field theory, 374
   generalised, 75
Non-degenerate Hamiltonian system, 350
Non-integrability, 316
   criterion, 316
   Yoshida's theorem, 316
Norm, 412
   canonical, 412
Normal coordinates, 172
Normal modes of vibration, 166
   completeness, 177
Null vector, 194
Nutation, 143

Observable (Problem 8.26), 287
Old quantum theory, 329
Operator
   adjoint, 416
   d'Alembertian, 196
   finite rotation, 117
   Hermitian conjugate, 416
   Lie, 264
   linear, 408
   time evolution, 265
Orthogonal complement, 414
Orthogonal matrix, 89
Orthogonal system of vectors, 413
Orthogonal transformation, 89
Orthogonal vectors, 413
Orthonormal basis, 413
Orthonormal system of vectors, 413

Parabolic coordinates (Problem 9.2), 332
Parallel axis theorem, 121
Passive point of view, 90

Path integrals, 306
Pendulum
   cycloidal (Problem 1.6), 38
   damped, 31
   double, 8, 27
   elastic (Problem 1.11), 39
   exact solution (Problem 5.2), 178
   Foucault, 107
   as a perturbed system, 343
   spherical (Problem 1.4), 37
Periodic function, 440
Perpendicular axis theorem, 122
Perturbation theory
   coupled harmonic oscillators, 346
   fundamental equation, 349
   fundamental problem, 341
   generating function method, 341
   Lie series method, 354
   one degree of freedom, 342
   pendulum, 343
   several degrees of freedom, 345
   small denominators, 348
   small divisors, 348
Phase space, 217
   extended, 229
   reduced, 230
Phase space Lagrangian, 227
Photon, 206
Physical laws in manifestly covariant form, 196
Poincaré and King Oscar II prize, 322
Poincaré lemma, 403
   proof for 1-forms, 426
Poincaré's recurrence theorem, 270
   and second law of thermodynamics, 271
Poinsot construction, 136
Point transformation, 23
   as a canonical transformation, 248
Poisson bracket, 254
   algebraic properties, 257
   on a differentiable manifold, 433
   free rigid body, 434
   in field theory, 372
   invariance under canonical transformations, 255
Poisson brackets
   angular momentum, 263
   canonicity criterion, 256
   fundamental, 257
      in field theory, 373
Poisson manifold, 257, 433
Poisson structure, 433
Poisson's theorem, 261
Polhode, 138
Potential
   central, 32
   effective, 33
   generalised, 28
   scalar, 29

Potential (Cont.)
  vector, 29
  velocity-dependent, 28
Precession, 143
  regular, 144
Primary constraints, 272
Principal axes, 123
  cube, 127
  triangular plate, 125
Principal axes of inertia, 123
Principal moments of inertia, 123
  cube, 128
  triangular plate, 124
Principle of virtual work, 15
Principle of least action, 52
  teleological interpretation, 78
Problem of small denominators, 348
Problem of small divisors, 348
Products of inertia, 119
Proper acceleration, 200
  Problem 6.13, 213
Proper time, 199

Quadratic form, 164
  positive, 164
Quadratures
  solution by, 141
    Liouville-Arnold theorem, 314
Quantum cosmology, 232
Quartic oscillator, 155
Quasi-integrable Hamiltonian system, 340

Rayleigh's dissipation function, 30
Real displacements, 12
Recurrence theorem, 270
Reduced phase space, 230
Regular motion, 318
Relativistic energy, 202
Relativistic collisions, 205
Relativistic free particle, 231
Relativistic momentum, 200
Relativistic oscillator, 359
Rest energy, 203
Rigid body, 86
  angular momentum, 112
  Euler's equations of motion, 131
  free, 132
    geometric phase, 431
    stability of uniform rotation, 134
  number of degrees of freedom, 86
  possible displacements, 93
Rolling coin, 129
Rotation, 67
  finite, 116
  formula, 117
  infinitesimal, 68
    angular momentum conservation, 70

Rotation group, 93
Rotational invariance and angular momentum
    conservation, 70
Rotations in the plane, 90

Scalar field, 194
Scalar potential, 29
Scalar product, 412
  in Minkowski space, 192
Schrödinger field, 365
  as a constrained system, 382
Schrödinger's equation, 331
  time-independent, 331
Schwarz inequality, 413
Second class function, 278
Secondary constraints, 275
Secular terms, 358
Self-adjoint operator, 417
Sensitive dependence on initial conditions, 318
Separable Hamiltonian system, 306
  multiply periodic, 307
Separatrix, 308
Similar matrices, 411
Similarity transformation, 411
Simultaneous diagonalisation of two
    quadratic forms, 178
sine-Gordon equation (Problem 11.8), 387
Skate, 60
Small denominators, 348
  problem of, 348
Small divisors, 348
  problem of, 348
Small oscillations, 149
  about stable equilibrium, 151
  about stationary motion, 161
  general case, 165
  one-dimensional case, 149
    anomalous, 154
Solitary wave, 378
Soliton, 381
Space cone, 138
Space dilation, 77
Spectral theorem, 418
Spectrum of a linear operator, 411
Spontaneous symmetry breaking, 153, 379
Stäckel's theorem, 296
Störmer's problem, 317
  non-integrability, 318
Stability, 150
  criterion, 150
  eternal, 353
  in the sense of Lyapunov, 422
    asymptotic, 422
  long term, 353
  uniform rotation, 134
  vertical top, 145
Star-shaped set, 427

Stationary motion, 161
Strong equality, 273
Super-Hamiltonian, 229
Swinging Atwood's machine, 37
    constant of the motion (Problem 2.8), 80
    Hamilton-Jacobi theory (Problem 9.12), 334
    non-integrability (Problem 9.12), 335
Symmetric top, 139
    with one point fixed, 139
        constants of the motion, 140
        effective potential, 142
        nutation, 143
        precession, 143
        regular precession, 144
        solution by quadratures, 141
        stability, 145
    on a smooth horizontal surface
        (Problem 4.15), 148
Symmetries and principal axes of inertia, 126
Symplectic form, 254
Symplectic manifold, 254
Symplectic matrix, 252
Symplectic transformation, 252
System with parameterised time, 230
System with parameterised time
    Einstein's gravitation theory, 232

Tautochrone, 49
Tensor, 118
    anti-symmetric, 198
        Problem 6.6, 212
    energy-momentum, 377
    field, 195
        transformation law, 195
    inertia, 119
        diagonalisation, 123
        symmetric, 119
    symmetric, 192
        Problem 6.6, 212
    transformation law, 118
    unit or identity, 115
Tensor product, 114
Theorem of Liouville, 268
Theorem of Liouville-Arnold, 314
Theorem on averaging, 328
Three-body problem, 319, 322
    restricted, 338
    Sundman's exact solution, 319
Threshold energy, 205
Time as a canonical variable, 228
    in quantum cosmology, 232

Time evolution operator, 265
Time rate of change of a vector, 98
Torus, 315, 320
    deformation in invariant tori, 348
    destroyed, 353
    invariant, 315
    non resonant, 320
    resonant, 320
Total partial derivative, 375
Translation, 67
    infinitesimal, 68
        momentum conservation, 69
Translational invariance and linear momentum
        conservation, 69
Triangle inequality, 413
Triatomic molecule, 173
    normal modes, 175
    zero mode, 174

Unitary operator, 416

Van Allen belts, 317
Van Dam–Wigner no-interaction theorem, 211
Variation
    of a function, 49
    of a functional, 49
    strong, 56
Variational principles and metaphysical issues, 78
Vector associated with an infinitesimal rotation, 98
Vector potential, 29
Virial of Clausius, 222
Virial theorem, 222
    equation of state of ideal gas, 223
Virtual displacements, 12
Virtual work, 13
    principle of, 15

Wave mechanics, 330
Weak equality, 273
Weber's electrodynamics
    Problem 1.10, 39
    Problem 7.14, 239
Wedge product, 399
    properties, 400
Wilson-Sommerfeld quantisation rules, 329
Worldline, 189

Yoshida's non-integrability theorem, 316

Zero frequency mode, 174
Zero mass particles, 204